The Book of CHAC

Programming Studies for Mexican Agriculture

CHAC
(the name of
a Mayan rain god)
is a programming model for
the study of Mexican agriculture.

A WORLD BANK RESEARCH PUBLICATION

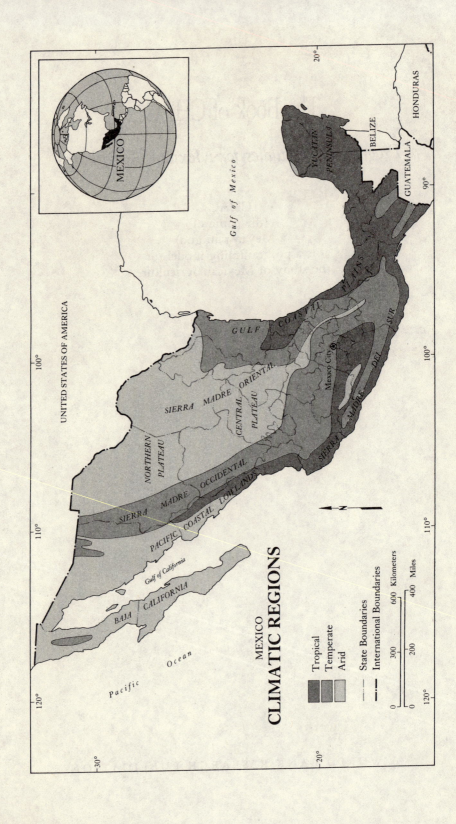

MEXICO
CLIMATIC REGIONS

Tropical
Temperate
Arid

State Boundaries
International Boundaries

| 0 | 300 | 600 Kilometers |

| 0 | 200 | 400 Miles |

UNITED STATES OF AMERICA

MEXICO

Gulf of Mexico

Pacific Ocean

Gulf of California

BAJA CALIFORNIA

NORTHERN PLATEAU

SIERRA MADRE OCCIDENTAL

SIERRA MADRE ORIENTAL

CENTRAL PLATEAU

PACIFIC COASTAL LOWLANDS

GULF COASTAL PLAINS

SIERRA MADRE DEL SUR

Mexico City

YUCATÁN PENINSULA

BELIZE

GUATEMALA

HONDURAS

120°   110°   100°   90°

30°   20°

The Book of CHAC

Programming Studies for Mexican Agriculture

edited by

Roger D. Norton and Leopoldo Solís M.

PUBLISHED FOR THE WORLD BANK

The Johns Hopkins University Press

BALTIMORE AND LONDON

The Johns Hopkins University Press
Baltimore, Maryland 21218, U.S.A.

The views and interpretations in this book are the authors' and should not be attributed to the World Bank, to its affiliated organizations, or to any individual acting in their behalf. The maps have been prepared exclusively for the convenience of readers of this book; the denominations used and the boundaries shown do not imply, on the part of the World Bank and its affiliates, any judgment on the legal status of any territory or any endorsement or acceptance of such boundaries.

Editor   James E. McEuen
Figures   World Bank Graphics Unit
Maps   Julio Ruiz and Larry Bowring
Book design   Brian J. Svikhart
Cover design   Joyce C. Eisen

Library of Congress Cataloging in Publication Data

Main entry under title:

The Book of CHAC.

(A World Bank research publication)
Bibliography: p.
Includes index.
  1. Agriculture—Economic aspects—Mexico—
Mathematical models—Addresses, essays, lectures.
I. Norton, Roger D., 1942–   . II. Solís M.,
Leopoldo.   III. Series: World Bank research publica-
tion.
HD1792.B66   338.1'0972   80-29366
ISBN 0-8018-2585-7   AACR1

Contributors

Roger D. Norton	*Professor of Economics, University of New Mexico*
Leopoldo Solís M.	*Director General, Banca Confía, formerly Deputy Governor, Banco de México*
Jock R. Anderson	*Faculty of Economic Studies, University of New England, New South Wales, Australia*
Luz María Bassoco	*Coordinación General del Sistema Nacional de Evaluacíon, Oficina de Asesores del C. Presidente, México*
Carlos A. Benito	*Department of Agricultural Economics, University of California at Berkeley*
Wilfred V. Candler	*Eastern Africa Projects Department, The World Bank*
R. G. Cummings	*Department of Economics, University of New Mexico*
John H. Duloy	*Economics and Research Staff, The World Bank*
Peter B. R. Hazell	*International Food Policy Research Institute*
Hunt Howell	*Inter-American Development Bank*
Gary P. Kutcher	*Western Africa Projects Department, The World Bank*
H. C. Lampe	*Department of Resource Economics, University of Rhode Island*
J. W. McFarland	*Business School, University of Houston*
Alexander Meeraus	*Development Research Department, The World Bank*
Anthony Mutsaers	*United Nations Food and Agriculture Organization*
Gerald O'Mara	*Development Research Department, The World Bank*
Malathi Parthasarathy	*Programming and Budgeting Department, The World Bank*
Carlos Pomareda	*Instituto Interamericano de Cooperación para la Agricultura*
Teresa Rendón	*Centro de Estudios Económicos y Demográficos, El Colegio de México*
Pasquale L. Scandizzo	*Agriculture and Rural Development Department, The World Bank*
Richard L. Simmons	*Department of Economics, North Carolina State University*
José S. Silos	*Secretaría de Hacienda y Crédito Público, México*

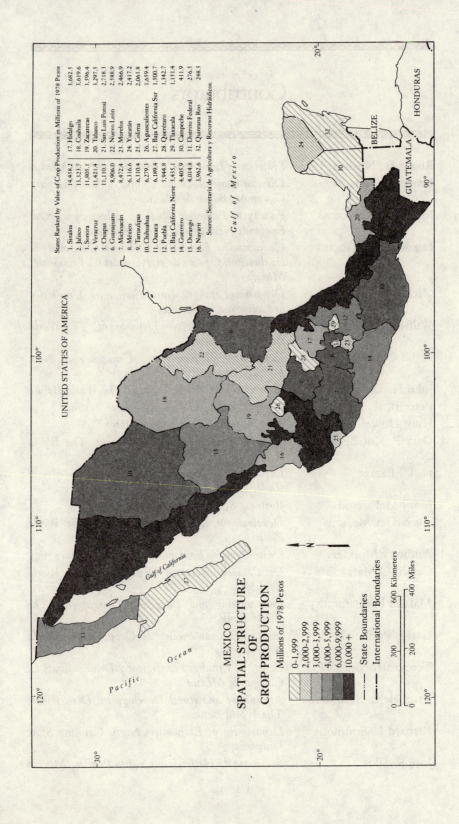

MEXICO
SPATIAL STRUCTURE
OF
CROP PRODUCTION

Millions of 1978 Pesos

0-1,999
2,000-2,999
3,000-3,999
4,000-5,999
6,000-9,999
10,000+

— · — · — State Boundaries
— · · — International Boundaries

0   300   600 Kilometers
0   200   400 Miles

N

States Ranked by Value of Crop Production in Millions of 1978 Pesos

1. Sinaloa	14,458.2		17. Hidalgo	3,682.5
2. Jalisco	13,323.7		18. Coahuila	3,619.6
3. Sonora	11,805.1		19. Zacatecas	3,596.4
4. Veracruz	11,421.4		20. Tabasco	3,297.5
5. Chiapas	11,110.3		21. San Luis Potosí	2,718.3
6. Guanajuato	8,906.0		22. Nuevo León	2,588.9
7. Michoacán	8,472.4		23. Morelos	2,466.9
8. México	6,316.6		24. Yucatán	2,417.2
9. Tamaulipas	6,310.4		25. Colima	2,063.8
10. Chihuahua	6,279.3		26. Aguascalientes	1,659.4
11. Oaxaca	6,189.4		27. Baja California Sur	1,500.7
12. Puebla	5,944.8		28. Querétaro	1,342.7
13. Baja California Norte	5,435.1		29. Tlaxcala	1,151.4
14. Guerrero	4,405.9		30. Campeche	411.9
15. Durango	4,014.8		31. Distrito Federal	276.5
16. Nayarit	3,962.6		32. Quintana Roo	248.5

UNITED STATES OF AMERICA

Pacific Ocean

Gulf of California

Gulf of Mexico

GUATEMALA

BELIZE

HONDURAS

Foreword

ROGER NORTON AND LEOPOLDO SOLÍS have performed a useful service for economists interested in development and quantitative applications of improved planning models. They have brought together in this volume analyses of a range of programming models that have been applied to Mexican agricultural policies—analyses made by themselves and twenty other authors. Some of these authors participated in the development of the original CHAC model, which has been applied for numerous purposes in Mexico and is now well known in other countries as well.

The current volume adds a rich set of opportunities for analyzing economic and developmental problems of a nation's agriculture. Programming models in their many modifications can be used at the level of sector, region, commodity, or resource and for analysis of other dimensions of agriculture. Models incorporating these opportunities are specified in the book. In addition to linear and nonlinear sectoral and regional models, the authors deal with problems of risk, rural development, vegetable exports, irrigation, and investment projects.

The initial CHAC models, having been used for a considerable time, provide a rich stock of capital assets to which the other models can be related. It is hoped that these additional models will also be extended and put to continued use for determining agricultural policy in Mexico. The nation has broad goals in agricultural development and plans to invest in and use quantitative methods to reach those goals.

Programming models are especially useful for analyzing the potential effects of change on resource use, productivity, and the generation of employment and income. In this sense, they are an important set of the quantitative techniques needed to assess development and policy alternatives in all countries. They generally can incorporate, in a complete model, greater regional, spatial, resource, and commodity details than can other types of quantitative models. These possibilities are well illustrated by the range of models presented in this book. Econometric models based on time series data (where available), simulation models, and programming models all have unique roles to perform in the evaluation of policy outcomes and alternatives. Programming models have an advantage when it is important to indicate "by how much and where" investments and change should be made, but when time series data are not generally available. The authors illustrate how the normative aspects of programming models can be combined effectively with demand functions and other positive relationships.

The book should illustrate the potential of programming analysis for both developing and developed countries. The World Bank has been extremely helpful in promoting the development and use of these models. I hope that these efforts can be extended to many other countries and the models put to use in policy analysis and selection.

EARL O. HEADY
*The Center for Agricultural
and Rural Development
Iowa State University*

Contents

Part Three: Regional Programming Models

Part Four: Data Processing for Agricultural Models

Definitions

Models and Submodels

CHAC
The multilevel linear programming model for Mexican agriculture; the name of a Mayan rain god

ALPHA
Regional submodel for production of annual irrigated crops in northwest Mexico; comprises eight irrigation submodels in CHAC

BAJIO
Regional submodel of CHAC for El Bajío area of Mexico's Central Plateau

BAJITO
Smaller version of BAJIO

BETA
Regional submodel (with ALPHA) for production of annual irrigated crops in northwest Mexico; a revision of the model for export vegetables (see chapter 12)

MEXICALI
Static (one-period) cropping submodel for Mexicali area of northwest Mexico; an extended version of RIO COLORADO regional submodel in CHAC

PACIFICO
Regional submodel for Mexico's Pacific Northwest

TECATE
Dynamic (investment) version of MEXICALI regional submodel; name of a smaller town near Mexicali in northwest Mexico

TOLLAN
Regional submodel for machinery-labor substitution in Tula irrigation district, state of Hidalgo, in Mexico's Central Plateau; name of the ancient Toltec capital at Tula

Operational and Spanish Terms

BOD
Biochemical oxygen demand; a measurement used in aquaculture to assess the decay rate of biomass

CDF
Cumulative distribution functions

c.i.f.
Cost, insurance, and freight (for import pricing)

CS	Consumer surplus
Culmaya	Aggregative name for the irrigation districts of Culiacán, Humaya, and San Lorenzo in Mexico's Pacific Northwest
ejido (ejidal, ejidatorio)	Mexican system (and unit thereof) of public ownership of smallholdings of farmland; the *ejido* farmer is granted lifelong farming rights but cannot sell or lease the holding
f.o.b.	Free on board (for export pricing)
FSD (SSD, TSD)	First-degree (second-, third-degree) stochastic dominance
FSE (SSE, TSE)	First-degree (second-, third-degree) stochastic efficiency
GESL	General equilibrium substitution locus
GDP	Gross domestic product
GNP	Gross national product
GASA	Grupos de Análisis Sectorial [Sectoral Analysis Task Forces] established by the Mexican government; precursors of COCOSA
grupos solidarios	Solidarity groups (of farmers, for example)
IRR	Internal rate of return
LHS	Left-hand side of an equation
MAD	Mean absolute differential
MEC	Marginal efficiency of capital
minifundio	Smallholding (as opposed to *latifundio*, large estate) of land
MPS	Mathematical programming system
MRS	Marginal rate of substitution
PDF	Probability density functions
PS	Producers' surplus
RHS	Right-hand side of an equation
SD	Stochastic dominance
TDM	Total dry matter; a measurement used in assessing feed requirements of livestock
TDN	Total digestible nutrients
temporal	"Rainfed" or "dryland" (as opposed to "irrigated")

Formats, Languages, Programs

APEX	Commercial linear programming package available from Control Data Corporation
Basic Cobol Fortran PLI	User's languages
MPS/360	Mathematical programming system for IBM 360 computer; the standard format for data input in linear programming
MODGEN	Reference term for class of commercial "table-driven" matrix generators (DATA-FORM, GAMMA, MAGEN, MGRW, and the like)
SECGEN	Matrix generator (written in Fortran) used in CHAC

Institutional Data Sources

CAADES	Confederación de Asociaciones Agrícolas del Estado de Sinaloa [Confederation of Farmers Associations of the State of Sinaloa]
CIMMYT	Centro Internacional de Mejoramiento de Maíz y Trigo [International Maize and Wheat Improvement Center]
COCOSA	Comisión Coordinadora del Sector Agropecuario [Coordinating Commission for the Agricultural Sector]; later CONACOSA
CONASUPO	Compañía Nacional de Subsistencia Popular [National Company for Public Food Supplies]
IBM	International Business Machines Corporation
IDRHEM	Instituto de Desarollo de los Recursos Humanos del Estado de México [Human Resources Institute of the State of Mexico]
IMCE	Instituto Mexicano de Comercio Exterior [Mexican Institute for Foreign Trade]
ISWYN	International Spring Wheat Yield Improvement Nursery

PAR	Programa de Altos Rendimientos [High Yield Program]; administered by the Mexican National School of Agriculture in Chapingo
SAG	Secretaría de Agricultura y Ganadería [Ministry of Agriculture and Livestock]; later combined with SRH to form SARH
SARH	Secretaría de Agricultura y Recursos Hidráulicos [Ministry of Agriculture and Water Resources]
SRH	Secretaría de Recursos Hidráulicos [Ministry of Water Resources]; later combined with SAG to form SARH
UNPH	Unión Nacional de Productores de Hortalizas [National Union of Vegetable Producers]

Currency Equivalents

IN THE ORIGINAL VERSION OF CHAC, all prices were defined in 1968 Mexican pesos. In later versions, for 1972 and 1976, prices were given in pesos of those years.

The yearly average exchange figure for pesos per U.S. dollar remained unchanged, at 12.49 pesos to the dollar (rounded to 12.5 pesos in the text), from 1966 through 1973. Yearly average currency equivalents for subsequent years were:

	Pesos per U.S. dollar
1974	12.500
1975	12.500
1976	19.950
1977	22.736
1978	22.724
1979	22.803
1980	23.256

Currency equivalents are from appropriate issues of the International Monetary Fund's *International Financial Statistics*. All dollar figures in this study are in U.S. dollars.

Preface

IN THE MOST GENERAL SENSE, the topic of this book is the economic structure of Mexican agriculture and how that structure responds to different kinds of policy initiatives. The book presents both methodological investigations and policy-oriented research results. A number of the analyses presented here have been incorporated into policy documents of the Mexican government, and they have made their mark on policy decisions over the years, as illustrated by the material in chapters 5 and 6.

The book can be regarded as a second progress report on an applied research project of long duration. The first progress report took the form of the seven chapters of Part IV in *Multi-Level Planning: Case Studies in Mexico*, edited by Louis M. Goreux and Alan S. Manne (Amsterdam and New York: North-Holland/American Elsevier, 1973). As of this writing, the CHAC model of Mexican agriculture has been functioning for ten years. After the papers in this volume were assembled, the government (in this case, the Coordinación General del Sistema Nacional de Evaluación) updated and revised all aspects of the structure of the model, and further analyses of pricing, investment, and crop insurance strategies have been conducted very recently.

Although this class of model, like any other, has definite limitations, various factors explain why CHAC has been so productive of applicable results. A decisive institutional factor has been the willingness of the Mexican government—in the form of various agencies, beginning with the Banco de México—to provide continuous support to the project. At times, this has required patience while new data were being assembled to update the tens of thousands of coefficients in CHAC. The key roles of two of our colleagues in the government, José Silos and Luz María Bassoco, certainly must be mentioned.

On the methodological side, one of CHAC's strengths is that it is a descriptive, not a prescriptive, model. Solutions of CHAC do not indicate "best" policies, but rather trace in an approximate way the multiple consequences of the policy alternatives being considered. In the work with CHAC considerable attention has been given to the specification of meaningful policy experiments.

Another advantage of the model is that it reflects, at least to a reasonable degree, the interdependencies which characterize the agricultural sector. There is interdependence on the supply side as different crops compete for scarce resources and different regions compete for access to markets. The model also reflects demand-side interdependence inasmuch as there is substitution of crops in consumption, substitution between foreign and domestic markets, and factor substitution accomplished by switching cropping technologies and locations of production. And of course the levels of factor use, output, and price are interrelated.

On the data side, the design of CHAC emphasizes use of cross-sectional field-level production coefficients, and it relies only minimally on aggregate time series data. In this way, not only does the model use the relatively more reliable kinds of data, but also the coefficients are, for the most part, susceptible to direct verification. Especially in the early stages of the model work, dialogues were established with field specialists, and CHAC was used to apply consistency tests to the data before the coefficients were finalized. In some instances, this process induced the data suppliers to revise their information.

Finally, it should be pointed out that CHAC is a flexible instrument, in the sense that different specifications of producer and consumer behavior and different possible policies can be incorporated in a straightforward manner without necessitating a revision of the mathematical solution procedure. In practical terms, this is a very useful property, for a model has to be solved many times if it is to be applied meaningfully.

Many of the chapters of *The Book of CHAC* refer not to the sectoral model but to regional models and related analyses. The regional models also have a policy role, particularly in investment analysis, but most of the chapters of Parts Two, Three, and Four are directed to issues of improvement in methodology. The analyses in Part Two already have borne fruit in that the latest version of CHAC (not reported here) contains specifications which represent farmers' aversion to risk. This now permits its application to questions of crop insurance, among other issues. We hope that the work reported in Parts Two, Three, and Four will help stimulate others to continue to improve these kinds of models. While their potential for usefulness is high, clearly there is substantial room for methodological improvements. Throughout the volume, and specifically at the end of chapter 5, the reader will find suggestions regarding priority areas for further work.

In the course of this study a large number of individuals in the Mexican government, the World Bank, and the academic world have contributed in a variety of ways. They are not listed as authors, but the study could not have been carried to conclusion without their help. They are too numerous to list individually, but we are grateful to them all.

ROGER D. NORTON
LEOPOLDO SOLÍS M.

The Book of CHAC

1

Introduction and Summary

ROGER D. NORTON AND LEOPOLDO SOLÍS M.

THIS BOOK GREW out of a collaborative research project on Mexican agriculture sponsored jointly by the Mexican government and the World Bank. For the Mexican agencies involved, the principal aim was to develop a comprehensive, quantitative framework that could assist in formulating policy guidelines for a sector marked by great diversity—both in the kinds of production conditions encountered and in the kinds of programs spread across a large number of administrative units.

Institutional Framework

For two decades, until the mid-1960s, Mexican agriculture had expanded rapidly and had played a crucial role in generating foreign exchange for the economy and in helping limit rural-urban migration to manageable proportions. In the late 1960s, however, the sector's dynamism began to falter. The opportunities for constructing massive irrigation works, which had contributed in important ways to the period of rapid growth, were coming to an end, and policymakers were seeking new sources of stimulus for the sector in coming decades. In the early 1970s, as the dimensions of the problem of rural underemployment became more apparent, the generation of agricultural employment per se moved to the head of the list of policy concerns.

The World Bank shared these concerns for the creation of agricultural growth and employment, but, in the context of this project, it was interested primarily in the development of usable methodologies for policy planning in developing countries. In the initial phase, the agricultural study was a component of a multilevel planning study that examined economy-wide behavior, the energy and agricultural sectors, some specific investment projects within those sectors, and the linkages among the three levels of analysis (Goreux and Manne 1973). As the first study was nearing completion, the Mexican government invited the World Bank to continue work on its agricultural component and, in particular, to refine and apply the sectoral model that had been developed, CHAC (from the name of the Mayan rain god). A second joint project was therefore launched, and most of the work in this volume is the direct outcome of that cooperative undertaking. While the work was being carried out, frequent exchanges of ideas were held with other research groups involved in related analyses of Mexican agriculture, and some of the chapters in the volume developed out of those contacts.

During the period when this work was being carried out, the Mexican government was experimenting with different means of coordinating its disparate public programs under way in agriculture. A great number of official agencies were involved in the sector, and elimination of their overlapping and conflicting functions was a concern of the government throughout the 1970s. By its nature as an integrative, numerical framework, the sector model complemented this effort and, in some respects, reinforced it.

The earliest efforts at coordination involved several sectors. "Programming units" were established in the ministries, and they were to serve as liaison to the central budgeting authorities (the Ministry of Finance and the Ministry of the Presidency).[1] Special task forces for developing sectoral policy guidelines (GASA, for Grupos de Análisis Sectorial) were then founded. Coordination was carried farthest for agriculture. The agriculture task force evolved into a Coordinating Commission for the Agricultural Sector (Comisión Coordinadora del Sector Agropecuario, COCOSA; later CONACOSA), which was staffed by all the public institutions that participate in the agricultural sector and was headed by the minister of the presidency. This commission sponsored the applications of the sector model reported in this book. The coordination embodied in COCOSA was carried one step further in 1977 when the two main public agencies for agriculture, the Ministries of Agriculture and Livestock (Secretaría de Agricultura y Ganadería, SAG) and of Water Resources (Secretaría de Recursos Hidráulicos, SRH) were merged to form a single ministry (Secretaría de Agricultura y Recursos Hidráulicos, SARH). COCOSA's sectoral coordination then passed to the new ministry and, for linkages with other kinds of programs, to the Ministry of Programming and Budgeting.

Policy Objectives and the Function of Models

In part, this book constitutes a report of an experience in using models for policy planning, but it also presents purely methodological material.

1. The Ministry of the Presidency is now called the Ministry of Programming and Budgeting (Secretaría de Programación y Presupuesto).

Many potentially useful, quantitative approaches were generated during the course of this study, and not all of them could be applied, in part because of limitations on government staff resources (as chapter 17, on an incomplete livestock model, well illustrates). The common theme running through the chapters is the use of quantitative methods—mainly, linear programming models—to simulate potential reactions of farmers to changes in policy. No attempt has been made to derive "optimal" policies with the models; rather, the power of analysis is applied to assist in understanding the múltiple consequences that would follow from specified policy options. Although growth in employment and sectoral output have been dominant policy concerns for agriculture, they have by no means been the only ones. The sector's potential as an earner of foreign exchange has been important in program formulation, and from the mid-1960s to the mid-1970s a substantial restructuring of agricultural exports occurred. Also, production growth and income growth are not the same thing, and income levels and their distribution within the sector have been a continuing concern. In particular, public programs have endeavored to foster higher productivity in the nonirrigated regimes— temperate rainfed (in Spanish, *temporal*) and tropical—wherein the bulk of the poorer farmers reside. Another continuing direction of policy has comprised the dual objectives of land redistribution and removal of tenure uncertainty. The emphasis given to one or the other varies by region.

When alternative policy instruments are arrayed against these goals, tradeoffs of many kinds have to be considered. The various components of the sector are tightly interdependent, and the policy aims are not always mutually compatible. The issue of the price support level for maize is a good illustration. Classically, there is the question of by how much the price must be increased if a given increase in production is to be attained and, therefore, the question of what is the tradeoff between consumer and producer welfare. But within the sector other choices arise. Increased maize production may be attained at the expense of other crops, some of which are destined for export. This tradeoff differs over regions, and, in the aggregate, policymakers have wanted to know whether the foreign exchange saved on maize imports is totally or only partially lost through reduced exports. And, although maize is a more labor-intensive crop than wheat and oilseeds, it requires less labor than many others, and so the net employment effect of raising the maize price is not obvious beforehand.

Most sectoral programs involve complex tradeoffs, and when alternative programs are considered against each other for their cost effectiveness in promoting certain aims, an additional dimension is added to an already difficult task of analysis. These circumstances have favored the use of a comprehensive quantitative framework that, although it might be too aggregative for many specific questions, would at least permit evaluation of the principal effects of the main policy instruments that could be utilized.

The principal tool of analysis in this study, the CHAC model, is sector-wide in its coverage of short-cycle crops and their markets and inputs, and it is decomposable for more detailed analyses into models for particular localities. Part Three of the book mostly comprises local models of this kind. A few of the CHAC model's salient characteristics are as follows.

• Specificity: Although it is reasonably aggregative, CHAC nevertheless contains enough detail for decisionmakers readily to interpret a solution relevant to their concerns. There are thirty-three crops, twenty producing areas, thirty-one representative farms, and monthly requirements for some of the inputs.

• Agronomic data base: The vast majority of the model's coefficients are derived from field-level data on production conditions. An advantage of this characteristic is that most of the model's parameters are in principle verifiable, and the model can be constructed via a dialogue between model builders and field technicians.

• Sectoral behavior: Rather than giving "frontiers" or maximal technological possibilities, the CHAC solutions are intended to be representations of particular kinds of behavior on the part of the sector's agents. Farmers are assumed to be profit maximizing, subject to meeting their subsistence consumption requirements, and in some cases averse to risk, and their individual production volumes are too small to influence market prices. Spatial price differentials persist, partly as a reflection of imperfections in the transport network, but, subject to these conditions, markets clear nationally in the competitive manner. (The possibilities of local monopolies in some crops are explored in chapters 12 and 13.) Consumers' behavior is described by price-responsive demand functions. Alternative behavioral hypotheses could be accommodated in the model, but the point is that it simulates some recognizable form of behavior—subject, of course, to limitations of data, aggregation biases, and other errors of specification.

• Interdependence: Efforts have been made to reflect in the model some of the pervasive interdependence that characterizes the agricultural sector: crops compete for the same locally fixed resources; regions compete in the supply of the same crop; domestic and imported sources of supply compete with each other; substitution effects are allowed for in domestic demand; and there are multiple markets for some crops.

• Comparative statics: CHAC is a static model and, with appropriate revision of policy assumptions or exogenous variables, can be solved in the comparative statics manner. The revised solutions do not, however, necessarily correspond to any particular year; they represent a new equilibrium toward which the sectoral system would tend, but no assumptions are made regarding the speed of adjustment. Truly dynamic effects regarding the time path of adjustment are beyond the scope of this kind of model. This is, of course, a weakness, but empirical knowledge about adjustment mechanisms is deficient. In this study the comparative statics framework has been chosen: it does help differentiate policy packages according to the different states of equilibrium toward which they push the sectoral system.

Overview

The book is organized in four parts. Part One is devoted to sector-wide studies: descriptions of the elements of CHAC and its applications. Two chapters of this section are reprinted, with minor revisions, from the earlier study (Goreux and Manne 1973) for the sake of completeness. Chapters 2–4 discuss the model, and chapters 5 and 6 examine its applications. (Each chapter is summarized in more detail in the next section.)

Part Two is devoted to the issue of producers' risk, an area in which the original CHAC was weak. The methods of Part Two were developed subsequently and then incorporated into some of the district-level models of Part Three. Given the importance of risk for agriculturalists, Part Two goes beyond linear programming models and attempts to measure farmers' attitudes toward risk and to develop for projects screening procedures that effectively account for the stochastic element in farmers' incomes. Chapters 7 and 8 present and test a modification of the deterministic CHAC market-equilibrium specification (which is given in chapter 3). Chapter 9 reports measurements of subjective attitudes toward risk and subjective assessments of new technologies for a sample of smallholding Mexican farmers. Chapter 10 develops for projects decision rules that in many circumstances enable unambiguous project rankings on the basis of the probability distributions of their outcomes, and the chapter is particularly relevant for extension programs involving new plant varieties and cultivation practices.

Part Three contains the district-level models. They are applied to a variety of concerns, including: employment generation, export supply responsiveness, rates of farmers' participation in extension programs, evaluation of irrigation investment, management of livestock herds, and the social tradeoffs arising from agrochemical pollution of rivers. Most of these models (see chapters 11, 12, 13, 15, 16, as well as chapter 8) were derived from CHAC district models, and some were enlarged with further local detail. Chapter 13 was applied to issues of pricing policy, along with CHAC, but these models were generally not as directly linked to the policy process. They were developed later than CHAC, and the staff limitations did not permit a full program of use of local models, although it was contemplated in COCOSA at one point. Nevertheless, their potential usefulness is high, and the methodological lessons of this section are many.

The volume concludes with Part Four, a single chapter on data management and computation. The amount of space allocated to these matters does not reflect their importance in an applied study, which is indeed considerable. The field is evolving rapidly, and generalizations are difficult to come by; chapter 19 is intended primarily to sensitize the economist or policymaker to some of the necessities for model management procedures as regards computational issues.

Chapter Summaries

A description of the CHAC model is provided in chapter 2, and most of its applications are then discussed in chapters 5 and 6. Even though these applications were made well after the material of chapter 2 was written, the model's structure did not have to be changed much. At a crop-specific and region-specific level, CHAC attempts to portray the major sources of interdependence among product supply, domestic demand, international trade, factor inputs, and government policies for agriculture. Although it is a maximizing model in mathematical terms, in economic terms it is a simulating model: it is designed to compute the effects of specified policy packages, and changes in purely external factors, on the sector's behavior as defined by production and employment levels, prices and incomes, and other variables.

On the product supply side, CHAC uses an activity analysis representation of alternative production possibilities, which means that most of its data requirements are met by agronomic information on cropping practices. In this way, the model implicitly embodies hundreds of cross-supply elasticities (or functions), and it would have been impossible to estimate them by econometric methods. As chapter 2 explains, considerable attention was given to the representation of factor supply possibilities, and production costs were specified via the individual factor supply activities, and not through aggregate cost coefficients in the cropping activities. This convention has facilitated experimentation regarding changes in both factor costs and factor market structures.

On its treatment of product supply, CHAC follows in the tradition of Heady and others (1963, 1975). For the market structure and consumer demands, the major predecessors were the single-product, nonlinear optimization models of Takayama and Judge (1964, 1971). Overall, perhaps CHAC's closest antecedent was the French national agricultural planning model (Farhi and Vercueil 1969). For the information of interested readers, models embodying elements of the CHAC market-simulating approach have been constructed for other countries in recent years: for Central America (five countries together, Cappi and others 1978), Nicaragua (Fajardo 1977), Brazil (Kutcher and Scandizzo 1981), Zambia (Candler and Pomareda 1978), Tunisia (Condos and Cappi 1976), and the Philippines (Kunkel and others 1978).

An analysis of CHAC's demand structures and market-simulating character is found in chapter 3. The treatment of demand is rather basic to the entire study, for it is the key to the model's market-simulating character. By maximizing the sum of consumer and producer surplus, CHAC simulates the market outcome in prices and quantities. Consumer surplus is defined with respect to given demand functions, and producer surplus with respect to crop supply functions, which are unknown beforehand. Chapter 3 discusses systematically the CHAC maximand, starting with the single-crop case and then introducing interdependence in demand. An earlier version of the chapter was published in the *American Journal of Agricultural Economics* (Duloy and Norton 1975). This version has been revised and expanded to include a discussion of the linear programming dual solution.

The ways in which the coefficients of CHAC were constructed from a large amount of raw data are explained in chapter 4. As explained, in virtually all cases the existing data could not be utilized directly but had to be transformed before they could enter the model. One of the major achievements of the chapter is the definition of a set of technology options for farming that encompasses more than the observed practices. This was accomplished by disaggregating the steps involved in mechanization and by specifying a crop-specific and location-specific sequence of those steps as mechanization proceeds. To convert agronomic data into an economic form for the model, it was necessary to convert field tasks to economic input categories, and then to differentiate input coefficients according to whether they are crop specific, location specific, or both. An immense amount of systematic work went into the effort reported in this chapter, but there is no substitute for it if the process-analysis representation of supply behavior is to be realistic.

Chapters 5 and 6 perform most of the task of policy interpretation of the CHAC analyses. Chapter 5 may be regarded as an interpretive, background piece, and chapter 6 is a policy document. In chapter 5 the major issues analyzed are sector-aggregate supply responsiveness, capital-labor substitution, international and interregional comparative advantage, pricing policies for grains, and model validation. Capital-labor substitution is studied by using the model to construct isoquants of varying types, and some conclusions are offered about the effects of wage policies. Some static and comparative static computations of the sectoral rates of income distribution and employment are also made with the model to delineate the initial situation more clearly. Pricing policy is addressed on the product supply side, and CHAC is utilized to compute effects on competing crops and exports of the grain price increases, as well as the magnitude of stimulus to grain production. Comparative advantage is discussed both in sectoral and regional terms and as it affects export strategies. The CHAC ranking of crops by comparative advantage for export turns out to be fairly stable over time, and the actual pattern of agricultural exports has in fact evolved in the direction indicated by the model. Finally, validation and further research issues are discussed. Although standardized, comprehensive procedures for validation of programming models such as this do not exist, specific aspects of the model's behavior can be used to evaluate its realism.

In chapter 6 a policy planning document of the Mexican government is reproduced. The document attempts to lay out guidelines for a coordinated agricultural strategy, and its principal concern is the generation of additional agricultural employment. The CHAC model and other analytical methods are used in a sector-wide review of principal policy programs. For employment, effects on job creation are quantified (for irrigation, land clearing, agricultural extension, export promotion, and a possible limiting of the pace of mechanization). One central conclusion is that it will not be easy to prevent further increases in rural underemployment, and that sector programs will have to be strengthened and modified to generate the required number of new jobs. Throughout the discussion, agricultural exports and irrigation stand out as the major factors favorable to increased employment. Chapters 12 and 13 provide additional evidence at the district levels on the employment question; they also conclude that exports are crucial. Chapter 6 does suggest, however, that there is a natural temporal dimension to the employment issue, that eventually the dynamics of rural and urban growth will begin to alleviate the problem. Other issues—such as agricultural extension, price support programs, and new institutional structures—also are discussed in the chapter. Some of its recommendations have been put into effect.

Chapter 7 sets the foundation for the treatment of risk in models such as CHAC and in several of the volume's district models (those of chapters 8, 12, 13, and 15). It shows how to modify the market equilibrium structures of chapter 3 to incorporate the assumption that farmers are averse to risk associated with their production decisions. Once this extension of the model is made, the concept of a market equilibrium is no longer as simple as it was before. In particular, the nature of the equilibrium now depends on how farmers form their anticipations of prices and yields. The empirical fruit of this chapter is found in other parts of the volume, and those results show that this way of representing behavior in response to risk substantially improves the models' realism.

Chapter 8 presents the most systematic tests of the theory developed in chapter 7. Two models are solved with varying assumptions about the degree of farmers' risk aversion, including, at one extreme, the hypothesis that they are in their attitudes neutral to risk. One of the models is a truncated version of CHAC and incorporates eight irrigation districts; it therefore may be considered illustrative of complete sectoral models. For both models, the inclusion of risk-averse behavior markedly modifies the results toward a better replication of observed cropping patterns. The position and curvature of crop-specific, supply response functions also are shown to be affected by the specification of risk. The procedure deals with two sources of variations in farm income: fluctuations in yield and market price. It requires for data a location-specific time series on prices and yields. Although this is a modest requirement, unfortunately it often is not met for the poorer, nonirrigated zones where risk is likely to be of greater concern. In these areas, choices of cropping patterns are somewhat less important, and packages of new technology are correspondingly more important. Consequently, for analysis of the role of risk for these areas, the procedures of chapters 9 and 10 are likely to be more relevant.

Whereas chapter 8 uses methods of model simulation to assign a numerical value to farmers' degree of risk aversion in the aggregate, chapter 9 reports on an ambitious attempt to measure directly, by means of field surveys, individual farmers' utility functions with respect to risk. In the area near Chapingo in the Mexican Central Plateau, eighty-four farmers were interviewed to determine both their degree of risk aversion (preferences) and their subjective assessments (beliefs) of the riskiness of a new package for higher-yield maize production that the extension service was promoting. By and large, the farmers of this sample correctly believed that mean yields from the new technique would be much higher than those of the existing technique, and their degree of optimism about the new technique increased with the length of time they used it. In most cases, according to the subjective probability distributions, the new technique was first-degree stochastically dominant over the existing technique; that is, monetary returns from the new technique were higher at all probability levels. Adopters tended, however, to have larger landholdings and to be wealthier. Nevertheless, it is likely that the new, high-yield maize package will ultimately be adopted by most of the farmers in the area.

In assessing preferences, utility functions with respect to wealth were measured, and some of these are shown graphically. The hypothesis of a common utility function for all farmers in the area is rejected by the evidence. For the few cases in which the new technique was not perceived to dominate the other one stochastically, the utility functions were utilized to predict the farmers' decisions on adoption. In most cases, the prediction was correct; that is, the farmer appeared to act as if the rule for his decision were the maximization of risk-averse utility.

The study in chapter 9 is bold and rewarding; for purposes of evaluating agricultural programs, however, it would be expensive and difficult to carry out. In chapter 10 it is pointed out that, worldwide, preference functions have been mapped out only for a few hundred farmers to date. Therefore, chapter 10 discusses a method of ranking risky programs, without knowing farmers' attitudes, by means of stochastic dominance rules. The rules are defined and are shown to have a natural relation to the definition of the utility function for the decisionmaker who is facing risky prospects. First-degree stochastic dominance (FSD) discriminates the least—it is the coarsest sieve—and it applies when one prospect's expected outcome is better than the other's at any probability level. In effect, it is useful when risk is not of concern in the preference function or when any attitude other than one that is risk preferring is admissible. Second-degree stochastic dominance (SSD) is a stronger rule that rejects more options than does FSD. Its use corresponds to risk-averse, maximization of expected utility. Third-degree stochastic dominance (TSD) is the finest sieve, and it is applicable when the decisionmaker is averse to risk and utility maximizing, and when he becomes decreasingly risk averse as he becomes wealthier. Applications are made to these rules in several cases: a pest-control program, a choice among wheat varieties, the high-yield maize program studied in chapter 9, choices of fertilization rates for wheat, and alternative packages for rice production. The examples demonstrate rather persuasively that useful and easy-to-apply rules exist for screening risky agricultural innovations. The chief limitation to more widespread use of these rules would appear to be lack of sufficient information on the variability of returns from new techniques under differing ecological conditions and husbandry practices.

Part Three, exclusively comprising regional and district-level models, begins with chapter 11, in which issues pertinent to regional (as opposed to sector-wide) analyses are discussed. The chapter addresses two major methodological issues: formal representation of the reactions of the rest of the sector to a policy change in one region and procedures for the validation of a linear programming model of this kind. The first issue obviously arises in the context of local policies that influence sector-wide prices through effects on local output levels. But it also arises in the case of sector-wide policies, such as interest rate changes, that are evaluated through use of a regional model that may or may not be "representative" of the entire sector. When the region-sector interdependence is expressed in variations in product prices, it can be captured by the specification of parameters for product demand, and chapter 11 spells out the procedure for each circumstance.

On the question of validation, the second issue, chapter 11 proposes two kinds of model tests: one that checks whether the model's feasible space in fact includes the observed values (the "capacity test") and another that tests the validity of the competitive market assumption (the "marginal cost test"). In the latter, it is found that the competitive assumption is not rejected for twelve of the fifteen crops and that the deviations for the three crops were explainable by model misspecification or special circumstances in reality. The regional model (PACIFICO) used in these analyses also is applied in some policy issues. It is found that credit subsidies are a cost-effective way of raising farm incomes, but that their effectiveness declines rapidly as the subsidy rate is increased. Also, a derived demand curve for labor is traced with the model, and it is found that as wages increase labor income rises. This result implies that at the regional level the elasticity of capital-labor substitution is less than 1—not along an isoquant but along a total response surface with output free to vary.

Chapter 12 applies the model techniques of this volume to the topic of vegetable exporters' responses to risk and market factors. Net import demand functions for the United States and Canada are given, and the model generates crop-specific and seasonal export levels and prices and sales to the domestic market. The region studied is the Pacific Northwest of Mexico, which also is the subject of chapters 11, 16, and 18. The farmers of the region are found to be risk averse, with tests conducted according to the methods of chapters 7 and 8, and the hypothesis of a monopolistic position in the export of winter vegetables is rejected by the experiments. The model is applied first to the question of increases in export demands, which could arise either from a reduction in U.S. tariffs or from other causes. It is found that the output and employment response is quite large, and also that domestic tomato prices in Mexico would fall because of certain complementarities in production. Mexican growers have flexibility in their seasonal scheduling of plantings. Therefore, the model is next applied to the question of optimal timing of vegetable exports, and some recommendations in timing changes—toward improving producers' revenues—are made. (The growers' associations in the Northwest adopted some of these recommendations and have requested updating of the model for continued use.) Finally, the allocation between domestic and export markets and wage-policy issues are analyzed. It is found that vegetable export levels are quite responsive (negatively) to increases in the wage rate, but that increases in the wage rate nevertheless lead to a larger total wage bill (as was found in chapter 11).

The possibilities for the creation of agricultural employment are examined in a systematic manner in chapter 13 for an irrigation district in which plot sizes are small and there is substantial underemployment. A district-level model (TOLLAN) with detailed choices for farming technology and with three classes of farm size was constructed for the analysis. One finding is that, given the small plot sizes, factor price variations do not lead to significant employment gains. It is found, however, that shifts in the composition of product demand (via increased vegetable exports, for example) do generate appreciable employment gains. Also, careful review of the existing production technologies suggests that it may be useful to make available to farmers some smaller-scale options for mechanization than those they now have. Finally, some conflicts between the policy goals of employment and rural income distribution are discussed for the case of this producing area. In addition to the policy conclusions, chapter 13 will be of methodological interest to some readers for its computation of machinery-labor use isoquants. It also contains the book's most thorough investigation of existing options at the local level for farming technology.

The subject of chapter 14 is the rural family's response to projects providing extension with credit, and the Plan Puebla project is studied as an example. It develops a theoretical model of a family's allocation of labor—between on-farm and off-farm activities, and also over time—and then a linearized version of the model is solved under parameter values appropriate for the Puebla area. The model includes a "social" variable representing the amount of the farmer's time spent in activities related to the organization of social groups. In Plan Puebla, these groups are means of gaining access to additional financial resources and, thus, of relaxing the credit constraint. For poor farmers with insufficient financial means to adopt the modern technology on their own, the Plan Puebla structure offers, to a certain extent, an opportunity for farmers to use their own time as a substitute for wealth (collateral). The model, however, posits off-farm employment as an option, and so the new farming technologies must have a labor productivity that is greater than the wage farmers can earn in temporary jobs. Risk considerations also are specified, but in a different form than in chapters 7–10. Whereas those chapters assume that the farmer is averse to the *variability* of income per se, in chapter 14 farmers are concerned with the possibility of their incomes' falling below a certain (subsistence) level. Other interesting features of the model include the effects of learning by doing in farming activities and a difference between the actual and perceived variance of yields from the package of new technology. In the empirical results, the "typical" farmer of the Puebla area is found to divide his time among farming, off-farm work, and organizational activities and to apportion his land to the traditional and modern farming technologies. The percentage of his land devoted to the modern technology is found to be especially sensitive to the gap between the actual and perceived variance of returns from it, to the off-farm wage rate, and to the level of requirements for subsistence consumption.

Chapter 15 discusses ways in which the district-level models may be used, directly and indirectly, in the process of evaluating investment projects. Special emphasis is given to the interdependence among policy choices: between investment outlays and policy decisions on pricing, between different kinds of investment projects, and between local and sector-wide decisions. The exposition is developed through a series of examples—a series of computer experiments conducted with models for the Río Colorado (Northwest region) and El Bajío (Central Plateau) producing areas.

This theme is treated in greater detail in chapter 16, which reports a detailed exercise in investment identification for the Río Colorado irrigation district. The objective is to define the level and composition of an investment program for the district, under different rates of interest and under different assumptions regarding relevant parameter values. Fifty-eight distinct investment activities are considered in a district-level programming model. The procedure generates estimates of the future expansion of the market for agricultural products from this particular area, estimates that differ from the national rate of agricultural market expansion. The model defines an investment package that maximizes the benefit-cost ratio for a given rate of interest (cost of capital). For this particular case, the results show that both the volume and the composition of the investment program are sensitive to the interest rate. For example, raising the interest rate from 8 percent to 16 percent reduces the justifiable amount of investment by 47 percent and reduces the share of canal lining in the total from 49 percent to 34 percent. The methodological contribution of chapter 16 is the demonstration of a workable procedure by which economic criteria can be employed in project design—that is, in defining the composition of a project in the context of variable output prices. Traditionally, economic guidelines are used only to accept or reject a project after its design has been completed. Often, the more important decisions are made at the design stage, especially regarding investments that comprise many small-scale, on-farm components.

In chapter 17 the study's only livestock model, for the dairy industry in the north-central part of the republic, is presented. It is an unusual chapter in that it presents an incomplete model, and it discusses much of the detailed labor that normally goes into data reconciliation and model construction but is typically not presented in final reports. The work on the model was suspended in midcourse for reasons that are, no doubt, familiar to other researchers: the Mexican government group (in the Ministry of Agriculture) responsible for application of the model was disbanded in a period of a few months because of promotions and transfers, leave for study abroad, and marriage. Nevertheless, we feel that this chapter offers many rewards to the reader who is interested in the livestock sector and is willing to persist through detailed, numerical reasoning on technical issues. It clearly shows that the art of building useful models requires the patient application of logic to initially inconsistent data and that, even without entirely reconciled data, a model structure can lead to the formulation of many interesting and specific policy issues. Should the study be continued where it was left off, the stage is set for a return visit to La Laguna to fill in questionnaires on a few, well-identified points.

Chapter 18 addresses the need to develop appropriate frameworks for the analysis of the ecological side effects of agricultural development. The specific concern of this chapter is the increased salinity of ocean lagoons caused by the greater amounts of water diverted by irrigation works. For Mexican agricultural products, the issue may be thought of as the interdependence of shrimp and wheat production, and it is becoming a serious economic concern for the fishermen along Mexico's northwest coast. In this chapter, the first quantitative analysis of the problem is presented through dynamic rather than linear programming. The authors show how to quantify the tradeoffs in irrigation-lagoon exploitation, and they then explore the social and institutional structures for which the model results are potentially applicable. They also discuss the institutional circumstances under which a corrective policy of taxation might be feasible.

Computational issues are fundamental to an applied study. The ease with which a model can be revised and solved determines the analysts' ability to respond to policy issues. Many steps are involved in translating the data into a form for computer processing and then into summary economic results, and the possibilities of errors, and delays because of them, mount up geometrically if the computational aspects are handled inefficiently. The CHAC model's having a matrix generator, for example, meant the difference between a few days and several weeks to a few months for the formulation and solution of a significantly revised version of the model. It is not too strong to say that, in a practical setting, an inflexible model that cannot be readily altered to address different hypotheses and policy issues is virtually useless. Chapter 19 discusses approaches to managing the computational aspects of sector models. It attempts to make the economist aware of the importance of systematizing data management and computations, and it discusses some critical steps in this process. This is a field that is changing rapidly in technology, so rigid rules should be avoided. The chapter offers, however, many useful guidelines to those embarking on construction of large-scale models.

Concluding Observations

When a multifaceted study such as this is codified in print, with the intent of presenting those aspects of it that might be applicable in other circumstances, it is natural that the more technical and quantitative side should receive greater emphasis. This aspect is more easily described and is capable of greater generalization than are the institutional aspects. It should be evident, however, that, without the appropriate institutional setting, a study such as this could not be useful and probably should not be attempted. A few reflections of a broader nature about this study would, therefore, seem in order.

As the experience of this study indicates, a favorable environment does not require massive amounts of manpower devoted to technical matters of model construction, but it does require effective communication between four kinds of persons: policymakers, quantitative economists, systems analysts or other computational specialists, and technical specialists (such as agronomists). In some respects it is preferable that the model team be small, in order to facilitate effective communication within the model group and with other groups.

Another requirement is that the model be a flexible instrument that can be revised readily and addressed to different issues that arise. In this respect, the CHAC model has been only barely adequate in our judgment. It has been possible to use it for varying purposes, but the revision and solution process has not been as easy as would have been desirable. (As of this writing, however, a completely new version has been built and used in a new presidential administration, and the computational management has been improved.) A model system also should be sufficiently flexible that components of the model may be used for some other purposes. The development of the district models from CHAC submodels is one illustration of this flexibility; in addition, the model's use as an organized data bank should not be overlooked. CHAC contains a large set of organized information on crop requirements for labor, water, credit, and the like, and this information is interesting in its own right. On some occasions, CHAC's employment norms and other input coefficients were used for advising, on short notice, on the implications of emergency plans for crop shifts involving wheat, soybeans, cotton, and other crops.

An exercise in applications such as this one is more a process than an isolated event. The model evolves over time, but at any moment it represents—notwithstanding its inadequacies—a best available picture of the sector in numbers. At minimum, the model provides a dimension of consistency to sectoral policy planning, and, merely by its presence, it sometimes induces more rigor in other analyses of sectoral issues. Even without being solved for response functions, CHAC was able to quantify the degree of seasonality in agricultural employment and to answer questions such as the ranking of crops by their competitiveness in export markets. Sometimes the simple solutions of a model are as useful as the sophisticated experiments.

The problems of Mexican agriculture have intensified during the 1970s, but many of the lessons of this study still appear valid. In retrospect, the early forecasts of economy-wide growth in GNP, which were important exogenous inputs to some of the sectoral model projections (see chapter 5, the sections "Basic Macroeconomic Results" and "Measuring the Aggregate Sectoral Supply Function"), were far too optimistic. Nevertheless, the restructuring of exports indicated by CHAC has taken place. Grains accounted for a large proportion of agricultural exports in the mid-1960s, but they required official subsidies. Their disappearance from the export bundle by the early 1970s exerted a strong downward pressure on aggregate agricultural exports, but now the export composition is heavily weighted in favor of crops with better market prospects—and, incidentally, with higher employment intensities.

A similar restructuring away from grain production has occurred for the internal market. CHAC results were used in making the revisions of the maize and wheat prices in 1973–75,[2] but unexpectedly high, overall inflation overtook these increases, so that by 1978 the real prices of grains had not increased over their 1970 level. In fact, they had declined slightly. The model's results serve as a reminder of the importance of price incentives, however—not only for grains in particular but for any kind of cultivation, especially cultivation on the more marginal lands, which appear to have been abandoned in some instances in the 1970s because of considerations of profitability.

CHAC's quantification of potential sources of agricultural employment (chapter 6) stands as the most definitive treatment of that issue in Mexico thus far and, in approximate terms, will remain valid as a policy guideline for some time to come. With the passage of time, several of the recommendations in chapter 6 have been implemented, especially those regarding the overall level of sectoral investment, the extension system, and the institutional reform embodied in the creation of the agricultural programming districts for nonirrigated zones.

Although this kind of model clearly has played a productive role in Mexico and has been adapted to other environments, the need for continuing methodological improvements (as discussed in the concluding part of chapter 5) must be emphasized. The material of Parts Two and Three of this volume already provides the basis for a better sectoral model, and no doubt other researchers will carry on the process of further refinement.

2. CHAC results also figured in the set of analyses used to establish the agricultural support prices in 1979–81, in the context of the production-incomes program of the Sistema Alimentario Mexicano.

References

Candler, Wilfred V., and Carlos Pomareda. 1978. "The Zambian Agricultural Policy Model—Progress Report." Washington, D.C.: World Bank, January 1978. Restricted circulation.

Cappi, Carlo, Lehman Fletcher, Roger Norton, Carlos Pomareda, and Molly Wainer. 1978. "A Model of Agricultural Production and Trade in Central America." In *Economic Integration in Central America*. Edited by W. Cline and E. Delgado. Washington, D.C.: Brookings Institution, pp. 317–70 and Appendix G. Reprinted as World Bank Reprint Series, no. 82. Washington, D.C.

Condos, Apostolos, and Carlo Cappi. 1976. "Agricultural Sector Analysis: A Linear Programming Model for Tunisia." Rome: Food and Agriculture Organization of the United Nations, Policy Analysis Division (October).

Duloy, John H., and Roger D. Norton. 1975. "Prices and Incomes in Linear Programming Models." *American Journal of Agricultural Economics* (November), pp. 591–600.

Fajardo, Daniel A. 1977. "Policy Analysis of Nicaraguan Agriculture: A Mathematical Programming Approach." M.S. thesis. Lafayette, Ind.: Purdue University (August).

Farhi, Lucien, and Jacques Vercueil. 1969. "Recherche pour une planification cohérente: Le modèle de prévision du Ministère de l'Agriculture." Monographies du Centre d'Econometrie, VI. Paris: Éditions de Centre National de la Recherche Scientifique.

Goreux, Louis M., and Alan S. Manne (eds.). 1973. *Multi-level Planning: Case Studies in Mexico*. Amsterdam/New York: North-Holland/American Elsevier.

Heady, Earl O., and Alvin C. Egbert. 1963. "Activity Analysis in Allocation of Crops in Agriculture." In *Studies in Process Analysis*. Edited by Alan S. Manne and H. M. Markowitz. New York: Wiley.

Heady, Earl O., and Uma K. Srivastava. 1975. *Spatial Sector Programming Models in Agriculture*. Ames, Iowa: Iowa State University Press.

Kunkel, D. E., G. R. Rodriguez, L. A. Gonzales, and J. C. Alix. 1978. "Theory, Structure, and Validated Empirical Performance of MAAGAP: A Programming Model of the Agricultural Sector of the Philippines." *Journal of Agricultural Economics and Development*. Vol. 8 (November), pp. 1–25.

Kutcher, Gary P., and Pasquale L. Scandizzo. 1981. *The Agricultural Economy of Northeast Brazil*. Baltimore, Md.: Johns Hopkins University Press.

Takayama, T., and G. G. Judge. 1964. "Equilibrium among Spatially Separated Markets: A Reformulation." *Econometrica*. Vol. 32 (October), pp. 510–24.

———. 1971. *Spatial and Temporal Price and Allocation Models*. Amsterdam: North-Holland.

Part One

Sectoral Programming

2

CHAC: A Programming Model for Mexican Agriculture

JOHN H. DULOY AND ROGER D. NORTON

THIS CHAPTER provides the basic description of the CHAC model. As the model was used over time, it underwent some modifications, which are discussed in chapter 5 in connection with the model's applications. The material of this chapter still stands, however, as a complete exposition of the basic structure of CHAC. Both chapters 5 and 6 discuss the ways in which CHAC was addressed to policy concerns; chapter 4 explains the data base and the development of model parameters; and chapter 3 formally analyzes the market-simulating aspects of CHAC.

Background

Agriculture in Mexico, as in most developing economies, is a major source of employment and foreign exchange earnings. In 1970, almost half the country's labor force was agricultural, and directly and indirectly through both raw and processed products, agriculture accounted for over half of export earnings. The crucial role played by agriculture in the

Note: This chapter first appeared as Duloy and Norton (1973*a*). Permission of the original publisher to use material here, with some editorial changes, is gratefully acknowledged. The authors are grateful to Leopoldo Solís for his encouragement and continuous support and his patience during the long gestation period of this model. We also gratefully acknowledge the support of the Banco de México, which devoted considerable resources to this study. The agricultural study was conceived by Louis Goreux and Alan Manne, who gave us useful comments and criticism throughout, and it has drawn upon the earlier work of Luciano Barraza. Dr. Barraza provided helpful comments on several aspects of the study. The enormous burden of constructing most of the 80,000 coefficients in the model has been borne with unflagging energy and goodwill by Luz María Bassoco, then of the Ministry of the Presidency, and Teresa Rendón, then of the Banco de México. Apostolos Condos and Donald Winkelmann provided helpful discussions of a number of aspects of the model's structure. Gary Kutcher became a (nearly prostrate) human link between the model and the computer and gave useful comments on the model's design.

economic development of Mexico is widely acknowledged.[1] It was partly because of this pivotal position that agriculture was singled out for detailed analysis in a multilevel planning study. The sector offers an opportunity to explore the relations between sectoral policies and economy-wide development strategies. Agriculture also provides an example of strong linkages between investment decisions and sector-level policies. Agricultural trade policies and policies on the pricing of inputs and products significantly affect the rates of return estimated for individual projects. Therefore, the initial objective in constructing the agricultural model, CHAC, was to formalize the major aspects of micro-level and sectoral decisionmaking.

The study also was designed to serve both the Mexican government's interest in analytical tools for its planning of sectoral policies, and the World Bank's interest in the methodology of project appraisal techniques and in general policy-planning models. As a tool for policymakers, CHAC is designed to be addressed to questions of pricing policies, trade policies, employment programs, and some categories of investment allocation. It is not particularly well suited for analyzing agricultural research and extension programs, crop insurance policies, or credit policies. It is structured so that it is a simple matter to change factor prices—including costs of labor, capital, water, and agrochemicals—and to represent subsidies to production by crop and geographical area. The prices received by farmers and paid by consumers for internationally tradable commodities also may be adjusted readily to reflect tariff, taxation, and exchange rate policies.

Commodity demand functions are included within the structure of CHAC; hence, prices are determined by demand as well as supply conditions.[2] Because relative product and factor prices are the dominant policy instruments in agriculture, this feature of the model gives scope for a wide variety of policy experiments. The production side of the model is decomposable into submodels for each of twenty geographical areas (referred to as "districts"). Under appropriate assumptions on prices, each submodel may be solved separately. This ability permitted checking each of the submodels prior to its inclusion in CHAC, and it also facilitated use of the submodels, in stand-alone fashion, for locale-specific analyses. The district models of chapters 8, 11–13, 15, and 16 were in fact based on CHAC submodels.[3]

1. See, for example, Solís (1970).
2. Some programming models yield marginal costs of production, or "supply prices," but, in the absence of demand functions, these do not yield market equilibrium prices. See, for example, Heady, Randhawa, and Skold (1966) and Piñeiro and McCalla (1971).
3. See chapters 15 and 16 for further discussions of models for investment evaluation.

The version of CHAC reported here covers only short-cycle crops.[4] Tree crops, livestock, forestry, and fisheries have been excluded. There is significant interdependence between the short-cycle crops and livestock through forage production and pricing and through allocation of labor and capital. There is also competition with some long-cycle crops for land and other resources. Nevertheless, it was decided to limit the scope of this version of the model in the interests of a more thorough treatment. (A companion model for livestock is reported in chapter 17 of this volume.) In total, CHAC contained some 1,500 equations and 80,000 nonzero coefficients in its first version.

Overview

CHAC is a sector-wide model in that it describes total national supply and use—production, imports, domestic demand, and exports—for the thirty-three principal short-cycle crops in Mexico (see table 2-1).[5] It is a one-period model that may be solved for the base year or a future target year. The timing of investment decisions cannot be studied, but investment choices may be included in the model. On the demand side, consumer behavior is regarded as price-dependent; thus, market-clearing commodity prices are endogenous to the model.

Basically, CHAC has been structured from the viewpoint of microeconomics rather than from that of macro theory. One reflection of this is the level of disaggregation: individual farm products are distinguished and factor inputs are disaggregated seasonally. A less trivial reflection of the microeconomic orientation is the model's description of a particular form of market equilibrium, in prices and quantities, with corresponding representations of producer and consumer behavior. In most of the solutions reported here, the market form is taken to be competitive, but the same programming structure could be used to represent the sector, or parts of it, as a monopolistic supplier of agricultural products. Purely as a descriptive matter, the competitive market mechanism is closer to the actual processes that determine production and prices in Mexican agriculture, and it has therefore been adopted as the basis for the model. Government policies—such as price supports, import quotas, and input subsidies—and their effects on producers' incomes, employment, and other variables are evaluated as interventions in a basically competitive market.

4. Annual crops plus sugarcane and alfalfa.
5. The thirty-three crops represent more than 99 percent of the value of production of short-cycle crops. A list of these crops and their production levels is given in table 2-1.

Table 2-1. *Area Harvested and Value of Production of the Crops in CHAC, 1966–67*

Crop[a]	Irrigated		Nonirrigated		Total	
	Area harvested (hectares)	Value (thousands of 1967 pesos)	Area harvested (hectares)	Value (thousands of 1967 pesos)	Area harvested (hectares)	Value (thousands of 1967 pesos)
Maize	441,939	326,407	7,844,996	7,681,953	8,286,935	8,508,360
Cotton fiber	415,997	2,521,700	279,382	887,876	695,379	3,409,576
Sugarcane	85,280	445,141	402,318	1,572,298	487,598	2,017,439
Beans	45,569	109,628	2,194,453	1,704,005	2,240,022	1,813,633
Wheat	421,685	1,113,180	304,910	304,050	726,595	1,417,230
Sorghum	268,037	487,479	307,823	414,040	575,860	901,519
Green alfalfa	24,784	151,465	83,347	602,982	108,131	754,447
Tomatoes	18,713	641,201	6,000	91,000	24,713	723,201
Rice (unhulled)	58,482	211,843	94,160	206,916	152,642	418,759
Sesame	36,349	71,281	215,760	278,877	252,109	350,158
Safflower	100,679	201,768	64,254	126,633	164,933	328,401
Strawberries	3,371	45,412	5,454	282,849	8,825	328,261
Tobacco	2,348	25,913	37,260	277,127	39,608	303,040
Watermelon	3,272	20,669	30,228	258,637	33,500	279,306
Potatoes	3,428	33,550	30,854	228,382	34,282	261,932
Green chile	5,975	57,337	36,527	150,417	42,502	207,754
Chickpeas	28,533	41,538	132,574	159,109	161,107	200,647
Barley	14,691	41,792	226,059	135,626	240,750	177,418
Pineapple	n.a.	n.a.	9,924	175,913	9,924	175,913
Dry chile	765	8,740	23,619	144,757	24,384	153,497
Dry alfalfa	31,989	137,819	n.a.	n.a.	31,989	137,819
Soybeans	42,601	122,771	11,642	12,635	54,243	135,406
Cantaloupe	7,167	68,689	8,567	65,272	15,734	133,961
Peanuts	5,452	16,473	57,229	103,982	62,681	120,455
Squash	n.a.	n.a.	16,150	110,279	16,150	110,279
Onions	1,725	12,306	15,281	72,014	17,006	84,320
Oats	963	1,993	74,457	54,131	75,420	56,124
Lima beans	1,307	2,283	46,393	51,422	47,700	53,705
Garlic	602	5,430	5,231	30,305	5,833	35,735
Flaxseed	4,369	10,917	14,133	22,755	18,502	33,672
Totals	2,076,072	7,434,725	12,578,985	16,206,242	14,655,057	23,640,967

n.a. Not available.

Source: Secretaría de Recursos Hidráulicos (SRH), Dirección de Distritos de Riego and Secretaría de Agricultura y Ganadería (SAG), Dirección de Economía Agrícola.

a. This list excludes cucumbers and cottonseed, and shows only one of the two forms of barley (that which is harvested as the whole plant, not that harvested only as the barley grain). There are, however, thirty-three crops altogether in the model.

Factor markets are specified in less detail. All purchased inputs and services of machinery and draft animals are priced at observed market prices, except for those experiments in which input prices are explicitly subsidized or taxed.[6] The supply of these inputs is assumed to be infinitely elastic. Water charges are included at the level of actual pumping costs for well water and administrative levies for the release of reservoir water. Because seasonal and annual limits to the availability of water are specified, the effective water price is augmented by a rental element. The rent is explicit in the dual solution. Cultivable land is specified in limited quantities. For land, the opportunity cost constitutes its entire valuation in the model.

The labor supply functions are based on observed wages by region, and the labor market equilibrium is viewed as competitive. It is assumed that the services of hired labor are offered at observed market wages, and that farmers offer their services at a positive reservation price below the market wage. A low but positive value has been taken for the farmers' reservation wage. Despite the underemployment that is characteristic of Mexican agriculture, farmers' time always has an opportunity cost. This cost may reflect either the production forgone or the opportunity to engage in traditional social activities. (Through this device, CHAC allows for fence-mending and other nonmarket production activities.)

This formulation means that the efficiency wage will be positive even during seasons in which there is underemployment. To this extent, the structure of CHAC follows the "subsistence" wage tradition of Ricardo, Marx, and W. Arthur Lewis. It is, however, the money wage and not the real wage that is fixed. Since product prices may change, the real wage is variable. Whereas the wage represents total earnings for hired labor, farmers in CHAC earn a rent in some seasons from land, water, and their labor. Thus, total annual earnings for farmers are greater than their wage earnings and are variable in the model.

In the mathematical sense, CHAC is an optimization model. In all solutions, the same maximand is used: the sum of producers' and consumers' surpluses. This use ensures that the optimal solution will be a competitive market equilibrium.[7] The model is not solved under, say, an employment maximand. Although greater employment may be a policy goal, it is not clear which policy instruments, if any, could implement an employment-maximizing solution. With CHAC, the implications for employment of specific policy changes are simulated. When policy changes are involved,

6. For the original version of CHAC, all prices were defined in 1968 pesos; for later versions for the years 1972 and 1976, prices were given in pesos of those years.

7. There may be departures from a competitive equilibrium when constraints are imposed to calculate the effects of government policy interventions.

the instruments are made explicit in the model; thus, the sectoral implications of a policy change can be estimated. This approach is amply illustrated in chapters 5 and 6; as these chapters indicate, employment has, in fact, been one of the major concerns of Mexican policymakers.

To address this and other policy questions, a wide range of technological choices in production were included in the model, along with domestic demand and trade activities. There are more than 2,300 different production techniques for thirty-three crops in twenty districts, ranging from completely nonmechanized to completely mechanized and including different degrees of efficacy for irrigation as well as nonirrigated techniques.

Basic Structure of the Model

Separation of sources of supply and demand, for both products and inputs, is the basic rule under which CHAC is specified. For each crop, there are production activities differentiated by location and technique; for twenty-one of the thirty-three crops, there are also importing activities. There are corresponding activities for sales on the domestic and export markets. In effect, the model contains multiple-step supply and demand functions for each crop, and these functions for different crops are interdependent. For most crops, the implicit sector-wide supply function contains dozens of steps, and in some cases there are more than 100 steps. The demand formulation is flexible and permits an arbitrarily close approximation to a nonlinear utility function.

The commodity balance equations require the clearing of markets, with simultaneous determination of equilibrium prices and quantities. Production for producers' consumption is given an imputed price equal to the price for commercial production. For nontraded agricultural goods, the prices are completely endogenous. For internationally traded crops, however, they must lie between the import and export prices. The assumed import and export prices may be varied in alternative solutions to reflect different world market conditions and tariff policies. Some sets of prices may be fixed in order to investigate the effects of price support policies. Export quotas are incorporated for a number of crops to reflect the realities of international markets.

The incorporation of demand functions (instead of exogenous product prices) provides a more realistic description of the aggregate market conditions faced by farmers. Moreover, it reduces the tendency of programming models to seek solutions with extreme crop specialization. It also opens the door to investigation of the effects of public interventions under different market structures. With appropriate modifications in the objective function, the same model may simulate a sector that behaves either as a monopolistic supplier of products or as a collection of competitive producers. By casting one of the objective functions as a constraint, it is possible to explore the tradeoffs and complementarities between producers' and consumers' welfare.

The cropping activities in the model also constitute factor demand activities. Factors are supplied by a separate set of activities, and there are balance equations to ensure equilibrium on the factor markets. Some factor supply functions are perfectly elastic (for example, chemicals and capital), and others are perfectly inelastic (for example, some categories of land). In the former category, factor prices are exogenous to the model; in the latter, they are endogenous. In intermediate cases, they are endogenous within limits, and labor falls in the intermediate category. When factor prices are exogenous, the factor is regarded as a national resource; that is, it has an opportunity cost in other sectors or in international trade. There are, however, factors in inelastic supply that have no economic use outside the sector in the short run. Agricultural land and water are placed in this category of sector-specific resources.

The demands for land, labor, and water are defined at seasonal intervals. All other inputs are treated on an annual basis, including services of farm machinery and draft animals. Virtually all farm machinery is in the form of tractors and is used in the irrigated areas of the Central Plateau and the arid northern zones. Because of the nearly uniform year-round climate in these areas, there is not a pronounced degree of seasonality in aggregate demand for machinery services. Hence, and to simplify an already complex model, the seasonal specification has been dropped for machinery.

Labor is divided into three classes: farm owners plus their family labor, hired (landless) agricultural labor, and machinery operators. Local and interregional migrations are taken into account for landless labor and for farmers on rainfed farms.[8] Machinery operators constitute less than 5 percent of the agricultural labor force, and their supply does not appear to be a constraint in Mexican agriculture. They are assumed to be supplied in fixed proportion to machinery services—with an infinite elasticity of supply. The wage for machinery operators is higher than the wage for hired labor. Both types of wages vary among regions, in accordance with observed behavior.

As noted, in any particular month farmers are assumed to be willing to work for an own wage, or reservation price, which is lower than the hired labor wage. Thus, in ascending order of marginal cost, the sources of labor supply are the following: (1) using the labor of the farmer and his

---

8. Throughout this volume, the English words *rainfed* and *dryland* and the Spanish word *temporal* are used interchangeably.

family; (2) hiring local landless labor; (3) hiring surplus, landless labor from other regions; and (4) hiring landless labor away from lower-productivity employment in other regions. The model is structured in a form that permits ready adjustment of all wages, so that various experiments—such as measuring capital-labor substitution—may be conducted (see chapter 5).

Spatial Disaggregation

On the product supply side, each of the twenty submodels represents either irrigated, rainfed, or tropical cultivation, and each covers a particular set of counties or districts, which are not necessarily contiguous.[9] In the case of rainfed and tropical agriculture, the submodels are defined on the basis of annual rainfall and altitude, which determine climatic conditions. Cropping activities are specified by submodel. The submodels are grouped into four geographical regions, and labor constraints are specified for each region to capture the differential labor mobility and wage rates that exist in Mexico. The four regions are the Northwest, the North, the Central Plateau, and the South. In the "subsistence-modern" dichotomy, the submodels for irrigated regions represent the modern, capital-intensive form of cultivation and the submodels for nonirrigated regions tend to represent the subsistence regimes. This is an oversimplification, however: within the submodels for irrigated regions, there is considerable variation in degrees of mechanization, depending on the factor endowments and ecological conditions. Similarly, some areas of nonirrigated agriculture, especially in tropical regions, cannot be properly called subsistence.

Although there are landless agricultural laborers who live in each region and gain a livelihood from part-time work on irrigated farms, the bulk of them reside in the Central Plateau. In this region there is closer access to the major urban centers for part-time work, and small rainfed plots may be cultivated. The dominant direction of seasonal labor migration is between the Central Plateau and the North and Northwest. There also is some movement from the South to the Central Plateau and the northern regions. Because of the distances involved, this is apt to be more permanent than seasonal migration. To help limit the size of the model, seasonal and permanent migration activities have been specified only for the three directions of significant, net interregional flow: from Central Plateau to Northwest; from Central Plateau to North; and from South to Central Plateau. Observed wages for hired labor are lowest in the South and

9. The submodels and the data sources are described in chapter 4, below.

highest in the Northwest. This reflects, at least in part, the relative abundance of labor in the tropical areas and the Central Plateau. Migration is a gradual process, and regional wage differentials have persisted for decades.

The constraints for each district submodel—primarily the annual and monthly bounds on land, water, and farmers—form a block in the block-diagonal production tableau (see figure 2-1). Because the constraints in one block are independent of all other constraints, additional submodels may be added to the system, with appropriate modifications in the coverage of the existing submodels. In this way, the model may be directed to the detailed choices in one geographical area, while treating other areas in a more aggregate fashion.

Demand functions are specified nationally rather than for each geographical submodel.[10] The only exceptions are a few food crops for which separate regional markets are introduced in the South and in the Northwest. (There is a high cost of transport between the tropics and the other parts of Mexico.) It is not assumed, however, that each submodel can supply the "national" market equally well. Spatial price differentials are used, and these reflect the differential transport costs faced by each submodel area, which are based on the historical patterns of transport. Thus, the farmers of the Northwest region receive a lower farm-gate price for vegetables than do the farmers of the Central Plateau, for the latter are located closer to the major urban markets of Mexico City and Guadalajara. For export crops, proximity to major ports determines the spatial pattern of price differentials.

The Production Technology Set

CHAC contains 2,345 cropping activities to describe alternative techniques for producing the thirty-three crops. The range of variation in these activities is described fully in chapter 4, below. Each cropping activity defines a yield per hectare, together with fixed proportions of the following inputs: land (monthly); water (monthly and annual); labor (monthly); machinery services; draft animals; chemicals; purchased seeds; and short-term, institutional credit. The relations between inputs and outputs are those which have been observed (and projected) in each locale, and not necessarily the biological or profit-maximizing optima. In princi-

10. This simplification was adopted because there was insufficient information on the spatial distribution of demand. Moreover, the introduction of local demand activities would enlarge the programming model still further.

ple, the possibilities for movement toward more efficient input–output mixes could be represented through activities for extension services. The existing data, however, do not provide a reliable basis for estimating the costs and benefits of extension.

The ratio of each input and output varies in the submodel for every crop. Some localities have shorter growing seasons than others, and so the number of months of land differs. Fertilization practices vary, especially between irrigated and nonirrigated areas. For irrigation, the amount of gross water release required at the dam depends on the length and condition of the canals. This, too, varies from area to area. In addition, there are systematic variations within many of the submodels in the input–output ratios—particularly in the amounts of water, machinery services, and labor per unit of output.[11]

For some of the irrigation submodels, the land is grouped into four classes based on efficiency of gross water use.[12] For all of the submodels, alternative degrees of mechanization have been specified in CHAC: totally nonmechanized (all power operations done with draft animals), partially mechanized, and fully mechanized (no draft animals used). Obviously, there can be many degrees of partial mechanization, but, in actuality, the choices are discrete and few. For example, plowing is done either with mules or with tractors but not with both. To avoid overstating farmers' short-run flexibility, one-degree changes of technique were permitted between 1968 and 1976, but two-degree changes were not. If the farmers in the area covered by one submodel were using totally nonmechanized techniques in the statistical base year, that submodel contains nonmechanized and partially mechanized techniques only. Similarly, it is assumed that fully mechanized farms may revert to partial mechanization but not to nonmechanization, even under drastic changes in relative prices.

The major advantage of mechanization versus the use of draft animals lies in land savings. One crop can be harvested, and the ground prepared for the next crop, significantly faster with tractors than with draft animals.[13] In some cases, this time saving makes it possible to plant a second crop during the year. This saving is shown in the model by requiring fewer months of land with the more mechanized techniques. If 1

11. For a given crop, variations in water-output ratios occur within five of the ten irrigation submodels. The machinery and labor requirements per unit of output vary within all twenty submodels.

12. This, in turn, is because of terrain conditions, distance from the water source, and state of repair of the canals.

13. One might think that the same savings could be achieved by simply using more mule teams. As anyone who has worked with mules knows, however, there are limits to the number that one farmer can supervise.

hectare is required for ten instead of thirty days, the first month's land requirement is represented by a coefficient of 0.33 instead of 1.0.

Differential land (and labor) requirements also constitute the distinction between two forms of the same crop. For example, alfalfa may be sold green, at a lower price per ton, or left on the land longer and sold dry, at a higher price. In the case of barley, the farmer also faces a choice of harvesting the entire plant and selling it as forage or of using substantially less labor and harvesting only the grain. As in the case of alfalfa, there is a separate demand function for both types of the same crop, so that prices move in the model in response to these production choices. For grain barley, there is an additional component on the demand side, the demand for malt grain. There is a minor element of postharvest, on-farm processing for the grain that is destined for malt, but this is ignored here. There are then two domestic markets specified for grain barley, malt and non-malt.

CHAC contains two markets for cotton also. This arises from the dual product nature of cotton: separate demand functions are specified for both cotton fiber and cottonseed. In the case of cottonseed, the price depends partly on the volume of production of other oilseeds. Hence, in the model the profitability of growing cotton depends on (1) the demand schedule for cotton fiber; (2) the demand schedule for oilseeds; (3) the production surface for competing oilseed crops; and (4) the production technology for cotton.

There is another dichotomy on the production side that is not explicit in CHAC—the *ejido*-private classification of farms. The ejido, one of the products of the Mexican Revolution, is the institution of public ownership of farmland. An ejido farmer is granted lifelong rights to work his land, but he may not sell it or lease it.[14] In some locales, the ejido is associated with collective farming. There has been considerable discussion of the relative efficiency of ejidos and their private counterparts, but the available evidence is ambiguous on this point.[15] The relevant consideration for CHAC is that production costs and yields are defined as averages over geographical areas. These averages include both ejidos and private farms. Since the numbers of farms in each category are stable, their contributions to the averages are stable. CHAC is not addressed to an evaluation of the ejido as an institution, but rather to sector-wide problems of supply, employment, trade, pricing, and resource allocation.

14. The *ejido* has been the subject of innumerable treatises. Perhaps the definitive work is that of S. Eckstein (1966).

15. See the analyses of agricultural census data for ejidal and nonejidal tenures by Hertford (1971).

Figure 2-1. Schematic Tableau of CHAC Coefficients

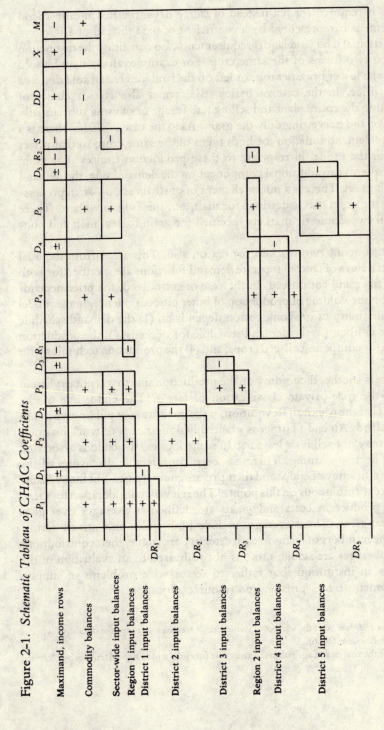

Note: P_i = production activities, district i; D_i = district-level factor supplies, investment, price differential activities, district i; R_j = regional input-supplying activities, region j; S = sectoral input-supplying activities; DD = domestic demand activities; X = export activities; N = import activities; and DR_i = district-level resource constraints, district i. See also the appendix ("Algebraic Statement") to chapter 2.

Factor Supply Activities

Three classes of factors may be distinguished in CHAC: those supplied at the level of each district submodel, those supplied at the regional level, and those supplied at the sector-wide level. At the submodel level, the fixed factors supplied are land, water, and the labor of farmers plus their families. Agricultural land is not priced, for it has no opportunity cost outside agriculture in the short run, but the dual solution of CHAC yields the value of rents which accrue to the land. Similarly, endowments of water are not priced, but the cost of tapping the water supply and providing it to farms is charged against the objective function. The reservation wage is charged for the labor of both farmers and their families; farmers may be fixed on the land in the short run, but their presence arises from long-run decision based in part on recognition of their opportunity cost. If it were assumed that farmers were willing to work for zero wages, cropping activities would enter the optimal basis that would not enter under more realistic assumptions. Hence, all of the supply functions in the model would be biased toward overestimation of the supply offered at a given set of product prices. Furthermore, unless extensive schemes for fiscal redistribution are to be considered, policy-oriented models must, if they are to provide solutions amenable to implementation, be based on wage assumptions not altogether different from actual wages.

Factors supplied at the regional level include hired labor, chemical inputs, and services of draft animals. Sector-wide factor supply activities in CHAC include those for credit, improved seeds, and machinery services. A sector-wide water pricing activity has been included in order to perform sensitivity analyses on the effects of systematic sector-wide variations in water charges. Most of the factor supply activities are straightforward. Except for labor, all regional and sectoral inputs are assumed to be supplied with infinite elasticity.

A schematic tableau for the entire matrix of coefficients is presented in figure 2-1. For simplicity, this schematic rendering shows only two regions and five districts, instead of the four regions and twenty districts actually contained in CHAC. The empty areas of the matrix represent blocks of zero coefficients. For blocks containing nonzero coefficients, the sign is indicated. An algebraic statement of the model, along with a listing of rows and columns, is found in the appendix to the chapter.

Details on the Treatment of Labor in CHAC

Labor activities and constraints constitute the most complex part of the factor supply set. One of the major purposes of CHAC is to measure the effect of various policies on employment patterns, and the labor components of CHAC have been designed accordingly. Some of the elements of the labor structure have been mentioned. Monthly labor demands are generated within each submodel, and these demands are met either with local labor or through interregional migration. Through the labor supply activities, regional wage differentials are incorporated. These activities also provide for a reservation price for farmers' own labor which is different from the wage for landless, or day, labor.

The number of farmers is fixed for each district, and the number of landless laborers is given for each region. That is, rural–urban migration is specified exogenously. Although farmers do migrate to cities, the number of farms in Mexico does not change very rapidly over time; in the short run, it therefore appears tenable to assume that the number of farmers in each locality is given. Farmers in nonirrigated areas in Mexico often work seasonally on irrigated farms, so this kind of labor transfer is allowed in the model. The reverse flow (farmers with irrigated land working on nonirrigated farms) virtually never occurs in Mexico, and so this is not included as an option. People leaving tropical areas are assumed to move permanently rather than temporarily because the distances are so great.

The landless labor force is divided into four regional pools. If one region employs all the members of its pool in a particular month, it may draw redundant laborers from another region.

Regional wage differentials are incorporated in CHAC by multiplicative factors, so that the proportional differences remain constant when experiments are conducted with different base wage rates. Official "minimum" wage rates exist, but in 1968 generally they were not fully enforced. Accordingly, they have been used as the maximum wages in parametric variations on the price of labor and capital. In 1968 the regional averages of official minimum wages were (in 1968 pesos daily): 19.5, 20.5, 24.0, and 26.0 for the South, Central Plateau, North, and Northwest regions, respectively. In the model's structure, the South's average, official minimum daily wage rate (19.5 pesos) is the base wage. Solutions have also been conducted with base wages of 13.5 and 16.5 pesos, maintaining the same proportional regional differences.[16]

The model is structured so that any ratio of farmers' reservation wage to day-labor wage may be employed. In the solutions reported here, it is assumed that the ratio is 0.5. This gives a reservation wage for farmers ranging from 7.8 to 13.0 pesos daily, depending upon the region and the assumed base wage. Recall, however, that the efficiency wage may exceed

16. At full employment (264 days annually), the daily wages of 13.5, 16.5, and 19.5 correspond to annual wages (in 1968 pesos) of 3,564, 4,356, and 5,148 pesos, respectively.

these levels in many months, and that farmers receive income from their property as well as from their labor.

The district submodels essentially reflect one or more "representative farms" in each district, since the production structures are taken from average data for the district or partial district. Hence, even within a fairly disaggregated model, there is a considerable degree of aggregation over farms. One consequence is an overstatement of resource mobility within the district. For example, because reservoir water is allocated centrally, it may be reasonable to assume that it can be reallocated in any manner. But, in general, this will not be possible for the water from private wells. In labor, too, there is an overstatement of mobility. Implicitly, the stock of farmers may be allocated in any manner among the farms in the district. In actuality, some farmers in a district may hire day labor during months when other farmers are idle. Farmers with irrigated land rarely work as seasonal laborers for other farmers. That is, the low reservation wage applies only to work on their own farms.

To overcome CHAC's bias toward labor mobility within a district, the model has been so specified that farmers with irrigated land may not offer their labor services on a monthly basis but only on a quarterly basis. Day labor, however, is available monthly. If both types of labor were supplied on a monthly basis, the lower reservation price of farmers would imply that day labor is hired only in the months when all farmers in the district are fully employed. With the quarterly contract device, this is not the case. For example, with a day-labor wage of 20 pesos and a farmer reservation price of 10 pesos, one-month peaks in labor demand would be met with hired labor, but two-month peaks would be met with farmers on quarterly contracts.

The effects of the quarterly contract assumption on labor hiring patterns are illustrated in figure 2-2. Of course, the quantity of labor demanded depends in part on the specification of crop supply, but, if it is assumed for the moment that the seasonal demand for labor is fixed, the seasonal pattern for labor hiring would look something like the solid line in the figure. If the reservation wage is half the day-labor wage, and if quarterly contracts are used for farmers, then day labor will be hired to meet the peak demands represented by the stippled areas. Farmers will satisfy the remaining labor requirements. The hatched areas show the number of slack days for which a cost is incurred when in fact farmers are idle.[17]

If farmers' availability were to be specified in the form of annual contracts, then hiring of farmers would correspond to the number of man-days that lie below both the solid line and the line AA'. Day-labor

17. These might be thought of as the "fence-mending" periods.

Figure 2-2. *Alternative Labor Hire Patterns*

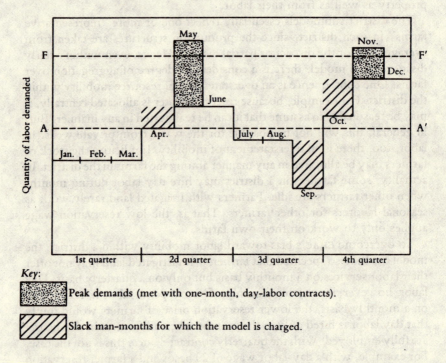

Key:

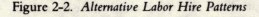

Peak demands (met with one-month, day-labor contracts).

Slack man-months for which the model is charged.

hiring would meet the remaining requirements. And, if farmers were available monthly, then all labor requirements up to the line FF′ (representing the total number of farmers in the district) would be met with farmers.

To summarize, the amount of labor hired in the model depends directly on four factors: (1) the wage rate for day labor; (2) the productivity of labor and other inputs to the various cropping activities; (3) the ratio of the farmer reservation wage to the day-labor wage; and (4) the length of the farmer contract. The last two factors are related. Whatever set of assumptions is adopted, it should be designed to offset the implicit assumption of complete farmer mobility within a district. This complexity is the price that must be paid to distinguish between the supply of labor from farmers and that from day laborers.

The reservation wage for farmers is clearly the most arbitrary element in the model. It should not be zero (the supply of labor would not, then, be positive), and yet it certainly is less than the day-labor wage. Since the reservation wage is seasonal, it does not measure farmers' income but, rather, the minimum return for which farmers would be willing to work in one season, realizing that the benefits will be reaped in another season. Hence, its value is difficult to assess a priori. Because of its arbitrariness, some sensitivity tests have been run. The results, reported in Duloy and Norton (1973b), are reassuring. The changes in employment, exports, and other variables are relatively slight when moderate changes are made in the reservation wage level, but extreme changes do distort the solutions significantly. Issues of validation are discussed more completely in chapter 5.

The Structure of Demand: A Summary

In its formulation of demand, CHAC differs from the conventional structure of sectoral planning models.[18] In most sectoral planning models, the problem is stated as either that of minimizing the costs of producing a fixed product mix or that of maximizing the sector's profits at exogenous input and output prices. In CHAC, demands and product prices are related endogenously through demand curves.

For a particular product, the demand function is illustrated in figure 2-3. It is assumed that all purchasers will pay the same price, and that all suppliers will receive this price. The import and export prices are indicated by p_m and p_x respectively. Transport costs account for part of the difference between p_m and p_x. This difference may be large for bulky agricultural products. Export and import prices are fixed exogenously.[19] Also, for convenience, all demand functions are assumed to be linear.

The purpose of this treatment of demand is threefold. First, it means that a programming solution will correspond to a market equilibrium. The effects of various policies—such as subsidizing or taxing product prices or varying the exchange rate—can then be investigated. Second, it allows the model greater flexibility. For instance, substitution between capital and labor (corresponding to different ratios of the wage rate to the rate of return on capital) can occur not only directly through the technology set or through changes in the commodity mix of exports, but also through substitution in domestic demand. Third, it enables a more realistic appraisal of the benefits (and particularly of the distribution of benefits between producers and consumers) accruing from an increase in agricultural output. Consider the (not unlikely) situation of agricultural production for the domestic market at prices between p_m and p_x. If the domestic

18. A more complete exposition of CHAC's demand structure is given in chapter 3.

19. For nontraded commodities, the demand function is specified between arbitrarily wide bounds that reflect the relevant range of potential prices and quantities. For some crops, there are bounds on exports to represent quotas.

Figure 2-3. *Demand Function for a Given Product in CHAC*

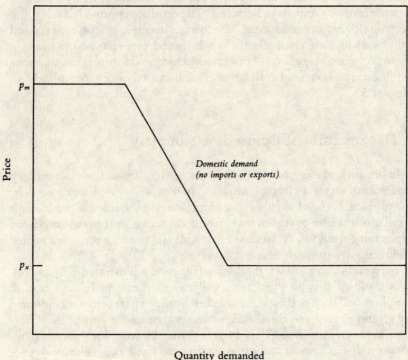

Note: p_m = import price of product; p_x = export price of product.

demand is price-inelastic, then the financial return to producers as a whole from an increase in output is negative. For consumers, the benefits are positive.

The maximand of CHAC is defined as the sum of areas under all demand functions, less costs of inputs purchased from outside the sector. This specification follows Samuelson (1952) and Takayama and Judge (1964, 1971), and it ensures that the model replicates a competitive equilibrium in which producers are price takers and in which, for each product, marginal cost is equated to price.

The derivation of the objective function coefficients is given in the next chapter, along with proofs of properties of the demand system.

In CHAC it has been assumed that the off-diagonal elements of the demand matrix (transformed matrix of own and cross-price elasticities) are unknown. The available information consisted of crude estimates of own-price elasticities for some commodities and commodity groups. (For the numerical values of the price and income elasticities, see table 2-2.) The demand group approximation procedure has the following properties.

Because of the lack of information on the off-diagonal elements, the system does not reflect a complete range of price interdependence among commodities.[20] Some (but not all) commodities enter into demand groups. Each group is specified so that demand substitution may occur at marginal substitution rates equal to the price ratios in the base period. The system is structured so that substitution is constrained within preassigned bounds on the commodity mix within each group. For example, the various vegetable oils constitute one such demand group, in which sub-stitution rates are determined by the relative prices of soybean oil, peanut oil, safflower oil, and so forth.

For any commodity group, both the consumers' surplus and producers' gross revenue are independent of the commodity mix in that group.

The system preserves the desirable property of the linearization of a concave functional, so that the function representing the area under the demand curve can be approximated to any desired degree of accuracy by adding activities without adding additional rows.

The revenue function is approximated so that the demand activities have coefficients in rows defining producers' profits and incomes.

Export selling activities are included as additional demand activities for individual products, and import activities are added as alternatives to domestic producing activities. Import activities would never enter the optimal solution with a monopolistic objective unless the model also included a social welfare constraint.

For the indifference curves within a demand group, we have employed the piecewise linear approximation shown in figure 2-4. There, in the case of a two-commodity group, the feasible area is the cone G0H. (The rays 0G and 0H define limits on the commodity mix within the group.) The indifference curves AB, CD, EF, and GH are parallel to each other, and the marginal rate of substitution is equal to the ratio of the base-period relative prices.

Time and Investment Choices

Since CHAC's size makes it expensive to obtain simultaneous mul-tiperiod solutions, it has been formulated as a one-period static model. It is, however, solved for different points in time with appropriate projec-tions of exogenous data. Investment activities may be included in the

---

20. This, of course, applies to the demand structure only. There is interdependence in product prices among all commodities arising from the interdependence of marginal costs on the supply side.

Table 2-2. *Crops, Types of Cultivation, and Price and Income Elasticities of Demand in CHAC*

Demand group	Crop[a]	Type of cultivation			Elasticity	
		Irrigated	Temporal	Tropical	Per capita income[b]	Own-price
1	Wheat	★	★		0.315	
	Maize	★	★	★	−0.453	} −0.10
2	Green chile	★	★		−0.119	
	Dry chile		★		−0.119	} −0.20
3	Sugarcane	★		★	0.117	−0.25
4	Beans	★	★	★	0.330	
	Rice	★		★	0.250	
	Potatoes	★	★		0.330	} −0.30
	Chickpeas	★	★		0.330	
5	Tomatoes	★	★		0.409	−0.40
6	Onions	★			0.598	
	Garlic	★			0.598	} −0.20
7	Cucumber	★			0.598	−0.60
8	Squash		★		0.330	−0.40
9	Lima beans	★	★		0.245	−0.40

Sources: For income elasticities, Banco de México, Secretaría de Agricultura y Ganadería, and U.S. Department of Agriculture (1965); for price elasticities, unpublished studies by L. Barraza and others.

a. Some products (maize, peanuts, chickpeas, and barley) appear in more than one demand group, and hence there are multiple domestic markets for these products. Malt barley and grain barley are the same product on the supply side.

sectoral model in principle, and in fact they have been incorporated in some of the stand-alone district models (see chapters 15, 16, and 17). Even though the timing of investment projects cannot be treated in this kind of model, the alternative projects can be ranked with respect to social profitability. For the CHAC policy experiments, the effects of possible investment programs are represented through exogenous increases in resource endowments (see chapter 5). The initial version of the model was based on data for 1968.[21] Solutions were obtained both for 1968 and 1976. The base-period solutions were used to check the model. Solutions for the later year constitute the policy experiments.

21. Because of short-term fluctuations, the average of 1965–69 was used for yields and the average of 1967–69 for production and other variables.

Table 2-2 *(continued)*

Demand group	Crop[a]	Type of cultivation			Elasticity	
		Irri-gated	Tem-poral	Trop-ical	Per capita income[b]	Own-price
10	Forage maize	★	★	★	0.500	
	Oats		★		0.500	
	Grain sorghum	★	★	★	0.500	
	Forage barley	★			0.500	−0.30
	Grain barley	★	★		0.500	
	Green alfalfa	★			0.500	
	Chickpeas	★	★		0.500	
11	Malt barley	★	★		0.460	−0.10
12	Cotton fiber	★	★		0.639	−0.50
13	Cotton seed	★	★		0.614	
	Sesame	★	★	★	0.614	
	Flaxseed	★	★		0.614	
	Safflower	★	★		0.614	−1.20
	Soybeans	★		★	0.614	
	Peanuts	★	★		0.614	
14	Peanuts	★	★		0.330	−0.20
15	Strawberries	★			0.330	
	Pineapple			★	0.330	
	Watermelon	★			0.330	−2.00
	Cantaloupe	★			0.330	
16	Tobacco			★	0.817	−0.10

b. The demand functions are shifted to reflect the combined effects of the rate of population increase (3.5 percent annually), per capita increase in gross domestic product (GDP; 2.5 percent annually), and the per capita income elasticity for the individual item. With these population and GDP shifts taken as exogenous data, the own-price elasticities are then applied for each demand group. In an optimal solution, a point is chosen endogenously along the price-quantity demand curve for the commodity group.

Endowments of labor are projected from 1968 to 1976. No attempt has been made to estimate rural-urban migration within CHAC. The labor force is projected at the natural increase rate, and rural-urban migration is exogenous. Export demand limits (for example, quotas) also are projected forward to the solution period. Disembodied technical progress is incorporated for purchased inputs and associated yield increases.

The major difference, in terms of effects on the solutions, between the 1968 and 1976 versions is the rightward shift over time of the domestic demand functions. These shifts are calculated through income growth and income elasticities of demand for agriculture products.

Figure 2-4. *A Family of Indifference Curves for a Two-commodity Demand Group*

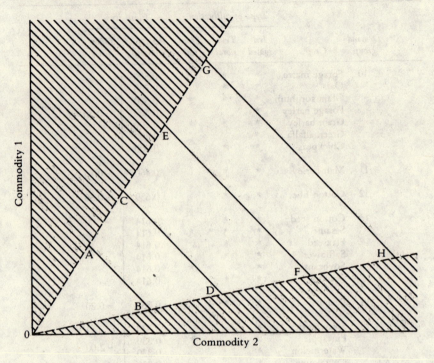

Risk and Dualism

Risk variables are the major omission on the production side of this version of CHAC. Perceived risk obviously plays an important role in farmers' decisions. Some early attempts were made to incorporate it, but for the first version the data and existing methods were insufficient to support the attempts. New methods have been tentatively formulated, however, and it is likely that this element will eventually be included.[22] In the meantime, it is instructive to discuss the reasons that earlier attempts failed. First, it should be noted that district-level changes in cropping patterns implied by the solutions of CHAC generally are not more severe than the historical, year-to-year changes observed in irrigated areas. Quite marked annual changes in planted hectarage per crop—often 50 percent or more—are observed in these areas. Hence, and even without allowing for

22. *Editors' note*: A rigorous risk formulation was in fact developed after this chapter was written. See chapters 7 and 8.

risk variables to ensure greater crop diversification, the model's results appeared satisfactory. The principal difficulty arises in the more traditional, nonirrigated farms. There it is optimal to shift substantially away from maize and into sorghum when it is assumed that import barriers are weakened.

The first thought on handling risk was to utilize the crop insurance premiums of the national agricultural insurance company. One formulation of risk leads to a quadratic objective function,[23] but, if it is assumed that the national crop insurance organization has made a linear approximation to the risk problem, insurance may be specified as a cost of production. The observed premiums may then be utilized as insurance input coefficients in the cropping activities. This approach was, unfortunately, vitiated by incomplete coverage of the sector in the insurance program and by inconsistency among premiums in areas where there was coverage. If nothing else, this inquiry indicates that it might be fruitful to examine the premium-setting rules in the insurance program.

As a second approach, the method of year-to-year "flexibility" constraints was examined. It was concluded that there is little objective basis for establishing appropriate parameters in such constraints, especially in circumstances of highly flexible cropping patterns. It was decided to impose such constraints only after the initial solutions, if they appeared warranted by the results. As things turned out, it was not necessary to do so.

Consideration of risk emphasizes the relative inflexibility of nonirrigated agriculture. This is reflected by CHAC in fewer crop choices for nonirrigated, traditional agriculture. In the solutions, the difference made by irrigation showed up strongly in responses to price subsidies.[24] In effect, farmers with irrigation face many alternatives of nearly equal profitability, and they will respond to minor perturbations of prices. Farmers without irrigation who grow maize, however, can grow little else easily. Stronger price changes are required to induce shifts in the traditional cropping patterns of these farmers, and the CHAC solutions show this circumstance. CHAC has therefore captured, at least to a degree, the dualistic aspects of Mexican agriculture.

Appendix. Algebraic Statement

In the case of large-scale programming models, matrix tableaus are often helpful in revealing the structure of the model. Nevertheless, the

23. See the classic early work on mean-variance analysis in an optimizing model by Freund (1956).
24. These should not be confused with a guaranteed price level.

Table 2-3. *Notation for Algebraic Statement*

Symbol	Description
Variables	
X_{hij}^{dz}	Crop production
I_n^d	Fixed investment
T_j^d	Total crop production at the district level
A^r	Regional supply of draft animal services
F^r	Regional supply of chemical inputs
C	Sectoral supply of short-term credit
K	Sectoral supply of machinery services (tens of days)
S	Sectoral supply of purchased seeds
D_{ms}^g	Domestic demand
E_j	Exports
M_j	Imports
P_j	Technical progress variable
K'	Sectoral supply of machinery services (10,000 pesos)
C^d	District-level counter for short-term credit
CP	Private long-term capital used
CT	Total long-term capital used
$\bar{W}_g^d$	Supply of gravity-fed water by district
$\bar{W}_p^d$	Supply of well water by district
W_g	Sector use of gravity-fed water
W_p	Sector use of pump water
$SALS$	Sectoral wage charging activity
$SALr$	Activities for charging regional wage differentials
$LMAN$	Sector annual employment counter (man-years at full employment equivalent)
$LMANt$	Sectoral monthly employment counter
$dDLt$	Monthly day-labor supply activities in each submodel
$dFLq, dFLt$	Farm labor supply activities by submodel, quarterly in irrigation submodels, monthly otherwise (q = quarterly index; t = monthly index)
$MDLrr't$	Migration activities for day labor from region r to region r' by months (rr' = 31, 32, 43)
$MA33t$	Migration activities in region 3 for farmers to the pool of day laborers
$MA44A$	Migration activities for region 4 farmers on annual basis
Parameters	
α_{mj}^g	Quantity of crop j demanded in mix m of group g
ω_s^g	Entry in maximand for demand group g and demand segment s (that is, weighted average price for segment s of all crops in the group)
ρ_s^g	Entry in income rows for demand group g and demand segment s (that is, weighted average marginal revenue for segment s of all crops in the group)
δ_m^g	Entry in the demand convex combination constraint for demand group g and mix m
κ	Ratio of farmer reservation wage to day-labor wage
γ_{ijt}^{dz}	Water input coefficients ($i = i'$ for gravity-fed water; $i = i''$ for well water; t = month)

Table 2-3 (*continued*)

Symbol	Description
Parameters	
σ_j^a	Purchased seed input coefficients
ϕ_j^d	Chemical input coefficients
μ_{hj}^d	Machinery services input coefficients
β_{hjt}^d	Labor input requirement
θ_{hj}^d	Draft animal services input requirements
η_n^d	Capital costs per unit of investment project (n = class of investment project)
τ_{hj}^d	Credit input requirements
λ_r	Ratio of region r wage to region 4 wage
γ_j^d	Yield per hectare
Prices	
p_j^e	Exports
p_j^m	Imports
p^l	Labor (region 4 hired labor wage)
p^k	Cost of machinery services, excluding interest cost and base wage component of machinery operators' wage
p^i	Long-term interest rate
p^c	Short-term interest rate
p_r^a	Regional unit cost of draft animal services
p_d^{wg}	Gravity-fed water
p_d^{wp}	Well water

Note: Super- and subscripts are:

d = district	r = region
z = zone (subdistrict)	g = crop group
t = type of irrigation	m = commodity mix
j = crop	s = demand segment.
n = class of investment	

In the section of the text that lists the equations, the vector α_j is the union over g and m of all coefficients α_{mj}^g.

algebra is also useful (particularly for writing instructions for matrix-generating computer routines), and so a statement is given here, in which algebraic quantities and symbols used are identified (the latter in table 2–3).

Special notation has been adopted for the algebraic statement. Capital italic letters represent vector unknowns or right-hand side (RHS) values, and small italic letters both parameters and sub- and superscripts. Greek letters denote vector and scalar coefficients. In raw form, some of the vector symbols are burdened with several superscripts and subscripts, but in most equations only part of the vector is relevant, so an abbreviated notation is used. For example, the typical production activity is denoted X_{hij}^{dz}. That set which corresponds to all the vectors for producing crop j in district r is written X_j^r. The total production of crop j in district r is abbreviated, $y_j^r X_j^r$, where y_j^r signifies the row vector of yields for those activities producing crop j in region r. Again, the set of symbols in full form is set out in table 2-3. Constraints are identified by the headings below, and the number of the constraints involved in each is given in the right margin. Accordingly, the equations of the system may be written as follows.

1. *Sectoral and district commodity balances*

(2.1a) $T_j^d + M_j - \alpha_j D^g - E_j + P_j \geq 0;$ (33 constraints)

$$\begin{bmatrix} \text{Domestic} \\ \text{production} \end{bmatrix} + \begin{bmatrix} \text{Imports} \end{bmatrix} - \begin{bmatrix} \text{Domestic} \\ \text{sales} \end{bmatrix}$$

$$- \begin{bmatrix} \text{Exports} \end{bmatrix} + \begin{bmatrix} \text{Adjustment for yield-increasing} \\ \text{technical progress} \end{bmatrix} \geq 0.$$

(2.1b) $y_j^d X_j^d - T_j^d = 0,$ each $d, j;$ (176 constraints)

$$\begin{bmatrix} \text{District-level} \\ \text{production in} \\ \text{various techniques} \end{bmatrix} - \begin{bmatrix} \text{Definition of} \\ \text{district total crop} \\ \text{production} \end{bmatrix} = 0.$$

2. *Sectoral and regional labor balances*

a. Sectoral wage accounting equation:

(2.2a) $- SALS + K + \sum_r \lambda_r SALr = 0;$ (1 constraint)

$$- \begin{bmatrix} \text{Wage charging} \\ \text{activity} \end{bmatrix} + \begin{bmatrix} \text{Accounting} \\ \text{activity} \\ \text{employment of} \\ \text{machinery} \\ \text{operators} \end{bmatrix} + \begin{bmatrix} \text{Regional wage} \\ \text{differentials} \times \\ \text{regional wage} \\ \text{accounting} \\ \text{activities} \end{bmatrix} = 0.$$

b. Regional wage accounting rows:

(2.2b) $- SALr + \kappa \left(\sum_{d \in r} dFLq + \sum_{d \in r} dFLt \right)$

$$+ \sum_{d \in r} \sum_t dDLt \leq 0, \text{ each } r;$$ (4 constraints)

$$- \begin{bmatrix} \text{Regional wage} \\ \text{accounting} \\ \text{activities} \end{bmatrix} + \begin{bmatrix} \text{Reservation wage ratio} \\ \times \text{ regional farmer} \\ \text{employment activity} \end{bmatrix}$$

$$+ \begin{bmatrix} \text{Sum over districts and} \\ \text{months of regional day-} \\ \text{labor employment} \end{bmatrix} \le 0.$$

c. Regional farmer employment accounting rows:

(2.2c) $- RESr + 3 \sum\limits_{d \in r} \sum\limits_{q} dFLq + \sum\limits_{d \in r} \sum\limits_{t} dFLt = 0$, each r; (4 constraints)

$$- \begin{bmatrix} \text{Regional farmer} \\ \text{employment} \\ \text{activity} \end{bmatrix} + 3 \begin{bmatrix} \text{Sum over districts} \\ \text{and quarters of} \\ \text{quarterly farmer} \\ \text{employment} \end{bmatrix}^{25}$$

$$+ \begin{bmatrix} \text{Sum over districts} \\ \text{and months of} \\ \text{monthly farmer employment} \end{bmatrix}^{26} = 0.$$

d. Total employment accounting row in man-years:

(2.2d) $- 12\, LMAN + \sum\limits_{t} LMANt = 0;$ (1 constraint)

$$- 12 \begin{bmatrix} \text{Total employment} \\ \text{in man-years} \end{bmatrix} + \begin{bmatrix} \text{Sum over months of} \\ \text{total employment} \\ \text{in man-months} \end{bmatrix} = 0.$$

e. Total monthly employment accounting rows in man-months:

(2.2e) $- 2.2\, LMANt + \sum\limits_{d} dDLt + \sum\limits_{d} dFLq + \sum\limits_{d} dFLt = 0,$

each t and q such that $t \in q$; (12 constraints)

$$- 2.2 \begin{bmatrix} \text{Total} \\ \text{employment} \\ \text{in month } t \end{bmatrix}^{27} + \begin{bmatrix} \text{Sum over districts of} \\ \text{day-labor employment} \\ \text{in month } r \end{bmatrix}$$

25. In irrigation districts, the quarterly contract device is used for farmers, but in nonirrigated districts farmers are assumed to be available on a monthly basis, so that seasonal migration to irrigated areas may occur.

26. See note 25.

27. The activities for hiring farmers and day laborers are stated in units of tens of man-days monthly (or quarterly), and there are twenty-two working days monthly; hence, the conversion factor of 2.2 is required in the first term of this equation.

$$+ \begin{bmatrix} \text{Sum over districts of} \\ \text{quarterly farmer} \\ \text{employment in the} \\ \text{quarter containing} \\ \text{month } t \end{bmatrix} + \begin{bmatrix} \text{Sum over districts} \\ \text{of monthly farmer} \\ \text{employment} \end{bmatrix} = 0.$$

f. Regional employment balances, by month:

(2.2f.1) $\sum_{d \in r} dDLt - MDL3rt \leq L_r,\ r = 1, 2,\ \text{each } t;$ (24 constraints)

$$\begin{bmatrix} \text{Total employment} \\ \text{of day labor} \\ \text{in region } r \\ \text{in month } t \end{bmatrix} - \begin{bmatrix} \text{Migration of day} \\ \text{labor from Central} \\ \text{Plateau to region} \\ r \text{ in month } t \end{bmatrix} \leq \begin{bmatrix} \text{Pool of} \\ \text{landless} \\ \text{labor in} \\ \text{region } r \end{bmatrix}$$

(2.2f.2) $\sum_{d \in r} dDLt + \sum_{r=1}^{2} MDL3rt - MDL43A - MA33t \leq L_3,$

$r = 3,\ \text{each } t;$ (12 constraints)

$$\begin{bmatrix} \text{Total employment} \\ \text{of day labor in} \\ \text{region } r = 3, \\ \text{month } t \end{bmatrix} + \begin{bmatrix} \text{Migration out} \\ \text{of region 3} \\ \text{in month } t \end{bmatrix} + \begin{bmatrix} \text{Migration from} \\ \text{region 4 to} \\ \text{region 3} \end{bmatrix}$$

$$+ \begin{bmatrix} \text{Movement of} \\ \text{dryland (\textit{temporal}) farmers} \\ \text{into day labor} \end{bmatrix} \leq \begin{bmatrix} \text{Pool of} \\ \text{landless labor} \\ \text{in region 3} \end{bmatrix}.$$

(2.2f.3) $\sum_{d \in r} dDLt + MDL43A - MA44A \leq L_4,$

$r = 4,\ \text{each } t;$ (12 constraints)

$$\begin{bmatrix} \text{Total employment} \\ \text{of day labor} \\ \text{in region } r = 4, \\ \text{month } t \end{bmatrix} + \begin{bmatrix} \text{Migration from} \\ \text{region 4 to} \\ \text{region 3} \end{bmatrix} - \begin{bmatrix} \text{Transfer of} \\ \text{tropical farmers} \\ \text{to day-labor pool} \end{bmatrix}$$

$$\leq \begin{bmatrix} \text{Pool of} \\ \text{landless labor} \\ \text{in region 4} \end{bmatrix}.$$

g. Migration constraints:

(2.2g.1) $\sum_{r=1}^{2} MDL3rt - MDL43A - MA33t$
$\leq M_{3t},\ \text{each } t$ (12 constraints)

[Bound on monthly migration out of region 3];

(2.2g.2) $\sum\limits_{t} \sum\limits_{r=1}^{2} MDL3rt - 12MDL43A - \sum\limits_{t} MA33t \leq M_3$ (1 constraint)

[Bound on annual migration out of region 3];

(2.2g.3) $12MDL43A + 12MA44A \leq M_4$ (1 constraint)

[Bound on annual migration out of region 4].

3. *Sectoral and regional input balances (excluding labor)*

   a. Short-term credit balance:[28]

(2.3a) $\sum\limits_{d} C^d - C \leq 0;$ (1 constraint)

$$\begin{bmatrix} \text{Sum of district} \\ \text{credit counting} \\ \text{activities} \end{bmatrix} - \begin{bmatrix} \text{Sectoral interest-} \\ \text{charging activity} \\ \text{for credit} \end{bmatrix} \leq 0.$$

   b. Machinery services balance:

(2.3b) $\sum\limits_{d} \sum\limits_{h} \sum\limits_{j} \mu_{hj}^{d} X_{hj}^{d} - K \leq 0;$ (1 constraint)

$$\begin{bmatrix} \text{Sum of demands for} \\ \text{machinery services} \\ \text{in cropping activities} \end{bmatrix} - \begin{bmatrix} \text{Activity supplying} \\ \text{machinery services} \end{bmatrix} \leq 0.$$

   c. Balance for charging interest component of machinery services:

(2.3c) $K - 2.308 \, K' \leq 0;$ (1 constraint)

$$\begin{bmatrix} \text{Machinery services} \\ \text{in tens of workdays} \end{bmatrix} - 2.308 \begin{bmatrix} \text{Machinery services} \\ \text{in 10,000 pesos annually} \end{bmatrix}^{[29]} \leq 0.$$

   d. Sectoral accounting row for use of gravity-fed water.[30]

(2.3d) $\sum\limits_{d} \sum\limits_{z} \sum\limits_{j} \gamma_{ij}^{dz} X_{ij}^{dz} - W_g = 0, \; i = i';$ (1 constraint)

$$\begin{bmatrix} \text{Total demands for} \\ \text{gravity-fed water} \end{bmatrix} - \begin{bmatrix} \text{Gravity-fed water} \\ \text{accounting activity} \end{bmatrix} = 0.$$

28. There are district-level credit balances that sum the demands for credit over cropping activities. There are also bounds on institutional credit allocations by crop that have been made nonoperative in the solutions reported here. Because there is no bound on C, credit is provided in the model in infinitely elastic supply.

29. The factor 2.308 converts from tens of days to 10,000 pesos annually, given the actual initial cost and lifetime of a typical piece of machinery in Mexico.

30. This and the subsequent row permit experiments with uniform sector-wide changes in the price of water.

e. Sectoral accounting row for use of well water:

(2.3e)
$$\sum_d \sum_z \sum_j \gamma_{ij}^{dz} X_{ij}^{dz} - W_p = 0, \ i = i''; \qquad \text{(1 constraint)}$$

$$\left[\begin{array}{c}\text{Total demands for}\\ \text{well water}\end{array}\right] - \left[\begin{array}{c}\text{Well water}\\ \text{accounting activity}\end{array}\right] = 0.$$

f. Sectoral balance for purchased seeds:[31]

(2.3f)
$$\sum_d \sum_j \sigma_j^d X_j^d - S \leq 0; \qquad \text{(1 constraint)}$$

$$\left[\begin{array}{c}\text{Total demands for}\\ \text{purchased seeds}\end{array}\right] - \left[\begin{array}{c}\text{Supply of}\\ \text{purchased seeds}\end{array}\right] \leq 0.$$

g. Regional balances for chemical inputs:[32]

(2.3g)
$$\sum_{d \in r} \sum_j \phi_j^d X_j^d - F^r \leq 0, \ \text{each } r; \qquad \text{(4 constraints)}$$

$$\left[\begin{array}{c}\text{Total regional demands}\\ \text{for fertilizers and}\\ \text{pesticides}\end{array}\right] - \left[\begin{array}{c}\text{Regional supply}\\ \text{of fertilizers}\\ \text{and pesticides}\end{array}\right] \leq 0.$$

h. Regional balances for draft animal services:

(2.3h)
$$\sum_{d \in r} \sum_h \sum_j \theta_{hj}^d X_{hj}^d - A^r \leq 0, \ \text{each } r; \qquad \text{(4 constraints)}$$

$$\left[\begin{array}{c}\text{Total regional demands}\\ \text{for draft animal}\\ \text{services}\end{array}\right] - \left[\begin{array}{c}\text{Regional supply}\\ \text{of draft animal}\\ \text{services}\end{array}\right] \leq 0.$$

i. Long-term private capital balances:

(2.3i)
$$\sum_d \eta_n^d I_n^d - CP \leq 0; \qquad \text{(1 constraint)}$$

$$\left[\begin{array}{c}\text{Cost of investment}\\ \text{activities financed}\\ \text{with private capital}\end{array}\right] - \left[\begin{array}{c}\text{Supply of}\\ \text{private capital}\end{array}\right] \leq 0.$$

31. Both terms in thousands of pesos.
32. The indexes d and j on ϕ_j^d indicate that rates of fertilizer use vary over district and crop, but not over other dimensions such as zones or degrees of mechanization. Both terms in this expression are in thousands of pesos.

j. Total long-term capital balance:

(2.3j) $$\sum_d \eta_n^d I_n^d + K' + CP - CT \leq 0,$$

those n not in equation (2.3i);   (1 constraint)

$$\begin{bmatrix} \text{Costs of investment} \\ \text{activities financed} \\ \text{with public capital} \end{bmatrix} + \begin{bmatrix} \text{Capital component} \\ \text{of machinery} \\ \text{services} \end{bmatrix} + \begin{bmatrix} \text{Private} \\ \text{capital} \\ \text{supplied} \end{bmatrix}$$

$$- \begin{bmatrix} \text{Total capital} \\ \text{supplied} \end{bmatrix} \leq 0.$$

4. *District-level input balances*

a. District labor balances:

(2.4a) $$\sum_h \sum_j \beta_{hjt}^d X_{hj}^d - dDLt - dFLt^{33} \leq 0,$$

each d, t;   (204 constraints)

$$\begin{bmatrix} \text{Demands for labor,} \\ \text{district } d, \text{ month } t \end{bmatrix} - \begin{bmatrix} \text{Day labor hired,} \\ \text{district } d, \text{ month } t \end{bmatrix}$$

$$- \begin{bmatrix} \text{Farmers employed in} \\ \text{district } d, \text{ month } t \end{bmatrix} \leq 0.$$

b. District credit balances:[34]

(2.4b) $$\sum_h \sum_h \tau_{hj}^h X_{hj}^d - C^d \leq 0, \text{ each } d;$$ (18 constraints)

$$\begin{bmatrix} \text{Demands for credit,} \\ \text{district } d \end{bmatrix} - \begin{bmatrix} \text{Total district} \\ \text{credit required} \end{bmatrix} \leq 0.$$

c. District gravity-fed water balances:

(2.4c) $$\sum_z \sum_j \gamma_{ij}^{dz} X_{ij}^{dz} - W_g^d \leq 0, \text{ each } d, i = i';$$ (10 constraints)

$$\begin{bmatrix} \text{Demands for gravity-fed} \\ \text{water, district } d \end{bmatrix} - \begin{bmatrix} \text{District } d \text{ activity} \\ \text{for charging costs} \\ \text{of gravity-fed water} \end{bmatrix} \leq 0.$$

---

33. Or $dFLq$, depending on the district.

34. The three El Bajío submodels are grouped together in measuring credit requirements, so there are eighteen balances instead of twenty.

5. *District resource constraints*

   a. Monthly land constraints:

(2.5a) $$X_t^{dz} \leq B_t^{dz}, \text{ each } d, z, t;$$ (348 constraints)

$$\left[\begin{array}{c} \text{Land requirements for} \\ \text{cropping (units of} \\ X \text{ are hectares)} \end{array} \right] \leq \left[\begin{array}{c} \text{Land availability} \\ \text{by district,} \\ \text{zone, month} \end{array} \right].$$

   b. Monthly gravity-fed and pump water constraints:

(2.5b) $$\sum_z \sum_j \gamma_{ijt}^{dz} \leq \bar{W}_{it}^d, \text{ each } d, t, i;$$ (168 constraints)

$$\left[\begin{array}{c} \text{Total month } t \text{ water} \\ \text{demands, district } d, \\ \text{water type } i \end{array} \right] \leq \left[\begin{array}{c} \text{Water delivery constraints} \\ \text{district } d, \text{ water type } i \end{array} \right].$$

   c. Annual gravity-fed and pump water constraints:

(2.5c) $$\sum_z \sum_j \gamma_{ij}^{dz} X_{ij}^{dz} \leq \bar{W}_i^d, \text{ each } d, i.$$ (14 constraints)

   d. District constraints on farmer and family labor:

(2.5d.1) $dFLq \leq A_d$, each q, each d with irrigation;   (44 constraints)

(2.5d.2) $dFLt - MA33t \leq A_d$, each t, each d
                in region 3 without irrigation;   (36 constraints)

(2.5d.3) $dFLt - MA44A \leq A_d$, each t, each d
                in region 4 without irrigation.   (12 constraints)

6. *Technical progress balances*[35]

(2.6) $$\alpha_j^g D^g + E_j - M_j - P_j = 0,$$
               each j, g such that $j \in g$;   (33 constraints)

$$\left[\begin{array}{c} \text{Total sales on} \\ \text{domestic markets} \end{array} \right] + \left[\text{Exports} \right] - \left[\text{Imports} \right]$$

$$- \left[\begin{array}{c} \text{Technical} \\ \text{progress factor} \end{array} \right] = 0.$$

---

35. These balances serve the purpose of adjusting total production to allow for exogenous changes over time in yields and associated inputs of seeds, chemicals, and credit.

7. *Income definitions*[36]

a. Farmers' profit:

(2.7a) $\sum_g \sum_s \sum_m \rho_s^g D_{ms}^g + \sum_j p_j^e E_j - \sum_j p_j^m M_j - p^\ell SALS - p^k K$

$- p^i K' - p^c CP - 0.1S - \sum_r 0.1 F^r - \sum_r p_r^a A$

$- \sum_d p_d^{wg} W_g^d - \sum_d p_d^{wp} W_p^d - (\Delta p^{wg}) W_g - (\Delta p^{wp}) W_p$

$+ \sum_d \sum_j (\Delta p)^d T_j^d - Y = 0,$ (1 constraint)

where $(\Delta p)^{wg}$ and $(\Delta p)^{wp}$ indicate the uniform sector-wide changes in water prices, and $(\Delta p)^d$ indicates the district price differentials by crop;

$$\begin{bmatrix} \text{Gross revenue} \\ \text{from domestic} \\ \text{sales} \end{bmatrix} + \begin{bmatrix} \text{Export} \\ \text{earnings} \end{bmatrix} - \begin{bmatrix} \text{Import} \\ \text{costs} \end{bmatrix} - \begin{bmatrix} \text{Total} \\ \text{labor} \\ \text{costs} \end{bmatrix}$$

$$- \begin{bmatrix} \text{Interest} \\ \text{on long-term} \\ \text{capital} \end{bmatrix} - \begin{bmatrix} \text{Interest} \\ \text{on short-term} \\ \text{capital} \end{bmatrix} - \begin{bmatrix} \text{Seed} \\ \text{costs} \end{bmatrix}$$

$$- \begin{bmatrix} \text{Chemical} \\ \text{input} \\ \text{costs} \end{bmatrix} - \begin{bmatrix} \text{Draft animal} \\ \text{service} \\ \text{costs} \end{bmatrix} - \begin{bmatrix} \text{Gravity-fed} \\ \text{water} \\ \text{costs} \end{bmatrix}$$

$$- \begin{bmatrix} \text{Well} \\ \text{water} \\ \text{costs} \end{bmatrix} - \begin{bmatrix} \text{Increments to} \\ \text{gravity-fed} \\ \text{water costs} \end{bmatrix} - \begin{bmatrix} \text{Increments to} \\ \text{well water cost} \end{bmatrix}$$

$$+ \begin{bmatrix} \text{District price} \\ \text{differences on crops} \end{bmatrix} - \begin{bmatrix} \text{Farmers'} \\ \text{profits} \end{bmatrix} = 0.$$

b. Farmers' income:

Equation (2.7b) is the same as (2.7a), except that the term

$$+ \sum_r a_r RESr$$

36. The seed and fertilizer supply activities are stated in units of thousand pesos, whereas the objective function and income rows are in 10,000 pesos—hence the factor of 0.1 in these rows.

is added, where a_r is the regional farmer reservation wage, to serve the purpose of adding farmers' wage income to profits in order to arrive at total farmers' income.

    c. Sector income:

Equation (2.7c) is the same as (2.7a) except that the term

$$- p^\ell \; SALS$$

is dropped and the price of machinery services, p^k, is reduced to take out labor costs. These adjustments result in an expression for total sector income, which is defined as farmers' income plus wage income of day laborers.

    8. *Objective function (maximand)*

(2.8) $\displaystyle \sum_g \sum_s \sum_m \omega_s^g \, D_{ms}^g + \sum_j p_j^e \, E_j - \sum_j p_j^m \, M_j - p^\ell \; SALS$

        $\displaystyle - p^k \, K - p^i CT - p^c CP$

        $\displaystyle - 0.1 \left[S + \sum_r F^r \right] - \sum_r p_r^a \, A^r - \sum_j p_d^{wg} \, W_p^d - \sum_d p_d^{wp} \, W_p^d$

        $\displaystyle - (\Delta p^{wg}) \, W_g - (\Delta p^{wp}) \, W_p + \sum_d \sum_j (\Delta p)^d \, T_j^d.$ (1 constraint)

There are differences between this and equation (2.7a) in the demand function term and in the role of long-term private and public capital. The first terms of the objective function are the sum of gross consumers' and producers' surpluses rather than gross revenue. Total long-term capital is costed, via CT, instead of just private long-term capital.

    In the above, the total number of rows for constraints (2.1)–(2.7) is 1,203. In addition, there are approximately 300 accounting rows.

References

Banco de México, Secretaría de Agricultura y Ganadería, and U.S. Department of Agriculture. 1965. *Projections of Agricultural Supply and Demand, 1965–75.*

Duloy, John H., and Roger D. Norton. 1973a. "CHAC, a Programming Model of Mexican Agriculture." In *Multi-level Planning: Case Studies in Mexico.* Edited by Louis M. Goreux and Alan S. Manne. Amsterdam/New York: North-Holland/ American Elsevier, pp. 291–337.

———. 1973b. "CHAC Results: Economic Alternatives for Mexican Agriculture." In Goreux and Manne, pp. 373–91.

Eckstein, S. 1966. *El ejido colectivo en México.* Mexico City: Fondo de Cultura Ecónomica.

Freund, R. J. 1956. "The Introduction of Risk into a Programming Model." *Econometrica.* Vol. 24 (July), pp. 253–63.

Heady, Earl O., Narindar S. Randhawa, and Melvin D. Skold. 1966. "Programming Models for Planning of the Agricultural Sector." In *The Theory and Design of Economic Development.* Edited by I. Adelman and E. Thorbecke. Baltimore, Md.: Johns Hopkins University Press, chapter 13.

Hertford, Reed. 1971. "Sources of Change in Mexican Agricultural Production, 1940–65." U.S. Department of Agriculture Foreign Agricultural Economic Report, no. 73. Washington, D.C.: U.S. Government Printing Office.

Piñeiro, Martin E., and Alex F. McCalla. 1971. "Programming for Agricultural Price Policy Analysis." *Review of Economics and Statistics.* Vol. 53, no. 1 (February), pp. 59–66.

Samuelson, Paul A. 1952. "Spatial Price Equilibrium and Linear Programming." *American Economic Review.* Vol. 42, no. 3 (June), pp. 283–303.

Solís, Leopoldo. 1970. *La realidad económica mexicana: retrovisión y perspectivas.* Mexico City: Siglo XXI Editores.

Takayama, T., and G. G. Judge. 1964. "Equilibrium among Spatially Separated Markets: A Reformulation." *Econometrica.* Vol. 32 (October), pp. 510–24.

———. 1971. *Spatial and Temporal Price and Allocation Models.* Amsterdam: North-Holland.

3

The CHAC Demand Structures

JOHN H. DULOY AND ROGER D. NORTON

THE USE OF MATHEMATICAL PROGRAMMING to simulate market behavior has been explored extensively in a number of studies since Samuelson (1952) first pointed out that an objective function exists whose maximization guarantees fulfillment of the conditions of a competitive market. Although some of the subsequent studies have been purely theoretical, Samuelson's basic idea has also proven fruitful in the realm of empirical economics—particularly in the construction of agricultural planning models, which may contain rather detailed supply-side specifications.

Overview

Nevertheless, in practice the existing empirical formulations of Samuelson's idea are incomplete and awkward to use in several respects. This chapter attempts to close some conceptual gaps and to make the idea of a maximizing objective function for a competitive market more practicable. Since the simplex algorithm is the most powerful computational programming algorithm available, linear programming is adopted as the context for the analysis. Grid linearization techniques, however, are used so that some classes of nonlinear programming problems can be approximated almost arbitrarily closely at very little increase in computational difficulty.

The procedures described in this chapter have been applied in many of the analyses for Mexican agriculture reported in other chapters of this book. To place this chapter in proper perspective, we begin with a brief review of the principal strands running through the published literature.

Note: This chapter is an extended version of Duloy and Norton (1975). Unrevised passages appear here with the original publisher's permission. The patient criticism of Peter Hazell is gratefully acknowledged (without implicating him in any remaining errors). Luciano Barraza, Richard Boisvert, Wilfred Candler, Hunt Howell, Gary Kutcher, and Alan Manne also contributed helpful comments.

Antecedents

The story began when Stephen Enke (1951) posed the problem of finding the competitive equilibrium levels of interregional product flows and prices. He ingeniously proposed solution by using an electrical analog, a system of interconnected batteries. Since the typical economist is not a competent electrician, the profession's interest was stirred more by Samuelson's mathematical formulation a year later, in which he pointed out the possibility of maximization.

Samuelson (1952) viewed his maximand, the "net social payoff," as an artifice whose usefulness lay simply in its driving of a programming solution to the point of competitive equilibrium with respect to quantities marketed at each location. He explicitly rejected any welfare interpretation of the maximand. Although Samuelson gave the first mathematical formulation, his suggestions on procedures for solution did not go beyond trial-and-error iterations. Also, and on the purely conceptual level, a decade later Vernon Smith (1963) showed that the same problem could be cast as the minimization of rents to fixed resources, which he felt was a more natural way of looking at the workings of a competitive market. Smith, too, dealt with the market for a single product, so questions of interdependence in demand were not considered.

Meanwhile, attempts at developing manageable, numerical procedures for solution were being made outside the optimization framework. Fox (1953) appears to have been the first to carry out a computational procedure for competitive equilibrium (for the case of a multiregional, feed-grain economy). Given estimated parameters of regional demand functions, Fox followed an iterative method to arrive at market-clearing prices and shipments. A few years later, Judge and Wallace (1958) and Tramel and Seale (1959) similarly proposed iterative, nonoptimizing tatonnement procedures for finding the equilibrium solutions for the interregional shipments problem.

Also in the numerical context, Heady and others had been concurrently developing spatial linear programming models in which interregional demands were fixed nationally or regionally (see, for example, Heady and Egbert 1959). In these studies the dual solution was sometimes used to discuss supply prices, but market equilibrium prices were not obtained. A more recent example of this approach is Piñeiro and McCalla's study of Argentine agriculture (1971).

The first attempts to explicitly introduce price-responsive demand functions into a linear programming model also led to iterative solution methods. In these cases a linear programming solution was calculated at each iteration. For example, Schrader and King's 1962 and 1963 works maximized producers' revenue at each iteration and successively revised prices to eventually attain the market-clearing solution. This procedure can be workable for some simple models, but, when the model becomes large or when many products are included, the iteration procedure becomes excessively time consuming.

Of course, iterations were adopted because the objective function is inherently nonlinear in this kind of problem: it is some function of price times quantity, where both factors are endogenous. Takayama and Judge (1964a and -b) were the first to solve the spatial equilibrium problem directly with quadratic programming and under linear, interdependent demand functions.[1] They also showed that, by appropriate definition of the maximand, the monopolistic equilibrium can be replicated with the model. In the Takayama-Judge formulation, the objective function for the competitive case is clearly identified as the sum of the consumer and producer surpluses. In the monopolistic case, it becomes net revenue to producers. For computational methods, Takayama and Judge showed that a modified simplex algorithm of Wolfe can be applied through reformulation of the problem as a primal-dual linear program. In practice, however, this algorithm is not nearly as powerful as the simplex algorithm itself.

Yaron, Plessner, and Heady (1965) distinguished between the following multiproduct cases: (1) independent demands; (2) interdependent demands with fulfillment of the integrability conditions; and (3) interdependent demands without fulfillment of the integrability conditions. For case (1), they showed how linear programming solution techniques could be applicable via stepwise approximation of the demand curves, and they made welfare interpretations of the objective function. For case (2), they established a quadratic programming formulation along the lines of Takayama and Judge and again stressed the welfare interpretation. For case (3), they used a primal-dual formulation and pointed out that the welfare interpretation breaks down, a point that lends force to satisfying integrability conditions. No computations were presented. Takayama and Judge also addressed the integrability problem in their 1971 book (chapter 12).

Hall, Heady, and Plessner (1968) actually applied a multiregional and numerical, quadratic programming model to U.S. agriculture and obtained estimated prices that were lower than actual prices.[2] Shortly

---

1. With nonlinear demand functions, the problem is of a higher order than quadratic.
2. They attribute the discrepancies to omission of fixed costs in the model and to the fact that supply-control and price support programs in the real world distort prices away from the competitive equilibrium levels. They appear to overlook the fact that both aggregation and

thereafter, Guise and Flinn (1970) used the quadratic formulation in an application to derive competitive allocations and prices of irrigation water for an agricultural district.

The work up to this point may be summarized as follows: the single-period (static) market demand structures had been fully worked out conceptually, in general optimizing models, for solutions treating prices and quantities as endogenous. On the computational side, however, although linear programming had been used for single-product models, little progress had been made on efficient solution procedures (that is, noniterative procedures). Hence, for multiproduct models it was not yet practical to use linear programming, which, by virtue of the computational advantages of the simplex algorithm, would permit complex and detailed formulations on both the supply and demand sides.[3] And, for multiproduct models with interdependence in demand, no simplex formulation had been established.

Yaron (1967) took up the question of income as an argument in the demand functions, which up to this point had been specified as dependent solely on prices. He established a lagged relation between demand and income and set up a two-period version of the model to show that the competitive equilibrium interpretation still holds. He made income exogenous, however, explicitly leaving out any effects on income of variations in the endogenous prices and quantities of the model. Thus, as Yaron pointed out, his approach is usable only when the portion of the economy represented by the model is small. And for linear programming solutions he recommended iterative procedures, ignoring the computational possibilities inherent in the earlier Yaron-Plessner-Heady specification.

Without taking into account interdependence in demand, Martin (1972) spelled out in detail the piecewise linear specification of product demand and factor supply functions in a linear programming model of competitive markets. Although Martin gave a systematic presentation of the procedure, national and regional multiproduct models that included stepped demand functions and the surplus-maximizing objective had already been developed in Europe. A national model was developed for France, as discussed by Farhi and Vercueil (1969) and Tirel (1971). About the same

---

specification biases could distort prices in a downward direction by overstating productive efficiency and, hence, placing the supply curve to the right of its proper location. Given the generally inelastic demands for agricultural products, this bias would show up more significantly in prices than in quantities, which is what their results display.

3. The relative ease of handling large-scale models by means of linear programming is especially relevant to the construction of sector-wide planning models, in which incorporation of considerable detail by location and product is almost unavoidable if useful results are to be obtained.

time, a regional model designed to yield the competitive market solution in prices and quantities was formulated for the Moldavian region of the U.S.S.R. and was reported by Mash and Kiselev (1971). These apparently were the first sectoral linear programming models to give the competitive equilibrium solution without iterations. Neither, however, considered interdependence in demand, and their stepped approximation technique led to a model much larger in row dimensions than does the procedure that follows in this chapter.

Given this background, this chapter provides five additions to the earlier contributions, all in the context of linear programming: (1) a workable method of handling interdependence among products in demand; (2) a method for defining an income variable for producers at endogenous prices; (3) a simple method for adjusting the linear programming coefficients of the demand system to reflect changes in population and per capita income, which permits comparative statics analysis; (4) an interpretation of the dual variables of the system as commodity prices and consumer and producer surplus; and (5) an application of the Miller (1963) grid linearization to permit arbitrarily close approximation to nonlinear demand functions without increasing the number of rows in the model. By this last technique, nonlinearities in both the constraint set and the objective function can be handled with linear programming algorithms. All of these elements may be viewed as steps toward making Samuelson's idea of market determination more usable for large-scale optimization planning models.

In CHAC, there are thirty-three agricultural commodities, including several in the oilseeds and forage groups. Consequently, it was necessary to develop some expression of interdependence in demand, and the number of crops ruled out an iterative, equilibrium-seeking procedure. The sheer size of the model—combined with the need to make many solutions to explore alternative, hypothetical policy packages—made mandatory the use of the simplex algorithm instead of a nonlinear algorithm.

Uses of Demand Structures in Planning Models

The entire stream of the literature charted above has had little effect on the analytic side of the national planning tradition. In the classic linear programming models of an economy or subeconomy, goods are assumed either to face infinitely elastic demand functions or to be traded in bounded quantities. Modifications sometimes are made for exports that constitute a significant share of the world market, and these lead to a formulation in which the optimizing unit equates marginal revenue and marginal cost on the export markets. But the existence of international markets normally is used to justify the price-taker assumption. In addition, for a large class of products, particularly agricultural commodities, the spread between c.i.f. and f.o.b. prices may be 20 percent or more, and for another group of products trading opportunities effectively do not exist. In these cases, domestic product demand functions are relevant in price determination.

Incorporating product demand functions into a planning model designed for the purpose of analyzing policy alternatives, rather than assuming exogenously determined product prices, has three principal advantages. First, it allows the model to correspond to a market equilibrium. The effects of various policies—such as subsidizing or taxing product or input prices or varying the exchange rate—can then be investigated. Second, it allows the model greater flexibility. For instance, substitution between capital and labor, corresponding to different factor price ratios, can occur not only directly through the technology set or through changes in the commodity mix of output but also through substitution in demand from changing relative prices of products that are more or less labor or capital intensive. Third, it permits an appraisal of the distribution between consumers and producers of benefits accruing from changes in output. For example, in the common situation of agricultural production for the domestic market in the face of demand curves with elasticities less than unity in absolute value, the returns from increased output are negative to producers as a whole and positive to consumers.[4]

Given these considerations, the approach of this chapter is intended to facilitate wider use of market equilibrium specifications in planning models.

The Basic Model

Throughout, the exposition that follows is developed as a static linear programming model. Experience with these models amply indicates by now that the linearity does not prevent the incorporation of significantly nonlinear behavior in the model.

The specification of the objective function follows from the choice of market form to be incorporated in the model. In the competitive case, producers act as price takers and equate marginal costs to the prices of products. In the monopolistic case, the sector maximizes its net income by equating marginal costs to the marginal revenues of products. For simplicity of exposition, the introduction of international trade is deferred to a later section.

4. None of these advantages accrues when a model is designed with fixed production targets and when marginal supply prices for products are derived from the dual solution.

In general terms, the static demand function may be written, in inverted form, as:

$$(3.1) \qquad\qquad p = \phi(q, Y),$$

where p is an $N \times 1$ vector of prices, q is an $N \times 1$ vector of quantities, and Y is an exogenous scalar representing (lagged) permanent income.

For an unconstrained model, the objective function for the competitive case may be written

$$(3.2) \qquad Z = \int_0^{q_n} dq_n \ldots \int_0^{q_1} \phi(q, Y) dq_1 - c(q)I \to \max_q,$$

where $c(q)$ is an $N \times 1$ vector of total cost functions,[5] I is the identity matrix, and $q \geq 0$. In this case, setting the derivative of equation (3.2) with respect to q equal to zero yields

$$(3.3) \qquad\qquad p - c'(q) = 0,$$

which is the equilibrium condition of price equals marginal cost.

If, in addition to the explicit costs $c(q)$, there are resources whose availability is constrained, the model is extended by adding the conditions

$$(3.4) \qquad\qquad Aq \leq b,$$

where A is an $M \times N$ constraint matrix and b is an $M \times 1$ vector of resource availability levels.

For the constrained maximization problem, the Kuhn-Tucker necessary conditions are equation (3.4) plus

$$(3.5a) \qquad\qquad p - c'(q) - \lambda A \leq 0,$$

$$(3.5b) \qquad\qquad [p - c'(q) - \lambda A] q = 0, \text{ and}$$

$$(3.5c) \qquad\qquad \lambda [Aq - b] = 0,$$

where λ is the vector of dual variables to the linear program.

Equation (3.5a) says that profits must be nonpositive. Profits per unit are defined as prices minus marginal costs, where costs now have two components: the explicit (market) costs of inputs whose behavior is subsumed in the vector of cost functions $c(q)$, and the economic rents that accrue to the use of the fixed factors represented by the vector b. Equation (3.5b) is the complementary slackness condition, which, together with the nonnegativity conditions, says that for every activity of nonzero level in the optimal basis profits are zero, and that an activity with nonzero profits at given levels of use cannot enter the optimal basis at any of those levels. Equation (3.5c) is the complementary slackness condition for the dual

---

5. The supply functions of a programming model usually are endogenous, in piecewise linear form, but this does not affect the generality of the exposition.

solution: either a resource's rent is nonzero or its slack is nonzero, but not both.

Taken together with Euler's Theorem, which guarantees equality in equation (3.5a) for activities that enter positively in the optimal basis, these conditions describe the characteristics of a system of competitive markets when fixed factors are used in the productive process. As such, these conditions constitute a generalization of equation (3.3) that is particularly relevant to agriculture. In the Mexican CHAC model, for example, the short-run fixed factors are land, irrigation water, and family labor on irrigated farms.[6] Hence, the economic rents to these factors are included in the sum of marginal costs, which give equilibrium product prices in this model.

The Linear Programming Formulation

In this section, the general model described by equations (3.2) and (3.4) is restated in linear programming form for the case of independent product demands. An approximation procedure is demonstrated that has the property whereby the nonlinear objective function of equation (3.2) can be approximated as closely as desired without increasing the number of rows in the model. An income variable is defined at exogenous prices, and, to handle the variable efficiently, the approximation procedure is extended to the constraint set. In subsequent sections, the linear programming specification is broadened to allow interdependence in demand, and some properties of the specification are demonstrated.

First, for the demand function (3.1), the area under the function is expressed as

$$(3.6) \qquad W = \int_{q_0}^{q_T} \phi(q, Y) dq,$$

where q_0, q_T are lower and upper limits, respectively, that are set to establish the relevant portion of the demand curve. This area measure becomes an important part of the objective function, but before this is demonstrated, some modifications are made in the basic model to cast it for linear programming.

To present an example of a linear programming formulation, the case of linear demand functions is used, although the procedure in general places no restriction on the shape of the demand function except that the Hessian matrix of detached coefficients of the joint demand functions be negative

---

6. Seasonal migration is assumed to take place only among landless laborers and family labor on the poorer (rainfed, or *temporal*) farms.

semidefinite to ensure convexity of the program.[7] The variable Y is dropped from the demand function since the model is static, or, if income effects are present, they are incorporated in lagged form. Equation (3.1) may be rewritten, then, as

$$(3.7) \qquad\qquad p = a + B_q,$$

where a is an $N \times 1$ vector of constants, and B is an $N \times N$ negative semidefinite matrix of demand coefficients. The objective function (3.2) then becomes:

$$(3.8) \qquad\qquad Z = q'(a + 0.5Bq) - c(q) \rightarrow \max_q.$$

The objective function[8] can be decomposed into components that correspond to consumer surplus (CS) and producer surplus (PS):[9]

$$(3.9a) \qquad\qquad CS = 0.5q'(a - p) = 0.5q'Bq;$$

$$(3.9b) \qquad\qquad PS = q'p - c(q) = q'(a + Bq) - c(q).$$

To describe the monopolistic equilibrium with the linear programming model, the appropriate objective function would be:

$$(3.10) \qquad\qquad M = q'(a + Bq) - c(q).$$

The Kuhn-Tucker conditions for this version of the model are equation (3.4) and

$$(3.11a) \qquad\qquad a + 2Bq - c'(q) - \lambda A \leq 0,$$

$$(3.11b) \qquad\qquad [a + 2Bq - c'(q) - \lambda A]q = 0, \text{ and}$$

$$(3.11c) \qquad\qquad \lambda[Aq - b] = 0.$$

The only difference between equations (3.5a–c) and (3.11a–c) is that the vector p is replaced by the term $a + 2Bq$, which is the vector of marginal revenues. Therefore, the previous interpretations of equations (3.5a–c) are

7. A slightly weaker statement of this (integrability) condition is found in Zusman (1969).

8. This objective function is essentially identical to Samuelson's "net social payoff" function, except that he includes interregional transport costs, whereas here only a single point in space is treated. See Samuelson (1952). The same objective function is elaborated in the multiproduct case by Takayama and Judge (1964b).

9. Of course, the function (3.2) may be interpreted merely as an equilibrium-seeking device, thus side-stepping the controversies surrounding the Marshallian surpluses. [See, for example, Mishan (1968).] If it is accepted as a social utility function, however, some interesting programming experiments are possible (as outlined later in this chapter). An alternative interpretation of the objective function is possible; it can be interpreted as the profit function of a discriminating monopolist. Such an interpretation, of course, is hardly tenable for a sectoral planning model, partly because of problems of separability of markets, but also because of the fact that the demand functions would require some reformulation to take into account income effects.

Figure 3-1. *Demand and Expenditure Equations for CHAC*

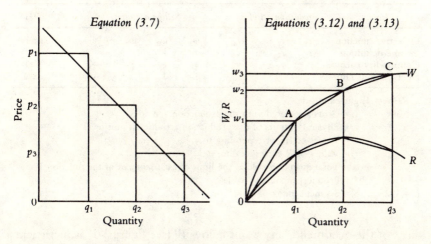

maintained subject to substitution of "marginal revenue" for "price." Hence, the model given by equations (3.10) and (3.4) guarantees the monopolistic equilibrium as the optimal solution.[10]

The final step required to set up the linear programming tableau is to rewrite equation (3.6) as

$$(3.12) \qquad W = q'(a + 0.5Bq)$$

and to write the revenue function as

$$(3.13) \qquad R = q'(a + Bq).$$

The demand function (3.7) and the counterparts (3.12) and (3.13) are shown diagrammatically in figure 3-1.

Even under the linear demand function, the competitive maximand, equation (3.8), and the monopolist's maximand, equation (3.10), both involve a quadratic form in p. Two linear approximation procedures have been developed; the first is for the case where estimates of the coefficients of B are available (interdependence among products in demand), and the second for the case where less information is known about the structure of demand (separability assumed). In this section the latter case is taken up.

Note that, for a linear demand function, both W and R are nonlinear. The approximation procedure, however, involves direct segmentation of W and R instead of the demand function. Since W is the positive compo-

---

10. For an interesting evaluation of the welfare loss implied by monopolistic agricultural markets, see Plessner (1971).

Table 3-1. *Linear Programming Tableau in the Single-Product Case*

Row	Production activities				Selling activities				M	RHS
Objective function	$-c_1$	$-c_2$	$\cdots$	$-c_m$	w_1	w_2	$w_3\cdots$	w_S		(max)
Income definition	$-c_1$	$-c_2$	$\cdots$	$-c_m$	r_1	r_2	$r_3\cdots$	r_S	-1	$=0$
Commodity balance	y_1	$y_2\cdots$		y_m	$-q_1$	$-q_2$	$-q_3\cdots$	$-q_S$		≥ 0
Demand constraint					1	1	$1\cdots$	1		≤ 1

RHS = Right-hand side.

Note: c_i = Costs associated with the production activities
y_i = Physical outputs of the production activities at unit level
w_s = Values of W corresponding to q_s
r_s = Values of R corresponding to q_s
q_s = Total quantities sold at the limit of each segment of the function W
M = Producer income variable

$(s = 1, 2, \ldots, S)$ = Segment index.

nent of the maximand, any point below W (see figure 3-1) is inefficient and, hence, nonoptimal. In the piecewise linear approximation to W, optimality guarantees that no more than two adjacent nodes (two points on the q-axis) will enter the optimal basis. The representation of the piecewise linear approximation in linear programming is shown in the following tableaus.[11]

In the single-product case, the linear programming tableau corresponding to the segmented approximation of the functions W and R for one product can be given as in table 3-1, taking equation (3.8) as the maximand and equation (3.10) as an income-accounting equation.

Note that, in table 3-1, no more than two activities from the set of selling activities (each corresponding to one segment in the approximation) will enter the optimal basis at positive levels. This may be seen by reference to the W-function in figure 3-1: a linear combination of more than two points is a line interior to the piecewise efficiency frontier, 0ABC in the figure's representation of equations (3.12) and (3.13).

Table 3-1 is a transformation (using elementary row and column operations) of an initial tableau that embodied additively separable segments with a separate bound for each segment. This initial tableau is shown in table 3-2. It will be noted that table 3-2 corresponds to the segmenting of the demand and marginal revenue functions as step functions, rather than to the linearization of the W and R functions that underlies table 3-1.[12] The

11. This is an application of the grid-linearization technique of separable programming. See Miller (1963) and Hadley (1964).

12. To the extent that nonlinear functions are incorporated into planning models, their inclusion as step functions, as in table 3-2, is a common procedure. For a recent example, see MacEwan (1971, pp. 66–9).

Table 3-2. *Single-Product Linear Programming Tableau in Stepped Form*

Row	Production activities				Selling activities					M	RHS
Objective function	$-c_1$	$-c_2$	$\cdots$	$-c_m$	p_1	p_2	p_3	$\cdots$	p_S		(max)
Income row	$-c_1$	$-c_2$	$\cdots$	$-c_m$	m_1	m_2	m_3	$\cdots$	m_S	-1	$= 0$
Commodity balance	y_1	y_2		y_m	-1	-1	-1	$\cdots$	-1		≤ 0
					1						$\leq K_1$
						1					$\leq K_2$
							1				$\leq K_3$
Demand segment constraints								.			.
								.			..
								.			.
									1		$\leq K_S$

Note: p_s, m_s = Prices and marginal revenues, respectively, corresponding to segments s of the demand function, and

$K_1 = q_1$
$K_i = q_i - q_{i-1}, \; i > 1$
M = Producer income variable.

principal advantage of table 3-1 over table 3-2 is that the demand function (or area function W) of the former can be approximated as closely as desired without additional constraints in the program. The number of selling activities increases as the number of linear segments increases, but the number of rows remains constant.

The approach is readily extended to two or more products that are additively separable in demand, with one commodity balance and one convex combination constraint per product.

Substitution in Demand

In the event that two or more products are not separable in demand, the nonlinear demand set can be linearized directly, to an arbitrarily close approximation, by the specification of activity vectors representing points on the demand surface and by the incorporation of an appropriate convex combination constraint. An example of the tableau in such a case, for two products and six segments per term in the objective function, is shown in table 3-3.

In the treatment of table 3-3, it is assumed that the elements of the matrix B, including off-diagonal elements, in equation (3.7) are known or can be estimated. Frequently, the available information consists only of estimates of own-price elasticities for a number of individual commodities and commodity groups, so an alternative approach is required.

The basis of the approximation procedure developed for this situation of limited information is the assumption that commodities can be clas-

Table 3-3. *Linear Programming Tableau for Two Products, Substitution in Demand*

Row	Production activities Good 1	Good 2	Selling activities						M	RHS
Objective function	$-c_{1j}$	$-c_{2j}$	w_{11}	w_{12}	w_{13}	w_{21}	w_{22}	w_{23}		(max)
Income row	$-c_{1j}$	$-c_{2j}$	r_{11}	r_{12}	r_{13}	r_{21}	r_{22}	r_{23}	-1	$= 0$
Commodity balance 1	y_{1j}		$-q_{11}$	$-q_{11}$	$-q_{11}$	$-q_{12}$	$-q_{12}$	$-q_{12}$		≥ 0
Commodity balance 2		y_{2j}	$-q_{21}$	$-q_{22}$	$-q_{23}$	$-q_{21}$	$-q_{22}$	$-q_{23}$		≥ 0
Convex combination constraint			1	1	1	1	1	1		≤ 1

Note: c_{ij} = Costs for the ith product in the jth activity producing it
   y_{ij} = Unit outputs of the ith product in the jth activity producing it
   q_{ij} = Quantities sold of the ith product corresponding to the end-point of the jth segment
   w_{ij} = Values of W for the ith commodity corresponding to the amount sold q_{ij}
   r_{ij} = Values of R for the ith commodity corresponding to the amount sold q_{ij}
   M = Producer income variable.

sified into groups, such that the marginal rate of substitution (MRS) is zero between all groups but nonzero and constant within each group. Clearly, this assumption is only an approximation to reality. A group may consist of one or more commodities, and limits are defined on the variability of

Figure 3-2. *Indifference Surface Depicting Limited Commodity Substitution*

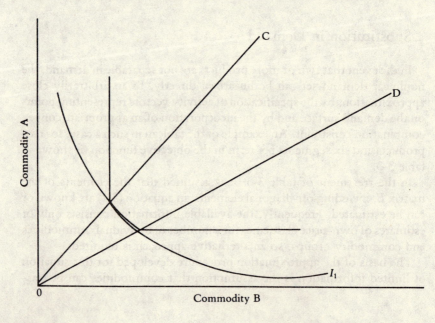

the commodity mix within each group. The relevant portions of the indifference surface with respect to two commodities in a group are shown in figure 3-2. The rays 0C and 0D in the figure define the limits on the composition of the commodity bundle.

If sufficient information is available, the approach can be extended to more linear segments per indifference curve, each segment representing a different value of the MRS. Consider a group consisting of C commodities. The appropriate linear programming tableau may be represented as shown in table 3-4. In the table, each of the block of activities $[W_s'\ R_s'\ -Q_s'\ 1]$ constitutes a set of "mixing" activities for one segment of the composite demand function for the commodity group. This block of activities can be written as:

$$(3.14)\quad \begin{bmatrix} W_s \\ R_s \\ -Q_s \\ \underline{1} \end{bmatrix} = \begin{bmatrix} w_s & w_s & \cdots & w_s & \cdots & w_s \\ r_s & r_s & \cdots & r_s & \cdots & r_s \\ -q_{s11} & -q_{s12} & \cdots & -q_{s1m} & \cdots & -q_{s1M} \\ -q_{s21} & -q_{s22} & \cdots & -q_{s2m} & \cdots & -q_{s2M} \\ -q_{sc1} & -q_{sc2} & \cdots & -q_{scm} & \cdots & -q_{scM} \\ \cdot & \cdot & & \cdot & & \cdot \\ \cdot & \cdot & & \cdot & & \cdot \\ \cdot & \cdot & & \cdot & & \cdot \\ -q_{sC1} & -q_{sC2} & \cdots & -q_{sCm} & \cdots & -q_{sCM} \\ 1 & 1 & \cdots & 1 & \cdots & 1 \end{bmatrix},$$

where the elements are as defined below.

The derivation of formulae for the elements of equation (3.14) is tedious because they take account of shifts both between and among segments. The starting point is a set of observed prices $\{\bar p_1, \ldots, \bar p_c, \ldots, \bar p_C\}$ and of quantities $\{\bar q_1, \ldots, \bar q_c, \ldots, \bar q_C\}$. Relative prices of commodities in the group are assumed fixed, both within and between segments, and are defined by

$$(3.15)\qquad \rho_C = \bar p_c / \sum_c \bar p_c.$$

Corresponding to the observed sets of prices and quantities is a quantity index

$$(3.16)\qquad \bar q = \sum_c \bar q_c\, \bar\rho_c$$

and a price index

$$(3.17)\qquad \bar p = \sum_c \bar p_c\, \bar q_c / \bar V,$$

where

$$(3.18)\qquad \bar V = \sum_c \bar q_c.$$

It is assumed that an estimate exists of a demand function for the group with a price index as a function of a quantity index, as in equations (3.16)

Table 3-4. *General Linear Programming Tableau with Substitution in Demand*

Row	Production activities	Selling activities			M	RHS
Objective function	$-\tilde{C}$	W_1 $\cdots$	W_s $\cdots$	W_S $\cdots$		(max)
Income row	$-\tilde{C}$	R_1 $\cdots$	R_s $\cdots$	R_S $\cdots$	-1	$= 0$
Commodity balances	Y	$-Q_1$ $\cdots$	$-Q_s$ $\cdots$	$-Q_S$ $\cdots$		≥ 0
Convex combination constraint		$\underline{1}$ $\cdots$	$\underline{1}$ $\cdots$	$\underline{1}$ $\cdots$		≤ 1

Note: $(s = 1, 2, \ldots, S)$ = Segment index
$\tilde{C}$ = Row vector of production costs
Y = C-rowed matrix of production coefficients entering the commodity balances
$W_s, R_s = 1 \times C$ vectors of areas under the demand function and gross revenues respectively
$Q_s = C \times C$ matrix of adjusted quantities as defined in equation (3.16)
$\underline{1}$ = Unit vector
M = Producer income variable.

and (3.17). Assume for a moment that no substitution occurs among commodities (that is, that they are consumed in the fixed observed proportions) and that the demand function is segmented in S segments. This case then corresponds to table 3-5, which is a simple extension of the single product case. Only the selling activities are shown in the table, and it is evident from the table that

$$(3.19) \qquad q_{sc} = \bar{a}_c V_s,$$

where $a_c = \bar{q}_c / \bar{V}_s$, the observed proportion in physical units of the cth commodity, and V_s is the total quantity sold in the sth segment in physical units. W_s and R_s are, of course, computed from the demand function with appropriate price and quantity indexes, although in table 3-5 the weights are all constant. The price-weighted total quantity is:

$$(3.20) \qquad \bar{q}_s^* = \sum_c \rho_c \, q_{sc} = V_s \sum \bar{a}_c \, \rho_c.$$

To extend the case of demand in fixed proportions within a group, it is supposed that, for C commodities, the set of feasible alternative mixes, as proportions in physical terms, *is given by the matrix A*, assumed for simplicity to be invariant across segments:

$$(3.21) \qquad A = \left[a_{cm} \right],$$

where $c = 1, \ldots, C$ commodities in the group; $m = 1, \ldots, M$ mixes of the commodities; and a_{cm} is the proportion in physical terms of the cth commodity in the mth mix, such that $\Sigma_c \, a_{cm} = 1$. The elements, a_{cm}, define the rays shown in figure 3-2.

Table 3-5. *Partial Linear Programming Tableau
with C Commodities in Alternative Fixed Proportions*

Row	Selling activities			RHS
Objective function	W_1 $\cdots$	W_s $\cdots$	W_S	(max)
Income row	R_1 $\cdots$	R_s $\cdots$	R_S	$= 0$
	$-q_{11}$ $\cdots$	$-q_{s1}$ $\cdots$	$-q_{S1}$	≥ 0
	$-q_{12}$ $\cdots$	$-q_{s2}$ $\cdots$	$-q_{S2}$	≥ 0
Commodity balances	.	.	.	.
	.	.	.	.
	.	.	.	.
	$-q_{1c}$ $\cdots$	$-q_{sc}$ $\cdots$	$-q_{Sc}$	≥ 0
Convex combination constraint	1 $\cdots$	1 $\cdots$	1	≤ 1

Note: See discussion of equation (3.21) in text.

The elements, in matrix Q_s in equation set (3.14) can now be defined as

$$(3.22) \qquad q_{scm} = a_{cm} V_s \sum_a \bar{a}_c \rho_c / \sum_c a_{cm} \rho_c,$$

which differs from the expression for q_{sc} (consumption in fixed proportions) in equation (3.19) by the factor $\sum_c \bar{a}_c \rho_c / \sum_c a_{cm} \rho_c$, which reflects the changing commodity weights. Using equation (3.20), equation (3.22) can be rewritten as

$$(3.23) \qquad q_{scm} = a_{cm} \bar{q}_s^* / \sum_c a_{cm} \rho_c,$$

and the price-weighted total quantity, q_{sm}^*, is given by

$$(3.24) \qquad q_{sm}^* = \sum_c \rho_c q_{scm} = \bar{q}_s^*.$$

That is, *the price-weighted quantity of the aggregate commodity is independent of the commodity mix*, and it can be written as q_s^*. Using this result, equation (3.22) can be simplified as follows:

$$(3.25) \qquad q_{scm} = a_{cm} q_s^* / \sum_a a_{cm} \rho_c.$$

This completes the definition of the elements of the matrix Q_s in equation set (3.14). By equation (3.24), q_s^* is invariant with respect to the commodity mix, so that the elements of w_s and r_s are invariant over the mixing activities. They are computed exactly as in the single-product case, using, however q_s^* in place of q_s. To recapitulate, if the demand function is linear, then:

$$(3.26) \qquad w_s = q_s^* (a - \frac{1}{2} b q_s^*), \text{ and}$$

$$(3.27) \qquad r_s = q_s^* (a - b q_s^*).$$

Table 3-6. *Partial Linear Programming Tableau for Limited Substitution with Two Commodities in a Segment*

Concept	Selling activity		RHS
Activity level	x_1	x_2	
Objective function	w_s	w_s	(max)
Income row	r_s	r_s	$= 0$
Commodity balances	$-q_{s11}$	$-q_{s12}$	≥ 0
	$-q_{s21}$	$-q_{s22}$	≥ 0
Convex combination constraint	1	1	≤ 1

Note: See text and equation (3.28).

The demand side of a planning model may be constructed to incorporate a number of product groups, some of which can consist of a single commodity. Between product groups, the MRS is zero; it is constant within; and it is given by the inverse of the price ratio. This last property leads to the constancy of consumer surplus $(w_s - r_s)$ and of consumer expenditure (r_s) within a commodity group.

The constancy of the MRS can readily be shown for the case of two products, shown in table 3-6, in which, again, only the selling activities are included. As is indicated in the table, and by the constancy of w_s and r_s, movement along a given indifference function requires changes in the activity levels, x_1 and x_2, which are equal but of opposite sign. Without loss of generality, consider the two cases $(x_1 = 1, x_2 = 0)$ and $(x_1 = 0, x_2 = 1)$. Then the MRS is given by equations (3.28), in which the subscript s, and q_s^* (which is common to all terms), are dropped:

$$(3.28) \qquad \text{MRS} = \frac{\Delta_1}{\Delta_2} = \frac{q_{11} - q_{12}}{q_{21} - q_{22}}$$

$$= \frac{a_{11}/\sum_c a_{c1}\,\rho_c - a_{12}/\sum a_{c2}\,\rho_c}{a_{21}/\sum_c a_{c1}\,\rho_c - a_{22}/\sum a_{c2}\,\rho_c}.$$

By expanding and rearranging equations (3.28), the following is obtained:

$$(3.29) \qquad \Delta_1/\Delta_2 = -\rho_2/\rho_1,$$

which is the required result.

Comparative Statics and International Trade

This specification of commodity demand structures incorporates one characteristic that makes it particularly convenient for obtaining compara-

Figure 3-3. *Transformation of a Demand Function*

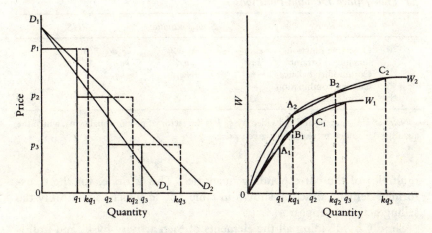

tive statics solutions. This property is that the demand function, for any commodity group, can be rotated merely by an appropriate change in the constraint value of the convex combination inequality; that is, the matrixes W_s, R_s, Q_s are invariant under this class of transformations of the commodity demand function.

The transformation of the demand function, for a single product, is illustrated in figure 3-3, in which it is assumed that the function is linear. The original demand function and corresponding W function are shown as D_1D_1 and $0W_1$, respectively, and the rotated demand function and corresponding W function by D_1D_2 and $0W_2$, respectively. If the original demand function is

$$(3.30) \qquad p = f(q),$$

it is required that the transformed function can be expressed as

$$(3.31) \qquad p = f(kq),$$

Such a formulation readily accommodates shifts in the demand function caused, for example, by changes in population or per capita incomes or both. The rotation upward of the demand function is expressed as a proportional lengthening of the segments, with price held constant. For the segmented W function, the slope of the linearized function in each segment, which is the approximation to price within that segment, is equal for both W_1 and W_2 for corresponding segments. A similar condition holds for the linearized R function, where the slopes are approximations to marginal revenue within the segment. Given linearity, and the constancy of the slopes of the segmented functions within each segment, the coefficients in the W_s and R_s matrixes can be expressed as simple

Table 3-7. *Linear Programming Tableau
for Transformed Demand Functions*

Row	Selling activities			RHS
Objective function	$kq_1w'_1$ $\cdots$	$kq_sw'_s$ $\cdots$	$kq_Sw'_S$	(max)
Income constraint	$kq_1r'_1$ $\cdots$	$kq_sr'_s$ $\cdots$	$kq_Sr'_S$	$\geq Y^{\star}$
Commodity balance	$-kq_1$ $\cdots$	$-kq_s$ $\cdots$	$-kq_S$	≥ 1
Convex combination constraint	1 $\cdots$	1 $\cdots$	1	≤ 1

Note: w'_s and r'_s are w_s and r_s divided by q_s; k is the factor of proportionality by which the quantity demanded increases at a given price.

multiples of the corresponding quantities. This is done, for the transformed demand function shown in table 3-7, in which, again, only the selling activities appear.

Simply by dividing all the elements of each activity by k, and multiplying through the convex combination constraint by k, the program with the transformed demand function in table 3-7 reduces to a program with coefficients in the constraint matrix identical to those before the demand transformation, but with k replacing unity on the right-hand side of the convex combination constraint. This result is readily extended to the commodity group case, as can be seen by replacing q_s with $q_s^{\star}$ (in the objective function and income constraint); by replacing w'_s and r'_s by the corresponding vectors W'_s and R'_s;[13] and by recalling that the matrixes Q_s in the commodity balances can be written as scalar multiples of $q_s^{\star}$. This characteristic of the demand structure permits computationally simple parametric variation of the position of the demand function. It also opens the possibility, in a larger system, of endogenously determining both the position of, and the position on, the demand functions.

A representation of international trade can readily be incorporated into the structures developed in this chapter in the usual way in which it is incorporated into planning models; that is, by adding commodity-specific importing activities as additional "production" activities and, similarly, by adding exporting activities as additional selling activities. Again, as usual, it is possible to specify import supply (export demand) as being infinitely elastic, but bounded, or as being represented by an upward-sloping supply (downward-sloping demand) schedule. In this last case, it is possible to approximate the nonlinearities involved by the methods developed above. Notice, however, that it is only possible to specify a monopolistic formulation of export supply (or a monopsonistic formula-

---

13. Note that $q_s^{\star}$, being invariant over mixing activities, is a scalar.

tion of import demand) if the objective function and the scope of the model represents multicountry welfare.

When trading opportunities are included as outlined above, the model captures the different trading positions posited by price theory. The trading positions depend on relative domestic and foreign supply and demand functions and on whether the objective function is chosen to reflect competitive or monopolistic behavior. For example, in the monopolistic case, final product importing activities never enter the optimal basis, and the model reproduces the expected two-price behavior when the foreign marginal revenue function lies above the domestic marginal revenue function.[14]

Shadow Prices

An essential feature of this approach is that commodity prices may be derived from either the primal or the dual solution. The structure ensures that they will be equal; if they are not, it is a signal to the model builder that something has gone wrong. Having the prices available in the primal solution can be useful for the purpose of introducing policies that fix or bound the levels of prices. If the demand activities for commodity i are represented by X_{is}, and where s is the segment index, then equation (3.7) may be entered directly in the model, in slightly modified form, as follows:

$$(3.32) \qquad p_i - b_i \sum_s X_{is} \, q_{is} = a_i.$$

In a detached coefficients tableau, equation (3.32) is transformed as follows (RHS = right-hand side):

		Selling activity				RHS
Activity level	p_i	X_{i1}	X_{i2}	$\cdots$	X_{iS}	
Coefficients	1	$-b_i q_{i1}$	$-b_i q_{i2}$	$\cdots$	$-b_i q_{iS}$ =	$a_i.$

This tableau defines a new price variable p_i, which may then be constrained if desired.

The equivalence of p_i and certain dual variables may be shown by reference to the structure of the primal and dual programs. The essential elements of the tableau in table 3-4 may be rewritten, in slightly different notation, as shown in table 3-8. It is assumed that the farm income

---

14. One case that the structure will not handle is the monopolist case in which either of the demand functions is of the double-log form and in which the elasticity of demand is less than unity in absolute value. In this case, marginal revenue is negative, but increasing (that is, the function is nonconvex).

Table 3–8. *Primal and Dual Programs*

Primal problem

$$-\sum_i c_i \; h_i \; + \; \sum_{is} w_{is} \; d_{is} \quad \text{(max)}$$

$$-y_i \; h_i \; + \; \sum_s q_{is} \; d_{is} \; \le 0$$

$$\sum_i a_{ji} \; h_i \qquad\qquad\qquad \le b_j$$

$$\sum_s d_{is} \le 1$$

Dual problem

$$\sum_j b_j \; \lambda_j \; + \; \sum_i \mu_i \quad \text{(min)}$$

$$-y_i \; \pi_i \; + \sum_j a_{ji} \; \lambda_j \qquad\qquad \ge -c_i$$

$$q_{is} \; \pi_i \qquad\qquad\qquad + \quad \mu_i \; \ge w_{is}$$

Note: Variables and parameters are defined in the text.

constraint is not set at a binding level. In the statement of the linear programming problem made in table 3-8, the first inequality in the primal version is the set of commodity balances, the second is the set of resource constraints on production, and the third is the convex combination constraint in demand. The h_i's denote production activities (acreages) and the d_i's are the levels of demand activities by commodity and segment.

By the Kuhn–Tucker theorem, the dual constraints are equalities for those dual variables that have nonzero values.[15] Thus, the second row in the dual problem yields

(3.33)
$$\pi_i = \frac{c_i + a_i \lambda_i}{y_i},$$

where the λ_i are marginal returns to fixed resources [as in equations (3.5a–c), above] and the c_i are explicit costs for purchased inputs. Thus, π_i is the product price.

It then follows that the dual variables μ_i are measures of consumer surplus, since from the last equation in the dual the μ_i are defined as

(3.34)
$$\mu_i = w_i - q_i \pi_i,$$

in the event that only one segment of the demand function enters the optimal basis. If two segments are optimal (there will not be more than two, as shown earlier in this paper), then equation (3.34) becomes

(3.35a)
$$\mu_i = w_{is} - q_{is}\pi_i \text{ and}$$

(3.35b)
$$\mu_i = w_{i,\,s+1} - q_{i,\,s+1}\pi_i.$$

15. For an enlightening exercise in interpretation of a linear program's dual solution, see Johansen (1967).

These two equations do not lend themselves directly to interpretation. Multiplying by d_{is} and $d_{i,\,s+1}$ yields

(3.36a) $\mu_i d_{is} = w_{is} d_{is} - q_{is} d_{is} \pi_i$ and

(3.36b) $\mu_i d_{i,\,s+1} = w_{i,\,s+1} d_{i,\,s+1} - q_{i,\,s+1} d_{i,\,s+1} \pi_i,$

and summing obtains

(3.37) $\mu_i = w_i^\star - q_i^\star \pi_i,$

since $d_{is} + d_{i,\,s+1} = 1$. The parameter $w_i^\star$ is the interpolated value of the W-function, which is applicable to the quantity $q_i^\star$. Thus, equation (3.37) is equivalent to equation (3.34).

These characteristics of the dual solution have been checked numerically with CHAC and have proven quite useful. Commodity prices can be read directly from the linear programming solution's information on the commodity balances, and, similarly, the total consumer surplus is found quickly by summing the shadow prices on the convex combination constraints.

The three types of dual variables—λ_i, μ_i, and π_i—in fact account for total product, as Euler's theorem requires. In heuristic terms (for a single product), this may be seen diagrammatically in figure 3-4. The variables λ_i define the supply function, and they therefore measure the area $0q^eED$ when multiplied by the various levels of resource use.[16] The variables π_i, when multiplied by quantities, measure the larger area $0q^eEp^e$. The difference between these two areas is DEp^e, which is the producer surplus. The remaining part of the maximand is the area p^eEA, or consumer surplus, which is registered in the variable μ_i.

Conclusions

This chapter has discussed a practical procedure for enforcing both competitive and noncompetitive market structures by means of the optimization inherent in linear programming. The procedure has the property that arbitrarily close approximations to nonlinear forms—in both the objective function and constraint set—can be made without much loss of the computational efficiency of the simplex algorithm. In this respect, the chapter applies the ideas of Miller (1963) and integrates them with the work on market forms in mathematical programming by Samuelson (1952) and Takayama and Judge (1964a and b, 1971).

---

16. By the same reasoning used above, λ_i is equal to the input price for inputs that are available in infinitely elastic supply.

Figure 3-4. *Interpretation of the Dual Solution*

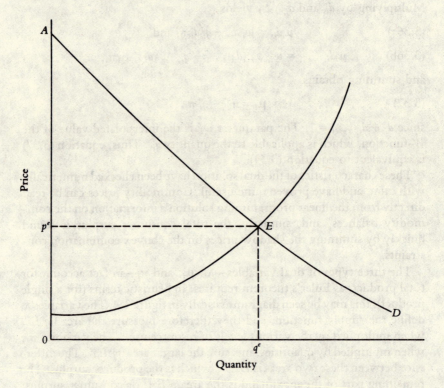

Furthermore, the chapter has shown that the noncompetitive market structure may be used for measuring income at endogenous prices in a competitive model. It has also developed a procedure for approximating product substitution effects in demand in a linear program. Alternative procedures are presented for the cases of full and partial information on the matrix of own- and cross-price elasticities. The demand structure can be simply transformed to take account of any shift in demand that can be represented by a rotation of the demand function. In addition, the chapter has shown that international trade can be integrated rather easily into the structure. Finally, it has been shown that the dual variables lend themselves to convenient interpretation.

References

Duloy, John H., and Roger D. Norton. 1975. "Prices and Incomes in Linear Programming Models." *American Journal of Agricultural Economics* (November), pp. 591–600.

Enke, Stephen. 1951. "Equilibrium among Spatially Separated Markets: Solution by Electric Analogue." *Econometrica*. Vol. 19 (January), pp. 40–47.

Farhi, Lucien, and Jacques Vercueil. 1969. *Recherche pour une planification cohérente: Le modèle de prévision du Ministère de l' Agriculture*. Monographies du Centre d'Econometrie, VI. Paris: Éditions du Centre National de la Recherche Scientifique.

Fox, Karl A. 1953. "A Spatial Equilibrium Model of the Livestock Feed Economy in the U.S.." *Econometrica*. Vol. 21, pp. 547–66.

Guise, J. W. B., and J. C. Flinn. 1970. "The Allocation and Pricing of Water in a River Basin." *American Journal of Agricultural Economics*. Vol. 52 (August), pp. 411–21.

Hadley, G. 1964. *Nonlinear and Dynamic Programming*. Reading, Mass. Addison-Wesley.

Hall, Harry H., Earl O. Heady, and Yakir Plessner. 1968. "Quadratic Programming Solution of Competitive Equilibrium for U.S. Agriculture." *American Journal of Agricultural Economics*. Vol. 50 (August), pp. 536–55.

Hazell, Peter B. R., and Pasquale L. Scandizzo. 1974. "Competitive Demand Structures under Risk in Agricultural Linear Programming Models." *American Journal of Agricultural Economics*. Vol. 56 (May), pp. 235–44.

Heady, Earl O., and Alvin C. Egbert. 1959. "Programming Regional Adjustments in Grain Production to Eliminate Surpluses." *Journal of Farm Economics*. Vol. 41 (November), pp. 718–33.

Johansen, Leif. 1967. "Regional Economic Problems Elucidated by Linear Programming." *International Economic Papers*. No. 12. London: Macmillan.

Judge, G. G., and T. D. Wallace. 1958. "Estimation of Spatial Price Equilibrium Models." *Journal of Farm Economics*. Vol. 40 (November), pp. 801–20.

King, G. A., and Lee F. Schrader. 1963. "Regional Location of Cattle Feeding: A Spatial Equilibrium Analysis " *Hilgardia*. Vol. 34 (July), pp. 331–416.

MacEwan, A. 1971. *Development Alternatives in Pakistan*. Cambridge, Mass.: Harvard University.

Martin, N. R., Jr. 1972. "Stepped Product Demand and Factor Supply Functions in Linear Programming Analyses." *American Journal of Agricultural Economics*. Vol. 54 (February), pp. 116–20.

Mash, Vladimir A., and V. I. Kiselev. 1971. "Optimization of Agricultural Development of a Region in Relation to Food Processing and Consumption." In *Economic Models and Quantitative Methods for Decisions and Planning in Agriculture*, Proceedings of an East-West Seminar. Edited by Earl O. Heady. Ames, Iowa: Iowa State University Press.

Miller, C. 1963. "The Simplex Method for Local Separable Programming." In *Recent Advances in Mathematical Programming*. Edited by R. L. Graves and P. Wolfe. New York: Wiley, pp. 89–100.

Mishan, E. J. 1968. "What Is Producer's Surplus?" *American Economic Review*. Vol. 48 (December), pp. 1269–82.

Piñeiro, Martin E., and Alex F. McCalla. 1971. "Programming for Argentine

Price Policy Analysis." *Review of Economics and Statistics*. Vol. 53 (February), pp. 59–66.

Plessner, Yakir. 1971. "Computing Equilibrium Solutions for Imperfectly Competitive Markets." *American Journal of Agricultural Economics*. Vol. 53 (May), pp. 191–96.

Samuelson, Paul A. 1952. "Spatial Price Equilibrium and Linear Programming," *American Economic Review*. Vol. 62 (June), pp. 283–303.

Schrader, Lee F., and G. A. King. 1962. "Regional Location of Beef Cattle Feeding." *Journal of Farm Economics*. Vol. 44 (February), pp. 64–81.

Smith, Vernon L. 1963. "Minimization of Economic Rent in Spatial Price Equilibrium." *Review of Economic Studies*. Vol. 30, pp. 24–31.

Takayama, T., and G. G. Judge. 1964a. "Spatial Equilibrium and Quadratic Programming." *Journal of Farm Economics*. Vol. 46 (February), pp. 67–93.

———. 1964b. "Equilibrium among Spatially Separated Markets: A Reformulation." *Econometrica*. Vol. 32 (October), pp. 510–24.

———. 1971. *Spatial and Temporal Price and Allocation Models*. Amsterdam: North-Holland.

Tirel, J. C. 1971. "General Design of French Model." In *Economic Models and Quantitative Methods*. Edited by Earl O. Heady. Ames, Iowa: Iowa State University Press.

Tramel, Thomas E., and A. D. Seale, Jr. 1959. "Reactive Programming of Supply and Demand Relations—Applications to Fresh Vegetables." *Journal of Farm Economics*. Vol. 41 (December), pp. 1012–22.

Yaron, Dan. 1967. "Incorporation of Income Effects into Mathematical Programming Models." *Metroeconomica*. Vol. 19, fasc. 3, pp. 141–60.

———. Yakir Plessner, and Earl O. Heady. 1965. "Competitive Equilibrium and Application of Mathematical Programming." *Canadian Journal of Agricultural Economics*. Vol. 13, no. 2.

Zusman, Pinhas. 1969. "The Stability of Interregional Competition and the Programming Approach to the Analysis of Spatial Trade Equilibria." *Metroeconomica*. Vol. 21, fasc. 1 (January–April), pp. 45–57.

4

The Technology Set and Data Base for CHAC

Luz María Bassoco and Teresa Rendón

A SET OF AGRICULTURAL SUPPLY functions is included in CHAC. These functions are represented implicitly by a series of fixed-coefficient production activities that are differentiated by crop, by technique, and by location. In each location, the model for activity analysis approximates a variable-coefficient production function at the district level. The same is true at the sectoral level for each crop and for the total value of agricultural output. An econometric specification of the production side would have confronted a number of deficiencies in the existing agricultural data series. For example, the sector-wide time series of production and prices are not very reliable. The spatial breakdown is less reliable, except for the irrigated areas. Beyond these problems, the time series do not include information on labor and other inputs.[1]

In these circumstances, an activity-analysis approach was adopted. It is based on estimates of discrete production alternatives, but the alternatives are sufficiently numerous so that aggregative behavior in CHAC is virtually continuous and nonlinear.

Overview

This chapter sets out the procedures used in constructing the spatial disaggregation scheme, the alternative input-output vectors, the resource availabilities, and other parameters for CHAC. Initial equilibrium conditions for product prices and quantities are also discussed, along with the prices of inputs that enter the productive process. In virtually all cases, the

Note: This chapter first appeared as Bassoco and Rendón (1973). Original material is used here, with minor editorial changes, by the kind permission of the publisher. The authors wish to express their appreciation to Roger Norton for helpful comments on development of the material in this chapter.

1. The decennial agricultural censuses include information on a few basic inputs, but only for aggregate production and not by crop. For an econometric analysis of supply based on the census data, see Hertford (1971).

existing data could not be used directly, but rather were subjected to a series of transformations so that they conformed to the accounting concepts in the model.

A major aim was to develop procedures sufficiently general so that the production side of CHAC could be altered readily to incorporate more district-level detail or less, or to selectively aggregate some portions and disaggregate others to shift the focus of investigation. Another intent, concerning the technology set in particular, was to describe feasible technological alternatives other than the set of farming practices observed in the base period.

Many of the individual district models (see Part Three of the book) also utilize the CHAC technology set (or extensions thereof), and, for those that do not, procedures similar to the ones of this chapter were followed.

Definition of Districts and Regions

There are twenty submodels on the product supply side of CHAC. Each represents either rainfed (*temporal*), irrigated, or tropical cultivation, and each covers a particular set of counties or districts, which are not necessarily contiguous. Cropping and investment activities are specified by submodel. The submodels are grouped into four major geographical regions, and labor constraints are specified for each region. This treatment reflects the different regional wage rates and the different degrees of interregional labor mobility.

For some submodels, the spatial building blocks are the administrative irrigation districts of the former Secretaría de Recursos Hidráulicos (Ministry of Water Resources, SRH). Some submodels represent individual irrigation districts, and others represent multiple districts. All but one of the single-district submodels are for the northwestern part of the country, where most of the export crops are produced. The production matrixes were designed, however, so that it would be a relatively simple matter to add submodels for individual districts in other areas.

In the case of *temporal* and tropical agriculture, the submodels are defined on the basis of altitude and annual rainfall rates, which together determine climatic conditions. In Mexico, crops are cultivated at altitudes ranging from sea level to 2,700 meters and under annual rainfall conditions of 400 millimeters to more than 1,500 millimeters. The kinds of crops cultivable, and their yields, vary considerably over climatic zones. Figure 4-1 shows the basis for defining the five *temporal* submodels (*Temporal A* through *E*) and the three tropical submodels (*Tropical A* through *C*).

The basic regions into which the submodels are grouped are described below. A schematic map of the regions is given in figure 4-2, and more exact descriptions of the submodels can be found in table 4-1.

Figure 4–1. *Climatic Definition of Temporal and Tropical Submodels*

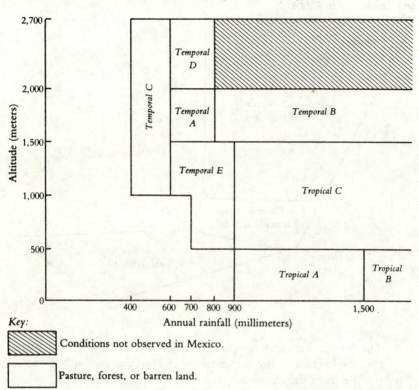

Key:

Conditions not observed in Mexico.

Pasture, forest, or barren land.

Note: Temporal A–E and *Tropical A–C* denote corresponding submodels in CHAC.

1. The Northwest: An arid zone of large-scale irrigation along a thousand-mile coastal strip between the Gulf of California and the Sierra Madre Occidental, plus Baja California. Agriculture is more extensively mechanized here than in any other region.
2. The North: The rest of the northern part of the country, this region is also extremely arid and cultivable only with irrigation except for the eastern portions near the Gulf of Mexico.
3. The Central Plateau: An area of mixed rainfed and irrigated farms, concentrated along the course of the Lerma River; the farms are generally smaller than in the North and Northwest; twenty years ago this was the most productive region in Mexican agriculture, but it has been surpassed by the northern regions.

Figure 4–2. *Schematic Map of Regions of Mexico and Submodels in CHAC*

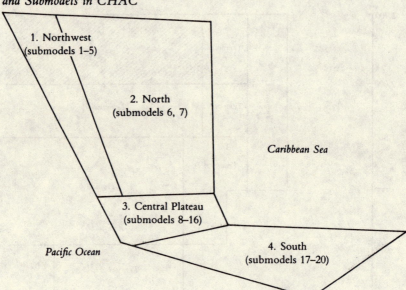

1. Northwest
(submodels 1–5)

2. North
(submodels 6, 7)

Caribbean Sea

3. Central Plateau
(submodels 8–16)

Pacific Ocean

4. South
(submodels 17–20)

4. The South: Tropical agriculture with very few systems of water control; because of the mountainous terrain, this region is the most remote from the major urban markets.

Production Alternatives

For each of the twenty submodels, various production alternatives have been identified. Each alternative describes a production process that embodies a fixed combination of resource inputs for a given level of output. There are a total of 2,345 column vectors representing such production alternatives in the model (see table 4–2). The production alternatives are functions of: (1) regional cropping patterns; (2) calendars of cultivation practices by crop; (3) classes of land by soil types, efficiency of water use, and climate; (4) modes of irrigation; and (5) degrees of mechanization.

Regional cropping patterns

In each of the submodels, crops were identified for production sets on the basis of the cropping patterns observed in the corresponding district during the 1960s. Because yields, fertilizer requirements, and other elements of the production vector are dependent on local soil and climate conditions, activities cannot be specified for crops that have not been grown previously in that district. This does not appear to be a serious omission in the Mexican context. On the one hand, producers in irrigated areas already cultivate a wide variety of crops; on the other, there are only a limited number of crops that are well adapted to conditions of rainfed agriculture in Mexico.

The basis for crop selection in irrigation submodels is the time series of statistics published for irrigation districts by the SRH.[2] The information utilized for *temporal* and tropical submodels was provided by the Secretaría de Agricultura y Ganadería (Ministry of Agriculture and Livestock, SAG).

Calendars of cultivation practices

Several Mexican agencies compile cost of production estimates by crops and location, but these compilations are based on the sequence of cultivation tasks and not on economic inputs. After identifying the crops to be included in the model, the next step was to establish the agricultural calendar for each of the 2,345 production activities. The calendar specifies the dates of planting, irrigating, fertilizing, crop tending, and harvesting.

The vector of production coefficients is derived from the agricultural calendar. For a given crop and location, the number of irrigation applications is not constant but varies with the month of planting. For a crop in a particular irrigation district, there are as many as four alternative planting dates in the model: two summer months and two winter months. For example, according to the activities for the Culmaya area (comprising Culiacán, Humaya, and San Lorenzo), it is possible to cultivate maize either in summer or winter. Winter maize may be planted in December and harvested in June, or planted in January and harvested in July. Summer maize may occupy the land from May to November or from June to December (see figure 4-3, and its discussion in the subsection "Land and Water," below).

For irrigation submodels, the alternative agricultural calendars were taken from information supplied by SRH. For each *temporal* or tropical submodel, there exists only one planting date—and, hence, one calendar

---

2. See SRH (1969*a*, -*b*). Since this study was undertaken, SRH and SAG have merged to form the Secretaría de Agricultura y Recursos Hidráulicos (Ministry of Agriculture and Water Resources, SARH). In the references of this chapter, the former names are used. Also, throughout the chapter the word "*temporal*" is used in reference to rainfed, nontropical agriculture.

Table 4-1. *Spatial Components of CHAC*

Region	Location[a]	Farm type[b]	Number	Name
			Submodels	
Northwest	Río Yaqui	I	1	Río Yaqui
	Culiacán } Río Humaya } San Lorenzo }	I	2	Culmaya
	Río Colorado	I	3	Río Colorado
	Comisión del Fuerte	I	4	El Fuerte
	Remaining irrigation districts in the states of Baja California, Sonora, and Sinaloa	I	5	Residual Northwest
North	Irrigation districts in the states of Chihuahua, Coahuila, and Durango	I	6	North Central
	Irrigation districts in the states of Nuevo León and Tamaulipas	I	7	Northeast
Central Plateau	Rainfed portions of the 17 *municipios* in Guanajuato, which include the irrigation districts of Alto Río Lerma and La Begoña }	LR	8	El Bajío A
		SR	9	El Bajío B
	Alto Río Lerma } La Begoña }	I	10	El Bajío irrigated
	Mostly parts of the states of Puebla, Guanajuato, Hidalgo, and Querétaro	R	11	*Temporal A*
	Mostly the states of Jalisco, Michoacán, and Morelos	R	12	*Temporal B*
(see figure 4-1)	Northern part of Central Plateau plus states further north	R	13	*Temporal C*
	Mostly the states of México, and Tlaxcala	R	14	*Temporal D*
	Mostly portions of the states of Oaxaca, Guerrero, Colima, Michoacán, and Tamaulipas	R	15	*Temporal E*
	The irrigation districts of 10 Central Plateau states	I	16	Central irrigated
South	Mostly the states of Campeche, Yucatán, Quintana Roo, and Nayarit	T	17	Tropical A
(see figure 4-1)	Mostly the states of Tabasco, and Veracruz	T	18	Tropical B
	Mostly part of the states of Puebla, Chiapas, Veracruz, and San Luis Potosí	T	19	Tropical C
	The irrigation districts in the tropical zones	I	20	South irrigated

a. For irrigation submodels, the location is defined by the administrative irrigation districts of the Secretaría de Recursos Hidráulicos (Ministry of Water Resources, SRH). For rainfed and tropical areas, altitude and rainfall define the submodels, and each submodel's precise coverage is stated in relation to *municipios* (counties). Each municipio is assigned wholly to one submodel.

b. The farm types are as follows: I, irrigated; LR, rainfed, large farms (10 hectares or more); SR, rainfed, small farms (less than 10 hectares); R, rainfed; T, tropical. In many of the irrigation submodels there are additional distinctions among farms that are based primarily on efficiency in water use.

of cultivation activities—depending on the month in which the rains begin. Planting dates were taken from information supplied by SAG.

Classes of land and water

The *temporal* and tropical submodels are defined on the basis of climatic conditions and, hence, are not necessarily contiguous. The irrigation submodels refer to the administrative irrigation districts of SRH.[3] Within each district, there are as many as four zones demarcated by that ministry to represent varying degrees of efficiency in use of reservoir water. In each zone, a different amount of gross water release at the dam is required to achieve the same net amount of water on the field. The water losses depend upon the length of canals and their state of repair.

Irrigation water is specified in two forms: gravity-fed water and well water. The former includes water from reservoirs and river pumps, and its allocation is controlled by SRH; the latter is supplied by private tubewells. In some of the irrigation submodels, both water sources are specified.

The water input norms differ for gravity-fed and well water because of larger transmission losses in the gravity-fed reticulation system. The prices also differ. For gravity-fed water, the administrative levy is entered as a cost in the objective function. For well water, the pumping cost is entered. Since both types of water are available in limited quantities, CHAC determines a shadow cost, which normally exceeds these direct costs.

Degrees of mechanization

The irrigated-nonirrigated distinction is one of the ways in which CHAC distinguishes more capital-intensive and more labor-intensive agriculture. Individual crops also vary enormously in their unit labor requirements.[4] In addition, alternative degrees of mechanization have been specified for each crop and location.

In CHAC, there are in entirety three degrees of mechanization for each crop: mechanized, partially mechanized, and nonmechanized. In some locations, depending on the observed techniques in the base period, only two degrees are specified. To account for the time lapse inherent in the adoption of new techniques, only one-degree changes of technique are permitted during the eight-year period studied (1968–76). That is, for districts in which only nonmechanized techniques were observed in the base period, CHAC includes partially mechanized as well as nonmechanized activities.

3. In some cases, the area in these districts is augmented to provide coverage of lands irrigated by dispersed wells.

4. See tables 4-3 and 4-4 in the next section of this chapter.

Table 4-2. *Production Activities in CHAC by Crop and by District*

Crop	Río Yaqui	Cul-maya	El Fuerte	North Central	North-east	Río Colorado	Residual Northwest	El Bajío temporal[a]	El Bajío irrigated
				Geographic submodel					
Garlic						32	4		32
Dry alfalfa	16		24	12		32	6		
Cotton	16		24	12		32	12		
Green alfalfa		12	24	12			6		32
Rice		12	24				6		
Oats									
Sugarcane		12	24		6		6		
Squash									
Safflower	16	12	24	12		32	6		
Peanuts					6				32
Onions									32
Forage barley	16			4		32	4		
Grain barley	16			4		32	4		48
Dry chile									
Green chile		12	24				12		32
Strawberries									32
Beans		12	24				9	8	48
Chickpeas		12	24				6	8	48
Lima beans									32
Tomatoes		12	24				12		32
Sesame	16	12	24				12		
Flaxseed	16		24				6		
Maize	16	22	48	12	12	32	22	12	48
Cantaloupe		12	24				12		
Potatoes			48				12		
Cucumber		12					6		
Pineapple									
Watermelon		12	24	12			9		
Sorghum	16	24	48	12	12	32	20	12	48
Soybeans	16	12	24				10		
Tobacco									
Wheat	16	12	24	12	6	32	8		48
Number of activities	176	214	528	110	36	288	210	40	544

a. Two submodels.

The alternative degrees of mechanization do not affect yields per hectare. The totally mechanized technique is defined so that operations requiring traction power are done with machinery rather than animals. These operations include land preparation, harvesting, and some intermediate cultivation steps. There are crop-specific variations. For example, in Mexico cotton is always harvested manually, no matter how capital-intensive the other operations. In this case, the mechanized cotton production technique includes manual harvesting.

| | | | | | Geographic submodel | | | | | |
Temporal A	Temporal B	Temporal C	Temporal D	Temporal E	Central irrigated	Tropical A	Tropical B	Tropical C	South irrigated	Number of activities
					2					70
					2					92
			3		4					103
					2					88
					4	2			4	52
		3	3							6
					4	2	2		4	60
	3									3
				3						105
	3				2					43
					2					34
					2					58
			3		2					109
	3									3
	3				2				2	87
					2					34
3	3	3	3	3	2	2	2	2		124
3	3				2					106
			3		2					37
	3				2					85
				3	4	2		2		75
				3						49
3	3	3	3	3	2	2	4	2	2	251
					4					52
			3							63
					4					22
							1			1
					4					61
3	3		3		2		2	2	2	241
						2				64
					1					1
		3	3		2					166
12	27	12	21	21	60	9	13	10	14	2,345

The partially mechanized technique refers to the practice of using mechanical power for land preparation and seeding, while using draft animals for all other operations. In the nonmechanized technique, draft animals are used for all traction operations. These discrete alternatives are the major ones observed in Mexico. When a farmer adopts only partial mechanization, he is very likely to use it at the beginning of the crop calendar in order to facilitate the process of getting the crop in the ground. Machinery operators' time is one of the inputs for the mechanized techniques. When draft animals are used, a (much larger) input of un-skilled agricultural labor is required. The input norms in CHAC reflect both kinds of labor. Since machinery operators do not appear to be a produc-tion constraint in Mexico, the supply of their services is assumed to be perfectly elastic at a given price. Hence, their services are not explicit inputs in CHAC, but they are reflected in the machinery cost entries in the objective function.

The input requirements for plowing, harvesting, and other power operations depend only on the degree of mechanization and, for harvest-ing, on the crop. They do not vary over districts. Plowing requirements per hectare are standard, and harvest requirements per ton are standard by crop. Through published data[5] and field surveys by one of the authors (Bassoco), it was possible to estimate these standard norms for each degree of mechanization. In this manner, activities were formed that represent degrees of mechanization other than those observed in a particular district.

Technical Production Coefficients

Information on agricultural production costs typically comes in the form of estimates of total expenses by operation, such as plowing, irriga-tion, and fertilizer application. These estimates include costs of materials, labor, draft animals, and machinery services. There are also estimates of the number of distinct irrigation releases, fertilizer applications, weedings, and the like, by crop and district. To form activities for CHAC, the problem was to convert this information into statements of required economic inputs such as labor, fertilizer, and credit. To facilitate this conversion, the unit activity level in all cases was defined to be cultivation of 1 hectare, rather than 1 metric ton of output.

Labor, machinery, draft animals

For each crop and degree of mechanization, standard inputs of labor and services of machinery and draft animals have been defined for each opera-tion in the agricultural calendar. These operations include both those which involve traction power and those which do not. They range from land preparation and seeding, through plant tending and application of water and chemical products, to the harvest. The standard inputs for each

---

5. The basic published series on costs of production are those of SRH (1969b), the Asegur-adora Nacional Agrícola y Ganadera (1969), the Instituto Nacional de Investigaciones Agrícolas (Gonzalez and Silos 1968), and the Banco Nacional de Crédito Ejidal (for the last two, annual statistics from various years).

operation are constant over districts. But the number of required operations varies over districts in some cases (plant tending, application of chemicals, water release), and the yield per hectare also varies over districts. Hence, the total labor requirement per hectare varies over districts for a given crop and a given degree of mechanization. The number of operations and yields by district are taken from data published by the four institutions mentioned above.

The assumption of standard inputs by operation, regardless of location, is not exactly true, but it is a close approximation to reality. The number of tractor-hours required to plow a hectare varies somewhat, depending on the average soil conditions in a district, but it does not vary greatly. To carry out the standard operation concept, machinery use requirements have been normalized for a tractor of 60 horsepower. Inputs of labor, animal power, and machinery services are expressed in days of labor. This concept is the bridge between technical agronomic information and the cost estimates of CHAC. It also permits ready identification of those portions of input packages which vary over crops, districts, planting dates, and degrees of mechanization.

Apart from the differing degrees of mechanization, some crops are simply more labor intensive than others. The mechanized form of cotton cultivation requires almost twice as much labor per hectare as the non-mechanized form of wheat cultivation. This may be seen from tables 4-3 and 4-4, which show the labor, machinery, and animal power inputs into the standard operations for two major crops (cotton and wheat).

The range of techniques in tables 4-3 and 4-4 implies certain elasticities of substitution of labor for capital. Calculated at the midpoints of the relevant range, they are as follows:

	Degree of mechanization	Elasticity
Cotton	Mechanized to partially mechanized	− 0.178
	Partially mechanized to nonmechanized	− 0.231
Wheat	Mechanized to partially mechanized	− 1.603
	Partially mechanized to nonmechanized	− 0.264

Grains offer more scope for factor substitution than cotton and many vegetables.

Land and water

The unit level of operation of the production activities is 1 hectare. Land inputs are specified monthly. Hence, the normal land input coefficient is 1.0, which signifies use of an entire hectare in a particular calendar month. An exception is made at the beginning of the cultivation cycle, when land preparation may require less than a full month. Plowing with draft ani-

Table 4-3. *Sequence of Standard Operations for Cotton Cultivation*
(days of unskilled labor, machinery services, and draft animal services required per hectare monthly)

Month	Operation	Mechanized		Partially mechanized			Nonmechanized	
		Unskilled labor	Machinery	Unskilled labor	Machinery	Animals	Unskilled labor	Animals
1st	Preparatory tasks		0.12		0.12		1.0	2.0
	Fallow		0.5		0.5		3.0	6.0
	Cross-plowing						2.5	5.0
	Harrowing		0.2		0.2		0.5	1.0
	Land leveling		0.25		0.25		1.0	2.0
	Canal cleaning	1.0		1.0			1.0	
2d	Irrigation ditches	1.0	0.2	1.0	0.2		2.0	2.0
	Forming borders[a]		0.2		0.2		2.0	
	Linking borders[b]	1.0		1.0				
	Water application	2.0		2.0			2.0	
	Harrowing		0.2		0.2		2.0	4.0
	Seeding and fertilization	0.2	0.2	0.2	0.2		4.0	
	Maintenance of field works		0.2	0.2			2.0	
3d	Thinning plants	4.0		4.0			4.0	
	Cultivation		0.2	2.0		4.0	2.0	4.0
	Weeding	6.0		6.0			6.0	
	Applications of insecticides (2)[c]		0.4	6.0			6.0	

94

4th	Fertilization		0.2	2.0		2.0	
	Cultivation		0.2	2.0	4.0	2.0	4.0
	Weeding	6.0		6.0		6.0	
	Water applications (2)	4.0		4.0		4.0	
	Applications of insecticides (2)			6.0		6.0	
5th	Cultivation		0.4	2.0	4.0	2.0	4.0
	Weeding	12.0		6.0		12.0	
	Water applications (2)	4.0	0.2	4.0		4.0	
	Applications of insecticides (2)			6.0		6.0	
6th	Weeding	6.0	0.4	6.0		6.0	
	Water application	2.0		2.0		2.0	
	Application of insecticides						
7th	Harvest (per metric ton)[d]	11.0		11.0		11.0	
	Transport to farm gate		0.12	2.0	4.0	2.0	4.0
8th	Harvest (per metric ton)[d]	11.0		11.0		11.0	
	Transport to farm gate		0.12	2.0	4.0	2.0	4.0
9th	Harvest (per metric ton)[d]	11.0		11.0		11.0	
	Transport to farm gate		0.12	2.0	4.0	2.0	4.0

a. *"Bordeo."*

b. *"Pegar bordos."*

c. For cotton, insecticide applications are made by airplane.

d. Normally the harvest covers three months, with 30 percent, 50 percent, and 20 percent occurring in each successive month.

Table 4-4. *Sequence of Standard Operations for Wheat Cultivation*
(days of unskilled labor, machinery services, and draft animal services required per hectare monthly)

| | | Requirement | | | | | | | |
| | | Mechanized | | Partially mechanized | | | Nonmechanized | | |
Month	Operation	Unskilled labor	Machinery	Unskilled labor	Machinery	Animals	Unskilled labor	Machinery	Animals
1st	Canal cleaning	1.0		1.0			1.0		
	Fallow		0.5		0.5		3.0		6.0
	Cross-plowing						2.5		5.0
	Harrowing		0.2		0.2		0.5		1.0
	Land leveling		0.25		0.25		1.0		2.0
2d	Irrigation ditches	1.0		1.0			2.0		2.0
	Bordering or bench terracing						2.0		
	Linking borders	1.0		1.0					
	Seeding and fertilization	0.2	0.2	0.2	0.2		4.0		1.0
	Harrowing		0.2		0.2		0.5		
3d	Water application	2.0		2.0			2.0		
	Applications of herbicides		0.2	3.0			3.0		
	Fertilization	2.0		2.0			2.0		
4th	Water application	2.0		2.0			2.0		
	Application of insecticides		0.2	3.0			3.0		
	Application of herbicides		0.2	3.0			3.0		
5th	Water application	2.0		2.0			2.0		
	Application of insecticides		0.2	3.0			3.0		
6th	Water application	2.0		2.0			2.0		
7th	Combine harvesting (per ton)		0.27		0.27				
	Hand cutting						4.0		
	Hand threshing							0.13	
	Transport to farm gate	2.0	0.12	2.0		4.0	2.0		4.0

96

mals, together with associated activities, usually requires a month's work. With mechanized techniques, however, the same operation requires about ten days. Hence, in the mechanized technique, a coefficient of 0.33 is used instead of 1.0. These land savings can be important when double cropping is feasible. Another exception is made in the case of the harvest, also to differentiate techniques.

In some cases, land savings are related to the nature of the product. In the case of alfalfa and barley, the form of the crop harvested is reflected in the length of time for which the land is used. Green alfalfa occupies the land longer in the drying process.

From figure 4–3 it is possible to derive the monthly land coefficients for various cropping activities in the Culmaya submodel; from the figure it may be seen which double-cropping combinations are feasible and which are not. For example, a wheat-soybean or wheat-sorghum rotation pattern is feasible, but wheat-sesame is not. In some cases, a particular rotation is prevented only because a mechanized harvest is not possible for one crop. This is most often true of fruits and vegetables. If it were possible to mechanize the harvest of cantaloupe, for instance, a soybean-cantaloupe rotation would be feasible. Tomatoes require a sufficiently long growing season to prevent rotation with any other crop. Hence, the cost of tomato cultivation includes the lost opportunity for double cropping.

The irrigation requirements are given by the irrigation schedules formulated by the SRH for each crop and district. The coefficients are measured in cubic feet of gross water released from the irrigation source. In some of the submodels, there are as many as four zones defined with respect to efficiency in gravity water use. Gross well-water coefficients differ from those for gravity water because of different rates of water loss in the reticulation systems.

The sets of production vectors that are based on these monthly land and water requirements permit, in many cases, two crops during a fifteen- to eighteen-month period. Seasonal temperature variations in Mexico are generally not severe enough to prohibit such rotations.

Credit, fertilizer, improved seeds, yields

Short-term credit requirements are related to total production costs; hence, they vary over crop, district, and technique. The basis for estimation of credit requirements is the set of credit norms used by the Banco Nacional de Crédito Ejidal.

Inputs of fertilizers and pesticides are grouped together in a row defining chemical inputs. The coefficients are given in pesos rather than physical units. Inputs of improved seeds also are given in pesos. Both sets of coefficients reflect prevailing practices and yields in each submodel area.

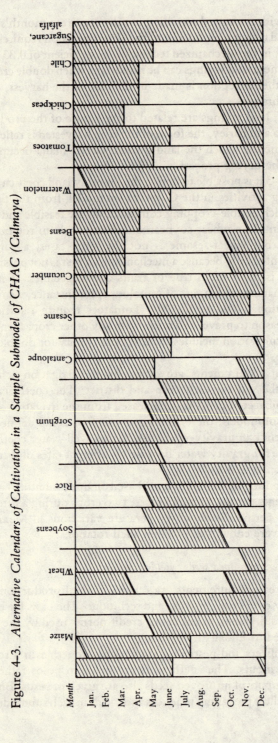

Figure 4-3. *Alternative Calendars of Cultivation in a Sample Submodel of CHAC (Culmaya)*

Note: Partial shading indicates that one-third of a month is required under mechanized techniques, a full month under nonmechanized techniques. "Culmaya" is the aggregative name for the irrigation districts of Culiacán, Humaya, and San Lorenzo in Mexico's Pacific Northwest.

Yields vary with the submodel (reflecting soil and climate conditions) and local practices regarding application of chemical inputs. For submodels of irrigation districts, these yields are five-year averages of yield statistics compiled by SRH. For submodels representing nonirrigated agriculture, yields are compiled as appropriate weighted averages of state-level data in the national agricultural plan of SAG (1968, 1969). Table 4-5 shows the range of variation of yields in CHAC.

Sources of Variation in the Technical Coefficients

In summary, variations in the technical coefficients in CHAC arise from two sources: geographical differences that give rise to variations in cultivation calendars, practices of fertilization and crop cultivation, irrigation requirements, and yields; and alternative degrees of mechanization and efficiency in water use within the same submodel area.

Table 4-6 shows the effect of mechanization on the coefficients for labor, machinery services, and draft animal services. Coefficients for two sample submodels are shown. Table 4-7 summarizes the sources of variation in all types of coefficients, which are grouped by basic agricultural operations. In labor, for example, there are two kinds of coefficients: those which vary over submodel districts and those which do not. The latter include labor inputs for land preparation activities. For the harvest, labor inputs per ton are constant, but yields vary among districts. Coefficients that are district-specific are those related to cultural operations: weeding, fertilization, applications of insecticides, and the like.

Restrictions on Resource Availability

The following section examines the accounting CHAC makes of two major restrictions on the availability of agricultural resources in Mexico: the labor force and land and water.

Labor force

There are two sources for estimates of the agricultural labor force in Mexico: the decennial population censuses (Secretaría de Industria y Comercio 1964) and the agricultural censuses (Secretaría de Industria y Comercio 1965). More resources have been invested in the population censuses, and as a result they are widely considered to be more reliable. They are, however, deficient in that they virtually ignore family labor. The agricultural census for 1960, on the other hand, lists 1.5 million unpaid family

Table 4-5. *Yields by Crop and Submodel in CHAC*
(metric tons per hectare)

					Crop		
Submodel	Garlic	Dry alfalfa	Cotton fiber	Cotton-seed	Green alfalfa	Rice	Oats
Culmaya					44.910	3.100	
Río Yaqui		12.154	0.828	1.421			
Río Colorado	7.000	7.962	0.972	1.667			
El Fuerte		8.679	0.828	1.421	43.897	2.600	
Residual Northwest	6.800	9.029	1.013	1.737	45.000	2.960	
North Central		16.270	0.833	1.428	72.000		
Northeast							
Central irrigated	6.750	20.300	0.935	1.600	87.630	$\begin{Bmatrix}5.667\\3.300\end{Bmatrix}$	
South irrigated						2.781	
Temporal							
A							
B							
C							0.500
D							0.800
E			$\begin{Bmatrix}0.357\\0.924\end{Bmatrix}$	$\begin{Bmatrix}0.612\\1.584\end{Bmatrix}$			
Tropical							
A_1							
A_2							
B_1						1.500	
B_2						2.500	
B_3							
B_4							
C_1							
C_2							
El Bajío *temporal*							
1							
2							
3							
El Bajío irrigated (nonlevel land)							
1							
2	5.500				80.000		
3	7.000				100.000		
El Bajío irrigated (level land)							
1							
2	5.780				84.000		
3	7.350				105.000		

					Crop				
Sugar-cane	Saf-flower	Squash	Pea-nuts	Onions	Forage barley	Grain barley	Dry chile	Green chile	Straw-berries
74.320	0.750							5.130	
	1.646				13.471	2.400			
	1.493				10.986	2.234			
145.920	1.256							4.179	
76.944	1.674				13.044	2.600		8.265	
	1.360		2.300		13.000	1.870			
52.912									
95.840			2.200	9.200	10.370	3.101	1.700	10.393	11.701
64.778								6.945	
		8.333	1.277				1.250	2.775	
	0.750					0.709			
45.000									
68.000									
43.000									
68.000									
						2.500			
			3.000	10.000		3.300		5.000	12.500
			4.000	15.000		4.000		7.000	12.500
						2.620			
			3.150	10.500		3.460		5.250	13.100
			4.200	15.750		4.200		7.350	15.700

(*Table continues on the following page.*)

Table 4-5 (*continued*)

				Crop			
Submodel	Beans	Chick-peas	Lima beans	Toma-toes	Sesame	Flax-seed	Maize
Culmaya	0.970	0.840		20.320	0.760		1.720
Río Yaqui					0.750	1.646	3.685
Río Colorado							1.744
El Fuerte	1.322			30.483	0.790	1.998	2.343
Residual Northwest	1.760	1.470		10.830	0.720	1.364	2.363
North Central							1.806
Northeast							2.600
Central irrigated	1.080	1.820	2.210	12.502	0.720		2.530
South irrigated							1.535
Temporal							
A	0.378	0.606					0.690
B	0.380	0.800		6.600			1.695
C	0.439						0.534
D	0.353		0.722				0.875
E	0.550				0.684	0.700	0.785
Tropical							
A_1	0.500				0.450		0.850
A_2	0.800				0.650		1.800
B_1	0.800						1.600
B_2	1.200						3.000
B_3							1.600
B_4							3.000
C_1	0.800				0.450		1.600
C_2	1.2000				0.650		3.000
El Bajío *temporal*							
1	0.500						0.700
2	0.500	0.600					1.200
3	0.600	0.800					1.500
El Bajío irrigated (nonlevel land)							
1	0.900	1.500					2.500
2	1.400	2.000	2.000	14.000			3.500
3	1.800	3.000	2.500	18.000			5.000
El Bajío irrigated (level land)							
1	0.950	1.570					2.620
2	1.470	2.100	2.100	14.700			3.680
3	1.890	3.150	2.630	18.900			5.250

				Crop				
Canta-loupe	Pota-toes	Cu-cumber	Pine-apple	Water-melon	Sor-ghum	Soy-beans	To-bacco	Wheat
7.610		11.120		6.850	3.060	1.370		2.782
					4.957	1.879		3.862
					3.552			3.600
4.223	14.000			6.540	3.843	1.913		3.166
7.400	14.000	10.176		7.590	3.600	1.840		3.320
				15.000	3.400			2.270
					3.700			2.662
9.750	14.000	9.500		11.590	2.791			1.830
					1.661		1.856	
					1.946			
					2.687			
								0.500
	11.000							0.861
					1.200			
						1.000		
			25.000			1.500	1.430	
					2.000			
					3.500			
					2.000			
					3.500			
					1.200			
					2.000			
					2.500			
					3.500			2.500
					5.000			3.500
					7.000			4.500
					3.680			2.625
					5.250			3.675
					7.350			4.725

Table 4-6. *Variation of Cotton Input Coefficients with Degrees of Mechanization, in Two Sample Districts*[a]
(days)

| | Input coefficients | | | | |
| | El Fuerte | | | Río Yaqui | |
Input	Mecha-nized	Partly mecha-nized	Non-mecha-nized	Mecha-nized	Partly mecha-nized
Machinery services	3.63	2.07	0	3.23	2.07
Mule services	0	32.00	58.00	0	32.00
Labor					
January	8.0	12.0	12.0		
February	8.0	10.0	10.0	1.0	1.0
March	8.0	10.0	10.0	4.2	4.2
April	8.0	10.0	10.0	12.0	14.0
May	10.2	12.2	12.2	8.0	12.0
June	13.7	15.7	15.7	12.0	14.0
July	5.5	7.5	7.5	8.0	8.0
August				8.2	10.2
September				13.7	15.7
October	1.0	1.0	9.0	5.5	7.5
November	4.2	4.2	14.0		
December	10.0	10.0	14.0		
Total labor	76.6	92.6	108.4	72.6	86.6
Yield (metric tons per hectare)					
Cotton fiber	0.868	0.868	0.868	0.828	0.828
Cotton seed	1.488	1.488	1.488	1.421	1.421

a. With different planting dates, the months and, in some cases, the values of coefficients change.

workers on ejidal farms alone (versus 0.1 million family workers in the entire sector, according to the population census, for the same year). In spite of this problem, the population census of 1960 has been taken as the basis for the CHAC labor force, with suitable augmentation made for family labor. The figures have been projected forward to 1968 in accordance with regional labor-force growth rates calculated from the 1960 census and the preliminary tabulations of the 1970 census. These regional growth rates add up to about a 1.5 percent growth rate for the sector as a whole.[6] Total population has been increasing at about 3.5 percent, but annual rural-urban migration has amounted to about 2.0 percent of the rural population.

The population census itself contains two kinds of estimates of the

6. Two percent is the growth rate assumed by Keesing and Manne (1973).

Table 4–7. *Sources of Variation in the Technical Coefficients*

	Source of variation					
Coefficient	Crop	Plant-ing date	Land class within a submodel	Type of irrigation (well or gravity-fed)	Degree of mecha-nization	Sub-model
Unskilled labor						
Land preparation	X				X	
Harvest	X				X	X
Other	X				X	X
Machinery services						
Land preparation	X				X	
Harvest	X				X	X
Other	X				X	X
Draft animal services						
Land preparation	X				X	
Harvest	X					
Other	X				X	X
Land	X	X	X			X
Irrigation water	X		X	X		X
Chemical inputs	X	X				X
Improved seeds	X					X
Short-term credit	X			X		X
Yield	X	X				X

"agricultural" labor force. One is by occupational category and the other is by sector. For the sectors of agriculture (that is, crops, livestock, forestry, hunting, and fishing), the total labor force is estimated at 6,143,530 in 1960. For the three occupational categories field workers, ejidal farmers, and nonejidal farmers, the total is 4,642,453 persons.[7] This latter total is engaged in crop agriculture. Of this figure, 2,671,852 persons are listed as heads of farm households (*propietarios*), and 1,970,601 are, essentially, field workers.

To estimate the family labor component, recourse is made to the demographic figures on family composition. The average farm household has about five and one-half persons, and about half the population is under fifteen years of age. Since a wife is occupied in the house most of the time, it is assumed that she contributes one-tenth the field work of her husband. These and other equivalence factors yield table 4–8. On this basis, table 4–8 indicates that there are 1.5 male adult-equivalent laborers per family. This implies that the total agricultural family labor available in Mexico in 1960 comprised 1,335,926 persons. This is less than the family labor estimated for *ejidos* alone in the agriculture census, but recall that the reliability of the

7. See Table 27 of the 1960 population census (Secretaría de Industria y Comercio 1964).

Table 4-8. *Calculation of Family Labor Component for CHAC*

Family member	(a) Number per household	(b) Labor equivalence factor	(c) Labor equivalent (c = a × b)
Household head	1.00	1.00	1.00
Spouse	0.95	0.10	0.10[a]
Children under fifteen years	2.75	—	—
Children over fifteen years	0.80	0.50	0.40
Total family labor equivalent			1.50

— Not applicable.
a. Rounded to nearest tenth.

latter is doubtful. Also, the CHAC estimate of the unskilled labor force is substantially higher than that of Keesing and Manne (1973), so it was deemed better to err on the conservative side than in the other direction.

Hence, the total labor force engaged in crop agriculture in 1960 is estimated as follows:

	Number
Heads of farm households	2,671,852
Family laborers	1,335,926
Day laborers	1,970,601
Total	5,978,379

Because CHAC excludes long-cycle crops, the model's figure for the labor force is correspondingly reduced. The total labor force that appears in CHAC for 1968 is 5,181,945. This reflects both the 1960–68 labor force increase and an allowance for the labor engaged in long-cycle crops.

Farmers and family labor are specified by submodel in CHAC and day laborers by region. To obtain the figures for each spatial entity, county-level data from the 1960 census were aggregated. As noted, to arrive at 1968 estimates, the annual growth rate for the regional labor force during 1960–70 was utilized.[8]

Land and water

The monthly land restrictions in CHAC are based on estimates of cultivable land by Secretaría de Agricultura (SAG) (1968; 1969) and SRH (1969c). For nonirrigated submodels, the building blocks are counties. Each

8. Based on preliminary data of the 1970 population census. Because the processing of the 1970 census was incomplete when the CHAC data were being compiled, the 1960 census (Secretaría de Industria y Comercio 1964) was used as the base for the labor force estimates.

county is assigned to a submodel according to its altitude and rainfall (see figure 4–1, above). For irrigation submodels, the building blocks are the administrative irrigation districts. In some submodels, additional land is included to represent scattered irrigation sites that lie outside the jurisdiction of the administrative districts.

Water restrictions are specified annually and monthly in CHAC. For gravity-fed water, the annual restrictions represent limitations on the annual rate of replenishment of reservoir water, whereas the monthly restrictions represent limitations on the capacity of the canal system for water delivery.[9] For pumped water from wells, the annual restrictions represent legal limits designed to maintain the level of the water table, and the monthly restrictions refer to pumping capacity.

Product Prices

For the 1968 solutions of CHAC, the domestic demand curves are passed through a point representing actual base-period prices and quantities. For the 1974 solutions, this point is shifted to represent income and population growth and per capita income elasticities.

For prices, 1967–69 averages were used to minimize the effect of short-term fluctuations. Rather than using existing sector-wide price estimates, it was deemed better to construct new sector-wide estimates from micro-level data. Weighted averages of local prices were constructed. SRH (1969c) collects extensive information on local crop prices every year, so these were used as the basis for the sectoral estimates. It was assumed that neighboring irrigated and nonirrigated plots face the same price for a given crop. But, as reported in the statistics, prices vary substantially among regions and, to a lesser extent, among major areas within a region. This procedure permits application of the SRH price data to all producing areas, irrigated and nonirrigated. Prices were weighted with local production statistics for both kinds of agriculture. Table 4–9 presents production estimates and computed average prices by crop for irrigated, *temporal*, and tropical areas and for the sector as a whole.

In the case of a few crops, the operations of the national price support agency, Compañía Nacional de Subsistencia Popular (CONASUPO), have resulted in a gap, after accounting for processing and transport costs,

---

9. In two submodels, the monthly restrictions also represent water availability, with opportunities for intertemporal water transfer by holding it in the reservoir, at the cost of evaporation loss. For the Río Colorado submodel, which represents an area on the U.S. border, the monthly and annual restrictions are in accord with an international treaty on water use.

Table 4-9. *Base-period Domestic Prices and Production in CHAC, 1968*

Crop	Farm-gate price (pesos per metric ton)[a]			Production (metric tons)			National average farm-gate price (pesos per metric ton)	National production (metric tons)
	Irrigated	Temporal	Tropical	Irrigated	Temporal	Tropical		
Garlic	2,213			31.0			2,213	31.0
Cotton fiber	2,459	2,381		1,270.8			2,447	1,504.1
Alfalfa (dry)	354						354	
Alfalfa (green)	126			10,932.0			126	10,932.0
Rice	1,134		1,358	368.6		131.8	1,219	500.4
Oats	653	805		22.0	87.2		774	109.2
Sugarcane	64		63	14,216.9		14,926.0	64	29,142.9
Safflower	1,544	1,552		173.4	75.3		1,546	248.7
Squash		576			108.1		576	108.1
Peanuts	1,593	1,298		24.5	50.7		1,391	75.2
Onions	637			122.7			637	122.7
Forage barley	86						86	
Grain barley	1,014	862		135.3	191.9		925	327.2
Dry chile	7,554	8,153		17.7	4.1		7,677	21.8
Green chile	1,413	1,651		121.5	50.9		1,496	171.5
Strawberries	1,977			109.3			1,977	109.3

Crop								
Beans[b]	2,202	2,070	1,726	104.5	720.7	119.4	2,040	944.6
Chickpeas	1,153	·957		24.4	159.2		992	183.6
Lima beans	855	865		14.2	31.9		862	45.4
Tomatoes[c]	1,998	1,255		586.2	78.8		1,906	665.0
Sesame	1,500	2,382	2,438	30.2	140.6	22.7	2,407	193.5
Flaxseed	1,701	1,661		8.8	10.1		1,680	18.9
Maize[b,d]	940	908	935	1,750.5	5,425.6	2,059.1	920	9,255.2
Cantaloupe	682			142.5			682	142.5
Potatoes	973	865		239.2	164.3		929	403.5
Cucumber[e]	1,301			20.5			1,301	20.5
Pineapple						297.6	513	297.6
Watermelon	777		513	187.6			777	187.6
Sorghum[b]	625	657	671	1,280.3	1,173.2	20.2	641	2,523.5
Soybeans	1,600	1,640		259.7		2.7	1,600	272.4
Tobacco		7,722	7,722			75.7	7,722	75.7
Wheat[b]	857	895		2,138.6	105.8		859	2,244.4

a. For prices, a 1967–69 average is used.

b. For maize, beans, sorghum, and wheat, the farm-gate prices in the table reflect the subsidies of Compañía Nacional de Subsistencia Popular (CONASUPO). For CHAC, prices were adjusted to remove the influence of the subsidy. Thus, the base-period prices used for these crops are as follows (in pesos per metric ton): maize, 861; beans, 1,834; sorghum, 633; wheat, 800.

c. The average price of tomatoes is strongly influenced by the export prices. In CHAC, a base-period domestic price of 1,150 pesos per metric ton is used.

d. In the case of maize, it is assumed on the basis of information contained in the National Agricultural Plan (SAG 1968, 1969) that 63 percent of production goes to human consumption and 37 percent to forage uses.

e. The average price of cucumbers is strongly influenced by the export price. In CHAC, a base-period domestic price of 586 pesos per metric ton is used.

between the farm-gate price and the corresponding price to consumers. For these crops, CONASUPO incurs a budgetary deficit. For CHAC, it was necessary to reduce the farm-gate price of these crops by an amount that reflects the subsidy to consumers. This yields a market-clearing price and quantity, and the CHAC demand curves are passed through that point.

Export markets are specified independently of the domestic markets. For products which Mexico exports, it is assumed that the Mexican share of the world market is sufficiently small so that the country is a price taker. In some cases, the quantity exported is limited by international agreement or by import quotas in other countries.[10] The fixed-price assumption is also made for imports.

For exports, farm-gate prices are used. These are less than f.o.b. prices. For imports, prices appropriate to consumption in Mexico City are required; these are higher than c.i.f. prices. This puts imports on the same price basis as domestic sources of supply.

Factor Prices

For sector-wide inputs, CHAC is based on market prices. These include hired labor, for which the wage varies over regions.[11] For land, a district-level resource, prices are completely endogenous. For water, the pumping costs for wells and the administrative charges for release of reservoir water are registered in the objective function, but, since quantities of available water are limited, the model also computes a shadow cost. Table 4-10 shows the calculation of the market price of tractor services. Similar calculations were made for pumping costs of wells and services of draft animals.

10. In the case of sugar, two export markets have been introduced in CHAC: one reflecting the Mexican quota for U.S. imports and the other, at a lower price, reflecting the free international market.

11. The pricing of labor, including imputation of farmers' reservation wages, is explained in chapter 2.

Table 4–10. *Costs Associated with the Operation
of One 60-horsepower Tractor*[a]
(1968 pesos)

Acquisition costs	
Tractor of 60 horsepower	75,364
Reversible plow with three-furrow disc	14,200
Eighteen-disc harrow	9,250
Land-leveling blade	6,000
Broadcasting seeder	20,000
Cultivator	10,500
Total	135,314
Useful life	10,000 hours
Summary of hourly operating costs	
Depreciation	13.50
Gasoline and oil consumption[b]	3.64
Maintenance[c]	2.38

a. Apart from the operator's salary.

b. For gasoline and oil consumption, the following is assumed: diesel fuel, fifty liters each 8 hours at 0.4 pesos per liter; oil, eight liters each 125 hours, at 8.0 pesos per liter; and grease, 1 kilogram each 8 hours at 5.0 pesos per kilo.

c. For maintenance, the following is assumed: transmission oil change, thirty liters each 1,500 hours at 6.5 pesos per liter; filter change each 250 hours at 30 pesos per change; tire change, one set each 3,500 hours at 4,500 pesos per set; tune-up each 1,000 hours at 300 pesos per tune-up; and cylinder change at 600 pesos per change.

References

Aseguradora Nacional Agrícola y Ganadera. 1969. *Programas de aseguramiento, ciclo primavera-verano 1968–69 y ciclo invierno 1968–69.* Mexico City.

Bassoco, Luz María, and Teresa Rendón. 1973. "The Technology Set and Data Base for CHAC." In *Multi-level Planning: Case Studies in Mexico.* Edited by Louis M. Goreux and Alan S. Manne. Amsterdam/New York: North-Holland/American Elsevier, pp. 339–71.

Bassoco, Luz María, and Donald L. Winkelmann. 1970. "Programación de la producción agrícola de la parcela ejidal en tres obras de regadío en el estado de Quintana Roo." Chapingo, México: Escuela Nacional de Agricultura.

Barrera Islas, Daniel, and Donald L. Winkelmann. 1969. "Análisis económico del uso del agua y la mano de obra en el sector ejidal de la Comarca Lagunera." Chapingo, Mexico: Centro de Economía Agrícola.

Conklin, Frank S., and Earl O. Heady. 1968. "Uso optimo de los recursos agropecuarios en el distrito de Riego de la Begoña." Chapingo, México: Centro de Economía Agrícola.

Gonzalez, Vicente, and José Silos. 1968. *Economía de la producción agrícola en el Bajío.* Mexico City: Instituto Nacional de Investigaciones Agrícolas.

Hertford, Reed. 1971. "Sources of Change in Mexican Agricultural Production, 1940–65." U.S. Department of Agriculture Foreign Agricultural Economic Report, no. 73. Washington, D.C.: U.S. Government Printing Office.

Keesing, Donald B., and Alan S. Manne. 1973. "Manpower Projections." In *Multi-level Planning: Case Studies in Mexico*. Edited by L. M. Goreux and A. S. Manne. Amsterdam/New York: North-Holland/American Elsevier.

Secretaría de Agricultura y Ganadería (SAG). 1968, 1969. *Plan Nacional Agrícola, 1968–69* and *1960–70*. Mexico City.

Secretaría de Industria y Comercio. 1971. *Anuarios de comercio exterior de 1955 a 1970*. Mexico City.

——. 1965. "Características de las personas ocupadas en el predio." In *IV censo agrícola ganadero y ejidal, 1960*. Mexico City.

——. 1964. *VIII censo de población, 1960*. Mexico City.

Secretaría de Recursos Hidráulicos (SRH). 1969a. *Características en los distritos de riego*. Mexico City.

——. 1969b (and earlier years). *Costos de producción de los principales cultivos en los distritos de riego*. Mexico City.

——. 1969c (and earlier years). *Estadística agrícola para los ciclos 1966–67, 1967–68, 1968–69*. Mexico City.

5

A Quantitative Framework for Agricultural Policies

LUZ MARÍA BASSOCO AND ROGER D. NORTON

IN THIS CHAPTER, the basic numerical results from the CHAC model that were used during 1972–76 in the Mexican government's discussions of agricultural policy options are discussed. Most of the material of this chapter was issued in one form or another by the Comisión Coordinadora del Sector Agropecuario (Coordinating Commission for the Agricultural Sector; COCOSA, later CONACOSA), and some of the results appeared in the policy document that is reproduced in the next chapter.[1] CHAC was used in ways similar to those reported in this chapter by the Ministry of Programming and Budgeting in 1980–81.

Overview

In the context of this volume, this chapter and the following one constitute the translation of analysis into policy applications. This chapter is primarily concerned with analytical issues, except for the section on pricing policies, and chapter 6 is concerned with applications, but both examine aspects of the process of interpreting analytic work for policy making. In a sense, the principal contribution of this chapter is to illustrate how a sector-wide model may be used to address certain kinds of issues; that is, how the basic version may be modified and solved to obtain meaningful "laboratory experiments" concerning the possible consequences of economic policy.

The two COCOSA documents that form the backbone of this chapter were issued as a working paper (background document) and a policy paper. The first one was distributed as COCOSA (1974a).[2] The second one (COCOSA 1974b) was issued as a policy note to reflect the considerations behind the new pricing policy for grains.

1. Ministry of the Presidency (1973). The Ministry of the Presidency is now called the Ministry of Programming and Budgeting (Secretaría de Programación y Presupuesto).
2. A shorter, English version was later published as Bassoco and Norton (1975).

The major issues treated in this chapter are the responsiveness of aggregate sectoral supply, capital-labor substitution, international and regional comparative advantage, pricing policies for grains, and questions of model validation.

An Approach to Sectoral Policy Planning

Agriculture is traditionally a baffling sector for policy planners in all parts of the world. In developing nations, the problem is often exacerbated by conflicting goals. Agriculture is expected to carry many burdens: principally, to satisfy national food requirements, to provide employment, and to generate foreign exchange. In addition, the data base in the developing world is often inadequate for estimating the appropriate response parameters of the sector.

The usual approach to agricultural policy planning involves setting production targets by commodity in physical units and then attempting to trace the input requirements for the target level of production of each commodity separately. Several criticisms may be made of this procedure.[3] First, the traditional framework does not permit assessment of sector-wide aggregates, such as an aggregate supply function or an aggregate elasticity of factor substition. Such measures are important for the evaluation of alternative sectoral programs to meet national development goals. Although individual commodity production targets may satisfy the food needs, and perhaps the foreign exchange goal, it is unlikely that they represent the best program for, say, employment purposes.

Second, and even from the viewpoint of food requirements, efficient resource allocation may require that certain product prices be allowed to rise while others decline in relative terms. In other words, the sector faces not point demands but demand schedules. The position on the schedule should be found as a result of a constrained resource allocation problem. Third, proper planning in the face of balance of payments constraints may require varying mixtures of imported and domestic supply for each product, and this cannot be handled properly without considering all products simultaneously.[4]

A fourth criticism of the usual approach is that attempting to add up

3. See, for example, Mellor (1966, pp. 382–84) for a critical view of the usual practice.
4. An iterative procedure could be envisaged, in which the cumulated production/import programs were revised in each round, but it would be cumbersome, especially if it were to allow for the effect of changing cropping patterns on the opportunity cost of land and other fixed resources.

resource requirements crop-by-crop ignores the substitution that may take place among crops on the supply side and, hence, can give quite biased aggregate estimates of resource needs. In monocultural zones this is not a problem, but in other areas the effects of substitution on the supply side can be important for short-cycle crops.

The quantitative procedures used in this study of Mexican agriculture meet these four criticisms, but, of course, they still leave many thorny questions unanswered. To overcome the usual limitations of data, the sectoral model relies heavily on the use of cross-sectional farm-level data on production costs instead of aggregate production time series. In this respect, it may be thought of as a procedure for translating micro-level data into macro-level (sectoral) statements.

The policy problem as treated here involves both traditional macroeconomic policy instruments (interest rates, foreign exchange rates, and the like) and also crop-specific and input-specific policies. In some cases, the instruments are identified by region, but they do not go so far as particular investment projects in particular localities (although the aggregate sectoral budget for investments is treated indirectly).

The Mexican policy document reproduced in chapter 6 (Ministry of the Presidency 1973) contains a good many measures of institutional reforms and other nonquantifiable programs, and it is unusual in that it specifies many concrete steps that subsequently have been fully implemented. To confine this discussion within reasonable bounds, we do not discuss these aspects but only elaborate the quantitative framework of the plan.

Policy planning in the Mexican context has meant the coordinated use of available policy instruments to attain the plan's objectives. Specifically, there are six major categories of quantitative instruments for influencing sectoral performance:

- Investment programs in physical resources (for example, land and irrigation facilities)
- Investment programs in research and extension
- Policies for pricing factors and products
- Trade policies (for example, tariffs and export incentives)
- In limited cases, factor allocations over crops or areas or both (for example, short-term credit)
- Policies on land tenure (for example, determination of farm size).

In addition, the overall rate of GNP growth—which may be influenced by fiscal, monetary, and other policies—affects sectoral performance through shifting the demand functions for agricultural products.

Each of these policy instruments, and the rate of GNP growth as well, is represented by a set of parameters in the model. For Mexican planning, the principal modeling procedure was to solve CHAC under alternative assumptions regarding the values of the policy parameters. In many cases, solutions were made for two points in time (1968 and 1976), and the rate of growth of each target variable was calculated ex post. Thus, the planning analysis may be regarded as an exercise in comparative statics. The different policy assumptions were reflected in alternative values of selected parameters for 1976. In the case of policies whose effect is cumulative over time, annual rates of change were hypothesized and projected to form values for the year 1976. The model thus represents a simulation of the various effects of these hypothetical policies. In agriculture, with all its interrelations on both the supply and demand sides, a fairly detailed model is required in order to make a reasonably realistic solution. For other issues, such as support price levels, comparative statics solutions were conducted with the base-year version of the model, which was updated to 1972.

Since CHAC is fairly disaggregative for crops, technologies of production, and producing locations, it has been possible to trace the potential consequences of hypothetical policies at a reasonably concrete level, where the judgement of agronomists and other specialists is applicable. This has been helpful both in model validation (see the section of that title, below) and the interpretation of projections.

A Summary of CHAC

In the following, the sectoral model is summarized briefly in a manner slightly different from the exposition of chapter 2. This summary also indicates some of the ways in which the model was modified for applications subsequent to development of the original version discussed in chapter 2.

Although CHAC is a mathematical programming model in its solution technique, it is best described as a behavioral simulation model. It attempts to describe how farmers will react, in the aggregate, to certain classes of economic policies that influence their cost-price structure and resource availabilities. The main elements of CHAC may be summarized by the following subsections.

Sectoral coverage

CHAC includes all sources of supply (domestic and imported) and all demands (domestic and export) for the thirty-three short-cycle crops analyzed. The model does not include livestock, forestry, or long-cycle crops.

Interdependence of supply

Supply is described as a process-analysis technology set for each of twenty spatial entities. Alternatives in mechanization, planting dates, fertilization, and irrigation are included.[5] The total set of alternative technologies for the thirty-three crops and twenty spatial submodels is 2,348. Because each submodel contains a large number of crops that compete for use of the same local resources (land, water, and farm family labor), the implicit cross-elasticities of supply are generally nonzero. This is the process-analysis manner of capturing extensive interdependence within the supply set. In addition to the local resources, other agricultural inputs included in the model are day labor, chemical inputs, improved seeds, agricultural machinery services, draft animal services, short-term credit, and miscellaneous cost items. Land, labor, and water are treated on a monthly basis. For the treatment of labor, as explained below, the twenty submodels are grouped into four major regions.

Interdependence of demand

As noted previously, price-elastic demand functions are incorporated in CHAC, and, when projections are made, income elasticities of demand are used to make appropriate shifts of the static demand functions. This structure permits varying crop portions in aggregate production, with corresponding variations in relative prices. This variation amounts to indirect substitution in demand. To permit direct substitution, crop groups are specified, within which limited substitution may take place at a constant marginal rate of substitution. For lack of more precise information, export demands are typically specified as perfectly elastic up to a bound.[6] The interdependence both in supply and demand is an important aspect of the agricultural sector, and capturing it in the model has helped considerably to increase the realism of the model's results.

Simulation of market equilibria

The incorporation of demand structures permits specification of alternative market forms; for example, a competitive, monopolistic, or quasi-monopolistic supply-control regime. For the bulk of the CHAC

5. This is basically an agronomic specification of supply conditions. Heady and various associates were pioneers in developing this kind of supply treatment; see, for example, Heady, Randhawa, and Skold (1966).

6. For some crops, such as cotton, export demands were specified as perfectly elastic at the going price.

solutions, the competitive market form was assumed because, with a few possible exceptions of fruits and vegetables, no producer or association of producers can influence the market price through production decisions. The optimization feature of the model is not used in a normative sense (that is, to maximize some goal set) but, rather, in a descriptive sense (that is, to simulate the behavior of the competitive market). This is achieved by the maximization of the sum of the Marshallian surpluses for each product's market.[7] Qualifications to the purely competitive assumption are made for the case of some of the producers in nonirrigated areas, where participation in the market is not as widespread. These are discussed in the next subsection.

Elements of dualism

Dualistic concepts are contained in CHAC in the technology sets and in the parameters of market participation. One explanation for the lower elasticities of crop supply that are often obtained for more traditional farmers is simply that these farmers have few alternative crops to consider in making their planting decisions. Farmers who have access to irrigation, and who have enough land to be able to afford to take some risks, can contemplate growing a wide variety of grains, vegetables, oilseeds, fruits, and other crops that are nearly equal in profitability per hectare. A small shift in relative crop prices is therefore more likely to induce such a farmer to change cropping patterns than it would in the case of the farmer who has a smaller array of choices.[8] In CHAC, the nonirrigated areas have fewer alternative crops and technologies than do the irrigated areas.

The second way in which traditional farmers are differentiated in the model is by the specification of constraints from home consumption. Many producers tend to satisfy their families' consumption needs in the basic food crop (maize) before marketing it or producing another crop. Several possible explanations, not all of them strictly economic, can be adduced for this behavior, but the following simple assumption for the model sufficed to explain the observed behavior. If a farmer meets his family's food requirements through market purchase of maize the year round, the average price he pays for a kilo will be higher than the price he could get for a kilo of his own maize crop at harvest time. This price differential arises from both the normal buying-selling margin—perhaps exaggerated by market imperfections—and the seasonal price move-

---

7. For a full exposition of the CHAC demand structures and their properties, see chapter 3.

8. This holds true as long as resource endowments are fixed in each case. As mentioned below, some solutions of CHAC have underscored the importance of idle, marginal lands in nonirrigated zones. Price increases may bring these lands under cultivation and therefore show a rather high supply elasticity for nonirrigated areas.

ments. For the model, it was assumed that this differential is paid when a farmer does not devote enough of his land to maize to meet consumption needs. In the solutions, the differential proved sufficient to enforce production for own consumption; that is, to allow crop diversification only after family consumption needs were satisfied.

Labor supply functions

Labor in CHAC is specified in three basic categories: farmers and family workers, day laborers, and machinery operators. The stock of farmers is divided into twenty parts, one corresponding to each spatial submodel on the production side. Farmers with irrigation are assumed not to migrate or work on other farms in the short run, but farmers without irrigation are assumed to be available for hire as day laborers in slack months.[9] The pool of day laborers is divided into four regional components, and interregional migration may occur in the model if the day laborers in a given region are fully employed in at least one month. Thus, hiring of day laborers and farmers in nonirrigated areas is specified on a monthly basis. Hiring of farmers in irrigation submodels is stated in annual terms: a farmer makes a commitment (to himself) to see his farm through the entire crop year. Machinery operators are assumed to be freely available at their going wage, and, thus, no quantity restriction is imposed. In practice, machinery operators form a tiny fraction of the labor force, and the lack of their availability has not been cited as an obstacle to agricultural undertakings in Mexico.

Day-labor wages are set at the going market levels for each of the four regions; the Northwest has a wage nearly twice that of the South—a reflection of the slow pace at which interregional wage differentials adjust. The labor of farmers is priced at a monthly "reservation wage" that is greater than zero but less than the day-labor wage. In narrow terms, the reservation wage may be regarded as the measure of the disutility of work; in broader terms, it is the minimal productivity at which farmers will undertake additional tasks on their farms. It is sometimes observed that farmers will not adopt new techniques that promise minimal additional returns per unit of additional work. In other words, at a zero wage the labor supply function is zero. On the one hand, time is simply too valuable (for noneconomic activities, also) to waste it in unproductive labor; on the other hand, farmers clearly undertake some low-productivity tasks on

9. This assumption follows from the less intense cycle of work observed on rainfed farms, where the most labor-intensive crops and double cropping are not feasible. There obviously are exceptions: small-scale farmers with irrigation may be found who work off the farm seasonally, and large-scale rainfed farmers may stick to their farms the entire year. On the whole, the assumption describes the actual degrees of labor mobility.

their farms, secure in the knowledge that their annual income will flow in at a higher rate. Over the course of a year, they gain not only the sum of monthly "reservation wages," but also the economic rents that accrue to their land, water, and their labor and management skills. In fact, the reservation wage payments in CHAC typically amount to one-third to one-fifth of a farmer's total income.

The empirical question confronted in building the model was the appropriate level of the reservation wage. Simulations were made with the model (and also with submodels solved in isolation) under varying reservation wage rates to see which figure gave the more appropriate cropping patterns and labor-hiring patterns. For irrigated areas, the answer fell consistently in the neighborhood of 40 to 50 percent of the day-labor wage; and values in the ranges 0–30 percent and 60–100 percent gave quite distorted results. For nonirrigated areas, the appropriate value appeared to be somewhat lower, around 30–40 percent of the market wage. Values in these ranges were therefore adopted for the planning solutions.

Comparative statics

CHAC is an annual model that may be solved for any given cropping cycle. Validation runs were made for the base year of 1968,[10] with the resource endowments of that year entered as constraints (see the section "Model Validation," below). Subsequently, solutions were made for 1976 under alternative assumptions along the following parameters:

- The rate of expansion of arable land, irrigation supplies, and the labor force
- The rate of change of yields per hectare for all crops
- The rates of GNP growth (which determine the degree of shift in the demand functions)
- The rate of change of upper *bounds* on crop exports (which is not the same as export *levels* in the solution) to reflect changing world market circumstances.

For each 1976 solution, 1968–76 annual rates of change were calculated and are reported in the following section. These solutions constitute the bulk of the planning runs, for they permit assessment of the sensitivity over time of several variables (including employment and income distribution) with respect to policies that would be designed to influence the parameters above. A number of other solutions were carried out to explore the static behavior of the model for the year 1968. In particular, a series of capital-labor substitution isoquants and response surfaces were traced out by varying relative factor prices and by making appropriate

10. For stochastic parameters such as yields and prices, three-year averages for the years 1967–69 were used.

assumptions about the constancy of output or other variables. These "static" experiments, of course, are also useful for the planning of employment-oriented policies.

Basic Macroeconomic Results

In the preceding discussion of the procedures for comparative statics, it was noted that four kinds of exogenous information define the solution. In numerical terms, the following assumptions were made to establish the "basic case" for 1976:

- The endowments of cultivable land and irrigation supplies increase by 2 percent annually from 1968 to 1976. This implies a corresponding annual increase of 2 percent in the number of farm families.[11]
- Real GNP increases at 8 percent annually, as does disposable income.
- Crop yields and the upper limits on exports by crop were increased in accordance with the judgments of specialists.

These sets of assumptions defined the solutions for 1976 that are discussed in the remainder of this section. The macroeconomic results shown in table 5-1 demonstrate, first of all, that the difference between 7 and 8 percent growth in GNP is important for the agricultural sector.[12] To avoid increased imports, sectoral production grows at 4.7 percent in the one case and 5.4 percent in the other.[13] Even with this increase in sectoral production, production is not keeping up with demand increases. This lag may be seen by the projected increases of agricultural prices relative to the economy-wide price level: 1.5 percent annually in the case of 7 percent GNP growth, and 2.0 percent annually in the case of 8 percent GNP growth. These rates of relative price increase constitute one of the measures of sufficiency of the development program for the agricultural sector in an economy-wide program. As noted above, they are based on certain rates of increase of cultivable land, irrigation water, and yields per hectare— which in turn are determined, in part, by the magnitude and composition of the public program of investment in agriculture. The implication of these results is clear: the assumed rates of expansion of the agricultural

11. With a continuation of the historical rates of urban-rural migration, the first assumption would imply that the absolute number of landless laborers neither increases nor decreases.

12. *Editors' note*: These projected growth rates turned out to be overly optimistic. Hence, even the pessimistic employment projections of this chapter and the next probably were not attained in reality.

13. By assumption, import levels were held constant in 1968 and 1976. Solutions could have been designed to accommodate changes in the import structure, which in fact occurred.

resource base are not sufficient to meet expanding needs for agricultural products,[14] and an increase in the agricultural terms of trade would be necessary to prevent increased imports.

Although the rates of relative price change are useful indicators, care must be taken in interpreting the CHAC prices at the overall sectoral level. They reflect changes in agricultural prices *relative* to the rest of the economy's prices, but agricultural prices are one of the main determinants of the economy-wide price level, and the second-round effects are not included in the analysis. Hence, there is a lack of closure in CHAC that cannot be overcome without enlarging it into an economy-wide model. Nevertheless, it is possible to use CHAC prices in the following two ways:

- The overall sectoral price index may be compared from one solution to another to see how inflationary each alternative program is relative to the other programs.
- The individual commodity prices in each CHAC solution may be examined to see which commodities are likely to be the most (or least) stable in price.

Another interesting aspect of the macroeconomic results concerns employment. Measured in total man–years, employment increases at 1.0 to 2.5 percent annually in the various solutions. Given that the sector's labor force increases at more than 3.0 percent, this implies continuing rural-urban migration at a significant rate. In comparing these employment growth rates with the production growth rates, it is seen that the "employment elasticity of agricultural output"[15] is about 0.40 (from 0.38 to 0.46 in the four solutions).

If man–years are considered instead of elasticities, a 4.7 percent growth rate in agricultural production creates about 55,000 man–years of employment annually, and a 5.4 percent growth rate creates about 73,000 man–years, given present relative prices of capital and labor.[16] Increasing the export growth rate (in varying proportions by crop) from 5.0 to 7.1 percent overall adds about 3,000 man–years annually.

If jobs are considered, the results are different because the sector's labor force is a mixture of day laborers who may work as little as one month a year and farmers who may work several months or more a year. The effect of this variability on various jobs is best seen through the changes in the monthly patterns of employment shown in the section "Income Distribution and Derived Demand for Inputs," below.

14. As of this writing, the public investment programs in Mexico are expanding the agricultural resource base at a slightly more rapid rate.

15. "Employment elasticity" is defined as the annual percentage change in employment divided by the annual percentage change in sector output, in a given solution.

16. For these calculations, it is assumed that the sector's labor force is roughly 7 million now and that the average laborer in the sector, including the full-time, part-time, and unemployed worker, works about five months of the year.

Measuring the Aggregate Sectoral Supply Function

The aggregate supply response of a sector may be measured in several ways. First of all, there is the simple "elasticity" of sectoral production with respect to GNP.[17] The results of table 5-1 show that this elasticity is of the order of 0.67 to 0.71 for annual GNP growth in the 7–8 percent range. The higher elasticity applies for the case of accelerated export growth. This, however, is not the same as a supply elasticity, for it measures the aggregate response of the moving supply-demand equilibria for all crops in the sector.

A version of the aggregate supply elasticity can, however, be measured from these results. Figure 5–1 illustrates the procedure for the case of a single crop. The curve D^{68} is the price-elastic demand curve for the base year, 1968, and the curves D^{76-1} and D^{76-2} are the corresponding curves for 1976, under 7 and 8 percent annual GNP growth, respectively. In CHAC, they have been shifted by an amount determined by both the rate of GNP growth and the magnitude of the income elasticities of demand. Hence, the amount of shift is different for each commodity. The implicit supply curve in CHAC (which is nonlinear, as shown) is represented by S^{68} for year 1968 and by S^{76} for the year 1976. It may be seen from the figure that the arc elasticity of supply between points A and B is readily calculated ex post as follows:

$$(5.1) \qquad \varepsilon = \frac{(q_2 - q_1) / (q_2 + q_1)}{(p_2 - p_1) / (p_2 + p_1)}.$$

Because the model provides both price and quantity estimates for all crops, the calculation of ε is in a straightforward matter, using the production and price indexes of table 5-1. The cases of 7 and 8 percent growth for 1976 are used because they jointly identify different points on the same short-run supply curve. Thus, for example, it is not possible to use pairs of points defined by the cases of faster and slower technological change, because these define different supply functions.

Taking the cases of 7 and 8 percent growth, then, the calculations yielded an aggregate supply elasticity of 1.383.[18] How is this figure to be interpreted? First of all, in one sense it represents a long-term supply

---

17. The elasticity is measured here as the percentage change in sectoral production divided by the percentage change in GNP. It should be pointed out that the values might be different outside the 7–8 percent range of GNP growth.

18. For reference, the weighted-average income elasticity of demand over all crops in CHAC is $+0.545$, using as weights the quantity produced in the model in the base solution for 1968.

Table 5-1. *CHAC: Principal Macroeconomic Results
and Annual Rates of Change*
(millions of 1968 pesos and percent)

		1976			
Concept	1968 (1)	7 percent GNP growth (2)	8 percent GNP growth (3)	7 percent GNP growth; slower technological change (4)	7 percent GNP growth; higher exports (5)
Objective function	66,822	97,589	102,934	96,349	97,952
Producers' income	12,250	17,891	18,295	18,295	17,706
Sectoral income	13,491	19,857	20,640	19,698	20,510
Value of production	25,692	37,110	39,065	37,071	37,958
Total employment (man-years)	2,016	2,350	2,451	2,334	2,357
Income (per man-year)	6,693	8,451	8,422	8,441	8,701
Exports	3,479	5,152	5,152	5,036	6,036
Price index	100.0	112.8	117.0	121.5	120.5
		Annual rate of change			
Objective function		4.9	5.6	4.7	4.9
Producers' income		4.9	5.2	5.0	4.7
Sectoral income		5.0	5.7	4.9	5.4
Value of production		4.7	5.4	4.7	5.0
Total employment		1.9	2.5	1.8	2.0
Income per man-year		3.0	2.9	3.0	3.4
Exports		5.0	5.0	4.7	7.1
Price index		1.5	2.0	2.5	2.4

elasticity because it refers to behavior between two *equilibrium* points (after all adjustment processes have worked themselves out). In another respect, however, it is a short-run concept, for it does not allow investment in expansion of the sector's resource base (land and water). That expansion is taken care of in the *shift* of supply function from S^{68} to S^{76}. Thus, we may call the CHAC elasticity an "equilibrium short-run elasticity."

Second, as noted before, the elasticity refers only to the supply of short-cycle crops. It is clear that the number would be smaller if it treated perennial crops and if it were a purely short-run (nonequilibrium) elasticity concept. Looked at in this light, the magnitude seems reasonable in relation to existing international studies.[19]

19. See Behrman (1968) for an extensive discussion of both estimation procedures and numerical results. Purely short-run supply elasticities are typically about half the value reported from CHAC. Part of the difference may be because of factors mentioned in the text, but part may also be because of the fact that most of the results cited by Behrman refer to

Figure 5-1. *Procedure for Measuring the Sectoral Supply Function for a Single Crop*

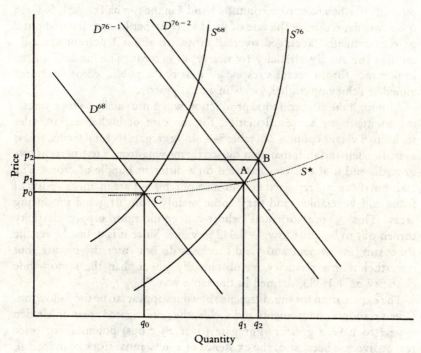

Note: D^{68} = price-elastic demand curve for base year 1968; D^{76-1} and D^{76-2} = corresponding demand curves for 1976 under 7 and 8 percent annual growth in GNP, respectively; S^{68} and S^{76} = implicit supply curves in CHAC for 1968 and 1976; $S^{\star}$ = long-run equilibrium supply function.

Alternatively, it is possible to redefine the supply function as a "long-run equilibrium supply function" that includes the effects of fixed investments and yield changes over time. This would be curve $S^{\star}$ in figure 5-1, which passes through points C and A. Using the same measurement rule, this long-run arc elasticity (between points C and A) is calculated at the value +3.030.[20] It may be asked what is the contribution of technological progress (changes in yields per hectare) to this value? Here it is necessary to

---

monocultural zones. Crop substitution effects do contribute somewhat to the overall supply response in Mexico. Over the 1930–60 period, substitution alone accounted for about 0.5 percent annual output growth (see Solís 1970).

20. If one wishes to view as a single function a line joining c, a, and b, then the arc elasticity between points c and b turns out to be +2.631 in value.

treat explicitly the cases of different rates of technological progress. With a slower rate of progress, the long-run equilibrium arc elasticity is computed as + 1.865 (compare columns 1 and 4 in the top half of table 5-1). In other words, reducing the rate of yield increase per hectare from about 2 percent annually (averaged over all crops) to about 1 percent annually reduces the supply elasticity by nearly 40 percent. This underlines the importance for the sector's responsiveness of the public sector programs aimed at achieving higher yields in actual practice.

Although the effect on total production was quite small, it was interesting to make the same calculation for the case of higher export sales (column 5 versus column 2 in table 5-1). In aggregate sectoral terms, this is a smaller demand shift than that caused by moving from 7 to 8 percent GNP growth, and it also is concentrated on a different bundle of crops. The additional exports reported in column 5 of table 5-1 are mainly exports of fruits and vegetables, and they come mainly from irrigated producing areas. This "export-oriented" short-run equilibrium supply elasticity turned out to be quite low: + 0.342 in value. Since it is defined over the short run, investment and yield increases do not enter the picture, but nevertheless this elasticity is substantially lower than the sector-wide elasticity of + 1.383, defined in the same way.[21]

The explanation for this difference in values appears to be the following. Given resource endowments and yields, the nonirrigated areas of Mexico appear to have a greater aggregate (over all crops) potential for price responsiveness because of the existence of a substantial stock of marginal, uncultivated land that will gradually be brought under cultivation as price incentives rise. In contrast, virtually all the cultivable irrigated land is already cultivated because of its higher levels of profitability.[22] A confirmation of this explanation is provided by an interesting set of figures from CHAC in the 1968 solution: 29.3 percent of the available nonirrigated land was uncultivated even at peak periods of field labor. In the 7 percent growth solution for 1976, this degree of slack was reduced to 8.3 percent, and in the 8 percent growth case for 1976 it was further reduced to 0.2 percent.

This explanation coincides with the observation of students of Mexican agriculture that in the postwar period the terms of trade, and hence the incentives to cultivate marginal land, have steadily worsened from the sector's viewpoint (Solís 1972, p. 40).

21. These elasticities refer to the *aggregate bundle* of crops produced in irrigated and nonirrigated areas. For individual crops, the elasticities tend to be higher in irrigated areas because of the crop substitution possibilities. (See "A Summary of CHAC," above.)

22. This does not include areas affected by land tenure disputes and other problems that are not responsive to price inducements.

Thus, at this particular point in Mexican history price incentives should have powerful stimulating effects on private expansion of the cultivated land and may be considered a necessary condition for achieving the results of table 5-1.[23] This result underscores the importance of the plan's prescriptions for utilization of price incentive tools. The obverse deduction for practical programs may also hold: to the extent that imperfect markets, sociocultural barriers, tenure disputes, and the like impede transmission of price signals to the majority of nonirrigated farmers, the sector's supply response may continue to be weak.

Measuring Factor Substitutability

CHAC also has been used to estimate the sector-wide elasticity of capital-labor substitution. In general, there are three types of capital in the sector: the physical availability of land, irrigation systems, buildings, and other forms of fixed capital; agricultural machinery; and working capital. With regard to the financing of investment, the first type of capital typically corresponds to long-term investments of ten years or more in duration. The second type corresponds to medium-term financing from two to five years, and the third type corresponds to short-term loans of no more than one year.

In agriculture, long-term capital in general is a complement and not a substitute for labor. Increases in cultivable land directly increase the possibility of employment. Increases in the availability of irrigation per hectare expand the employment possibilities by permitting higher yields, double cropping, and cultivation of crops that are intensive in the use of labor, such as fruits and vegetables. Similarly, increases in the stock of buildings augment storage capacity and therefore increase sales and production prospects. Medium-term capital—that is, capital incorporated in agricultural machinery—is normally a direct substitute for field labor. Short-term capital can be either a complement or a substitute with respect to the use of labor, depending upon the particular field tasks that it supports.

In most econometric studies of factor substitution, the first two kinds of capital are lumped together, and sometimes all three classes are grouped. Both positive and negative substitution effects are thus aggregated, and the sign that dominates, and by how much, depends on the strength of the two opposing effects and the relative weights of the different classes of capital within the total capital stock of the sector. For example, Behrman

23. *Editors' note*: In fact, the terms of trade for agriculture did not rise, and imports of agricultural goods rose in the 1970s.

(1972) in his estimates of capital-labor substitution for Chile used time-series data that group several forms of capital. In his study, the value of all of the estimated sectoral elasticities of substitution is less than unity, and for the agricultural sector the value is 0.31.

Estimations with CHAC refer solely to the second type of capital—that is, machinery—and therefore they measure solely the substitution effect without any admixture of effects of the opposite sign. Thus, it could be expected that the elasticities of substitution measured by CHAC would be of higher absolute value, and in fact they are: they range from around 1.0 to more than 3.0, in accordance with the different isoquant definitions that are presented below (table 5-2). Given that the financing of investment in machinery is generally of a different term than investment in land and other long-term works, the conceptual separation of types of capital for the elasticity calculations is consistent with a distinction between different instruments of policy.

The experiments with the model were carried out by specifying proportional salary increases for all types of labor as a means of inducing movement along the isoquant. The total cost of labor, which includes the farmers' returns to their land and water, always increases by a lower proportion than the nominal salary. This occurs because, as the cost of production (which includes the salary) increases, farmers lose part of their fixed factor returns. Also, the higher salary levels tend to encourage the substitution of family labor for day laborers, since the reservation wage for family laborers is less than the market wage for hired labor.

Two important characteristics of the elasticities of factor substitution that come out of CHAC summarize the foregoing: the elasticities refer solely to medium-term capital (machinery), and labor as defined for these measurements is not a homogeneous factor. Another important characteristic is that the isoquant is derived from a sectoral production function, or envelope of production functions, that is defined over multiple factors. Land and irrigation supplies are two factors whose availability is specified in monthly form in each locality. The actual amounts of land and water used in the model are endogenous, but their availability is fixed. In formal terms, this multiple-factor production function corresponds rather closely to the process-analysis model described and analyzed by Georgescu-Roegen (1972, 1969).

A fourth important characteristic of the CHAC estimate is that the model's sectoral production function is a multiproduct function. Because of this characteristic and to define the isoquant, users must decide which concept remains constant. The solutions that are presented here are based upon three different definitions: (1) the economic rent of producers (profits) is maintained constant; (2) nothing is maintained constant; and (3) the total value of production is maintained constant. (Results under the three definitions are presented as cases 1, 2, and 3, respectively, in table 5-2. Cases 1 and 2 are also shown in figure 5-2.) Given that the income of labor is composed in part by the economic rent, it is to be expected that the first definition would allow the least factor mobility and hence would give the lowest elasticity of substitution between factors, and that is exactly what occurs. This definition is the isoprofits curve. The second definition does not give an isoquant but rather a locus of equilibrium points associated with changes in factor prices. Although this is not an isoquant, it is perhaps more interesting from the viewpoint of decisionmakers because it constitutes a complete estimate of the set of multimarket reactions to hypothetical changes in prices. It is a type of response surface. Among other things, it is interesting to see how closely the response surface approximates the isoquant. The third definition given above is very close to that of the isoquant itself because, as will be explained below, it ensures that production, measured by a quantum index, is maintained approximately constant.

The production function of CHAC is also specified with respect to the flows of various current inputs that are used in the production process. These inputs have a price in the model, but they are not restricted in any way in the versions used for these solutions.

The results, then, are that the sectoral elasticity of factor substitution—measured as an arc elasticity over the longest arc—has a value of 0.956 when producers' profits are held constant, a value of 1.395 in the case of unrestricted equilibrium points, and a value of 3.341 along the isoquant.[24]

If cases 1 and 2 are compared, the locus of equilibrium points shows a greater degree of factor substitutability than the isoprofits curve. In other words, the isoprofits curve underestimates the degree of factor response in the sector as a whole; this is the relevant point for the formulation of agricultural policy. Furthermore, both curves have elasticities that vary substantially over the different segments, and in some cases they are not even convex. This behavior was foreseen by Georgescu-Roegen (1972). The nonconvexity arises from the fact that CHAC is a model with multiple products and multiple factors and that the "isoquants" are projections of a multidimensional hyperplane onto Euclidean two-space. The following question arises from these results: if in fact the process-analysis production model is a reasonable representation of reality, how useful are substitution parameters estimated by imposing on the data a production model that includes the implicit assumption of constant elasticities of substitution and by using a production function of two factors and one product?

---

24. For an application of these results to the possibility of employment generation via a machinery purchase tax, see chapter 6.

Table 5-2. *Elasticities of Factor Substitution*

Case	Segment of the curve					
	0	1	2	3	4	5
1. Isoprofits curve						
X: Value of production[a]	2311.5	2342.6	2374.3	2403.6	2439.3	2478.9
P: Producers' profits[b]	927.71	925.71	927.71	927.71	927.71	927.71
W: Wage payments[c]	512.40	554.01	558.67	595.31	621.72	653.38
Y: Total labor income[d]	1440.11	1481.72	1486.38	1523.02	1549.43	1581.09
E: Employment[e]	2015.59	1988.64	1884.04	1856.56	1818.29	1799.33
Y/E: Income per man-year	0.7145	0.7451	0.7889	0.8203	0.8521	0.8787
K: Use of machinery[f]	865.55	858.04	911.67	913.58	942.16	941.64
R: Rate of interest	0.12	0.12	0.12	0.12	0.12	0.12
K/E	0.4294	0.4315	0.4839	0.4921	0.5182	0.5233
(Y/E) ÷ R	5.9542	6.2092	6.5742	6.8358	7.1008	7.3225
Elasticity (a)[g]		+0.116	+2.008	+0.430	+1.359	+0.319
Elasticity (b)[g]		+0.116	+1.206	+0.987	+1.067	+0.956
2. Locus of market equilibria						
X[a]	2311.5	2276.9	2356.4	2391.2	2390.7	2403.7
P[b]	927.71	847.80	902.90	901.21	856.27	840.25
W[c]	512.40	560.96	560.24	596.75	623.18	653.58
Y[d]	1440.11	1408.76	1463.14	1497.96	1479.45	1493.83
E[e]	2015.59	2009.31	1890.18	1863.54	1829.00	1802.30
Y/E	0.7145	0.7011	0.7741	0.8038	0.8089	0.8288
K[f]	865.55	861.72	915.60	929.55	963.01	952.29
R	0.12	0.12	0.12	0.12	0.12	0.12
K/E	0.4294	0.4289	0.4844	0.4988	0.5265	0.5284
(Y/E) ÷ R	5.9542	5.8425	6.4508	6.6983	6.7408	6.9067
Elasticity (a)[g]		+0.062	+1.223	+0.778	+8.543	+0.149
Elasticity (b)[g]		+0.062	+1.504	+1.272	+1.639	+1.395

3. *Value of production constant*

X^a	2311.5	2311.5	2311.5	2311.5	2311.5	2311.5
p^b	927.71	880.55	853.36	807.48	767.59	729.15
W^c	512.40	561.36	560.92	600.12	622.51	653.70
Y^d	1440.11	1441.91	1414.28	1407.60	1390.10	1382.85
E^e	2015.59	2010.58	1894.16	1873.75	1832.14	1810.25
Y/E	0.7145	0.7172	0.7467	0.7512	0.7587	0.7639
K^f	865.55	863.27	919.96	942.59	975.57	972.61
R	0.12	0.12	0.12	0.12	0.12	0.12
K/E	0.42943	0.42936	0.4857	0.5031	0.5325	0.5373
$(Y/E) \div R$	5.9542	5.9767	6.2225	6.2600	6.3225	6.3658
Elasticity (a)^g		−0.043	+3.053	+5.857	+5.710	+1.314
Elasticity (b)^g		−0.043	+2.792	+3.157	+3.573	+3.341

Note: As specified in the text, in case 1 producers' profits are constant; in case 2 nothing is constant; and in case 3 the total value of production is constant. See also figure 5–2.

a. The value of production is defined at endogenous prices. The units are tens of millions of 1968 pesos.

b. Producers' profits are the sum of economic rents that accrue to land, water, and family labor. The units are the same as above.

c. Wage payments include both wages paid to day laborers and "payments" of the reservation wage to the farmer and family laborers. The units are the same as above.

d. Total labor income is the sum of producers' profits and wage payments.

e. Employment is measured in man-years and includes employment of both hired labor and family labor.

f. The use of agricultural machinery is measured as the flow of machinery services in units of ten million pesos.

g. "Elasticity (a)" is the arc elasticity measured between contiguous endpoints of the linear segments of the curve. For example, in case 1 its value is +2.008 from the end of segment 1 to the end of segment 2 "Elasticity (b)" is always measured from segment 0 to the end of the segment indicated. Thus, the longest arc is that from segment 0 to segment 5, and it has an elasticity of +0.956 in case 1 and +1.395 in case 2. "Segment 0" is not a segment but rather a point that corresponds to the base solution. The elasticity is always measured as the percentage change in factor proportions divided by the percentage change in the ratio of factor prices. See Ferguson (1971) regarding methods for calculating elasticities along an isoquant of piecewise linear segments.

Figure 5-2. *Sector Isoquants for Capital and Labor in CHAC*

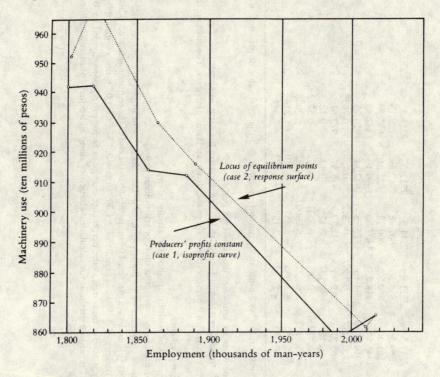

Another interesting aspect of the capital-labor substitution results is that the isoprofits curve gives levels of net labor income that are always higher than those of the response surface. Correspondingly, the levels of employment are always lower along the isoprofits curve than along the response surface. The reason for this can be seen clearly in table 5-2. With profits held constant, the producers' economic rents are not permitted to fall as nominal salaries rise. And the only way that profits can be maintained constant while production costs are increasing (through the salary increases) is through sufficient rises in product prices. Therefore, the physical levels of production are lower in the isoprofits case than in the response surface case. Given that agricultural products in the aggregate have a price elasticity of demand less than unity (in absolute value), the reduced levels of production tend to raise producers' profits slightly so that the higher costs of production are exactly compensated. This chain of reactions is thus reflected in a lower production index for the isoprofits curve, in comparison with the response surface, and a higher value for production at endogenous prices for the isoprofits curve.

In sum, imposing constant producers' profits on the model stimulates a series of compensating changes in production levels and in product prices. These changes are completely different from the case of the unrestricted market response surface. In the latter case the value of production rises neither as rapidly nor as uniformly as salaries are raised. Because of these production effects, the isoprofits curve not only underestimates the elasticity of substitution, but it also underestimates absolute levels of utilization of capital and labor in all segments, in comparison with the response surface curve.

Although it may be preferable to use the response surface instead of the isoprofits curve for policy purposes, it must be recognized that neither of these concepts permits the measurement of a pure subsitution effect. Both case 1 and case 2 include output effects as well as substitution effects.[25] For this reason, and to isolate the substitution effect alone, CHAC was formulated for a third set of results by maintaining the value of production constant at endogenous prices. These results are presented as case 3.

Case 3 by definition does not permit the physical levels of production to fall as factor costs increase. Although this case has been generated with CHAC's holding constant the value of production, that procedure implies that the quantum index of sectoral production also must remain approximately constant[26] (permitting compensating changes among individual products), given that the average price elasticity of demand for agricultural products is not equal to unity. Case 3 therefore gives physical levels of production that are higher, and higher levels of utilization for both factors, than does either case 1 or 2.

As expected, the pure elasticity of substitution in this case is significantly higher than in cases 1 and 2. Though this measure is simpler conceptually than either of the other two, to calculate it has required the imposition of restrictions on the market response in the model and these restrictions have forced the aggregate value of production to differ significantly from its full equilibrium level along the unrestricted response surface. For this reason, case 2 is likely to be more useful for policy purposes. If the response surface of case 2 is the relevant concept for program formulation, then using the isoquant results would appear to be misleading since they overestimate more than twofold the percentage response of employment with respect to changes in labor income levels.

As a final point of interest, table 5-3 shows the income elasticity of

25. Over some segments, the fall in physical production is also accompanied by a fall in value of production because a few of the crops face demand curves that are relatively elastic with respect to price, and in some segments of the isoquants these products are the ones that register greater movement.

26. The quantum index of production will not be exactly constant because of index number problems.

Table 5-3. *Elasticities of Employment*
with Respect to Total Labor Income

Case	Segment of the curve				
	1	2	3	4	5
1. Isoprofits curve					
Elasticity (a)	−0.321	−0.946	−0.376	−0.548	−0.342
Elasticity (b)	−0.321	−0.682	−0.596	−0.586	−0.550
2. Locus of market equilibria					
Elasticity (a)	+0.165	−0.617	−0.377	−2.958	−0.605
Elasticity (b)	+0.165	−0.802	−0.666	−0.783	−0.754
3. Value of production constant					
Elasticity (a)	−0.660	−1.480	−0.1803	−0.2260	−1.760
Elasticity (b)	−0.660	−1.409	−1.456	−1.589	−1.606

Note: The three cases are defined as in table 5-2. The elasticity is defined as the percentage change in employment divided by the percentage change in total labor income (variables E and Y in table 5-2. As before, "elasticity (a)" refers solely to the arc of one segment, whereas "elasticity (b)" refers to the arc that reaches from segment 0 to the end of the indicated segment. The elasticities are always calculated about the midpoint of the arc.

employment for all segments of the curves in all three cases. This concept is measured as a percentage response in all types of employment divided by percentage change in total labor income (salaries plus producers' profits). As the table shows, there is a substantial variation along the course of each isoquant and among definitions of isoquant. Once again case 3, in which production is held constant, shows the greater degree of response. The limiting value of elasticity is − 1.606 in case 3, whereas it is − 0.754 in case 2, and − 0.550 in case 1.

A curiosity is that in one segment of case 2 the sign of the income elasticity of employment is positive; that is, both employment and total income per man-year fall. This is attributable to the complex structure of labor income determination in the model. When the salary alone is taken into account (and not producers' profits), the data in table 5-2 show that the salary elasticity of employment for this segment has the usual sign and that its value is − 0.033.

Chapters 11 and 13 both report results that give an elasticity of machinery-labor substitution that is less than unity at the regional level, in contrast with the above results for the entire sector. As discussed in chapter 13, the differences are attributable to the fact that the complete sector model allows for indirect factor substitution via substitution in domestic demand and agricultural foreign trade. Although the differences are explainable, they do imply that caution is needed in attempting to make sectoral policy inferences from region-specific studies.

What do these factor substitution measures imply about rural wage policies? For one thing, a clear distinction must be made between farmers and day laborers; because farmers earn an imputed wage while working their own lands, wage policy cannot directly affect their returns in that activity. Increases in day-labor wage rates generally, but not always, do lead to higher wage bills for hired field workers. Farmers' profits generally are reduced, however, so that total rural income rises by a lower proportion than the day-labor wage bill does. Employment falls, and machinery use tends to rise, depending on how strong the output effects are. In summary, enforcement of higher rural wages would lead to higher income levels for most day laborers; it would hurt some because they would lose their jobs, and it generally would hurt farmers who use hired labor. Output would fall, and product prices would rise. The magnitudes of these effects depend on the precise magnitudes of the changes being considered.

Income Distribution and Derived Demand for Inputs

In a model such as CHAC there are basically two ways of specifying an income distribution: by including various size classes of farms and by specifying various producing areas. The latter can be delineated, of course, to capture important distinctions such as that between dryland and irrigated farming. In CHAC, farm sizes are incorporated only for one submodel (El Bajío), and the pattern of income over those size classes was reported earlier (Bassoco and others 1973). Hence, for the sectoral distributional measures the regional income results are reported here.

Income distribution

Use of average regional income levels as points on an income distribution entails the well-known disadvantage that each point represents a group whose range of individual income levels may overlap the income ranges of other groups. Nevertheless, the regional measure is of some interest, in part because many kinds of policies may be pursued on a regional basis. To conform to widely accepted regional designations in Mexico, the CHAC results for the submodels were aggregated to a basis of seven regions: five representing irrigated agriculture and two representing nonirrigated agriculture. The 1968 CHAC results for net producer income[27]

27. Net producer income is calculated as gross sales at endogenous prices less the value of purchased inputs. Here the services of day laborers are regarded as purchased inputs. The results of this section are an example of the use of the model essentially as an accounting device to provide statistics, such as net income levels, which are simple functions of other parameters in the model (such as yields, prices, input costs) but for which there do not exist direct survey data.

Table 5-4. *CHAC Estimates*
of the Agricultural Income Distribution, 1968

Region	Annual net income per farm (pesos)	Number of farms	Cumulative percentage of net income	Cumulative percentage of farms	Average farm size (hectares)	Net income per hectare
Dryland	1,393	1,579,174	17.2	51.2	3.5	398
Tropical	3,886	792,217	41.3	76.9	2.9	1,340
North	5,270	81,882	44.7	79.6	2.1	2,510
Central Plateau	8,825	407,665	72.9	92.8	2.4	3,677
South	9,806	47,541	76.6	94.3	3.0	3,269
Northeast	10,530	40,396	79.9	95.6	6.5	1,620
Northwest	19,220	133,299	100.0	100.0	8.4	2,280
Nonirrigated (total)	2,226	2,371,391	41.3	76.9	3.3	675
Irrigated (total)	10,527	710,783	100.0	100.0	3.7	2,845
Total	4,140	3,082,174			3.4	1,218

Note: Dryland (rainfed or *temporal*) and tropical are both nonirrigated regions; the rest are irrigated.

for the seven regions are shown in table 5-4. The coverage of the model excludes farms which are primarily dedicated to tree crops and livestock; nevertheless, the typical farmer of annual crops earns a small amount of supplementary income from fruit trees and small-scale livestock. CHAC does not include these sources of supplementary income, and to that extent it understates farm income levels.

Table 5-4 shows a wide divergence in farm incomes. At one extreme, the rainfed farms constitute 51.2 percent of the population (as defined here) and yet earn only 17.2 percent of the income. At the other extreme, irrigated farms in the Northwest represent 4.4 percent of the population and earn 20.1 percent of the total income. In the Northwest the average farmer with irrigated land earns 13.8 times as much as his dryland counterpart. Yet less than half the difference is accounted for by higher productivity per unit of land. The irrigated farms of the Northwest produce 5.7 times the income per hectare of rainfed farms, but the irrigated farms in the Northwest are more than twice as large.

In productivity per hectare, irrigated farms of the Central Plateau are the most efficient: 3,677 pesos per hectare versus 2,280 pesos per hectare in the Northwest. The irrigated farms in the South and the North are also more productive per hectare than those in the Northwest. Part of the explanation for this is found in the cropping patterns: the Central Plateau produces proportionally more high-value fruits and vegetables than any other part of the country. The South has tobacco, and the North has cotton.

But it is also true that the farmer of the Central Plateau uses fewer purchased inputs and relies more on his own labor and hence has a higher ratio of net income to gross income. Having a smaller farm makes it economic for the farmer to use much less machinery and hired labor. From tables 5-4 and 5-5, the ratios of net to gross income for the regions are as follows:

	Ratio		Ratio
Dryland (*temporal*)	0.29	South	0.50
Tropical	0.52	Northeast	0.45
North	0.62	Northwest	0.46
Central Plateau	0.78		

Relative to nonirrigated agriculture, irrigated agriculture as a whole generates 4.7 times as much net income per farm, 4.2 times as much net income per hectare, and 2.8 times as much employment per hectare. These figures reveal that a man-year generates 1.5 times as much net income with irrigation as without. The employment comparison is striking for policy purposes. Adding water increases enormously the employment absorption capacity of agriculture, even though the typical irrigated farm also uses machinery more intensively than does the typical nonirrigated

Table 5-5. *Employment and Production by Region in CHAC, 1968*

Region	Employment			Value of production		
	Per hectare cultivated[a]	Per farm[b]	Per unit of water[c]	Per hectare cultivated[d]	Per farm[e]	Per unit of water[f]
Irrigated						
Northwest	2.08	16.95	1.69	5,143	41,904	4,171
North	2.07	10.50	2.48	4,120	8,524	2,009
Northeast	3.36	21.78	2.50	3,615	23,467	2,694
Central Plateau	4.52	11.06	7.90	4,652	11,383	8,129
South	7.06	21.46	5.89	6,402	19,436	5,341
Dryland	1.11	3.88	—	1,377	4,826	—
Tropical	1.65	4.77	—	2,609	7,527	—
Irrigated (total)	3.58	13.41	3.42	4,811	18,003	4,600
Nonirrigated (total)	1.26	4.18	—	1,737	5,729	—
Total	1.85	6.31	3.42	2,517	8,559	4,600

— Not applicable.
a. Man-months per hectare.
b. Man-months per farm.
c. Man-months per ten thousand cubic meters.
d. Gross value in pesos per hectare.
e. Gross value in pesos per farm.
f. Gross value in pesos per ten thousand cubic meters.

Figure 5-3. *Lorenz Curves for the Sectoral Income Distribution in CHAC*

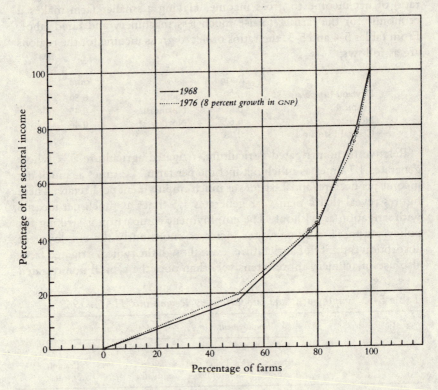

farm. The additional sources of employment with irrigation are from higher yields (higher harvest labor requirements), double cropping, and the ability to grow labor-intensive fruits and vegetables that need controlled water. The typical fruit and vegetable crop in Mexico needs four times as much labor per hectare as the typical grain crop (for example, maize or wheat).

In every respect—production, income, and employment—irrigation is clearly the factor of critical importance in Mexico. The uneven distribution of water over farms is clearly the major determinant of the skewness of the sectoral income distribution. Figure 5-3 shows the Lorenz curves for the sector's income distribution for both 1968 and 1976 under 8 percent growth. The curves are very similar except that the lowest-income groups appear to gain somewhat over time. In numbers, farmers in the rainfed areas receive 17.2 percent of total producers' income in 1968, 18.5 percent in 1976 under 7 percent growth, and 19.2 percent in 1976 under 8 percent growth. In all cases, these farms represent 51 percent of the farms.

Higher growth clearly makes the sectoral income distribution some-what more uniform. The reason for this is the same as the reason for the higher aggregate supply elasticity in nonirrigated areas: the nonirrigated farmers have more idle, marginal land, and hence they respond more to price incentives. Higher growth means more favorable terms of trade and therefore induces the nonirrigated farmers to put a higher proportion of their land under cultivation; the consequence is an improved income position for them. Conversely, slow growth in the Mexican context brings about an increasing skewness in the incoming distribution.

These results are, of course, conditional with respect to the hypotheses established regarding rates of increase of yields and the agricultural resource base in each region. To present the problem of income distribution in its simplest profile, we have used the income results from the same solutions reported earlier, which contain the assumption of equal rates of yield and resource increase for both irrigated and nonirrigated agriculture. Unfortunately, the historical time-series evidence on this is not very reliable, but it does seem to indicate roughly equal rates of technological progress and resource expansion in both regimes of agriculture.

To pursue the matter further, it would be interesting to alter these assumptions for additional CHAC solutions; that is, to see what would be the effect on the income distribution of a research and extension program that favored nonirrigated areas.

Seasonal employment patterns

As mentioned in the earlier section on macroeconomic results, it is difficult to evaluate the rate of employment increase in the sector only in total man-years of employment. Seasonality is the essence of the agricultural employment problem.

Figure 5-4 shows sectoral employment, by month, for the three solutions for 1968, 1976 at 7 percent growth, and 1976 at 8 percent growth. The first characteristic that stands out is that employment is highly seasonal in the sector. In the peak month there are about five times as many jobs as in the least busy month. Each of these seasonal curves is, of course, an aggregate of the corresponding curves for irrigated, dryland, and tropical farming. Both irrigated and tropical farming generate fairly smooth seasonal demands for labor; that is, for dryland areas alone the seasonality is even more marked than in figure 5-4.

A comparison of the three curves shown in the chart reveals that the increased demands for employment do not occur uniformly over seasons. Rather, employment is increasing more rapidly in the peak months than in the base months: the degree of seasonal variation is becoming more pronounced. This is an inevitable consequence of a trend pointed out earlier: the area cultivated is expanding most rapidly in dryland regions as

Figure 5-4. *Seasonal Employment in CHAC*

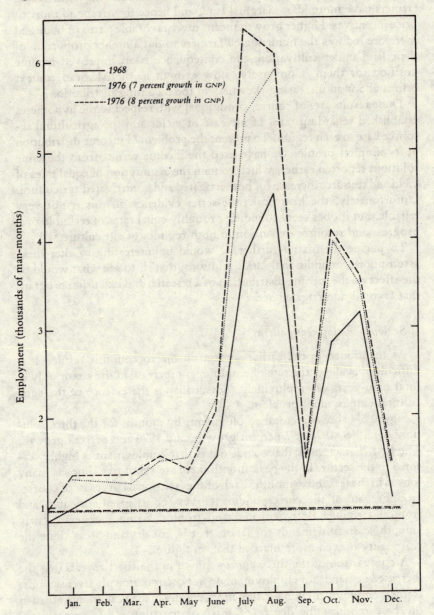

greater price incentives bring·more marginal lands under the plow. Although expansion of area cultivated is one of the objectives of sectoral policy, the increasing seasonality of employment is an unfortunate by-product.

The following numerical comparisons may be made. As shown in table 5-1, total sectoral employment, measured in man-years, grows by 2.5 percent per year when GNP grows by 8 percent annually. "Steady" employment, however, as measured by the man-years worked in jobs that last ten, eleven, or twelve months per year, is growing at only 2.0 percent annually in that case. In contrast, highly seasonal employment, as measured by time devoted to jobs that last only one, two, or three months per year, is increasing by 3.5 percent annually in that case. The lowest rate of increase is registered for the six- and seven-month jobs: 1.6 percent annually.

Similar results are available for each submodel and region. Here the aim is simply to offer a numerical example of the seasonal results that flow from CHAC.

Derived demands for other inputs

As with employment, input use can be tabulated on a regional basis from CHAC solutions. Here we present only sectoral aggregates. Table 5-6 shows the percentage response of the use of various inputs relative to the percentage change in production for 1968–76. It can be seen that demands for credit, improved seeds, and fertilizer grow substantially faster than does production itself.[28] In other words, 5 percent annual output growth requires about 8 percent annual credit and fertilizer expansion, and 11 percent annual increases in improved seeds.

When GNP growth is at 7 percent, labor-intensive techniques, as represented by the use of draft animals, grow more rapidly than do capital-intensive techniques (machinery), and the reverse is true under higher GNP growth.

As a final note on this point, it is interesting to see how the marginal productivity of irrigation water responds to GNP growth: under 7 percent growth when agricultural prices increase 1.5 percent in relative terms, the value of water grows by 1.8 percent annually. Under 8 percent growth, however, with prices increasing at 2 percent, the value of water goes up even faster—by 3 percent annually. These kinds of calculations are relevant to benefit-cost evaluations of irrigation projects.[29]

---

28. CHAC, however, is a cross-sectional model, and it does *not* include historical estimates of the relations in table 5-6.

29. See chapter 6 for other kinds of aggregate evaluations of irrigation investment programs.

Table 5-6. *CHAC Estimates of Income and Input Elasticities with Respect to Production, 1968–76*

	Growth in GNP	
	7	8
Concept	percent	percent
Sectoral income	1.063	1.056
Short-term credit	1.829	1.796
Improved seeds	2.319	2.185
Agricultural chemicals	1.765	1.740
Agricultural machinery	0.894	1.000
Draft animals	1.297	0.981

Analysis of Price–support Programs

The basic issues associated with the rationale for maintaining some crops' prices above the market level are discussed in chapter 6. There the conclusion is reached that in Mexico certain prices should indeed be supported by fiscal intervention, in the short run at least, and so the CHAC model was used to study some possible supply-side consequences of the support programs. Specifically, the model was used to quantify the amount of price increase for grains required to achieve domestic self-sufficiency in grains, and also to calculate approximately the short-term sacrifices in other crops and in export earnings associated with attainment of the goal.

After being maintained constant for twelve years, guaranteed maize and wheat prices were raised significantly in the early 1970s. A succession of annual price increases were instituted in an attempt to keep ahead of inflation in real terms, but from 1970–77 the wheat price in fact declined relative to the general price index, and the maize price increased annually about 1 percent more rapidly than did the general index. Both crops' prices declined relative to the overall agricultural price index in that period. Because of this factor and some uncertainty about land tenure, food grain production has increased rather slowly in the 1970s, and throughout the decade substantial quantities of grains have been imported annually.

The way in which the CHAC model contributed to discussions of price supports was by computation of supply response coefficients and emphasis of the interdependence of crop choices on the supply side. Although the analysis was conducted with a 1968 version of the model, it was assumed that crop-specific supply response elasticities are not likely to change much over a few years' time.

Also, district-specific models were used to indicate probable responsiveness on a local basis to revised national pricing policies. In particular, the TOLLAN model of chapter 13 (for the Tula irrigation district) and an unpublished model for the Zamora irrigation district were used to analyze supply response possibilities in the Central Plateau area. This region was of especial concern in view of government investment programs for the construction of new marketing and storage facilities there.

The district models contain risk specifications along the lines suggested in chapters 7 and 8. They otherwise look very much like a detached regional component of CHAC. As in all the policy analyses, care was taken not to interpret CHAC too literally at the district level, for its lack of risk structures means that it tends to give overly specialized cropping patterns by district. Yet it was felt that its results in the aggregate and by the four broad regions (chapter 4) were representative of sectoral behavior. For district-specific analyses of several types, models such as those of chapters 11–18 were used. Of course, in many cases the construction of those models was facilitated by use of CHAC technology vectors.

In general, the bias of both CHAC and the district models can be said to lie on the side of an overstatement of supply response elasticities, because of a failure to incorporate all factors that inhibit changes in cropping patterns. This being the case, the policy bias would be toward an understatement of the amount of price increase required to achieve a given increment in production.

In both the district-level and sector-wide experiments in pricing policy, the procedure for tracing out the implicit supply response function was the same: to conduct successive model solutions under rightward rotation of the demand function of the product in question, and to calculate ex post supply response elasticities along the sequences of points revealed in this procedure. For the aggregate sectoral supply response, the method is illustrated in figure 5-1.

In interpreting the results, some thought was devoted to identifying the relevant range of the supply response functions, for it has been noted that the arc elasticity varies significantly from segment to segment of a given response function. As in the case of isoquants for measuring capital-labor substitution, it has been found that an assumption of constant elasticity is not very realistic.

The remainder of this section illustrates these kinds of model applications by citing (in translation) selections from one of the price-support working papers (COCOSA 1974b). Similar model analyses had formed the underpinning of the initial revisions to support prices a year earlier.

In order to evaluate the price response of national maize production, a sector-wide linear programming model was used in which the production levels of all crops are competing for the use of land in all regions.

The results of the analysis permit an evaluation of the supply response elasticity of maize at the national level and also by irrigated and rainfed subsectors. This concept is the relation between the percentage increase in maize production and the percentage increase in its price.

Also it is possible to evaluate the relationships of substitution or complementarity between maize and other agricultural products. . . .

The results of the analysis indicate that the supply response elasticity of maize, for the segment of the supply response that is currently relevant, is around 0.5 (0.486). The value for irrigated areas is 0.636 and for rainfed areas it is 0.4. . . . In order to identify the relevant portion of the supply response function, that segment was selected which contained the current [and proposed] price levels, deflated to 1968 prices. . . .

This value of the elasticity implies that—if it is desired to increase national maize production by 10 percent above the trend to reduce substantially the imports running at 1.4 million tons annually—the producer price of maize must be increased by 20.6 percent [in real terms]. At the current level of the support price of 1,200 pesos per ton, this increase signifies that the price must be raised to 1,447 pesos per ton.

Studies carried out for the irrigation districts of Tula and Zamora indicate that the value of the maize supply response elasticities for those districts are 0.4 and 0.2, respectively. The explanation of the lower elasticity obtained for these districts is that they have cropping patterns that are intensive in highly profitable crops, such as alfalfa, tomatoes, garlic, and chile (in Tula) and strawberries, onions, and potatoes (in Zamora). Therefore the opportunity cost of land is higher in those districts than in the majority of the irrigation districts, which dedicate most of their land to crops such as wheat, sorghum, and soybeans. . . .

If allowance is made for the expected greater possibilities for export of certain crops [especially fruits and vegetables], then the required additional price stimulus for maize is even greater. In this case, the maize supply elasticities are 0.380 for all production, 0.459 for rainfed production, and 0.285 for irrigated production. . . .

Table 5-7 [tables 5-7 through 5-9 have been reproduced and renumbered as tables of this chapter] presents the calculations of cross-supply response elasticities for the same segment of the response function that gives an own-price elasticity of 0.486. It is noteworthy that the crops that are most sensitive to maize price changes are wheat, squash, lima beans, soybeans, and peanuts. . . . An increase of 1 percent in the maize price will give rise to decreases of 0.80 percent and 0.58 percent in irrigated production of wheat and soybeans respectively.

Table 5-8 presents the corresponding reductions in production levels of the different crops that would be attributable to each ton of increased maize production under higher support prices. . . .

Table 5-7. *Cross-supply Response Elasticities with Respect to the Maize Price, Computed from CHAC Solutions*

Crop	Substitutes	Complements	Independent
Garlic			X
Dry alfalfa	−0.2674		
Cotton	−0.2397		
Green alfalfa	−0.6800		
Rice			X
Oats	−0.2120		
Sugarcane			X
Peanuts	−8.6460		
Squash	−100.0[a]		
Safflower		+0.8660	
Barley			X
Forage barley	−0.2028		
Grain barley	−0.6178		
Dry chile			X
Green chile			X
Strawberries			X
Kidney beans	−0.1807		
Lima beans	−0.5791		
Tomatoes		+0.0240	
Flaxseed	−0.2121		
Cantaloupe			X
Potatoes			X
Cucumbers			X
Pineapples			X
Watermelon	−0.0245		
Soybeans	−0.5828		
Wheat	−0.8090		

Source: COCOSA (1974*b*).

a. Indicates total substitution for that crop.

From the foregoing analysis, it can be seen that—with increased maize production and if there were no increases in yields per hectare or in the area cultivated—there would be reductions in the domestic supply of other basic food products and forage crops. At the same time, the export crops are relatively insensitive to maize price increases [of the proposed magnitude], with the exception of cotton. . . . The reduction of cotton exports amounts to a loss of 56 dollars of export earnings per additional ton of maize production, but this is more than offset by the savings of 112 dollars, which is the foreign exchange cost of importing a ton of maize.

The impact of maize price increases differs among producing regions. The Northwest is the region that shows the strongest response to price variations. This region is composed entirely of irrigation districts that have high yields per hectare. . . . Its producers typify commercial

Table 5-8. *Reduction in the Production of Various Crops Attributable to Each Ton of Additional Maize Production under the Price Support Program, from CHAC*

Crop	Loss (tons)	Crop	Loss (tons)
Dry alfalfa	0.067	Kidney beans	0.007
Cotton	0.085	Lima beans	0.006
Green alfalfa	0.142	Watermelon	0.013
Oats	0.005	Soybeans	0.040
Peanuts	0.117	Wheat	0.336
Forage barley	0.254	Grain barley	0.005

Source: COCOSA (1974b).

agriculture, for they [extensively] use machinery and improved seeds, and they fertilize at appropriate levels. This group is more responsive to price changes, for both products and factors, than the traditional producers who are more hesitant to change their cropping patterns and technologies. In addition, in a good part of the Northwest irrigated area the cultivation patterns emphasize crops such as wheat, soybeans, safflower, and the like, and all these crops provide a similar level of net income per hectare to the producer; therefore even small changes in their prices, or in the price of maize, can give rise to a chain of crop substitution effects. Table 5-9 shows some of these effects. In the Northwest, the additional maize production is secured mainly at the expense of wheat, soybeans, barley, and forage crops. And, as a result of realignment of production patterns and seasonal resource requirements, there are increases in output of sorghum, tomatoes, garlic, and rice, along with the maize.

The irrigated districts of the North Central and Northeastern parts of the country do not respond strongly to changes in the maize price, a result which suggests that these districts (especially those of the state of Tamaulipas) have reached a practical maximum in maize production. . . .

Maize production in the irrigated districts of the Central Plateau responds to price incentives but to a lesser degree than in the Northwest. As was discussed already [in the context of the individual district models], these districts include in their cropping patterns very profitable crops such as fruits and vegetables, and substituting 1 hectare of these crops for maize would involve a high opportunity cost. . . .

Rainfed maize production responds strongly to price increases. It is worth noting that in these areas the substitution occurs principally with respect to squash, peanuts, sesame, sorghum, and potatoes. The incorporation of idle lands figures importantly in the rainfed production increase. The model results indicate that an 11.4 percent increase in the maize price would lend to a 1.37 percent increase in the area cultivated. . . .

Variations in the price of one agricultural product, via a chain of relations of substitution and complementarity with other crops, lead to modifications in the general price index for agricultural products. . . . [An 11 percent increase in the maize price, according to the model] gives rise to a 2.8 percent increase in the price index for all other crops . . . and also a 10.5 percent increase in producers' profits, but only a 1.5 percent decrease in consumer well-being [as measured by consumer surplus].

Calculations of Comparative Advantage by Crop

CHAC has been used to rank crops by their social profitability in export and also to obtain some indicators of interregional comparative advantage within Mexico. The initial computations of international comparative advantage were published in 1973 and were made on the basis of a version of CHAC that used 1968 data (Duloy and Norton 1973). Subsequently, those computations were repeated for 1976 and were made on the basis of assumptions regarding the GNP growth rate and the rate of expansion of the agricultural resource endowment. It turned out that the conclusions regarding the export structure of Mexican agriculture were quite stable over time.

The concept of comparative advantage has been quantified in terms of Bruno's "exchange cost" (Bruno 1967) or "domestic resource cost" measure: that is, the level of production costs in pesos required to earn a dollar of foreign exchange through export. The major empirical contribution of CHAC for making this computation, of course, is the marginal valuation of fixed resources, especially land and irrigation water. Because this valuation is endogenous, it depends on the production pattern, and hence it will vary with the level and composition of agricultural exports. For the present studies, it is desirable to know the exchange cost for incremental exports from the existing levels; therefore, for this particular experiment all exports were upper-bounded at base-year levels (1968 actual values and 1976 forecast values).

The procedure for using the model for the computations of the exchange costs is straightforward. With the export bounds set at the desired level, the model is solved, and the shadow prices on those bounds are tabulated. Shadow prices on export bounds signify the amount of producers' excess profits per incremental unit exported. The domestic resource cost—excluding profits but including the "normal" returns to producers'

Table 5-9. *Qualitative Changes in Regional Production Levels as a Consequence of a Change in the Maize Price (of 11 Percent)*

	Region				
	Irrigated			Nonirrigated	
Crop	North-west	North	Central Plateau	Rainfed	Tropical
Garlic	+ +		− −		
Dry alfalfa		−			
Cotton	−	+	− −		
Green alfalfa	− −				
Rice	+ +		−		− −
Oats			−		
Sugarcane	+ +				−
Peanuts		+		− −	
Squash			− −		
Safflower				+ +	
Barley					
Forage barley	− −	+ +			
Grain barley	+ +		− −		
Dry chile			− −		
Green chile	−		+		
Strawberries					
Beans	−			+	
Chickpeas					
Lima beans			−		
Tomatoes	+ +				
Sesame				− −	
Flaxseed			−		
Maize	+ +		+ +	+ +	
Cantaloupe					
Potatoes			+ +	− −	
Cucumbers	− −		+ +		
Pineapple					
Watermelon			− −		
Sorghum	+ +			− −	+ +
Soybeans	− −				+ +
Wheat	− −			−	−

Key:
+ + Increase of more than 5 percent
+ Increase of zero to 5 percent
− Decrease of zero to 10 percent
− − Decrease of more than 10 percent
(blank) No significant change.
Source: COCOSA (1974*b*).

labor, land, and water—is therefore found by subtracting the export-bound shadow price from the exogenous export price. "Normal" factor returns here refer to the rates of return accruing from production for sale on domestic markets. This measure of exchange cost obviously varies by crop, and it is interesting to see which are most profitable in export; that is, in which crops Mexico is most competitive on world markets. It also is worthwhile to compare the exchange cost for each crop with the prevailing exchange rate; crops whose exchange cost in pesos is greater than the exchange rate (in pesos per dollar) require subsidies for additional exports at the margin. A similar comparison may be made for an average of all export crops, using export volumes as weights, to see if agriculture as a whole is competitive on world markets and how it compares with industry.

For commodities that should by market criteria be imported, but that in reality are exported under subsidization, the procedure is to tabulate the "reduced cost" of their export activities in the model (the cost of forcing their exports into the optimal basis) and then to add this reduced cost to the peso export price to obtain the marginal exporting cost (which is greater than marginal export revenue).

Table 5-10 reports the CHAC exchange costs calculated in this manner, both for a 1968 version of the model and for a projected 1976 version.[30] Several conclusions are apparent from the table without assigning an unwarranted degree of accuracy to its figures. First, the ranking of crops by their export competitiveness is fairly stable over time, especially for those crops whose cost is less than the official exchange rate of 12.5 pesos to the dollar that prevailed when these calculations were made (in 1974; a horizontal space breaks the table at the value 12.5). Second, in many crops (especially fruits and vegetables) Mexico's production cost is significantly below the world market price. These are farm-gate production costs, but allowance for farm-to-port transport typically would not add more than 10 percent to the unit costs.

In the 1960s, Mexico's agricultural export structure emphasized some crops in which the nation does not appear to have a comparative advantage, especially maize and wheat. In fact, grain exports were realized only through Compañía Nacional de Subsistencia Popular (CONASUPO) subsidies, and in the 1970s those crops disappeared from the export bundle. The export structure has been reoriented along the lines suggested by these cost calculations (as shown in table 6-5 of the next chapter). Early results from CHAC, along the lines of table 5-10, were used in discussions with the official Mexican Institute for Foreign Trade (IMCE) regarding the

30. The list of changes made for the 1976 version is given in the section "Basic Macroeconomic Results," above.

Table 5-10. *CHAC Computations*
of Export Comparative Advantage by Crop over Time

	Exchange cost			
Crop	1968	Rank	1976	Rank
Garlic	3.13	1	3.28	1
Cantaloupe	3.69	2	3.94	3
Onion	3.89	3	4.28	5
Tobacco	3.99	4	4.10	4
Strawberries	4.65	5	3.47	2
Peanuts	5.45	6	6.50	8
Cucumber	5.84	7	6.31	7
Potatoes	5.86	8	6.73	9
Sesame	5.94	9	5.74	6
Tomatoes	6.22	10	6.89	10
Watermelon	6.49	11	7.51	12
Sugarcane[a]	6.61	12	8.27	13
Chickpeas	6.73	13	7.12	11
Green chile	9.48	14	10.83	15
Pineapple	10.53	15	9.88	14
Kidney beans	11.31	16	11.80	16
Dry chile	12.39	17	13.70	18
Cotton fiber	12.39	18	12.39	17
Squash	14.06	19	14.63	19
Sorghum	16.25	20	16.48	20
Safflower	16.58	21	18.16	21
Cottonseed oil	19.02	22	40.71	32
Sugarcane[b]	19.58	23	22.92	23
Wheat	20.33	24	21.92	22
Safflower	20.54	25	24.39	25
Maize	21.00	26	24.39	25
Soybeans	21.23	27	26.51	27
Green alfalfa	22.50	28	29.25	28
Dry alfalfa	24.43	29	22.94	24
Grain barley	25.48	30	32.20	29
Rice	28.25	31	34.38	30
Oats	30.39	32	34.72	31
Lima beans	33.46	33	44.98	33

Note: "Exchange cost" is the level of production costs, in pesos, required to earn a dollar in foreign exchange through export (see text for method of calculation). The horizontal break in the table indicates the official 1968 exchange rate of 12.5 pesos to the U.S. dollar.

a. On the U.S. market.

b. On the world market.

c. Assuming 7 percent annual GNP growth during 1968–76. The other assumptions for the 1976 solution are listed in the text (see the section "Basic Macroeconomic Results").

usefulness of expanding the program of export marketing for certain fruits and vegetables, but no doubt market forces alone have been responsible for most of the trends in table 6–5.

A third conclusion that emerges from table 5–10 is that Mexican agriculture as a whole is quite competitive in world markets; the weighted average 1968 exchange cost for all crops was 8.75 pesos per dollar earned, using actual 1968 export levels as weights. Compared with the then prevailing exchange rate of 12.5 pesos, this figure implies that agriculture may in fact be more competitive than the nonagricultural sectors.

Within Mexico, the diversity of regional ecological conditions gives rise to differing spatial patterns of comparative advantage. In the discussion of pricing policies above, reference was made to the comparative advantage that rainfed zones have in maize production, even though maize yields are lower there than in irrigated areas. Other similar instances of regional differences emerged from study of the model solutions. The procedure was simple: to solve the model with no restrictions on the spatial allocation of crops and then to study the resulting differences in regional cropping patterns, as compared with the actual situation. In this way, CHAC's character as a spatial equilibrium model is exploited. As noted in the section on validation, the base-year data were not reliable enough to permit exact appreciation of these differences; nevertheless, some points emerged clearly.

One spatial result that stood out was the comparative advantage of the tropical zones in sugarcane production, even though cane yields per hectare are more than twice as high in the irrigated areas of the Northwest and almost twice as high in the irrigated Central Plateau areas (state of Morelos). Given the existing capacities of sugar refineries in each region, it is not possible to contemplate relocating sugar production in the short run, but on the basis of these results a policy directive was adopted to prohibit construction of new refineries outside the tropics.

As an expression of official interest in development of tropical agriculture, the bulk of water control and irrigation investments is now taking place in those areas. A question that arose in this regard was which kinds of crops should be favored in the new tropical production areas in relation to official programs of input supply and credit authorization. In addition to sugarcane, the model indicated rice, soybeans, pineapple, and maize as being especially suitable for the tropics.

The Northwest irrigated zones can dedicate their land to crops such as cotton, tomatoes, watermelon, wheat, and certain oilseeds that, for the most part, are more profitable per hectare. The Northwest's advantage in the first three crops arises from the high yields that have been obtained as a result of cumulative experience and investment in human capital. For wheat and oilseeds, the comparative advantage is based on the relatively large size of farms and, hence, the greater advantage of mechanized cultivation techniques.

Another result that appears significant is the model's tendency to understate maize production and overstate wheat production (see table 5-11 in the next section). No doubt this distortion is in part because of the assumption of a constant marginal rate of substitution in consumption between these two grains (see chapter 3). But, rather than attempt to adjust the model to eliminate the effect, it was preferred to leave it as it is because, on grounds of principle, arbitrary adjustments to bring the solution closer to reality were avoided and because the maize-wheat pattern in the solution reveals an important fact—that it is not only changes in taste that are moving the national consumption basket away from maize products and toward wheat products, but also the relative production costs for the two grains. To the extent that dietary needs can be satisfied by substitution of wheat for maize, then it is a more efficient use of a sector's resources to do so.

As in the other cases of regional comparative advantage, it would have been possible to quantify the amount of efficiency loss involved in the present production patterns by imposing them as constraints on the model and then comparing the solution values (for farm income, consumer surplus, and so forth) with the corresponding values for the unrestricted solution. This, however, was not done.

For the sector as a whole, the basic CHAC solution gave a quantum index of production that was about 12 percent higher than the observed value for 1968 (1967–69 average). This difference may be attributed to three sources: aggregation bias in the model; misspecification in the sense of failing to incorporate some real restrictions, be they economic or otherwise; and the model's attainment of a more efficient resource allocation because of its assumption of perfect information transmission. Given that all three factors add up to a 12 percent overestimation of sectoral production capacity, it appears that the aggregation bias of linear programming models may not be as great as has sometimes been feared.

Model Validation

Validation of a large-scale linear programming model is very different from that of an econometric model. A number of diverse issues must be addressed to form a clear idea of the economic content of the model, but as yet no standard procedures exist for validation. In the remainder of this chapter, issues concerning validation are discussed and several pieces of evidence on the model's performance are presented, but no pretense of a systematic, definitive validation is made.

To place the CHAC model in broad perspective, we can start by noting that it does not fall into either of two common categories: one is the econometric model for which preferred parameter values are selected according to formal statistical criteria, and the other is the numerical "simulation" model for which parameter values are selected through a repeated process of trial and error for model solution and parameter revision. Almost all of the parameters in CHAC are agronomic coefficients derived from extensive collation of published surveys and small-sample supplementary field interviews (as discussed in chapter 4).

In addition to this agronomic base, the other pillar of CHAC's structure is provided by economic theory and accounting relations. The competitive market form was used in the model structure in accordance with the assumption that atomistic, price-taking, profit-maximizing behavior most accurately describes the production responses of Mexican farmers. Four important modifications were made to this assumption, however, as it is embodied in CHAC: (1) prices differ spatially to reflect differential access to markets; (2) subsistence farmers behave according to a modified home-retentions rule with respect to maize and kidney-bean production (see "A Summary of CHAC," above); (3) some factor prices (labor, irrigation water) are not determined competitively; and (4), in the case of the individual district models (see chapters 8, 12, 13, and 16), farmers are risk averse as well as profit maximizing.

In a sense, a programming model of this type is a set of "structural equations," whereas econometric estimation of, say, supply functions gives a "reduced form." (Validation of much of the CHAC structural representation of the sector takes place through field checks of the magnitudes of the agronomic coefficients and through comparisons of the results of different surveys.)

The risk factor clearly is quite important for explaining observed behavior, and its omission from the present version of CHAC constitutes the model's greatest weakness. The relevant theory for incorporation of risk into this kind of model (see chapter 7) was formulated after the construction of CHAC, but risk elements were added to many of the district models, some of which were used in the drafting of papers on pricing policy (see "Analysis of Price-support Programs," above). As of this writing, a new version of CHAC that incorporates risk has been implemented by Mexican government personnel.

Validation in the broader sense has implied four classes of concerns for the work on CHAC:

- How to deal with the situation of unreliable base-year, benchmark data against which CHAC should be validated
- The definition of principal variables—both primal and dual—for which validation is most revealing about the model

- The delineating of certain classes of "policy experiments" for which CHAC should *not* be used, given its known limitations, and the definition of the directions of some of its biases, to be taken into account in evaluating the policy results
- Compilation of an agenda for improvement of the model's reliability in future versions.

The first concern—for benchmark data—has been particularly serious. In principle, the structure of CHAC, which is based on micro theory and field-level data, can be tested against its macro results (that is, against its aggregate results for income, employment, and crop production levels). There are, however, problems of definition that prevent a direct comparison of model results with data from the census and national accounts. As regards crop production levels, the official statistical reporting system underwent revisions in the late 1960s and early 1970s, and the reported data for 1967–69 do not seem particularly reliable. (At one point, three units of the former Ministry of Agriculture gave estimates of 1968 barley production that differed from each other by as much as 50 percent.)

In view of these circumstances, two strategies were adopted: to use the model to indicate priority areas for improvements in data collection and to seek other, microeconomic benchmarks for validation purposes. Only rough validation with respect to aggregate production data could be expected (see table 5-11).

A major area of data improvement as a result of the CHAC experience was in the set of official irrigation norms by crop and district. These were included in the irrigation submodels of CHAC; when in a test solution actual cropping patterns were imposed on these submodels, the resulting total calculation for irrigation use turned out to be significantly different from that reported in the official statistics.

For validation, the individual submodels were tested before their inclusion in CHAC. The two submodels analyzed most thoroughly were those for the Bajío region (Bassoco and others 1973) and for the Pacific Northwest region (chapter 11). The former includes both irrigated and rainfed zones and two sizes of farms in each zone. It is called BAJIO; a smaller version called BAJITO was made by aggregating the twelve monthly land restrictions into three constraints representing seasons of four months each. The latter embraces all five of the CHAC submodels in the Northwest, and it is called PACIFICO.[31]

Chapter 11 provides a full explanation of validation procedures for PACIFICO, including a discussion of why in most instances prices should be endogenous even for a regional model. The contention is that most policy

---

31. Other validation experiences for submodels including risk are reported in chapters 8, 12, and 13.

changes are likely to influence several producing regions and not only one. Pricing policies have this characteristic, and even an investment project in one region should be regarded as part of a sector-wide investment program that influences production capacity in several areas (see Harberger 1974). In any event, in tables 11-5 through 11-8 of chapter 11 the goodness-of-fit in the output and price spaces is reported for PACIFICO under the alternative versions with prices exogenous and endogenous. The fit is much better for the latter, and that version corresponds more closely to the concept of PACIFICO as a part of CHAC when sector policy experiments are conducted.

Other validation tests were conducted on PACIFICO—specifically, a capacity test to see if in fact the model specification permits replication of observed output levels (the answer is yes) and a test for marginal cost. The latter is designed to be a test of the competitive market assumption—by solving the model for its marginal costs of production (by crop) and then comparing those costs with actual prices. Table 11-4 in chapter 11 shows the results, and it is seen that the comparison is generally quite close.

For the Bajío area, perhaps the most interesting test involved computation of the implicit price of land and comparing it with actual sales prices (Bassoco and others 1973, table 5, p. 412). The computed prices by class of land corresponded closely to the actual prices for a real discount rate of 18 percent, which probably reflected accurately the actual lending rates of the rural informal sector in 1968. This means that, at least in the aggregate, the technology vectors of the model give an adequate description of the process of transforming factor inputs into final output.

For other submodels that were not incorporated directly into CHAC, validation experiments were conducted with and without risk specifications. These are reported in chapter 8, tables 8-5 through 8-8; chapter 12, tables 12-6 and 12-7; chapter 13, table 13-1; and chapter 16, table 16-1. The ALPHA model of chapter 8 actually is only a slightly simplified variant of the part of CHAC that represents eight irrigation districts in the Northwestern, North-Central, and Northeastern parts of the country, and the MEXICALI and TECATE models of chapter 16 are close relatives of the RIO COLORADO submodel of CHAC.

In general, it may be said that the validation results of the individual models are reasonably acceptable, and sometimes good, but that for the cropping pattern in particular noticeable improvements are obtained through the inclusion of the risk specifications as spelled out in chapters 7 and 8.

For the irrigation districts, the official production statistics are regarded as more reliable than the aggregate production statistics, and therefore validation for output levels (and prices also) is more meaningful at the irrigation district level. Most of the statistical uncertainty concerns the nonirrigated producing areas.

Table 5-11. *A Comparison of CHAC Production Levels with Official Production Data*
(metric tons)

Crop	CHAC (1968)	Actual (1967–69 average)	Crop	CHAC (1968)	Actual (1967–69 average)
Garlic	46,023	29,580	Chickpeas	153,222	150,462
Green alfalfa	8,499,999	7,753,700	Lima beans	34,199	77,848
Dry alfalfa	435,526	n.r.	Tomatoes	835,192	679,828
Cotton fiber	477,143	487,800	Sesame	343,158	149,601
Rice	461,901	378,130	Flaxseed	20,391	12,282
Oats	90,375	43,472	Maize	7,229,853	8,704,788
Sugarcane	30,580,719	31,922,000	Cantaloupe	241,497	171,207
Squash	119,042	119,001	Potatoes	803,736	321,359
Safflower	227,743	105,446	Cucumbers	54,132	n.r.
Peanuts	78,873	75,698	Pineapple	400,000	265,095
Onions	113,151	154,463	Watermelon	319,831	159,009
Forage barley	4,761	n.r.	Sorghum	3,249,855	2,046,245
Grain barley	306,793	217,707	Soybeans	261,647	215,484
Dry chile	27,692	25,842	Tobacco	193,024	59,538
Green chile	289,276	229,805	Wheat	2,884,756	2,067,587
Strawberries	268,141	108,210	Cottonseed oil	788,177	832,725
Kidney beans	1,016,846	857,536			

n.r. Not reported.
Note: The average percentage of absolute deviation is 13.4 percent. The official data are from the Bureau of Agricultural Economics, Ministry of Agriculture and Livestock (Secretaría de Agricultura y Ganadería, SAG; now Ministry of Agriculture and Water Resources, Secretaría de Agricultura y Recursos Hidráulicos, SARH).

Taking the aggregate statistics as they are, a comparison with the CHAC production levels for 1968 is given in table 5-11. Given the data uncertainties, not many conclusions can be drawn from the table, but three points emerge clearly: the average discrepancy by crop between the CHAC results and the official statistics is 13.4 percent; CHAC consistently overstates the production of fruit and vegetable crops; and CHAC overstates maize production and understates wheat production. The first problem is because of the absence of risk structure, as discussed; both field experience and formal work with district models support this contention. It may also be partly attributable to the presence of some quasi-monopolistic elements in the actual organization of production and marketing of these crops; this point is explored in chapter 13 and discussed briefly in the recommendations of chapter 6.[32] For the state of Sinaloa in the Northwest, however, tests performed in chapter 12 reject the hypothesis of a monopoly. The maize-

32. Conscious control of production levels (for the domestic market) by producer groups is perhaps most commonly recognized to occur in the case of potatoes.

wheat problem may be a reflection of CHAC's exaggeration of substitution possibilities in demand, as mentioned.

There exists a study that also permits an approximate check on CHAC's results regarding producer income levels and the rate of underemployment in the sector. As reported in Norton and Duloy (1973), CHAC gives a net income figure for short-cycle crops in 1968 of about 4,000 pesos per person in the agricultural labor force and an average employment rate for all members of the labor force of 51 percent to 54 percent for the year as a whole.

With some adjustments, a figure for a comparable employment concept may be derived from the sample surveys of the Centro de Investigaciones Agrarias (1970). That figure is 59 percent, averaged over the year and over the entire labor force. It is not possible to derive a comparable income figure, but the Centro's study does report an (adjusted) 1968 income figure of 6,300 pesos, including livestock and tree crop activities as well as short-cycle crops.[33] Given the definitional differences, this income level appears to be approximately equivalent to the CHAC income level of 4,000 pesos.

The reservation wage rate, which was discussed at length in chapter 2, is the only CHAC parameter estimated by trial and error. As noted in an earlier work (Norton and Duloy 1973, p. 381), the model was not very sensitive to variations in the ratio of the reservation wage to the market wage, in the range of 0.40 to 0.70. At values approaching zero and 1.0, however, the model's cropping pattern became much less realistic. For the individual district models, similar results were observed, and a ratio of 0.5 gave the best fit on cropping patterns for irrigated areas and 0.3 to 0.4 for rainfed areas. This difference appears to reflect correctly the higher opportunity cost of farmers' time in irrigated zones.

Regarding the empirical implications of CHAC's biases, it already has been noted that interpretations of pricing policy should be made to reflect the model's tendency to overstate supply responsiveness. It so happened that CHAC's tendency to understate the required price increase (to attain a specified production target) was unimportant because other groups in the government were proposing even lower price increases. Had that situation been reversed, then CHAC's results would not have been so applicable.

Also, given CHAC's lack of fertilizer-yield response surfaces, care was taken not to use it for recommending fertilizer pricing policy, although that question did arise in governmental discussions. Data from experimental stations are available for a few crops on the fertilizer-yield relation, but their ranges of values typically lie well above actual average

33. The adjustments made to the Centro's figures are described in Duloy and Norton (1973, p. 379).

yields per unit of fertilizer. Hence, they are not usable in the model directly although, perhaps with some transformations, they could be.

Given the lack of precision on base-year output levels (table 5-11), no recommendations were drawn from the model that would have required more accurate prediction of those levels. For example, on the question of incremental capacity for sugar refining, exact amounts of new capacity were not prescribed on the basis of the model; rather, it was stated that all incremental capacity should be located in the tropics.

Similarly, given the model's tendency to overly specialize in production by district, care was taken not to use CHAC for specific recommendations at that level, but only at the sector-wide level. Moreover, with over 200 different activities for maize production, plus a large number of resource constraints by location and month and many alternative crops, it is not unreasonable to place some confidence in the sector-wide maize supply responses of CHAC.

As in all work with models, judgments must enter in making interpretations. For CHAC, whenever possible interpretations have been made in relative rather than absolute terms—for example, results on export comparative advantage were applied as crop rankings rather than as absolute measures of the exchange cost by crop.

Any model is most useful if revised and solved repeatedly at intervals over time. Once the fixed cost of basic model design, data collection, and model construction has been incurred, there is a relatively high payoff from model revision and solution to address different issues. For CHAC and models of its type,[34] a basic agenda for future improvements during this process is fairly clear, although other items no doubt could be added to the list. First, as mentioned a number of times already, risk considerations should be made explicit whenever possible. Unfortunately, the time series of data on prices and yields that are available for irrigated zones were not available for the nonirrigated (poorer) zones of Mexico when CHAC was constructed initially, but they are available now. With risk, the new CHAC fits the reported data better. More basic research on model techniques is needed for the nonirrigated areas, perhaps along the lines of chapters 9 and 10.

In Mexico great improvements have been made in recent years in the quality of the data base, especially as regards price information and crop reporting. This will ameliorate some of the validation difficulties discussed above. Also, a new regional disaggregation of CHAC, one that corresponds more closely to program administration, has been developed for nonirrigated areas since the studies of this volume were carried out.

34. For information on other examples of the CHAC family of models, see in particular Cappi and others (1978) for Central America, Kutcher and Scandizzo (1981) for Northeast Brazil, and Candler and Pomareda (1978) for Zambia.

On the purely methodological side, certainly more attention can be paid to cross-price and income effects (see Norton and Scandizzo 1981) and to labor supply behavior (Hazell 1979). Also, more formal methods of searching for optimal policy packages (Candler and Norton 1977) would substantially increase the model's policy usefulness. As indicated by the references, work is underway on these topics, but obviously more remains to be done.

CHAC's omission of long-cycle crops and livestock constitutes an important limitation to its usefulness. Chapter 17 reports a beginning on a model of the dairy subsector, but more work is needed in this direction. Very little has been done for the analysis of tree crops in optimization models (see the review by Kennes and Hazell 1977), and again this is an important area for further research.

Some issues of land tenure have been studied with other models of the CHAC kind (Kutcher and Scandizzo 1976), as have food processing industries (Cappi and others 1978), but the existing treatments are by no means complete. It would be important to include better representation of the linkages with the food distribution system.

In general, one of the more promising areas for use of agricultural programming models concerns the application of district-level models to the investment decision. This line of investigation is exemplified by chapters 14 through 18 of the present volume, and some follow-up work has been conducted in the World Bank (Husain and Inman 1977), but the possibility has not been exploited very extensively.

Finally, computational systems are basic to the applicability of models such as CHAC. A well-designed model-generating system permits easy model management, both in the control of inputs and the interpretability of outputs and in the facility of model revision. Absence of such a system can mean literally months of elapsed time wasted in manual revisions and elimination of the inevitable errors. The initial CHAC model was built in a very inefficient way in this respect, and then the team progressed to use of a custom-designed matrix generator system, as reported in chapter 19. Only a beginning was made, however, regarding this all-important issue of model management.

References

Bassoco, Luz María, and Roger D. Norton. 1975. "A Quantitative Approach to Agricultural Policy Planning." *Annals of Economic and Social Measurement.* Vol. 4 (October–November), pp. 571–94.

Bassoco, Luz María, John H. Duloy, Roger D. Norton, and Donald L. Winkelmann. 1973. "A Programming Model of an Agricultural District." In *Multi-Level Planning: Case Studies in Mexico.* Edited by Louis M. Goreux and Alan S.

Manne. Amsterdam/New York: North-Holland/American Elsevier, pp. 401–16.

Behrman, J. R. 1972. "Sectoral Elasticities of Substitution between Capital and Labor in a Developing Economy: Time Series Analysis in the Case of Postwar Chile." *Econometrica*. Vol. 40 (March), pp. 311–26.

———. 1968. *Supply Response in Underdeveloped Agriculture*. Amsterdam: North-Holland.

Bruno, Michael. 1967. "The Optimal Selection of Export-Promoting and Import-Substitution Projects." In *Planning the External Sector: Techniques, Problems, and Policies*. Report of the First Interregional Seminar on Development Planning. New York: United Nations, pp. 83–135.

Candler, Wilfred V., and Roger D. Norton. 1977. *Multi-level Programming and Development Policy*. World Bank Staff Working Paper, no. 250. Washington, D.C.

Candler, Wilfred V., and Carlos Pomareda. 1978. "The Zambian Agricultural Policy Model—Progress Report," Washington, D.C.: World Bank. Restricted circulation.

Cappi, Carlo, Lehman Fletcher, Roger D. Norton, Carlos Pomareda, and Molly Wainer. 1978. "A Model of Agricultural Production and Trade in Central America." In *Economic Integration in Central America*. Edited by William R. Cline and Enrique Delgado. Washington, D.C.: The Brookings Institution, pp. 317–70 and Appendix G. Reprinted as World Bank Reprint Series, no. 82. Washington, D.C.

Centro de Investigaciones Agrarias. 1970. *Estructura agraria y desarrollo agrícola en México*. Vols. I–III. Mexico City.

Comisión Coordinadora del Sector Agropecuario (COCOSA). 1974a. "Una metodología cuantitativa de la programación agrícola." Technical Note no. 2. Mexico City. (See Bassoco and Norton 1975.)

———. 1974b. "Producción de maíz y política de precios." Mexico City.

Duloy, John H., and Roger D. Norton. "CHAC Results: Economic Alternatives for Mexican Agriculture." In Goreux and Manne, Multi-Level Planning, pp. 373–99.

Ferguson, C. E. 1971. *The Neoclassical Theory of Production and Distribution*, New York: Cambridge University Press.

Georgescu-Roegen, Nicholas. 1972. "Process Analysis and the Neoclassical Theory of Production." *American Journal of Agricultural Economics*. Vol. 54 (May), pp. 279–94.

———. 1969. "Process in Farming versus Process in Manufacturing: A Problem of Balanced Development." In *Economic Problems of Agriculture in Industrial Societies*. Edited by U. Papi and C. Nunn. New York: Macmillan.

Harberger, Arnold. *Project Evaluation*. 1974. Chicago: Markham Publishing Company.

Hazell, Peter B. R. 1979. "Endogenous Input Prices in Linear Programming Models," *American Journal of Agricultural Economics*. Vol. 61 (August), pp. 476–81.

Heady, Earl O., Narindar S. Randhawa, and Melvin D. Skold. 1966. "Programming Models for the Planning of the Agricultural Sector." In *The Theory and Design of Economic Development*. Edited by I. Adelman and E. Thorbecke. Baltimore: Johns Hopkins University Press, pp. 357–81.

Husain, Tariq, and Richard Inman. 1977. "A Model for Estimating the Effects of Credit Pricing on Farm-Level Employment and Income Distribution." World Bank Staff Working Paper, no. 261. Washington, D.C.

Kennes, Walter, and Peter B. R. Hazell, 1977. "A Review of Tree Crop Models." Washington, D.C.: World Bank. Restricted circulation.

Kutcher, Gary P., and Pasquale L. Scandizzo. 1976. "A Partial Analysis of Share-tenancy Relationships in Northeast Brazil," *Journal of Development Economics*. Vol. 3, pp. 343–54.

————. 1981. *The Agricultural Economy of Northeast Brazil*. Baltimore: Johns Hopkins University Press.

Mellor, John W. 1966. *The Economics of Agricultural Development*. Ithaca, N.Y.: Cornell University Press.

Ministry of the Presidency. 1973. *Lineamientos de la politica economica y social del sector agropecuario*. Mexico City.

Norton, Roger D., and Pasquale L. Scandizzo. 1981. "Market Equilibrium Computations in Activity Analysis Models," *Operations Research*. Vol. 29, no. 2 (March-April) pp. 243–62.

Shumway, C. R., and A. A. Chang. 1977. "Linear Programming versus Positively Estimated Supply Functions: An Empirical and Methodological Critique." *American Journal of Agricultural Economics*. Vol. 59, pp. 344–57.

Solís M., Leopoldo. 1970. *La realidad económica mexicana: Retrovisión y perspectivas*, Mexico City: Siglo XXI Editores.

————. 1972. *Controversias sobre el crecimiento y la distribución*. Mexico City: Fondo de Cultura Económica.

6

A Program for Mexican Agriculture

MINISTRY OF THE PRESIDENCY
GOVERNMENT OF MEXICO

THIS CHAPTER ILLUSTRATES the link between analysis and policy by repro-
ducing one of the principal Mexican government agricultural planning
documents of recent years.[1] The document presented here, *Lineamientos de
la política económica y social del sector agropecuario* [Guidelines for economic
and social policy for the agricultural sector], was issued in 1973 by the
Dirección General Coordinadora de la Programación Económica y Social
[General Coordinating Bureau for Economic and Social Program Plan-
ning], a bureau of the Ministry of the Presidency.[2] It is oriented toward
employment and income distribution, although not to the exclusion of
considerations of efficiency.

Overview

As the text of the document shows, the CHAC model of the agricultural
sector was used frequently to provide the numerical backbone of the
policy arguments, but it also contains many sections that were developed
without reference to CHAC. Preparing the document with only issues that
were amenable to analysis by the model in mind would have made it much
too narrow.

The ways in which a model is used are not always direct, and in some
instances simple recourse to it as an organized data bank was helpful. For
example, there are no surveys of Mexican rural labor that can be counted
on to give a reliable picture of the amount of rural underemployment and
unemployment, much less an estimate of incomes by functional groups of
agricultural producers. Therefore, as is seen below, one early use of the
model was the derivation of static pictures of the amount of productive

1. The principal authors of this document are Leopoldo Solís M., José S. Silos, and Luz
María Bassoco, with the advisory assistance of Roger D. Norton. They gratefully acknowledge the assistance of the others, in and out of the Mexican government, who also contributed.
2. Now the Ministry of Programming and Budgeting (Secretaría de Programación y
Presupuesto).

employment by season and by locale and of the net income levels of producers in different zones. The model was used in other ways as well, and a more formal presentation of some of the model's policy-related results was given in the preceding chapter.

Minor editing has been performed on the document to improve its clarity and eliminate unnecessary repetitions, but the text that follows is presented essentially unchanged.[3]

Guidelines for Economic and Social Policy for the Agricultural Sector

1. *Introduction*

Agriculture has played a predominant part in Mexico's economic development. Over the thirty years from 1930 to 1960, agricultural production grew at an annual rate of 4.8 percent, a rate high enough not only to meet the demands of the domestic market, thereby avoiding inflationary pressure on food prices, but also to provide growing surpluses for export. From 1940 to 1960, the dollar value of agricultural exports grew by 13 percent annually while their share of Mexico's total exports rose from 25 percent to 50 percent.

This performance contributed substantially to economic development by providing the foreign exchange needed to import industrial inputs. Agricultural growth also created new rural employment opportunities, thus reducing migration from the land, which has tended to exceed the capability of urban industry to provide new jobs

During the past ten years, agriculture has lost some of its dynamism and a number of problems have arisen. The chief symptoms of this decline include the need to import large quantities of grains during the last two years, a fall in exports, an increase in rural unemployment, and the existence of unsatisfactory rural living conditions. Generally speaking, these problems stem from two main causes: (1) an actual loss of dynamism as measured by the behavior of certain indicators (such as the indexes of production and employment and the need to import basic grains) during recent years, and (2) the emergence of problems that had remained hidden during the period of agricultural expansion, connected with the uneven pattern of development of the sector. The hidden structural problems are exemplified by the existence and development of a class of commercial farmers who have received considerable government support and have

---

3. The translation is unofficial. Throughout, numbered footnotes were in the original, except those identified as containing comments added by the editors.

managed to achieve higher productivity than the majority of producers, who in turn have clung to their traditional methods of production and whose development has lagged behind the rest of the economy. The basic reason for this difference is that commercial agriculture has developed mainly in the irrigated areas and in rainfed [*temporal*] areas of favorable ecology.

The majority of producers (approximately 70 percent) have not yet received the benefits of irrigation and the farms they work are small. There are also a large number of day laborers whose living conditions are even more modest.

In view of this, the main focus of the new agricultural development strategy is on the solution of two sets of related problems: how to restore dynamism to the sector, and how to even out the inequalities in income distribution that have arisen within it. These two factors are closely interrelated. For example, rainfed agriculture produces the greater part of the national corn supply: therefore, in order to regain self-sufficiency in this cereal under these conditions, it is necessary to raise the productive efficiency and income levels of the majority of the traditional farmers. If, on the other hand, policy concerning irrigated agriculture should be turned toward the cultivation of basic crops (such as corn), this would limit the resources for export crops and the resulting losses in foreign exchange earnings from exports might outweigh the gains achieved by reducing imports. It could also mean a loss of income to both farmers and farm workers. Another example of the interdependence of the objectives of growth and equity derives from the nature of agricultural exports: not only do these provide an opportunity for significantly increasing farming output; at the same time the crops in which Mexico is able to compete favorably on the world market are those that require relatively more labor to produce—that is, that provide more employment for both farmers and farm workers. Tables 6-1 and 6-2 [for ease of reference, tables and figures in the document have been renumbered following the convention used in the other chapters of this volume] give a picture of the income and employment distribution in the sector.

2. *Antecedents*

To find out the causes of the recent loss of dynamism in agriculture, we must examine the factors that in the past helped speed up its development. In macroeconomic terms, three factors determine the growth of agricultural production: expansion of the area cultivated, improved unit yields, and changes in the cropping patterns in the direction of higher-value crops. Public investment contributes directly to increasing cropland and raising

Table 6-1. *Estimation of the Regional Distribution of Agricultural Income, 1968*

Region	Income per hectare (pesos)	Hectares per farm[a] (average)	Income per farm (pesos)
Irrigated			
Northwest	2,359	8.1	19,220
North Central	2,547	2.1	5,270
Northeast	1,621	6.5	10,530
Central Plateau	3,593	2.4	8,791
South	3,227	3.0	9,806
Nonirrigated			
Rainfed	397	3.5	1,393
Tropical	1,347	2.9	3,886
Subtotals			
Irrigated	2,808	3.7	10,508
Nonirrigated	675	3.3	2,226
Total	1,216	3.4	4,136

Note: Estimates include only the income from annual crops.

Source: CHAC solution I–10. This solution corresponds closely to reality as regards cropping patterns and other quantifiable variables.

a. Hectares actually cultivated.

yields. Favorable market conditions strengthen all components of sector activity.

The contributions of the above factors to the sector's growth were as follows: over the 1930–60 period the area of land under cultivation increased at an annual rate of 2.4 percent, bringing a proportionate increase in output; physical yields contributed the equivalent of an annual 1.9 percent; and changes in the cropping patterns contributed 0.5 percent annually.[4] The total of these three components constitutes the sector's annual growth over the period; that is, 4.8 percent. Over the decade of the 1960s, the contributions of all three factors declined: the area of land under cultivation expanded by only 1.7 percent annually, as did yields, while changes in cropping patterns contributed only 0.1 percent.[5] The sector's overall annual growth was thus 3.5 percent, much lower than in the preceding decades.

4. *Editors' note*: These figures were taken from Solís (1970).

5. In spite of the changes in physical production patterns in most agricultural areas, the increase in income per hectare attributable to crop pattern changes has not been very great.

Table 6-2. *Estimation of the Regional Distribution*
of Agricultural Employment and Value of Production, 1968

Region	Employment			Value of production		
	Per hectare cultivated[a]	Per farm[b]	Per unit of water[c]	Per hectare cultivated[d]	Per farm[e]	Per unit of water[f]
Irrigated						
Northwest	2.08	16.95	1.69	5,143	41,904	4,171
North Central	5.07	10.50	2.48	4,120	8,524	2,009
Northeast	3.36	21.78	2.50	3,615	23,467	2,694
Central Plateau	4.52	11.06	7.90	4,652	11,382	8,129
South	7.06	21.46	5.89	6,402	19,436	5,341
Nonirrigated						
Rainfed	1.11	3.88	—	1,377	4,826	—
Tropical	1.65	4.77	—	2,609	7,527	—
Subtotals						
Irrigated	3.58	13.41	3.42	4,811	18,003	4,600
Nonirrigated	1.26	4.18	—	1,737	5,729	—
Total	1.85	6.31	3.42	2,517	8,559	4,600

— Not applicable.
Source: CHAC.
a. Man-months per hectare.
b. Man-months per farm.
c. Man-months per ten thousand cubic meters.
d. Gross value in pesos per hectare.
e. Gross value in pesos per farm.
f. Gross value in pesos per ten thousand cubic meters.

Given the structure of consumption in the Mexican economy, when national product is expanding at approximately 7.0 percent, the supply of agricultural products should grow at a rate of 4.2–4.5 percent in order to satisfy the needs of the domestic market. In fact, agricultural production grew over the 1960s at only 3.5 percent annually, compared with 7.1 percent for the national product. This meant that agricultural exports had to fall, or imports had to rise, or both.

It is true that the penetration of foreign markets has been more difficult lately than during the 1940s and 1950s. However, the sector's recent loss of dynamism stems more from problems of supply than of demand. The determining factor has been the rate of increase in the area under cultivation, which over the past thirty years has fallen from 2.4 percent to 1.7 percent, even reaching zero in recent years. The increase in farmland has been adversely affected by three basic factors: (1) a succession of bad-weather years, which has affected harvests in the rainfed areas;[6] (2) the gradual increase, with the passage of time, in the amount of cultivated land in rainfed areas of less favorable ecology, which, together with terms of trade unfavorable to agriculture [see table 6-3],[7] has deterred any great progress in the cultivation of the remaining marginal land;[8] (3) a decline in the rate of growth of investment in agricultural projects, plus a rise in the amount of investment required per hectare to bring fresh land under cultivation. In fact, the lands suitable for rainfed farming are gradually becoming used up, as are also those that offer the best opportunities for building large-scale irrigation works, which means a higher real unit cost to provide additional irrigation.

The decline in agricultural production has affected subsistence farmers more than commercial farmers, since the slowdown in the expansion of cultivated land has been more marked in nonirrigated areas.[9] Temperate-zone, rainfed [*temporal*] farming and tropical-area farming are more vulnerable to changes in weather conditions than irrigated farming, with the result that it is the lowest-income groups that suffer financial losses from poor weather.

Although it is not possible, from the available information, to determine the precise role of each of these three factors in the slowdown of the increase in area under cultivation, there is indirect evidence of the part played by each. The fact is that no substantial public investments have been made to open up rainfed and tropical land; nearly all such investments having been made in irrigation works. In spite of this, from the beginning of the 1930s to the mid-1960s the increase in nonirrigated land under cultivation exceeded 2 percent a year. And then from 1966 to 1971 the area of nonirrigated land under cultivation fell slightly. Thus this recent decline is indicative of the importance of the first two factors: poor

6. The weather was so bad that a substantial part of the land sown was damaged and was not harvested in its entirety. Official statistics (Ministry of Agriculture and Livestock; Secretaría de Agricultura y Ganadería, SAG) record the area harvested, not the area sown; they do not, therefore, reflect the effect of the weather conditions.

7. In this context the terms of trade refer to the ratio between the indexes of change in agricultural prices and in nonagricultural prices. Throughout the 1950s agricultural prices rose at an annual rate of 5.8 percent in comparison with 7.1 percent for nonagricultural prices. The corresponding figures for the 1960s are 2.5 percent and 3.8 percent. This means that over a period of twenty years the terms of trade have worsened from the point of view of the agricultural sector (see table 6-3).

8. The first stage of Agrarian Reform played an important role in bringing idle *latifundio* land under cultivation through land redistribution, but land reform now has a much more limited field of action in increasing the total amount of arable land.

9. Although not all commercial farmers work irrigated land, it is clear that policies for irrigated areas have a greater effect on commercial farmers. In the same way, policies for rainfed areas have a greater effect on subsistence farmers.

Table 6-3. *Price Deflator Indexes for GNP, 1950–71*

| | | GNP in agriculture and livestock | | | | |
Year	GNP	Total	Agri-culture	Live-stock	For-estry	Fish-eries
1950	50.6	57.5	59.7	55.4	44.9	47.5
1960	100.0	100.0	100.0	100.0	100.0	100.0
1961	103.4	106.5	109.2	100.9	116.6	91.2
1962	106.5	111.5	114.2	104.5	127.6	100.5
1963	109.9	114.8	119.5	102.7	144.8	103.2
1964	116.0	120.9	124.8	110.8	146.7	94.0
1965	118.7	120.4	124.3	109.3	149.0	106.8
1966	123.4	120.8	123.8	111.7	150.2	112.5
1967	127.0	125.3	127.8	117.6	155.4	115.3
1968	130.0	125.3	129.3	114.3	156.3	130.2
1969	135.1	131.2	133.2	125.0	152.4	145.2
1970	141.2	137.4	140.6	127.2	170.8	169.6
1971	148.1	136.5	143.2	121.8	159.6	161.3

Note: The base year for the indexes is 1960, at 100.

Source: Ministry of the Presidency [now Ministry of Programming and Budgeting, Secretaría de Programación y Presupuesto], Information System for Economic and Social Programming.

weather and use of increasingly marginal lands where cultivation is sensitive to fluctuations in weather and the terms of trade.

Public investment in land still has the potential to play a critical role in the sector's development. A new impetus can be given to expansion of the area under cultivation by directing considerable amounts of public investment toward rainfed and tropical areas for soil conservation, land clearing, and drainage works. Similarly, the terms of trade between the farming sector and the rest of the economy can be improved so as to encourage private investment for the opening up of new land. Specific programs will be commented on later in the document.

The slowdown in agricultural investment can be seen from the following figures [see table 6-4]. From 1951 to 1961, agricultural investment grew at an annual rate of 10.7 percent (at current prices), whereas for 1961–67 the rate was only 8.6 percent. This situation is reflected also in a change in agriculture's share of that investment (public and private): from about 18 percent at the beginning of the 1950s, this ratio had already fallen to approximately 10 percent by the mid-1960s. Private investment has decelerated relative to public investment, a fact that is indicative of the importance of price incentives in the sector. However, the relative decline in private investment also was due to inadequate public expenditure on necessary infrastructural works, such as rural roads, research and extension work and dams. Although in recent years the growth of public investment has been more rapid in agriculture than in other sectors, from

Table 6-4. *Total Gross Capital Formation*
(millions of current pesos)

| Years | National total (1) | Agriculture and livestock | | | Relative sectoral participation (4)/(1) |
		Private (2)	Public (3)	Total (4)	
1950	5,962	655	526	1,181	0.198
1951	7,790	1,140	621	1,761	0.226
1952	10,089	733	568	1,301	0.129
1953	9,411	894	501	1,395	0.148
1954	12,659	1,739	615	2,354	0.186
1955	15,953	2,108	597	2,705	0.170
1956	20,051	1,388	639	2,027	0.101
1957	21,078	1,622	665	2,287	0.109
1958	21,565	2,575	696	3,271	0.151
1959	22,207	1,581	714	2,295	0.103
1960	30,205	3,772	524	4,296	0.142
1961	29,289	3,003	909	3,912	0.134
1962	29,260	2,738	1,661	4,399	0.150
1963	37,820	3,379	1,340	4,719	0.125
1964	46,295	4,233	2,105	6,338	0.137
1965	50,143	3,581	1,105	4,686	0.093
1966	61,189	4,129	1,629	5,758	0.094
1967	66,045	4,197	2,341	6,538	0.099

Source: *National Accounts*, Banco de México, table 145.

1951 to 1966[10] the reverse was true: public investment in the farming sector grew at an annual rate of 8.7 percent compared with 13.5 percent for total public investment (both figures at current prices).[11] Since public projects generally have a long lead time, this lag in public investment in farming has continued to affect the behavior of the sector up to the present time.

So far we have examined only the main factors to which the decline in expansion of the area under cultivation may be attributed. Turning to other factors determining the overall behavior of the sector, it may be seen that aggregate yield increases have been about as rapid during the 1960s as they were for the 1930–60 period. However, the contribution to growth given by changes in the cropping patterns declined. This is explained mainly by the behavior of cotton production, which increased rapidly up to the early 1960s and then declined. Cotton production was 17 percent lower in 1969–71 than during 1960–62; this change practically offset the rapid increase in the production of fruit and vegetables. The cotton

10. This calculation is based on the average of the years 1950–52 and 1965–67.
11. These figures relate to authorized investments. Obviously, with growth running higher in the nonagricultural than in the agricultural sector, a larger proportion of investment has to go to the former. However, the differences in sectoral growth rates do not match the very wide discrepancy in sectoral investment rates.

Figure 6-1. *Estimate of Sector-wide Seasonal Employment in Short-cycle Crops, 1968*

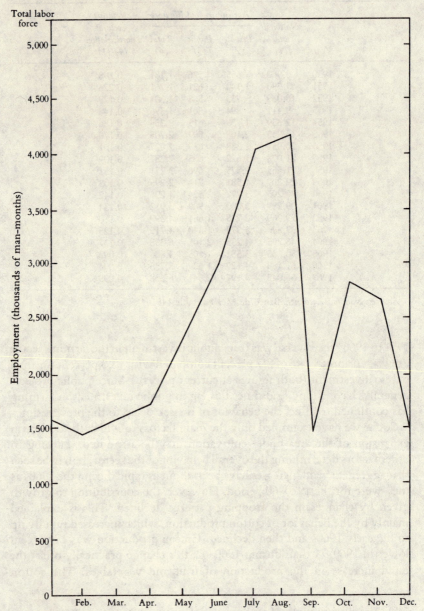

Source: CHAC.

problem is basically one of supply, since the world market could have absorbed larger quantities than were exported. Cotton growing has become less profitable, mainly because of the increase in labor costs, and also there have been serious problems of pest infestations in the northeastern growing areas. At the same time, the possibility of turning cotton lands over to other crops has become more attractive because their prices have risen in relative terms.[12] On the other hand, if the arable land expansion had been faster, the advantage to the farmer of replacing cotton by other crops would not have been as great; in other words, during a period of general scarcity of agricultural products brought about by insufficient increases in such basic factors as the amount of capital invested in opening up land to cultivation and irrigation, the real marginal values of the limiting factors rise substantially. This puts greater pressure on producers to choose crops that are relatively efficient in the use of the limiting factors. Thus, the problems of an individual crop cannot be separated from the behavior of the sector as a whole.

A review of employment in the sector reveals that: (1) during the 1950s employment opportunities increased, generally speaking, as rapidly as the working-age population, so that there was no increase in either unemployment or underemployment; (2) during the 1960s available jobs increased more slowly than the working-age population; and (3) at the present time about 40 percent of the potential work force is not being used effectively. A number of studies have yielded the same conclusion with regard to the rate of agricultural underemployment.

There are both seasonal and regional aspects to this problem. About 30 percent of the work force is productively employed more or less throughout the year—60 percent for only one to five months, and about 10 percent only occasionally [see figure 6-1]. The regional aspect of the problem is illustrated by the following figures: irrigated agriculture accounts for only about 16 percent of cultivated land, against 30 percent of total production. Studies based on data collected at farm level in all parts of the country show that employment per hectare triples when irrigation water becomes available;[13] in other words, irrigated land, with only 16 percent of the cultivated area,[14] generates 36 percent of employment.

12. The programs of guaranteed prices for some of the grains and oil-bearing plants have provided additional incentives to switch from cotton by eliminating the risk of price fluctuations in these crops.

13. This figure is derived from the CHAC sectoral model and is based on the monthly labor requirements of all main short-cycle crops, given the prevailing cropping patterns in all regions. This model is based mainly on microeconomic data; that is, farm-level production ratios by crop, technology, and region. With this structure, macroeconomic results can be derived for the sector from microeconomic data.

14. Irrigated agriculture also makes more intensive use of machinery per hectare and per unit of employment. Irrigated agriculture nevertheless generates more jobs per hectare than does rainfed agriculture.

In short, the sector's capacity to absorb the rural work force has declined. According to the census data, the product elasticity of the remunerated economically active population of the agricultural sector, which measures the percentage increase in that population attributable to a percentage increase in gross agricultural product, fell from 0.28 in the 1950s to 0.11 in the 1960s, reflecting the technological change in which capital replaced labor.[15]

The unemployment problem is reflected in a maldistribution of income. This can be seen from table 6-1, which shows the sector's income distribution by region and by farm type (irrigated or nonirrigated). These figures have been derived from the CHAC sectoral model and include only income from annual crops. Estimates by various sources agree that income is very unevenly distributed, mainly owing to the unequal distribution of physical resources. [See figure 6-2.] Because commercial farmers adapt more rapidly to new crops and new production methods, this unequal distribution of income appears to be worsening.

These income distributions relate only to farmers; if landless farm workers are included, the distributions are seen to be even more inequitable. The poorest segment of the agricultural labor force comprises two groups: about 1.5 million smallholding farmers (ejidatarios and private smallholders), whose resources are insufficient to enable them to earn an income above subsistence level; and about 2 million landless farm workers, who own no property at all and enjoy no common land rights. These people, living on or outside the fringe of the economy, together with their families make up a quarter of the country's entire population.

Studies carried out with the sectoral model indicate that, according to the main employment trends and unless great efforts are made, unemployment will not decline until the 1990s; however, this assertion cannot be made with certainty because of the unreliability of some of the census data relating to the number of people of working age. The sectoral model also shows that the disparity in income between farmers on irrigated and on nonirrigated land has continued to widen over the past twenty years. The challenge at the present time is how to increase manpower absorption and at the same time narrow the income gap between the two categories of farmers.

3. Development objectives

It may be concluded from this outline that agriculture's main areas of concern are four: employment, output, net foreign exchange earnings,

15. In view of the problems of incompatibility between the censuses taken in different years, it is believed that the product elasticity of the remunerated economically active population of the agricultural sector is not, in fact, as low as these figures suggest. The negative trend of this elasticity is, however, probably correct.

Figure 6-2. *Lorenz Curves for Distribution of Annual Net Agricultural Income from Short-cycle Crops and for Use of Irrigation Water*

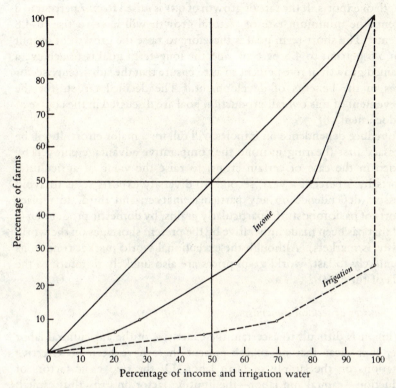

Percentage of farms

Percentage of income and irrigation water

Income

Irrigation

Source: CHAC.

and an adequate income for the least-favored stratum of the population. These areas of concern are adopted as the targets for the present program, for they fit within the economy-wide economic and social development strategy, which has as its principal objectives raising the level of productive employment, increasing the general level of income while improving its distribution, and reducing dependence on imports.

In agriculture, increasing employment basically means providing new jobs to absorb the annual increase in the rural labor force and gradually reducing the existing underemployment. Improvement of the income distribution will be sought mainly through an increase in the level of productive employment among smallholders and landless laborers. Reduction of dependence on imports calls for strengthening the sector's foreign trade structure to make it more viable and, above all, increasing agriculture's net export earnings to provide foreign exchange for the imports required for the country's development.

As already stated, the studies made indicate that to be able to meet domestic consumption needs, and to keep employment growing faster than the labor force, calls for a minimum production growth rate for the sector of 4.2 percent.[16] This rate also provides for an adequate rate of growth of exports. If the rate of growth of GNP is raised from 7 percent to 8 percent, the minimum rate of sectoral growth will have to rise to 4.8 percent.[17] The short-term goal is therefore to raise the growth of output from 3.5 percent to 4.5 percent, and the long-term goal is to achieve a sustained growth of five percent so as to ensure that the subsistence sector shares in the benefits of development.[18] The detailed targets for the achievement of this overall production goal are discussed in the course of the document.

To reduce dependence on imports will call for a major effort. It will be necessary first (bearing in mind the comparative advantage enjoyed by Mexico in the case of certain crops) to raise the value of agricultural exports by 5 percent a year; second, to diversify exports so as to avoid excessive dependence on any particular market; and third, to replace imports of basic foodstuffs, particularly grains, by domestic production (a need that has been made imperative by the present shortages on the world foodstuffs markets). Although the exceptional world market conditions are unlikely to last, world grain prices are also unlikely to return to the levels of the 1960s.

4. *Sources of sectoral dynamism*

While it is difficult to ascertain the exact size of the agricultural labor force, it is clear that up to the present there have been no effective limitations on the supply of farm labor. Of the two basic factors of production—capital and labor—the limiting factor on growth is capital; labor is not expected to become a limiting factor until the 1980s at least.[19] In agriculture, capital takes many forms: land improvements (clearing, leveling, and so forth); irrigation works, herds, fencing, silos, orchards, and the like; plus the capital invested in training research and extension workers and other agricultural technicians. The following sections contain comments on capital requirements and the importance of each category of capital.

16. The figure of 4.2 percent is, in fact, the *minimum* rate, since overall growth of 7 percent calls for an agricultural sector growth of 4.2 percent to 4.5 percent, as already mentioned.

17. These estimates have been derived from the sector model, CHAC, which takes into account income and price elasticities for all important crops.

18. Slow development of the farming sector is reflected in difficulty in meeting domestic market demand. This in turn tends to lead to higher prices for farm products, which benefit the commercial farmers. However, because weakness in growth of production results in inadequate increases in employment, subsistence farmers and landless laborers receive no benefit at all from the price increases.

19. The employment picture is discussed in greater detail in section 5, below.

GLOBAL ANALYSIS. Global analysis of the country's agricultural growth shows the sector's capital needs in terms of its share of total fixed investment. As already stated, this share has fallen in the last five years. Simple arithmetic from macroeconomics shows that, to achieve a sectoral growth rate of 4.8 percent, the sector's share of total fixed investment will have to rise to 11.5–12.5 percent. Since private investment in agriculture plays a relatively smaller role than in the other sectors of the economy, agriculture's share of public investment will have to be even higher, in excess of 20 percent. Such as increase appears to be necessary to enable agriculture to regain its former dynamism. In addition to channeling more public investment to the sector, it will clearly also be necessary to apply indirect policies to make returns on private investment in agriculture more attractive.

Priorities by investment category may be examined within the general context of the three basic components of sectoral development: expansion of the area under cultivation, increased yields per hectare, and changes in the cropping patterns. The projections show that no significant change is to be expected in the structure of value of production over the next five years. Because a number of incentives are planned, grain production will probably grow more rapidly than in the past, reducing the present trade deficit in grains and maintaining grain production's output share despite the rapid increase in the output of fruits and vegetables. This would leave as the only potential sources of development in the near future an expansion of the cultivable area and an increase in unit yields. These two factors will have to provide a combined annual growth rate of 4.5 percent.[20] In the short term, and even with a growing investment in research and agricultural extension, it would be difficult to raise the rate of growth of yields above the historical levels of 1.7–1.9 percent.[21]

Assuming that 2 percent annual growth in unit yields can be achieved, the area under cultivation will have to expand by 2.5 percent in the short term. Over the long term, to achieve the growth target of 5 percent, either land area or unit yields will have to grow more rapidly even though, by the end of the decade, the composition of production will again begin to make a slight contribution. Since the opening up of new land to cultivation is becoming more expensive, it seems that raising unit yields is inevitably going to be more important than in the past.

Efforts to improve marketing machinery, changes in price policy, and

20. To ensure that the minimum rate of 4.2 percent is achieved, it will be necessary to plan for 4.5 percent.

21. This figure refers to average yields for all agricultural areas of the country (both irrigated and rainfed). It therefore includes the contribution of increased irrigation to raising the overall average yield.

so on will have no impact on sectoral supply unless they bring about changes in one or another of the basic factors of production; that is, an increase in land under cultivation, a rise in yields, or changes in the crop composition.

These reflections suggest a two-stage strategy. In the first stage, over the next four or five years, the area under cultivation will have to be expanded rapidly by means of irrigation, drainage, land clearing,[22] and soil conservation. In the meantime, great efforts will have to be made to strengthen the agricultural research and extension program. This will ensure that production will not be affected by a diminution in the rate of farmland expansion after 1978, when increases in yields and changes in crop composition will play a more important role. This means that human resources will begin to assume increasing importance in the sector's development. Greater emphasis on human resource investment will foster higher yields from the investment in physical capital and thereby attract more private investment to the sector. Human resource investment belongs traditionally to the sphere of public investment, and the prevailing low returns on physical investment in agriculture may be attributed to the insufficient public expenditures on agricultural education, research, and extension over the past twenty years. This lack of proportion between physical capital and human resource capital (typified by the low rate of investment in extension services) has reduced the return on physical investment and has also led to a situation in which human resource investment should show a higher return than physical investment.

A 2.5 percent annual increase in cultivated land means a yearly expansion of about 375,000 hectares (though not all in the form of public investment programs). Until the late 1960s, private enterprise and the agrarian reform programs were increasing the area of nonirrigated farmland by about 2 percent a year (that is, about 250,000 hectares annually during the 1950s and early 1960s). As already stated, some deterioration in agriculture's terms of trade, plus bad weather conditions, have lately reduced this figure to zero. With a price policy that offers greater incentives, as proposed later in this document, private enterprise may be expected to bring into cultivation about 100,000 hectares of new land each year, though not all of this land will be of the best quality; this will reduce the public investment target to 275,000 hectares a year. (This figure does not include irrigation of land at present cultivated without irrigation. It relates solely to new land in the real sense;[23] that is, irrigation of arid land,

22. In programs of land clearing, it is essential to avoid destruction of woodlands where they are more valuable than the crops that will replace them.

23. Irrigation of existing rainfed land helps greatly to raise yields but does nothing to increase the area under cultivation.

clearing and leveling for cultivation, and reclamation by means of soil conservation and drainage works.) By the end of the 1970s this annual target can be permitted to decrease, in accordance with the parallel effort being made in research and extension. The importance of irrigation projects to both production and employment targets should not be underestimated. In the first place, irrigation and water-control projects are important in bringing under cultivation land that otherwise could not be cultivated. In the second place, irrigation greatly improves the productivity of land already in cultivation, making it possible to effect changes in the direction of higher-value crops, to increase yields, and in some cases to obtain two harvests within twelve to eighteen months. The CHAC model was used to evaluate these effects at the sectoral level, and it showed that the value of production per hectare can be increased threefold by irrigation, with a threefold increase also in employment and a fourfold increase in net income per hectare.[24] Average net annual income per hectare (gross value of production less cost of production) for the country as a whole is 749 pesos on nonirrigated land and 3,117 pesos on irrigated land (at 1971 prices).[25] This difference of 2,368 pesos per hectare annually is indicative of the net return on irrigation and ancillary investments, such as agricultural research centers. At the national level, using 8 percent as the true rate of discount, we arrive at a present value of income flow of 29,600 pesos per hectare higher in the case of irrigated land. Using a discount rate of 10 percent, this flow is greater by 23,680 pesos, whereas for 13 percent the excess is 19,733 pesos.[26]

Since these calculations are national averages, it is possible to lay down guidelines for cost-benefit analysis for irrigation projects. Investments, including irrigation works and other associated infrastructure works up to a conservatively estimated cost limit of 20,000 pesos per hectare, appear to be entirely justified by the size of the benefits they bring. Indeed, considering benefits arising from the increase in employment, even higher investment costs per hectare may well be justified. For employment, the reasoning is as follows: according to the model, the addition of irrigation to 1 hectare of land provides fifty-one additional man-days of employment a year. It is assumed that, if a person is not working, society has to pay out at least 10 pesos a day to meet his basic subsistence needs (food, medical services) and those of his family. This figure of 10 pesos a day may be regarded as money saved by society as a whole when the worker is

24. See table 6-2.

25. For the purposes of this calculation, it is assumed that taking on paid labor represents a cost to the producer. The effects on employment are estimated later.

26. These discount rates are in real terms (constant prices) and thus correspond to a higher rate in current terms.

provided with employment. In the case of irrigation projects, therefore, these savings amount to 510 pesos a year.[27] A discount rate of 8 percent means a benefit flow of 6,375 pesos. Adding these figures to the previously stated benefits, we obtain a total benefit per hectare under irrigation, at 1971 prices, of 35,975 pesos, 28,780 pesos, and 23,983 pesos, respectively, for the three real rates of discount. Again, estimated conservatively, the side effect on employment justifies projects costing up to about 24,000 pesos per hectare at 1971 prices,[28] discounted as of the date on which the irrigation works are brought into use.

Although the cost of irrigation projects has risen as the successive irrigation of the more suitable areas has proceeded, the national average cost of recent works per hectare[29] does not appear to have exceeded 20,000 pesos per hectare at 1971 prices. At this cost, irrigation projects must be carried out as quickly as possible, with the emphasis on small irrigation projects, which have a more marked effect on income redistribution.[30]

The main assumption underlying these cost calculations is that, on average, the crop patterns in the new irrigation areas would not emphasize high-value crops more heavily than do the present irrigated areas. This is a conservative assumption that tends to underestimate the real benefits of irrigation. However, in calculating the cost of the project, due regard must be paid to the fact that yields from newly irrigated lands are generally lower than the national average during the first three years of useful life of the project. This loss of potential profits has to be included in the project cost to allow the previous benefit figures to be compared for the purpose of a cost-benefit analysis. This increase in costs generally adds 2,000–5,000 pesos in discounted flow to the total costs of the project. On the other hand, it has also to be borne in mind that these calculations are derived from a model in which it has been assumed that the paid labor force

27. These are savings in transfer payments and are not, therefore, benefits of the same type as the increases in net income per hectare, which reflect an improvement in the productive efficiency of the sector. However, since the aims of the plan include both efficiency and equity, the benefits under the heading of employment are included in the analysis. From the point of view of methodology, the two effects in efficiency and equity cannot, perhaps, be added together.

28. All these calculations are based on the assumption that current inputs (fertilizer, seeds, and so forth) supplied to the new irrigation areas will be subsidized to the same degree as at present. However, this is not an important assumption in quantitative terms: changes in the rate of subsidy of inputs would represent, at most, 100–200 pesos per hectare (that is, 3–6 percent of annual net income flow).

29. The average cost of recent works represents the marginal cost of irrigation over time.

30. These recommendations imply that all ancillary inputs, such as extension credit, and the like, will be supplied in adequate quantities, a factor that may largely determine the success of the physical works. Similarly, we must not overlook the need to solve institutional problems that affect production in such areas as land tenure and organization of production.

receives the current wage at market price, and that the farmers receive a significant remuneration for their own labor in addition to an economic rent on their labor, land, and endowment of irrigation supplies. In other words, these calculations of the return on investment in irrigation are not inflated by an assumption of unrealistically low wages.[31]

As regards investments in research and extension, the balance between research and extension requirements varies according to crop and ecological area. In the case of maize, methods already known to the research centers in most parts of the country provide yields several times higher than do the more common farming practices. For this reason, the most pressing need is not for more research but for more extension facilities (together with the necessary ancillary agricultural inputs such as insurance, fertilizer, improved seed, and credit).[32] However, in order to adapt certain oil-bearing plants to rainfed farming conditions, intensive research is required in combination with agricultural extension. There is a priority need for research on the combating of cotton pests in the irrigation districts of the Northwest and on the development of suitable varieties of fruits and vegetables for export.

In general terms, what constitutes an adequate extension effort may be determined as follows. At the present time, agricultural extension adequately serves about 2.5–3.0 percent of the farmers each year. At this rate the entire sector receives attention only once every thirty-three or forty years. However, the introduction of new crops and new farming methods, the need to update the farmer's technical knowledge, and the fact that the composition of the work force is constantly changing mean that the sector should be completely "serviced" every ten years at most.[33] In other words, in budgetary terms the target should be a three- or fourfold expansion of the present program of investment in extension services.

Although it is difficult to measure the quality of the extension service, interviews with farmers and with various Mexican credit institutions leave little doubt that there are still a great many ways in which it could be

---

31. The commonly adopted assumption of a zero wage substantially overestimates investment returns, since it is difficult to sustain the argument that a supply of labor exists at a price (wage) of zero.

32. Input packages are discussed in greater detail in later sections; here the focus is on the requirements of human resource capital. It should be emphasized that little is yet known about the determinants of the behavior of the small farmers with respect to the adoption of new farming methods.

33. This does not mean that each farmer should receive only one visit every ten years, but that extension should be a more or less continuous process. The figures of ten years and thirty-three to forty years are used here to illustrate the need for expansion of the extension network.

improved. Specifically, this calls for: (1) the coordination of the work of
the research and extension professionals;[34] (2) better working conditions,
including the provisions of an adequate number of vehicles, to enable
extension workers to make more frequent visits; and (3) the allocation of
more funds to the training of extension workers. These improvements
will entail additional expenditure over and above that required to achieve a
three- or fourfold expansion of the extension service.

As regards the focus of the extension service, the present tendency to
place major emphasis on rainfed agriculture should be reinforced, for two
reasons: (1) the bulk of the grains are grown in these areas, so that small
increases in yields have a significant effect on national production, and (2)
this being the least prosperous segment of agriculture, the income redistribution will be achieved more easily if it is given priority.

It is difficult to quantify the requirements of the research program. It is
quite clear, however, that to maintain a suitable interrelation between
research and extension, the research program will have to be appreciably
expanded, though not as rapidly as the extension program. In view of the
enormous diversity of Mexican agriculture, the specific regional programs will have to be strengthened. A considerable increase is needed also
in the number of farming schools and technical study centers. These
schools exemplify that extension services can be provided in various ways.

To ensure that research and extension will be truly effective, a program
directed toward a substantial increase in investment in agricultural human
resource capital should pay due regard to the importance of a properly
coordinated distribution of agricultural inputs. Other specific input programs calling for more effective coordination at the regional level include
fertilizers, credit, agricultural insurance, and crop purchases at guaranteed
prices.

As noted previously, the change in crop composition was one of the
dynamic factors of agricultural development during the 1950s but had
relatively little impact in the 1960s. One of the objectives of agricultural
policy now is to structure production in such a way that the criteria of
comparative regional advantage predominates, or to eliminate factors that
tend to inhibit such a structure from emerging naturally.

The studies carried out using the sectoral model CHAC point to a possible
static increase of 11 percent in the value of agricultural production, given
more rational use of resources and a shift in production structures toward
crops that enjoy comparative regional advantages, as determined by regional resource availability and ecological conditions. The transfer of

34. This would mean providing extension workers with intensive courses on the results of
research in fields of experimentation and on methods of transmitting technical knowledge to
the small farmer.

sugarcane production from the central and northwestern irrigation districts to the tropical areas ranks high among such changes. The productivity of resources in those districts would be increased by more crops such as vegetables, fruit, and cotton. To encourage this change is a long-term policy direction that can be promoted by locating investment in the construction and expansion of sugar mills (in conjunction with improvements in crop yields) in the tropical areas and discouraging them in the above-mentioned irrigation districts. The temperate rainfed areas offer a comparative advantage for the production of sorghum, corn, beans, and oil-bearing plants, even though in rainfed farming absolute yields are lower than under irrigation.

Investment in research to identify the most suitable regions and varieties, and expenditure on agricultural extension, will be among the means used to help bring about these and other appropriate changes in crop patterns. Research into suitable microclimates in the irrigation districts of the central area will make it possible to move some cotton growing to regions where labor costs are lower.

To conclude this outline of the dynamic factors involved, mention should be made of the trends in the distribution of agricultural income between irrigated and rainfed areas. Using the sectoral model, projections for distributional analysis have been made, subject to certain assumptions that facilitate analysis; namely, that irrigated and rainfed lands expand at the same rate and that yields per hectare show similar increases in both cases. The results of these experiments indicate that the income of producers in irrigated areas increases about 35 percent more rapidly than that of the rainfed farmers, even with the above assumptions. This reflects the fact that irrigated areas take a larger share of the export markets (which tend to be more dynamic than the domestic markets) and that farmers of irrigated land respond more rapidly to an increase in both domestic and external demand because their unit costs of production are lower. These projections illustrate the magnitude of the problem confronting distributional policies, since they probably reflect actual distributional trends fairly accurately.

PRICE POLICIES AND AGRICULTURAL SUPPLY. The price policy for agricultural products has not changed significantly since the 1960s. Seven products are receiving support in the form of government-guaranteed prices but most of the price levels have not changed and the coverage of this price-guarantee program has been restricted mainly to commercial agriculture. In the latter half of the 1960s, CONASUPO's[35] net operating

---

35. *Editors' note:* CONASUPO (Compañía Nacional de Subsistencia Popular) is the national agency charged with administering the price-support program.

deficit was equivalent to only 6 percent of the aggregate net income from annual crop production.

In view of the difficulties faced by small farmers in transporting their products to the official receiving centers, CONASUPO has recently begun to place more emphasis on extending its price support programs to more remote localities. However, this new development has been limited to some extent by the insufficient increase in maize production, which has meant that market grain prices have risen above the guaranteed price in most regions and have thus tended to reduce the impact of the CONASUPO program there.

The guaranteed price policy is at present under review. It should be borne in mind that pricing policy can help to direct agricultural production along lines consistent with the patterns of long-term comparative advantage between regions and also in relation to the rest of the world. To encourage agriculture to follow different lines would cause distortions that would hinder development.

The idea of differential price support between irrigated and rainfed areas is consistent with a regional development strategy. Studies with the sectoral model have shown that maize, sorghum, beans, and most oil-bearing plants enjoy a marked comparative advantage in regions of natural rainfall (as opposed to irrigated regions), an advantage that is shared by wheat in some rainfed areas. This means that, even though unit yields of these crops are lower in the rainfed than in the irrigated areas, there is a net gain by freeing irrigated land for other crops.

Turning to maize, it is noteworthy that Mexico has ceased to be a large exporter (495,000 tons exported annually on average during the 1960s) and has had to import considerable amounts. Part of the need for imports can be explained by unusually unfavorable weather conditions and by rising demand, which is outstripping the growth of supply.[36] What is the reason for this trend? Why have market conditions not restored the balance between supply and demand through adjustments in relative prices? At least two principal factors have been responsible. First, during the first half of the 1960s the guaranteed price of maize was probably above the equilibrium market price, with the result that production was stimulated compared with other crops. But toward the end of that decade, the position changed in most of Mexico. Second, the slower rate of increase in the area of cultivated land after 1967 limited the supply position for all crops and in some areas encouraged farmers to switch to more profitable crops than maize. To sum up, relative prices within the production sector, and the general supply position, were the principal factors affecting production.

36. An increasing proportion of the demand for maize is because of its use as fodder; its overall elasticity of demand is therefore slightly higher than expected: approximately 0.6.

The quantitative importance of these two factors has been shown by projections made using the sectoral model. In the long term, if overall sectoral production increased by at least 4.2 percent annually, the natural market pressures would cause maize production to rise at a sufficiently rapid rate to supply the needs of the home market. If on the contrary overall production increased at a slower rate, the production of maize would be insufficient. (In the near future, maize production must increase even more rapidly, by approximately 5 percent annually, in order to make up for the recent shortages and to raise production again to the point where supply and demand are in equilibrium.) It therefore appears that in the long term a healthy rate of growth of the sector as a whole will allow the deficiency in maize to be made up.

Solely from the point of view of efficient allocation of resources in the short term, increasing maize production would mean sacrificing other crops of which Mexico is an efficient producer and would probably lead to a reduction in other potential exports. In any case, this argument loses its force in the long term, and the problem can be resolved by giving sufficient attention to the factors determining increases in land area, in water resources, and in investment in improved technology.

"Efficiency," however, is not the sole objective of national agricultural policy; job creation and redistribution of income are more important goals than efficient production. Since maize is practically the only crop grown by the majority of subsistence farmers, a higher price for it would certainly serve the aim of income redistribution,[37] particularly if a differential price system can be maintained, with higher prices in areas where there is a larger concentration of rainfed farming.

Similarly, the argument against self sufficiency in maize production is based on considerations of a static comparative advantage and, although Mexico at present probably does not offer a comparative advantage for maize production compared with other crops, recent experience proves that small changes in maize-growing methods could result in significant increases in unit yields. These techniques are already being applied in some parts of the country, and a higher support price for maize will act as a strong incentive to their diffusion throughout other farming areas.[38] If the improved techniques are widely adopted, Mexico will be able to reduce or even eliminate its comparative disadvantage in maize growing. Guaran-

37. Where reference is made to "new production techniques," this means new seed varieties, new methods of cultivation, fertilization, and so forth, but it does not include increased mechanization. As discussed later, although mechanization makes it possible for medium- and large-scale producers to increase their profits, it significantly reduces employment of labor.

38. It must be remembered that the new techniques sometimes involve greater risks to the farmer than the traditional methods. Where this is true, higher prices do not necessarily provide sufficient incentives to adopt the new techniques.

teed prices therefore will be used to encourage technological change for maize production in the rainfed areas, which are the slowest to adopt new techniques but which, at the same time, have a comparative advantage in relation to the country as a whole.

From the foregoing it may be concluded that, at least during the next few years, the price of maize will have to be maintained at a sufficiently high level to stimulate its production in comparison with other crops.

One of the points for consideration when fixing the new levels of guaranteed prices could be to make the near-term price increase only sufficiently great to overcome the shortfall in present production, allowing future growth needs to be satisfied by more rapid overall development in this sector. The present national shortfall in maize production is probably about 1.5 million tons. It is also proposed that the deficit should be made up by increasing production in rainfed areas only as far as this is possible. Studies using the sectoral model provide estimates of crop supply elasticities, both in rainfed areas and in nonirrigated zones, affording a basis for revising the guaranteed prices over time as circumstances change.

Guaranteed prices act as a powerful stimulus to crop substitution and obviously must be planned in a way compatible with the goals for long-term production patterns. At the same time, account must be taken of the attitude of producers to crop substitution in response to relative changes in prices. Although a more detailed analysis has not been made as yet, it would be useful to examine the possibility of applying a policy of guaranteed prices to oil-bearing plants, in the rainfed areas only, since they enjoy a comparative advantage in those regions. Similarly, consideration might be given to the use of a variant of the guaranteed price system to encourage production of export crops. It should be remembered that, while it is true that fruit and vegetables are showing the fastest rates of development on both the home and foreign markets, they are also the crops that carry the greatest risks for the producer of market price fluctuations, and yet they receive no support.

In view of the natural variations in production and in prices on the vegetable and short-cycle fruit markets, it would be inappropriate to try to maintain constant prices over a period of several years. However, it might be worth looking into the possibility of offering incentives to production through pricing that would vary from year to year but would be fixed for each year. The important thing is that farmers should be able to know for certain, before sowing time, what the harvest price will be, so they can make their production plans free from uncertainty about prices. To maintain supply and demand balance over time, the price would be adjusted each year, sometimes upward and sometimes downward. Uncertainty about prices, however, would no longer be a factor influencing the producer's decisions. It is stressed that this kind of price support is not intended to raise the general level of prices, but simply to eliminate excessive fluctuations. The goal of the program would be to stabilize prices in accordance with their natural trend and not to affect the trend itself.[39]

Farmers face three main kinds of uncertainty: the risk with regard to the actual physical yield, which is governed partly by variations in the availability of inputs, such as rain or irrigation water; market risks (prices); and uncertainty about legal provisions (for example, concerning land tenure). Farm insurance is designed to cover the first kind of risk, while the programs of the Department of Agrarian Affairs are directed toward eliminating the third. Guaranteed prices help to reduce the risks of the second category, but at present they affect only a very small number of crops.

For input prices, the complex decisions that a producer must make—in face of the varying yields from different crops and varying product and factor prices—may result, when the cost of resources is not in line with social conditions, in underutilization of resources or economic waste.

An example of such a policy is to be found in the rates charged for water in the irrigated areas. At the agricultural sector level, this policy has resulted in inefficient allocation of the most scarce resource, water. In some districts the rates charged per unit of water are much below the marginal productivity of this resource.

The results obtained from the CHAC model indicate that the water demand is rather inelastic at the present price of irrigation water, so that an increase in the price has no significant effect on production levels. Pricing policy for this input could be used as a means of redistributing income in cases where an increase in government revenues would make it possible to finance development programs in rainfed areas.[40]

In the case of other inputs such as fertilizer and improved seed, present prices do not appear to cause any problems. Rather, the limiting factors have been the unavailability of sufficient credit and, particularly, the lack of information about the proper use of chemical inputs, which could, of course, be overcome by strengthening the agricultural extension service.

AGRICULTURAL EXPORTS AND BEHAVIOR OF THE SECTOR. The rapid growth of agricultural exports during the 1940s and 1950s not only

39. *Editors' note*: This proposal regarding the use of support prices may appear quite conventional and obvious, but in Mexico price supports customarily have not been adjusted in a downward direction.

40. If the prices of gravity-fed water were raised, it would probably be necessary to make exceptions in favor of small farmers cultivating holdings of less than 5 hectares to ensure that the aim of income redistribution was achieved.

supplied the foreign exchange needed by the economy but also provided a strong stimulus to the growth of the agricultural sector itself. Foreign markets offer possibilities of expansion far beyond the limits of the home market. Moreover, production for export did not have the depressive effect on farm prices that is caused by increased production for the home market. The main exports during the 1950s were sugar, cotton, oranges, coffee, and tomatoes.

As already stated, the decade of the 1960s was a period of slow growth of exports, due mainly to supply problems but partly to the difficulties of breaking into new markets. Analysis of the recent changes in the composition of agricultural exports indicates that the decade of the 1960s was a period of transition leading toward a new structure of exports and that the behavior of agricultural exports will probably be more dynamic in the future. The first indication of possible future changes in the virtual disappearance of grain from total exports and the increase in the share of fruit and vegetables, coffee, tobacco, and cocoa; that is, the share of total exports occupied by crops with brighter prospects is now greater.

Although in the early and mid-1960s maize and wheat became very important earners of foreign exchange, their recent falling-off has contributed greatly to the overall decrease in agricultural exports. Table 6-5 shows their share of the total value of exports. One factor that encourages optimism is that the most active products, with the best prospects, are beginning to take a larger share in total exports, which is a hopeful sign that exports may regain the lost ground.

Another important structural change is that the composition of exports is moving toward products with a high yield (both in quantity and value) per hectare and per irrigation unit. This means that larger foreign currency earnings can be achieved today than in 1968 for a smaller area in export crops, and this in turn means less competition between the objectives of supplying the home market and of increasing foreign currency earnings.

Some of the "new" export crops have also shown a fairly high rate of increase in yield per hectare, considering that the farmer has had little past experience with such crops.[41] This trend is expected to continue in the future, so that a rapid expansion of exports will not be greatly dependent on increasing the acreage under export crops.

The studies using the sectoral model show that the behavior of the most dynamic exports is not attributable simply to fortuitous conditions on the international market, but to the fact that Mexico is an efficient producer of

41. For example, the percentages of annual increase between 1953–57 and 1965–69 in yields per hectare for certain selected crops are as follows: maize, 1.9; garlic, 4.7; melons, 6.8; tomatoes, 6.9; strawberries, 15.5; onions, 5.9; cotton, 4.3; sugar, 1.5; and eggplant, 5.6. The six crops with the highest rates of growth are among the most dynamic exports.

Table 6-5. *Value of Crop Exports*
(thousands of current pesos; farm-gate prices)

	1968		1971		1972	
Crop group	*Value*	*Percent*	*Value*	*Percent*	*Value*	*Percent*
Grains	1,014,446	19	314,840	8	442,700	8
Cotton	2,065,710	39	1,254,002	31	1,639,369	30
Coffee	726,651	14	767,761	19	1,214,809	22
Sugarcane[a]	561,021	11	546,185	14	579,512	11
Short-cycle fruits and vegetables	541,186	10	893,012	22	1,106,962	20
Plantation fruits	66,037	1	60,793	2	39,080	1
Tobacco, cacao	75,227	1	95,088	2	180,559	3
Others	213,282	4	85,093	2	275,584	5
Total	5,263,560	99	4,016,774	100	5,478,575	100

Note: The annual rate of increases for 1968–72 was 1.0 percent.

Source: Data from the Bureau of Agricultural Economics, Ministry of Agriculture and Livestock [SAG].

a. Valued in cane at the farm gate. [There is a large difference between the farm-gate cane price and the processed sugar price, and data for foreign trade earnings through export are based on the latter.]

these exports by international standards. The same study, based on data for 1968 updated to 1971, indicated that grain exports were not as competitive as those of vegetables, fruits, and certain oil-bearing crops; this is in line with the trend of exports from 1968 to the present time. However, there would be savings of foreign exchange by substituting local production for present imports of all grains.[42]

The study is based on calculations of the expenditure in pesos required to earn one U.S. dollar in foreign exchange [the "exchange cost"; see preceding chapter, especially table 5-10]; it shows that, for agriculture in general, and for certain kinds of crops in particular, Mexico enjoys a comparative advantage on international markets. Many agricultural exports are considered to have a very high rate of economic and financial return. The overall effective rate of exchange for all crops is between 12 and 12.5 pesos per dollar; that is, there is a slight gain from exporting these crops at the present rate of exchange. For in comparison, exports of manufactures receive subsidies of about 15 percent, and it is being proposed that this should be increased to 18 percent. The rank order of

42. Even when the study is updated in accordance with world prices for the first half of 1973, grains show less competitiveness than other crops. It is quite important that production of grain be stimulated to supply the domestic market, but the analysis suggests that investment for marketing other crops abroad yields higher returns.

competitiveness of agricultural products on the world market is more or
less as follows (for those crops that have a positive economic return in
exporting):

1. Pineapples, tomatoes, cucumbers, chile
2. Melons, eggplant, carrots, potatoes, grapefruit
3. Chickpeas, garlic, onions, peanuts, sesame, sugar, tobacco, coffee
4. Soybeans, cotton, strawberries, black beans.

None of these export products require subsidies to make them competitive on the international market. Those listed under categories 1 and 2
above enjoy significant advantages.

The majority of these crops are quite labor intensive, so that encouragement of their production for export is in line with the policy of expanding
agricultural employment. Marketing these products is, however, more
difficult than marketing grain, since the collection and distribution system
is designed primarily for goods such as cotton, coffee, and cereals. Many
of the vegetables and fruits are highly perishable, some require processing
and special methods of packaging, and others sell only in small quantities.
Moreover, as has already been said, these are crops that involve risks as
regards yields and market prices. Some efforts are being made to develop a
wider market for these exports, but there is a need to provide more
support for production and for the whole marketing process if these crops
are to be the main agricultural export lines in the future.

More attention should also be paid to the organizational procedures for
encouragement of production, particularly:

• Those that reduce the climate and market risks faced by the farmer
• Those regulating the flow of production among various uses; that is, between
  different markets or different forms of the product, which are constantly
  changing in response to fluctuations in supply
• Those designed to provide technical assistance concerning both the most
  suitable varieties of crops and methods of marketing and packing for export
• Those directed at exploring foreign markets and securing contracts on behalf
  of the producers.

The new organizational machinery should also take account of another
factor limiting the expansion of fruit and vegetable production: the prevailing semimonopolistic character of the growing and marketing process, which affects the home market more than the export market. If the
hold of monopoly by a few groups of farmers were weakened, production
(and thus employment) could be considerably expanded, considering that,
sector-wide, fruit and vegetable growing employs about four times as
much labor per hectare as grain farming.[43]

43. This figure is obtained from the analysis using the CHAC model.

This completes the general review of the sources of dynamism in agriculture. Because of its importance, employment is discussed more thoroughly in the next section, followed by a summing up of the implications of all the sections in the form of guidelines for the new agricultural policy.

5. *Agricultural Employment in Mexico*

The first section discussed the present level of unemployment in rural areas, which is very closely related to the seasonal nature of job opportunities. The inherent difficulty of measuring employment makes it difficult to ascertain how it has evolved over a given period. The sectoral model provides a general view of the regional and seasonal employment situation at a given moment and also some indications of the extent to which agricultural employment is responsive to increases in production, but it cannot provide information about the growth of the labor force or of working-age population in the rural areas. The agricultural census shows that, over the period 1950–60, the agricultural work force grew by 2.6 percent annually; the population census indicates for the same period only half this rate of growth. This difference can be explained by the fact that many rural workers switch from rural to urban employment according to season. For the period 1960–70, the population census (if certain adjustments in the data are taken in account) again shows that the potential agricultural work force (that is, the working-age population not attending school) increased by slightly over 1 percent annually. The results of other census surveys undertaken in 1970 are not available. These circumstances mean that there will be a wide range of uncertainty in the demographic variables used in planning. It may be assumed that at present the potential agricultural work force is increasing annually by 1.0–1.9 percent, taking into account the capacity of the urban areas to provide new jobs. It also appears that the potential agricultural work force at present numbers about 6 million. The annual increase in the work force is probably, on the basis of average values, around 90,000 workers plus or minus 34,000. It may therefore be assumed, as a guideline for government policy, that about 90,000 new agricultural jobs need to be provided each year. Given the dynamics of growth of rural and urban populations, the number of new workers each year will be significantly less in 1980, and by 1990 the rural work force will probably have stopped growing. The problem of absorbing manpower into the agricultural sector is therefore a ten- to fifteen-year problem.

The short-range question is how to provide permanent employment in agriculture.[44] There are two ways: first, by extending the area of cultivated

---

44. Here the emphasis is on permanent employment in agricultural activities and not on temporary employment directly associated with the construction of public works in rural

(*Note continues on following page.*)

land and at the same time creating new production units for the incremental labor force; second, by making it possible for farming units to employ more people. The second method means, in fact, reducing the average effective size of farms; that is, the number of hectares per person employed in agriculture. The creation of new production units involves expanding the cultivated area, particularly in the rainfed zones, since irrigation works have for the most part been constructed in already cultivated areas that lacked irrigation or drainage. For all practical purposes, it appears that the era when desert land was brought under cultivation by means of large-scale irrigation works has come to an end.

The suggested aim of an annual increase of 375,000 hectares in land under cultivation implies the establishment of about 37,500 new holdings each year, taking the average size of a new holding to be 10 hectares. Holdings of this size do not produce an income much above subsistence level: a rainfed holding of 10 hectares may be considered to generate ten man-months of employment a year, which, taking the seasonal nature of the work into account, represents full-time employment of one and one-half persons a year.[45] This means, for the proposed rate of expansion of cultivated land, the creation of about 56,000 jobs a year. If the remainder of the incremental work force—about 34,000 workers a year—remained on the existing production units, the labor force density per production unit would increase by 1.1 percent yearly, and the "effective size of the production unit" would decrease by 12 percent in ten years.[46] The question that now arises is by what means can employment per production unit be increased sufficiently to compensate for the reduction in effective size. Reference is made to the sectoral model projections for results on the effect of yields per hectare on employment. In terms of the entire sector, the historical rate of increase of yields would, under present conditions, call for about 9,000 additional man-years of work in each year's harvest, assuming no increase in the use of machinery. For landless laborers and farmers together (whether or not they are farming irrigated land), the average employment rate is estimated at five and one-half months a year.[47] This means that it is hoped that the normal increase in

---

areas. Although the latter may be useful in taking up a certain amount of underemployment, it cannot go on increasing year after year on a scale that contributes substantially to absorbing the annual increase in the labor force into productive work.

45. Assuming that the farmer himself works seven months in the year and an additional person (for example, his son) does the other three months' work during the heavy workload periods.

46. For the lowest and highest values in the range of estimation of the annual increase in the work force, the decrease during the ten-year period would be 8 percent and 34 percent, respectively.

47. This employment rate, which is the average for the sector, differs from that used in the calculation at the beginning of this section, which relates to a specific holding of 10 hectares, without irrigation.

yields will provide about 20,000 additional jobs a year; that is, that labor intensiveness per production unit will increase by 0.6 percent annually.

Expansion of the cultivated area and raising of yields would together provide approximately 76,000 jobs a year with a constant rate of mechanization; however, mechanization is displacing about 30,000 jobs a year,[48] so that the net annual increase in jobs is about 46,000. This means 9,000 jobs less than the estimated minimum incremental work force of 55,000 persons, with an even greater shortfall relative to the planning aim of 90,000 new jobs a year. While in fact there may be no way of achieving this aim, with certain policies it might be possible to approach it. For example, if mechanization could be slowed down for another ten years, the net increase in jobs would be considerable. The studies using the sectoral model indicate that a 10 percent reduction in the use of machines will yield a 3 percent increase in employment; in other words, avoiding a 10 percent increase in use of machines means avoiding a 3 percent drop in employment.

There are no accurate figures on the total inventory of agricultural machinery or the extent to which it is used, but the machinery use is said to be increasing by 5 percent annually. According to the above calculations, 120,000 jobs could be provided over 10 years by reducing this growth to 3 percent per year. Such a reduction could probably be achieved by policies that raised the effective price of the machinery.[49] During the coming decade, it will also be important to avoid excessive increases in rural wages, which have a marked effect on speeding up mechanization. Even with a policy of changing relative factor prices so as to foster higher employment, it does not seem that it will be easy to achieve the aim of 90,000 new jobs a year, and without strong measures the rate of rural unemployment will probably continue to grow.[50]

Given these somewhat unfortunate conclusions, it may be worthwhile to look outside the scope of normal policies for sources of employment generation. In particular, it may be in order to examine certain problems of livestock raising in the tropics. There may be some potential interrelation between measures to encourage stock raising and measures designed to increase employment, even though there is obviously no direct connection, because stock raising is generally much less labor intensive than farming. Farming in fact provides about ten times as many jobs per hectare.

48. Each tractor displaces ten to twelve jobs.

49. A change of this degree in relative factor prices would not alter the ranking of crops by comparative advantage on external markets. Any change would in fact tend to sharpen the advantage of fruit and vegetables.

50. Obviously, the use of agricultural machinery can mean high profits for large-scale commercial farmers. However, the social cost in terms of unemployment can more than nullify this if mechanization is allowed to be speeded up without restriction.

Stock raising in the tropics is at present of relatively low density—not more than one animal is carried per hectare—and the land used for livestock covers 9.4 million hectares of pasture land. Techniques for raising the livestock density are known: greater use of the new types of pasture grasses that have already been tested in practice, and the application of fertilizer, would substantially increase the fodder-producing capacity of the land. At present, the legally fixed limits on herd size and on the animal-land ratio do not encourage ranchers to put their operations on a more efficient footing. If these regulations were revised, ranchers would have incentives to increase the livestock density. A speed-up in the application of this policy would have a salutary effect on production, and at the same time, would allow more rational use of fertile land, which is the scarce resource.

If the stock-raising density could be increased in this way, significant amounts of land in tropical areas could be turned over to crop farming. In addition to providing more jobs per hectare, crop farming also generates more income per hectare. With present practices, a tropical cattle ranch earns 200–500 pesos net per hectare; in comparison, net income from the simplest annual crops, such as unfertilized maize, is 600 pesos per hectare, and from other crops net income is several times that figure. In view of the great pressure on land, a changeover to higher-density stock raising seems, in the long run, inevitable.[51]

Less than a fourth of the tropical ranch land would be suitable for such conversion: probably 2 million of the existing 9.4 million hectares. If 100,000 hectares a year could be converted to 10-hectare farms, this could yield 15,000 new jobs annually. While this would not solve the employment problem, it would help considerably to improve the position for the next ten years,[52] and, as stated earlier, the problem would begin to be alleviated after that.[53]

The cumulative effect of extension of the area under cultivation, increased yields, changes in the crop patterns, increased prices of farm machinery, and the program of conversion of pasture land would be a net increase in employment of the order of 70,000 jobs a year. Additional

51. A comparison can be made with Argentina, with its obvious advantages for low-density stock raising. The average farm size in Argentina is 80–100 hectares, compared with 6.8 hectares in Mexico. Moreover, Argentina has a good deal of unused land, in contrast with the enormous pressure on fertile land in Mexico. These considerations point to the need to switch from low-density to high-density operations in the tropics.

52. It should also be noted that some crop-farming zones are so unproductive that they should probably be converted to ranching.

53. The reason that no reference is made in this report to rural industries as sources of employment is that these have already been taken into account in calculating the drift from the country to the towns.

efforts to increase water supply and to open up export markets for fruit and vegetables would also contribute to higher employment.

6. *Principal lines of the new agricultural policy*

The foregoing diagnosis suggests that the orientation of present policy should be confirmed and strengthened in certain areas, whereas in others a change of approach is called for. The following are the four main lines of approach of the new agricultural policy:

- Investment in physical resources
- Investment in human resources
- New methods of organizing the production and distribution of inputs
- New forms of market organization and new price policies.

This listing does not indicate an order of relative importance, since the various aspects of supply and demand in the agricultural sector are so closely interrelated that the four lines of action will need to be applied simultaneously to achieve the national objectives for the sector.

INVESTMENT IN PHYSICAL RESOURCES. The estimates indicated earlier show that, to achieve the sector objectives, adjustments will be necessary in the allocation of public expenditure so as to raise agriculture's share of total investment to more than twice what it was at the end of the 1960s. This change of focus has already been operating since the beginning of the present administration but should be reinforced in the coming years. The biggest increase should be in human resource investment, though investment in physical capital should also be increased considerably.

Investment in physical resources relates to investment in expanding both land under cultivation and availability of irrigation water. So far as extension of irrigated land is concerned, the present programs are consistent with the general lines of the policy; in the case of expansion of rainfed land, execution of the existing projects will need to be speeded up. The historical rate of expansion of the area under irrigation of three percent a year will have to be sustained or exceeded for some years to ensure the national increase in production. The investment budget for the remainder of the six-year period provides for the bringing under irrigated cultivation of more than 100,000 hectares a year, a 3 percent annual increase. At the same time, a considerable effort is being made to rehabilitate irrigated districts, and this is effectively increasing the total availability of water and of irrigated land. When the rehabilitation areas are added to the new irrigation hectarage, the total is much more than 100,000 hectares a year.

As stated earlier, the total benefit of the irrigation programs appears to be of the order of 25,000 pesos per hectare or more. The projects should be executed rapidly, to the point where cost per hectare (discounted as of the date the irrigation works enter into service) is equal to the benefit, taking into account all related costs (such as investment in regional extension services and agricultural experimentation stations).

In the rainfed and tropical zones, the target of a further 275,000 hectares under cultivation will have to be pursued by means of three types of land programs: clearing, drainage, and reclamation through soil conservation. For these programs, it is of the greatest importance that research, agricultural extension, and supply of credit and other agricultural inputs be coordinated with the present programs of investment in physical works. Where the investment programs include land-settlement schemes, careful planning in collaboration with the settlers will be very important to their success.

INVESTMENT IN HUMAN RESOURCES. In this field the plan calls for greater divergence from past trends than in the case of physical resource investment. By means of a stronger human resources program, it will be possible to increase the return on physical investment and to encourage greater private investment in the rural sector. Some of the main points have been discussed in the preceding sections; they may be summarized as follows:

- Three- or fourfold expansion and improvement of the agricultural extension service
- A corresponding expansion of agricultural research
- Increased regional (that is, local) emphasis on research and agricultural extension, with particular stress on intensive training courses for farmers
- Increased emphasis on research and agricultural extension in rainfed and tropical areas
- Coordination of agricultural extension services with the supply of input packages, with the purchase of produce at guaranteed prices included among these "inputs."

To these five main points should be added two others:

- Greater emphasis on technical education at all school levels, particularly in rural areas
- Raising the technical ability of persons involved in the preparation and execution of development programs and projects—both to understand and take into account the outlook, views, and aspirations of the potential beneficiaries and to learn from them.

This last point means that all action programs must be the fruit of a two-way flow of ideas between the agents of change and the members of the communities concerned. A prerequisite for achievement of these objectives is the carrying out of successful literacy campaigns among the rural population.

All these lines of approach in the new policy call for some degree of adaptation on the part of the individual producer. This involves learning techniques that enable farmers to cultivate new crops, to adapt them to local conditions, and to participate fully in the existing agricultural institutions. Individuals adapt more easily when they have a good basic education, and improvement of the rural education system is essential to the long-term success of the new agricultural policy. At the present time, less than half the subsistence farmers have completed primary education; it is therefore highly desirable that this proportion should be substantially increased, in parallel with the introduction of educational reforms to shift the focus of rural schools toward agricultural technical training.

At the basic educational levels, instruction could be given, by way of illustration, in the role of the credit institutions, the use of fertilizers and new seeds, cultivation methods, and other practical aspects of farming.

Agricultural technical education should also be expanded at the high school level. It is discouraging to note that the overwhelming majority of students enrolled for agricultural studies at the professional and postgraduate level come from urban areas, owing to the lack of the secondary and "preparatory" facilities in rural areas.

The third point above (special training courses for farmers) refers to programs of the kind in which the farmers attend a school or workshop for a period of a month to a year, during which they exchange day-to-day experience with other farmers attending the same course. Practice has shown that the shyness and reserve displayed by the farmers to outsiders do not begin to disappear until after some days in these group classes. From that point on, the farmers begin to participate and to assimilate the new material at a much more rapid rate. The experience of the Human Resources Institute of the State of Mexico (IDRHEM) provides a good example of this approach. This type of course multiplies the effectiveness of extension work by turning the farmers themselves into extension "auxiliaries" who can help to promote change in their own communities.

NEW METHODS OF ORGANIZING THE PRODUCTION AND DISTRIBUTION OF INPUTS. Agricultural input packages have to be fairly flexible and adaptable to local production conditions, which vary more widely in rainfed than in irrigated areas; this calls for a major effort to design input packages for the rainfed farmers. The diversity of the conditions in rainfed and tropical regions normally means that the points of view of the farmers have to be reflected in the regional policy measures of the governmental agencies responsible for promoting agricultural production, since the individual farmers have little chance to exert any influence on the decisions of the banks and of the fertilizer-marketing companies. It follows that, if any significant change is to be brought about in yield trends in rainfed areas, new forms of organization are needed. These mechanisms must be set up by local initiative within the general planning process and out of them must come measures that reflect the local point of view as well as background information for the decision of the agencies that formulate overall policy measures for sectoral planning. Although examples of cooperative action in rainfed areas do exist, local participation in decision-making concerning the use of inputs and the planning of production has not been systematized in the way that it has in the majority of the irrigation areas, through the official irrigation districts.

It is suggested that a system similar to that in use in the irrigated areas should be introduced in the rainfed areas by setting up what might be called "rainfed-zone programming districts" (*distrítos de temporal*). Each district would have a directive committee composed of representatives of the farmers, of the credit institutions, and, through their various agencies, of the Department of Agrarian Affairs and the Ministry of Agriculture. The farmers should constitute a majority on the committee. The first task of each directive committee would be to regularize land tenure in its region. The obligations would be to set production goals for the region and, on the basis of these, to plan the requirements of productive inputs such as technical advisory assistance, credit, chemical products, and agricultural insurance. The committees could also determine the need for agricultural research to solve local problems and could act as an action group in relation to the official institutions—for example, to ensure that fertilizers reach the area when needed.

It is felt that each *temporal* district should comprise not more than 1,000 farms. This limit is necessary if the districts are to cater efficiently to the needs of all the farmers. This means that a total of about 1,000–2,000 districts would be set up. At the present time there are 80 agricultural directive committees operating in about 100 irrigation districts. For administrative reasons, therefore, the planned number of new districts could not be reached until the system had been under development for several years. The program would obviously have to be carried out gradually to enable advantage to be taken of the experience accumulated as the system expands. The districts should operate under the direction of the government agency that acts as a clearing house for information and local priorities for the formulation of national goals. As in the case of the irrigation districts, the *temporal* districts would be composed of both private farmers and *ejidatarios*. This type of organization would combine elements of the irrigation districts, of the various development "plans" in rainfed areas, and of the new population centers in the tropical parts of the country.

The *temporal* districts would be coordinated at state level and also, at the national level, in the Ministry of Agriculture and in the Coordinating Commission for the Agricultural Sector [Comisión Coordinadora del Sector Agropecuario, COCOSA].

The *temporal* districts would be an effective instrument for the continuation and expansion of the current program of regulation of land tenure; this will make it possible to increase the area under cultivation, since some cases of idle land are attributable to disputes about ownership rights.

It is felt that the input packages might include extension services, credit (fertilizers, improved seed, and other improved inputs), and also support prices. If the input packages are to be effective in increasing production, there must be a close link between administration of the support price and credit availability. Among other essential characteristics, agricultural credit must be timely and adequate.

With regard to the structure of credit, it must be borne in mind that it is being used with two aims: redistribution of income and economic efficiency. Unfortunately, in the present administration of the system these two aims of credit policy are confused, with the result that in the subsistence areas the granting of credit often does not operate as a strong incentive to productive efficiency: it is common for the representatives of the banking system to require payments against the loans only in cases of good harvests, an attitude that is hardly conducive to improvement of yields. A means must be found of making a clear separation, for any given producer, between the element of government subsidy and that of commercial credit.

Credit must stimulate improvement of the production process, but it cannot fulfill this role when it simultaneously serves as a subsidy vehicle. It would be worth investigating the possibility of granting each subsistence farmer, at the start of the annual cycle, a subsidy equivalent to 200 pesos, which would in fact be a higher subsidy than that implicit in the present system of bank credit,[54] simultaneously removing the subsidy element from the credit administration. This measure would also provide a means of "training" the farmers in the use of credit by requiring each farmer to handle his production credit in a businesslike way. The subsidy would also help to foster the introduction of new farming methods.

With regard to the prices of agricultural inputs, as mentioned earlier the price of irrigation water should be raised to a level that closely reflects its real productivity. It has also been stated that, if a way were found of significantly raising the price of agricultural machinery, this would help substantially to achieve the employment goal.

54. The implicit subsidy is estimated by means of the loan recovery rate.

NEW FORMS OF MARKET ORGANIZATION AND NEW PRICE POLICIES. The new features of price policy have been discussed in previous sections. Its main features are: (1) the need to maintain a guaranteed price for maize above the free market price for a few years; (2) the desirability of focusing support price programs on rainfed grain and oil-bearing crops to the greatest possible extent; (3) the need to investigate the possibility of introducing flexible support prices for high-value crops, with the object of smoothing out price fluctuations during the year; and (4) the importance of maintaining guaranteed price policies that reflect the patterns of regional ccmpetitive advantage. The remainder of this summary is devoted to a more detailed discussion of the points concerning the market for agricultural exports. The need has been mentioned for greater attention to institutional mechanisms directed toward the following objectives: (1) reducing the risks faced by producers; (2) regulating the flow of an agricultural product into its alternative processing or packaging forms; (3) providing technical assistance with respect to suitable varieties and forms of packing and marketing; and (4) exploring the possibilities of, and entering into, sales contracts with foreign buyers on behalf of domestic producers. For this last purpose, the possibility would be worth considering of setting up a system of futures purchases and sales that could be carried out with the assistance of the Perishable Products Marketing Trust, CONASUPO, or the like, thereby enabling the producer to share his commercial risks. It would also have the effect of contributing to the stability of prices to the producer and, indirectly, to control of supply by offering low prices in futures purchases when domestic production was high in relation to explicit world demand.

These functions could be performed through specialized marketing committees whose operating instrument could be a guaranteed price that would fluctuate from year to year to keep it continuously adjusted to the balance of supply and demand. Separate committees could be set up for each fruit or vegetable crop, or for a group of such crops, for a region supplying a consumer area, or for a group of regions, and could be composed of representatives of the agricultural authorities and the producers. These marketing committees would deal with matters concerned with both the domestic and the export market. They would be responsible for distribution to the markets for perishable products or to the industrial processors and, in years of surplus supply, for disposing of the surpluses either through export or through special programs such as those of the National Institute for Childhood Protection.

A policy for perishable products could also be linked through these committees to expansion and coordination of the capacity of refrigerated stores. Coordination is important because, through rotation of products, it can permit full utilization of storage capacity throughout the year. This storage capacity for perishable products means an expansion of domestic demand because it extends the purchasing period to the entire year.

Another limiting factor on expansion of fruit and vegetable growing is the semimonopoly characterizing their production. Weakening of this monopoly would lead to a substantial expansion of production and, therefore, of employment, since fruit and vegetable crops use about four times as much labor per hectare as grains.

With the elimination of the risks of price fluctuations, the producers would be more willing to reduce their unit profits in exchange for increases in sales volume.

The basic features of the four main lines of the new agricultural policy have been examined. In conclusion, attention is again drawn to the strong interdependence of all facets of the agricultural sector and the consequent need to seek a balance among these four policy lines, each receiving a matching effort, so as to ensure accomplishment of the economic and social development objectives.

To achieve the sectoral development goals, the economic and social development program for agriculture should be implemented through the annual preparation of national agricultural and livestock development plans that reflect regional growth targets (total and product-by-product), that refer to domestic and external demand expectations, that reconcile requirements and availabilities of operational inputs, and that provide for scheduling and coordination of specific policy actions during the various periods of the plan.

Part Two

Risk in Agricultural Models

7

Risk in Market Equilibrium Models
for Agriculture

PETER B. R. HAZELL AND
PASQUALE L. SCANDIZZO

LINEAR PROGRAMMING MODELS are gaining increasing acceptance as tools for analysis of agricultural supply response and agricultural investment programs at both the regional and sectoral levels. One difficult problem in specifying aggregate models, however, lies with ways of incorporating the considerable price, yield, and resource uncertainties that confront producers.

There is now considerable literature that attests to the fact that farmers are not risk neutral. Notable examples are Officer and Halter (1968) and O'Mara (chapter 9 of this volume), both of which include estimates of farmer's utility functions. These functions typically show risk aversion in the relevant range of values of farm income. A direct consequence of this risk aversion is that omission of risk considerations in programming models is likely to lead to an overestimate of the supply response for farm enterprises with high variance in yields, prices, or both. Furthermore, since these are often high-value enterprises, omission of risk is likely to lead to an overstatement of the returns to investment. These biases may be particularly large in models of low-income agriculture, in which risk aversion is likely to be greatest.

Overview

Methodologies for handling risk at the individual farm level are well developed in the literature for a wide range of decision criteria. Two of the more appealing of these, when information about the probability distributions of stochastic components is available, use the E, V decision criterion (Freund 1956; Heady and Candler 1958; Markowitz 1959; McFarquhar

Note: This chapter is a revised and extended version of Hazell and Scandizzo (1974). Permission of the original publisher to use unrevised material here is gratefully acknowledged.

1961; Stovall 1966*b*) or the related E, σ measure (Baumol 1963; Hazell and Scandizzo 1974).

Such models are generally more descriptive of individual farm behavior than linear programming models that maximize expected income, and therefore provide a more useful starting point for the construction of aggregate models. At the aggregate level, however, problems arise when market demand structures are interfaced with the supply model to obtain simultaneous determination of the equilibrium levels of production and prices under the usual assumption of a perfectly competitive agricultural industry. The problem then is to specify the objective function of the aggregate model in such a way as to simulate the results that would be obtained if each farmer operated in a competitive environment according to the E, V or E, σ decision criterion.

Methods of simulating competitive market equilibrium through linear programming are well developed for the deterministic case. Samuelson (1952) showed that the appropriate objective function in the aggregate model is the maximization of net social payoff (the sum of consumers' and producers' surplus). He developed this result in the context of spatial equilibrium models, and Takayama and Judge (1964, 1971) further developed this objective function to obtain a quadratic programming formulation for multiproduct models. Duloy and Norton (1971, 1973; chapter 3 of this book) subsequently applied the method to agricultural sector models using linear programming approximations.

The purpose of this chapter is to provide a modification of the Duloy-Norton method when production is risky and individual farmers maximize E, σ utility instead of expected profits. A crucial issue in the development of such modifications lies in appropriate specification of the equilibrium solution to be simulated. The nature of market equilibria under risk conditions is a complex subject that has received scant attention in the economic literature. Furthermore, results are known to depend very much on assumptions made about the dynamics of market adjustment and on the nature of the stochastic components involved. In this chapter we assume that the initial source of risk lies in yields. Stochastic yields also lead to stochastic costs for those costs which are related to production rather than numbers of hectares or livestock. Under these assumptions, and assuming lagged behavior in supply response, useful results can be developed about market equilibria (Bergendorff, Hazell, and Scandizzo 1974; Hazell and Scandizzo 1973; Turnovsky 1968).

Some systematic experience with numerical implementation of this approach are reported for various Mexican producing areas in the next chapter of this book. In addition, the aproach has been incorporated in the models of chapters 12, 13, and 15.

In the following sections, we briefly review methodology for the deterministic case, and pose the problem for the risk situation. Relevant market equilibrium results are then developed, and a methodology for obtaining market equilibrium solutions assuming an E,σ decision criterion at the farm level is developed. Finally, linear programming approximations are provided and illustrated with a schematic tableau.

The Deterministic Model

The deterministic model is premised on the assumption that individual farmers are profit maximizers, and that they compete in a perfectly competitive way. The latter assumption implies, in particular, that farmers plan on the basis of constant anticipated prices.

Define

$\hat{p}$ = An $n \times 1$ vector of anticipated product prices

c = An $n \times 1$ vector of unit costs

x = An $n \times 1$ vector of enterprise levels

M = An $n \times n$ diagonal matrix of enterprise yields with jth diagonal entry m_j

$y = Mx$ is the $n \times 1$ vector of total outputs.

Then the objective function for an individual farm problem is:

(7.1) $$\max \pi = \hat{p}' \, y - c' x,$$

and this is to be maximized over some set of constraints that are usually specified to be linear.

Now if the product markets attain an equilibrium, then regardless of the way in which the anticipated prices $\hat{p}$ are formed over time, the equilibrium is unique. Furthermore, the market equilibrium prices and outputs occur at the points where the demand and implicit model supply functions intersect. This fact provides the basis of the solution procedure.

Let X, Y, C, and W be some appropriate aggregates[1] of the individual farms x,y,c and M matrixes, and P be the vector of unknown market prices. Then, assuming the linear demand structure

(7.2) $$P = A - BY,$$

where B is a symmetric matrix of demand coefficients,[2] the Duloy–Norton aggregate model objective function is:

---

1. Aggregation should be exact to avoid biased results in the sector model. The usual approach to this problem is through appropriate classification of farms into homogeneous groups (Day 1963; Stovall 1966a). To simplify notation, however, it is assumed throughout this chapter that there is only one homogeneous group of farms in the sector.

2. The condition of symmetry is necessary to ensure the existence of the potential function in equation (7.3). An analysis of the implications of such an assumption is contained in Zusman (1969).

(7.3) $$\max \Pi = X'W\,(A - 0.5BWX) - C'X,$$

where it is understood that $Y = WX$.

The term $X'W\,(A - 0.5BWX)$ is the sum of areas under the product demand functions. For example, in the single product case this would be

$$\int_0^y (a - bt)\,dt = y(a - 0.5by) = wx(a - 0.5bxw).$$

The term $C'X$ is total production costs, or equivalently, the sum of areas under the product supply functions. Consequently, the difference between these two terms is the sum of producers' and consumers' surplus over all markets, and this reaches its maximum at the required intersections of supply and demand functions.

Introduction of Risk

The basic source of risk to be introduced is confined to yields. Thus, the vector of products for an individual farm now becomes

$$y = Nx,$$

where N is an $n \times n$ diagonal matrix of stochastic yields with jth diagonal element ε_j.

Stochastic yields imply stochastic supply functions, and hence lead to stochastic market prices. It is assumed, however, that input costs and the market demand structure remain nonstochastic, and that the farm linear programming constraints are not affected. The latter assumption can easily be relaxed, since several techniques are available to handle stochastic constraints that do not affect the farm model objective function (Charnes and Cooper 1959; Hillier and Lieberman 1967; Madansky 1962; Maruyama 1972).

It is further assumed that the individual farmers are risk averse, and that their behavior conforms to a single period E,σ specification. Consequently, the individual farm model objective function each year is

(7.4) $$\max_x u = E(p'y) - c'x - \phi V(p'y)^{1/2},$$

where E and V denote, respectively, the expectation and variance operators, and ϕ is a risk-aversion coefficient.

In order to enumerate equation (7.4) more precisely, it is necessary to make explicit assumptions about the nature of farmers' subjective expectations. These in part depend on the nature of perfect competition under risk.

Perfect Competition under Risk

As a natural generalization of the deterministic concept of perfect competition, it is assumed that farmers continue to expect their outputs to not have any effect on the market. A set of assumptions regarding behavioral anticipations that are consistent with this are:

(A.1) $$E(\varepsilon_j) = m_j$$

(A.2) $$V(\varepsilon_j) = \sigma^2_{\varepsilon_j}$$

(A.3) $$E(p_j) = \hat{p}_j$$

(A.4) $$V(p_j) = \sigma^2_{p_j}$$

(A.5) $$\mathrm{cov}(p_i p_j) = \sigma_{p_{ij}}; \quad \mathrm{cov}(\varepsilon_i \varepsilon_j) = \sigma_{\varepsilon_{ij}}, \quad \text{for all } i \neq j$$

(A.6) $$\mathrm{cov}(p_j \gamma_i) = x_i \, \mathrm{cov}(p_i \varepsilon_i) = 0, \quad \text{for all } i,$$

where all operators are now subjective expectations that may differ from the real-world parameters.

Assumption (A.3) states that farmers expect a constant mean price for each product, and by making the variance homoskedastic [assumption (A.4)] and the covariances between prices and outputs zero [assumption (A.6)], this implies that there is no expected relation between the output of the individual farm and the market.

It is important to note that these are a set of behavioral assumptions, and it is not required that farmers anticipate the true state of affairs. Market behavior is in part a reflection of what farmers anticipate, but this is basically no different from the deterministic case where, in the short run, the farmers' expectation of prices $\hat{p}$ can differ from the vector of market-clearing prices P.

Given the above set of assumptions, the components of equation (7.4) can be enumerated as follows:

$$E(p'\gamma) = \hat{p}'Mx,$$

where $M = E(N)$,

$$V(p'\gamma)^{1/2} = (x'\Omega x)^{1/2},$$

where Ω is an $n \times n$ covariance matrix of activity revenues with diagonal elements

$$\omega_{jj} = V(p_j \, \varepsilon_j)$$
$$= E(p_j^2 \, \varepsilon_j^2) - \hat{p}_j \, m_j^2$$
$$= E(p_j^2) \, E(\varepsilon_j^2) - \hat{p}_j^2 \, m_j^2$$
$$= \sigma_{p_j}^2 \, E(\varepsilon_j^2) + \hat{p}_j^2 \, \sigma_{\varepsilon_j}^2,$$

and off-diagonal elements

$$\omega_{ij} = \operatorname{cov}(p_i\varepsilon_i, \; p_j\varepsilon_j)$$
$$= E(p_ip_j\varepsilon_i\varepsilon_j) - E(p_i\varepsilon_i) \, E(p_j\varepsilon_j)$$
$$= E(p_ip_j) \, E(\varepsilon_i\varepsilon_j) - \hat{p}_i\hat{p}_j \, m_im_j$$
$$= E(p_ip_j) \left[E(\varepsilon_i\varepsilon_j) - m_im_j \right] + m_im_j \left[E(p_ip_j) - \hat{p}_i\hat{p}_j \right]$$
$$= \left[\sigma_{p_{ij}} + \hat{p}_i\hat{p}_j \right] \sigma_{\varepsilon_{ij}} + m_im_j \, \sigma_{p_{ij}}.$$

The farm problem objective function is then

(7.5)
$$\max_{x} u = \hat{p}'Mx - c'x - \phi(x'\Omega x)_{1/2}.$$

Obviously, alternatives to assumptions (A.1) through (A.6) are possible. However, assumptions (A.1), (A.3), and (A.6) are retained, then the only effect of changing the assumptions is on the elements of Ω. This will affect the market behavior and the estimation of Ω in a quantitative way, but does not deter the development of qualitative results about farm behavior.

Let the linear programming constraints for the farm model be denoted by $Dx \le b$, $x \ge 0$, then the Lagrangian function for maximizing equation (7.5) over this set is:

(7.6)
$$L = \hat{p}'Mx - c'x - \phi(x'\Omega x)^{1/2} + v'(b - Dx),$$

where v is a vector of dual values. An optimal solution to the problem is then a "saddle point," and necessary and sufficient conditions for any (x,v) to be this saddle point, are obtainable from the Kuhn-Tucker conditions. The necessary conditions are:

(7.7)
$$\frac{\partial L}{\partial x} \le 0, \quad \frac{\partial L}{\partial v} \ge 0,$$

(7.8)
$$x\frac{\partial L}{\partial x} = 0, \quad v\frac{\partial L}{\partial v} = 0.$$

Of these, the requirements in equation (7.8) are the complementary slackness requirements that an activity cannot be active and at the same time have a nonzero opportunity cost and that a resource cannot be slack and at the same time have a nonzero dual value. Sufficient conditions for a saddle point can be derived, but they reduce to the requirement that Ω be a positive semidefinite matrix (Takayama and Judge 1971, p. 19).

Applying the necessary requirements in equation (7.7) to equation (7.6) gives

$$(7.9) \qquad \frac{\partial L}{\partial x} = P'M - c' - \phi x' \Omega (x'\Omega x)^{-1/2} - v'D \le 0,$$

$$(7.10) \qquad \frac{\partial L}{\partial v} = b - Dx \ge 0.$$

Equation (7.10) is merely the feasibility requirement, but equation (7.9) contains the risk counterpart to the classical marginality rules for output determination in a deterministic firm. Taking the jth element of the vector $\partial L/\partial x$, and rearranging terms and dividing by m_j, we obtain

$$(7.11) \qquad \hat{p}_j \le \frac{1}{m_j} \left[\sum_k v_k\, d_{kj} + c_j + \phi (x'\Omega x)^{-1/2} \sum_i \omega_{ji} x_i \right].$$

This states that for each product the expected marginal cost per unit of output must be equal to or greater than the expected price. The expected marginal cost comprises expected own marginal cost c_j/m_j, plus a marginal risk factor

$$\frac{\phi}{m_j} \frac{\partial V(p'\gamma)^{1/2}}{\partial x_j} = \frac{1}{m_j} \phi (x'\Omega x)^{-1/2} \sum_i \omega_{ji} x_i,$$

plus expected opportunity costs

$$\frac{1}{m_j} \sum_k v_k\, d_{kj}$$

as reflected in the dual values of the resources used by that activity. This differs from the comparable requirements of a deterministic model primarily in that a risk term has been introduced. This is quite reasonable because the risk term is really nothing but a new cost; namely, 'the additional expected return demanded by farmers as compensation for taking risk. This is even clearer when farmers can participate in a crop insurance program, for then the risk term is the marginal premium a farmer would be willing to pay to insure against risk (that is, a certainty equivalent cost).

This result is not new and is consistent with the results obtained for single-product firms in economic analysis (for example, see Magnusson 1979). Further, the appearance of the risk factor as a marginal cost provides the rationale for the expectation that deterministic models overestimate the supply response of high-risk crops. This is because $\sum_i \omega_{ji} x_i$ will then be positive, hence the marginal cost curve must lie above the marginal cost curve that would be obtained from a deterministic (or risk-neutral) model.

Although equation (7.11) is a necessary condition, it is clear from the complementary slackness condition $x(\partial L/\partial x)$ in equation (7.8) that the condition will always be satisfied as an equality in an optimal solution for all activities that enter the basis; that is, $x_j > 0$. Consequently, for all basic activities the risk counterpart to the price-equals-marginal-cost rule can be written as:

$$(7.12) \qquad \hat{p}_j = \frac{1}{m_j} \left[\sum_k v_k \, d_{kj} + c_j + \phi(x'\Omega x)^{-1/2} \sum_i \omega_{ji} x_i \right].$$

The right-hand side of equation (7.12) is then the short-run supply function for the farm as implicitly embedded in the programming model. This is a basic behavioral relation, and expresses the farmer's determination of x_j given his expectations about yields and prices. That is, $x_j = f(M, \hat{p}, \Omega)$, with everything else constant. Multiplying by the mean yield m_j, a conditionally expected supply function is immediately obtained:

$$(7.13) \qquad E(y_j | x_j) = m_j x_j = m_j \, f(M, \hat{p}, \Omega).$$

Since all the expectations involved are subjective anticipations, it is useful to denote equation (7.13) as the "anticipated supply function" for the farm to distinguish it from a true statistical relation.

Market Equilibrium under Risk

By summing the anticipated supply functions over all farms, an aggregate anticipated supply function can be obtained as a basic behavioral relation in the market. Ignoring aggregation problems for now, the jth supply function can be written as

$$(7.14) \qquad E(Y_j | X_j) = w_j X_j = w_j \, g(W, \hat{\rho}, \Gamma),$$

where X_j, Y_j, W, $\hat{\rho}$, and Γ are suitable aggregates of x_j, y_j, M, $\hat{p}$, and Ω, respectively, and w_j is the jth diagonal element of W.

Given X_j, actual supply is

$$(7.15) \qquad Y_j | X_j = e_j X_j = e_j g(W, \hat{\rho}, \Gamma),$$

where e_j is a suitable aggregate of the farm ε_j's such that $E(e_j) = w_j$. Clearly, actual supply is stochastic with e_j and, furthermore, is of a specification in which the slope of the supply function is stochastic. Ignoring the unnecessary complication of a stochastic intercept term, and assuming that e_j is bounded on some positive interval $e_m \leq e_j \leq e_x$, the market situation can be portrayed as shown in figure 7-1.

Figure 7-1. *Market Situation under Aggregate Anticipated Supply Function*

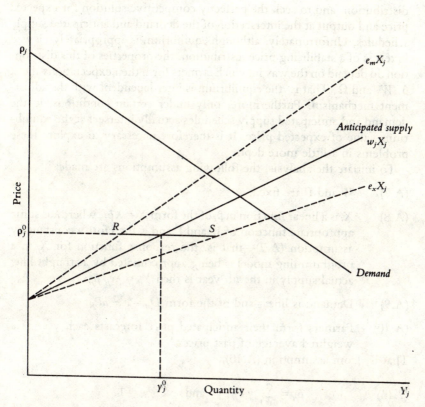

If W and Γ remain constant, the anticipated supply function is fixed, and aggregate anticipated supply is determined by $\hat{\rho}$. Thus for given $\hat{\rho}$ with $\hat{\rho} = \rho_j^0$, farmers will plan their farms so that aggregate expected output is y_j^0. Yet, because yields are stochastic over the range e_m to e_x, actual supply can take on any value between R and S. More generally, the actual supply function can rotate around the anticipated supply function to any position contained in the funnel defined by $e_m X_j$ and $e_x X_j$. It follows that market price must always be stochastic and will fluctuate with both e_j as well as with X_j if the latter does not stabilize to some equilibrium amount. Further complications arise when W or Γ is not fixed, for then the whole "supply funnel" may shift structurally over time.

The question of what is a perfectly competitive and equilibrium solution to the market now arises, and, since it is obviously not a point solution, the further question of what characteristics can usefully be derived with a mathematical programming model can be posed. The intuitive answer would be to view the market as stabilizing in its price distribution, and to seek the perfectly competitive solution for expected price and output at the intersection of the demand and anticipated supply schedules. Unfortunately, although equilibrium is appropriately viewed in terms of a stabilizing price distribution, the properties of this distribution do depend on the way in which farmers form their expectations about $\hat{\rho}$, W, and Ω. That is, the equilibrium is interdependent with the adjustment mechanism. Furthermore, only under certain conditions do the demand and anticipated supply schedules actually intersect at the equilibrium value of expected price. It is therefore necessary to explore these problems in a little more depth.

To initiate the analysis, the following assumptions are made:

(A.7) W and Γ are fixed.

(A.8) X_j is a linear function in $\hat{\rho}_j$ of the form $X = \lambda \hat{\rho}_j$, where λ is some appropriate function of W and Γ and is therefore a constant by assumption (A.7)—that is, the response function for X_j in a programming model is being approximated by a straight line; actual supply in the tth year is then $Y_{jt} = \lambda e_{jt} \hat{\rho}_{jt}$.

(A.9)   Demand is linear and of the form $D_{jt} = a - bP_{jt}$.

(A.10)   Farmers form their anticipated price forecasts each year as a weighted average of past prices.[3]

That is, from assumption (A.10),

(7.16) $$\hat{\rho}_{jt} = \sum_{i=1}^{m} \gamma_i P_{jt-1} \quad \text{and} \quad \sum_{i=1}^{m} \gamma_i = 1.$$

Note that a "naive cobweb" formulation and the Nerlove-type adaptive expectation models are special cases of equation (7.16), so that the formulation is quite general. A further assumption is that

(A.11) $$\text{cov}(e_{jt}, e_{jt-1}) = 0, \text{ for all } t.$$

That is, the yield of an individual activity is uncorrelated with itself over time.

Ignoring the jth subscript for convenience, we find that if the market clears each year, then market-clearing price is

---

3. It is to be noted that an E,σ decision specification states only which price and yield parameters are relevant each year; it does not state how anticipations about these parameters are formed over time.

(7.17)
$$P_t = \frac{a}{b} - \frac{\lambda}{b} \, e_t \, \hat{\rho}_t$$

$$= \frac{a}{b} - \frac{\lambda}{b} \, e_t \sum_{i=1}^{m} \gamma_i P_{t-i}.$$

Equilibrium can now be defined according to the convergent properties of P_t over time. There are a number of alternative properties to choose from (Bergendorff, Hazell, and Scandizzo 1974; Turnovsky 1968), but all basically are variants of the concept of convergence in the probability density function of p_t and its various central moments. Consider first the convergence of expected price:

(7.18)
$$E(P_t) = \frac{a}{b} - \frac{\lambda}{b} \, E(e_t \hat{\rho}_t)$$

$$= \frac{a}{b} - \frac{\lambda}{b} \, w \, E(\hat{\rho}_t)$$

$$= \frac{a}{b} - \frac{\lambda}{b} \, w \sum_{i=1}^{m} \gamma_i \, E(P_{t-i}).$$

This is an mth order difference equation and, if convergence occurs,[4] has the particular solution

(7.19)
$$\lim_{t \to \infty} E(P_t) = \frac{a}{b + \lambda w}.$$

Further, it is the same for any choice of γ weights satisfying equation (7.16).

Solving now for the intersection price (P^*) of demand and anticipated supply $(Y_t | X_t = \lambda w_t \, \hat{\rho}_t)$, we obtain $a - bP^* = \lambda w P^*$; hence $P^* = a / (b + \lambda w)$, which is identical to $\lim E(P_t)$.

Thus, under assumptions (A.7) through (A.11), the asymptotic expectation of market price is the same regardless of the specific price-learning model and, furthermore, corresponds to the desired intersection of the demand and anticipated functions. It is also clear from the derivation of anticipated supply that at this point $\lim E(P_t) = E(\text{marginal costs})$, which provides an acceptable equivalence to the equilibrium point of a competitive but deterministic market. These results also hold for the multiproduct case, as is shown in the appendix to this chapter.

Turning now to other properties of the equilibrium, we can see that.

---

4. Necessary conditions for convergence depend on the characteristic roots of equation (7.18), but a sufficient condition is simply $\lambda / b < 1 / w$, and which is also the necessary condition for the naive cobweb (Bergendorff, Hazell, and Scandizzo 1974).

even under assumptions (A.7) through (A.11), the variance and probability density function of price *do not* converge to the same limits for alternative choices of the γ weights (Bergendorff, Hazell, and Scandizzo 1974; Hazell and Scandizzo 1973). These properties of the equilibrium do depend on the way farmers form their price anticipations each year and, consequently, can only be enumerated through simulation-type procedures given explicit assumptions about the behavior involved. A similar result pertains for the stochasticity of input decisions, and hence X_j.

The results for $E(P_t)$ are clearly quite useful, but what happens if any of the assumptions (A.7) through (A.11) are relaxed?

- If either assumption (A.8) or (A.9) is relaxed to permit nonlinearities, then lim $E(P_t)$ usually becomes dependent on the γ weights in equation (7.16) and no longer corresponds to the intersection of demand and anticipated supply. Under these conditions the intersection price is only a linear approximation to the asymptotic expectation of price.

- Relaxation of assumption (A.7) leads to a situation in which the slope of the supply formation (λ) varies structurally over time, and the location of the anticipated supply function will no longer be fixed. However, if W and Γ converge,[5] then a stable equilibrium is attained with the above properties, though the length and nature of the adjustment path in attaining this equilibrium may be quite different. A similar result pertains when assumption (A.10) is relaxed providing lim $E(\hat{\rho}_t) = $ lim $E(P_t)$, otherwise approximations are again involved.

- Assumption (A.11) is only a simplifying assumption, and can be relaxed to consider autocorrelated yields providing the stochastic residual of the process satisfies assumption (A.11), that is, $e_t = \mu e_{t-1} + \mu_t$ where $cov(u_t, u_{t-1}) = 0$, and μ is a constant. In this case the basic equilibrium properties still hold, and the main effects are on the length and nature of the adjustment path.

If it is accepted that assumptions (A.7) through (A.11) and the possible modifications stated above are quite reasonable, then any mathematical programming model that could provide the intersection solutions for demand and anticipated supply would generate results that have a direct and relevant economic interpretation and that would be reasonably general with respect to the way in which farmers form their price and yield anticipations over time. The task of providing a modification to the Duloy-Norton objective function that solves this problem in an aggregate model remains.

---

5. Note that if assumptions (A.1) through (A.6) are a true description of competitive behavior, then Γ is unlikely to converge to its real statistical value, for there will almost certainly be nonzero correlations between individual farm outputs and market prices.

Solving for the Asymptotic Expectation of Price

X, Y, W, C, $\hat{\rho}$, and Γ have already been defined as aggregates of the farm x, y, M, c, $\hat{p}$, and Ω matrixes. Also, let Φ be an aggregate of the farm risk parameters ϕ.

There are definite problems in forming these aggregates that are closely related to the problem of establishing criteria of farm classification for exact aggregation in quadratic models. This problem lies beyond the scope of this chapter, and it is merely noted that the aggregate variables Γ and Φ must be chosen so that

$$(7.20) \qquad \Phi(X'\Gamma X)^{1/2} = \sum_k \phi_k (X'_k \, \Omega_k \, x_k)^{1/2},$$

where k denotes the kth individual farm. This equation states that the aggregate level of risk calculated for the model must be equal to the sum of individual risks over all farms. Without this condition, covariance relations between farms could be exploited in the aggregate model in seeking efficient diversification, and this would be inconsistent with assumed competitive behavior. For example, if all the farms were identical, a suitable choice of the aggregate variables would be $\Phi = (1/k)\phi$ and $\Gamma = \Omega$, so that

$$\sum_k \phi_k (x'_k \, \Omega_k \, x_k)^{1/2} = K \, \phi (x' \, \Omega \, x)^{1/2}$$
$$= \Phi(kx' \, \Gamma \, kx)^{1/2} = \Phi(X' \, \Gamma \, X)^{1/2},$$

where $X = Kx = \Sigma_k \, x_k$.

Given the necessary aggregate variables and parameters, it is now possible to modify the Duloy-Norton objective function to obtain the solution corresponding to the intersection of demand and anticipated supply schedules. This modified function is:

$$(7.21) \qquad \max U = X'W(A - 0.5BWX) - C'W - \Phi(X'\Gamma X)^{1/2},$$

where $X'W(A - 0.5BWX)$ is now the sum of expected areas under the demand curves and $C'X + \Phi(X' X)^{1/2}$ is a revised sum of areas under the supply curves.

To verify that equation (7.21) gives the desired market equilibrium solution, form the Lagrangian function

$$L = X'W(A - 0.5BWX) - C'X - \Phi(X'\Gamma X)^{1/2} + v' \, (b - DX),$$

where b and D denote the aggregate constraints and v is a vector of dual values. Apart from complementary slackness requirements, the necessary Kuhn-Tucker conditions are:[6]

$$(7.22) \qquad \frac{\partial L}{\partial X} = WA - WBWX - C - \Phi \, \Gamma X (X'\Gamma X)^{-1/2} - D'v \leq 0$$

and

$$(7.23) \qquad \frac{\partial L}{\partial v} = b - DX \geq 0.$$

Equation (7.23) is the feasibility requirement, and equation (7.22) can be rearranged as:

$$(7.24) \qquad (A - BWX) \leq W^{-1} \left[C + \Phi \, \Gamma X (X'\Gamma X)^{-1/2} + D'v \right].$$

Now WX is the vector of anticipated supplies $E(Y|X)$, and $A - BWX$ is the corresponding vector of market prices. Further, the right-hand side of equation (7.24) is the sum of expected marginal cost curves over all farms [the jth component is the sum of the right-hand sides of equations such as (7.11)]. That is, it is the vector of aggregate anticipated supply functions. The inequality in equation (7.24) states that, in aggregate, farmers must operate around some expected point on the anticipated supply functions that lie at or above the intersections with demand. Because of the complementary slackness condition $X(\partial L / \partial X) = 0$, optimality will occur at the intersection point for all nonzero activities in the solution, and then $A - BWX$ is the intersection price vector. Conditions, however, have already been established for this to be an approximation to the vector of asymptotically expected prices, and if these conditions are met, the perfectly competitive solution for $\lim E(P) = E$ (marginal costs) will have been obtained.

Linear Programming Approximations

The aggregate model with the objective function defined in equation (7.21) is a quadratic programming problem. Because of the large dimensions of any realistic sector model, and the difficulties that still exist with computer codes for quadratic programming, it is clearly desirable to linearize this problem.

---

6. In order for the Kuhn-Tucker conditions to be necessary and sufficient, it is required that the matrix $(WBX + \Gamma)$ be positive semidefinite. Since W is a diagonal matrix, it is sufficient that the individual matrices B and Γ be positive semidefinite. This condition is invariably satisfied for covariance (Γ) matrixes and is generally satisfied for demand coefficient (B) matrixes.

Duloy and Norton (1971, 1973) have shown how the term Y' $(A - 0.5BY)$, where $Y = WX$, can be linearized. To illustrate their method, consider the simplest case when B is diagonal, implying that the product demands are independent. Then, letting V_j denote the area under the demand curve (the definite integral) from 0 to Y_j for the jth product,

$$Y' (A - 0.5BY) = \sum_{j=1}^{n} V_j.$$

Now V_j is a quadratic, concave function when plotted against Y_j, and, since the programming model is a maximization problem, V_j can be approximated by a series of linear segments using conventional computer codes for linear programming. Duloy and Norton introduce segment weighting activities, V_{ij}, $0 \le V_{ij} \le 1$, $i = 1$ to k, for each V_j; assign segment upper bounds g_{ij} on Y_j, over which interval V_{ij} is relevant; and assign a single value of V_j—say, d_{ij}—which is to approximate V_j over the interval $Y_j \le g_{ij}$. They then suggest that the part of the programming problem involving Max $Y' (A - 0.5BY)$, with $Y = WX$, be replaced by the linear programming problem:

(7.25)
$$\max \sum_j \sum_{i=1}^{k} d_{ij} V_{ij},$$

such that

(7.26)
$$X_j m_j - \sum_i g_{ij} V_{ij} \ge 0, \quad \text{for all } j,$$

(7.27)
$$\sum_i V_{ij} \le 1, \quad \text{for all } j.$$

This method adds only two rows for each product, but permits inclusion of as many V_{ij} activities as necessary to increase the accuracy of the approximation to any desired degree of precision.

When Γ is estimated on the basis of time-series data, an efficient way of linearizing the remaining quadratic term $\Phi (X'\Gamma X)^{1/2}$ is to use the mean absolute deviation (MAD) method proposed by Hazell (1971).

Let $r_{jt} = P_{jt} m_{jt}$ denote the tth year's observation on the revenue of the jth activity X_j, $t = 1$ to T, and let $\bar{r}_j$ denote the sample mean revenue for the activity over the T years.[7] Then, the MAD estimator of the standard deviation of income[8] is:

(7.28)
$$\text{est}(X'\Gamma X)^{1/2} = \Delta \frac{1}{T} \sum_t \left| \sum_j (r_{jt} - \bar{r}_j) X_j \right|,$$

---

7. The raw data should first be analyzed for any trend and other systematic movements over time, and these components removed to obtain a random residual.

8. Since costs are nonstochastic by assumption, the variances of income and total revenue are identical.

where

$$\Delta = \left(\frac{T\pi}{2(T-1)}\right)^{1/2}$$

is Fisher's (1920) correction factor to convert the sample MAD to an estimate of the population standard deviation,[9] and π is the mathematical constant.

To obtain a linear programming formulation, define new variables $z_t \geq 0$ for all t such that

$$(7.29) \qquad z_t + \sum_j (r_{jt} - \bar{r}_j)X_j \geq 0,$$

where $\sum_j(r_{jt} - \bar{r}_j)X_j$ measures the deviation in total revenue from the mean, $\sum_j \bar{r}_j X_j$, for the tth set of revenue outcomes. Now, if the z_t variables are selected in a minimal way, then for each t either $z_t = 0$ when $\sum_j(r_{jt} - \bar{r}_j)X_j$ is positive, or z_t measures the absolute value of the negative deviation in total revenue when $\sum_j(r_{jt} - \bar{r}_j)X_j$ is negative. Consequently, $\sum_t z_t$ measures the sum of the absolute values of the negative deviations. Since the sum of the negative deviation around the mean is always equal to the sum of the positive deviations for any random variable, it follows that $2 \sum_t z_t$ is the sum of absolute deviations in total revenue, and hence

$$\text{est}(X'\Gamma X)^{1/2} = \left|\frac{2\Delta}{T}\right| \sum_t z_t.$$

In summary, the appropriate linear programming subproblem to minimize $\Phi (X'\Gamma X)^{1/2}$ is:

$$(7.30) \qquad \min K \sum_t z_t,$$

such that $z_t + \sum_j(r_{jt} - \bar{r}_j)X_j \geq 0$ for all t, where $K = \Phi (2\Delta / T)$ is a constant.

The reliability of this method, compared with using quadratic programming directly on $\Phi (X'\Gamma X)^{1/2}$, has been discussed elsewhere (Hazell 1971a; 1971b). It basically depends on the relative efficiency of the sample MAD compared to the sample standard deviation as an estimator of the population standard deviation.[10] Surprisingly, the sample MAD may actually be better than the sample standard deviation for skewed income distributions (Hazell 1971b), but it is less efficient for normal distributions (Hazell 1971a).

---

9. Strictly speaking, this form of Fisher's correction factor only holds when income is normally distributed (that is, $\sum_j r_{jt}X_j \sim N$). For other distributions, appropriate correction factors should be used (see Fisher 1920).

10. More favorable results about the ability of the sample MAD to rank farm plans efficiently (Thomson and Hazell 1972) do not hold in the current problem, because the MAD estimate of $(X'\Gamma X)^{1/2}$ is required to appear in the weighted objective function defined in equation (7.21).

In order to show how all these approximations fit together, a complete linear programming tableau is formulated in table 7-1, which approximates the solution to the original quadratic programming problem. If the quadratic variables V_j, $j = 1$ to n, are each approximated by k segments, the linearized problem adds an additional $kn + T$ activities and $2n + T$ rows to the problem, but this is highly efficient computationally when solutions are obtained through a revised simplex algorithm.

Conclusions

In this chapter a methodology has been developed for simulating, with a linear programming model, the market equilibrium of a perfectly competitive but risky agriculture in which producers behave according to an E,σ decision criterion. Such results are useful for comparative static analysis of policy problems, and, according to the degree of risk involved, should be more descriptive than existing models that ignore risk.

The development of this methodology has pinpointed a number of difficult issues with respect to both the design and implementation of aggregate risk models.

First, market equilibrium is considerably more complicated under risk than in a deterministic setting, with interactive effects between the way farmers form their anticipations about prices and yields and the properties of an equilibrium if attained. Consequently, it is difficult to design a general programming model that will always provide meaningful economic answers. In this chapter, the problem was resolved by specifying a set of plausible assumptions under which the proposed model is appropriate, albeit with an obvious loss in generality.

Second, an aggregate model must be defined with variables that are inherently difficult to measure. This is partly because such variables are based on individual farmers' utility functions and subjective expectations about stochastic variables and are therefore difficult to observe, but partly because such variables involve aggregation procedures that have not been adequately explored in the literature. Although practical ways of overcoming these problems are demonstrated in other chapters of this book, more refined procedures really ought to be developed.

Appendix. Derivation of Market Equilibrium Results for the Multiproduct Case

In the multiproduct case it is likely that there will be interrelations on the demand side as well as contemporaneous correlations between yields. With this in mind, the market structure can be written as:

Table 7-1. Layout of Tableau for the Linearized Problem

Constraint and equation number in text	Production activities			n sets of activities to linearize areas under demand functions				T negative deviation counters	RHS
	X_1	X_2	X_n	$V_{11}\cdots V_{1k}$	$V_{21}\cdots V_{2k}$	$\cdots$	$V_{n1}\cdots V_{nk}$		
Objective function, (7.21), (7.24), and (7.29)	$-c_1$	$-c_2$	$-c_n$	$d_{11}\cdots d_{1k}$	$d_{21}\cdots d_{2k}$		$d_{n1}\cdots d_{nk}$	$-K\cdots\cdots -K$	(Max)
n commodity balance constraints, (7.25)	m_1	m_2	m_n	$-g_{11}\cdots -g_{1k}$	$-g_{21}\cdots -g_{2k}$		$-g_{n1}\cdots -g_{nk}$		≥ 0
n convex combination constraints, (7.26)				$1\cdots 1$	$1\cdots 1$		$1\cdots 1$		≤ 1
T revenue constraints containing sample data, (7.28)	$r_{11}-\bar{r}_1$	....	$r_{1T}-\bar{r}_1$	$r_{21}-\bar{r}_2\cdots\cdots r_{n1}-\bar{r}_n$			$r_{2T}-\bar{r}_2\cdots\cdots r_{nT}-\bar{r}_n$	1 $\cdots$ 1	≥ 0
Resource constraints		D Matrix							$\leq b$ Vector

RHS = Right-hand side.

220

$$Y_t = \Gamma_t \, G_t \, (L) \, P_t$$

$$D_t = A - BP_t,$$

where Y_t and D_t are $n \times 1$ vectors of quantities supplied and demanded respectively; P_t is the $n \times 1$ vector of prices; Γ_t is the $n \times m$ matrix with ijth element $\gamma_{ijt} = \lambda_{ij} \, e_{ijt}$; $G \, (L)$ is a polynomial form in the lagged operator L; and A and B are the demand coefficient matrixes defined earlier.

It is again assumed that yields are distributed independently over time, so that $\text{cov} \, (e_{ijt}, \, e_{ijt-1}) = 0$, for all $i, \, j$, and that anticipated prices $\hat{p}_t$ are linear lagged functions of past prices. In particular, let:

$$\hat{p}_t = G_t \, (L) \, P_t = \begin{bmatrix} \sum\limits_{i=1}^{m} \delta_{1t-i} \, (L^i) & \cdots & 0 \\ & \ddots & \\ & & \\ & & \\ 0 & \cdots & \sum\limits_{i=1}^{m} \delta_{nt-1} \, (L^i) \end{bmatrix} \begin{bmatrix} P_{1t} \\ \\ \\ \\ P_{nt} \end{bmatrix},$$

where $\Sigma_{i=1}^{m} \, \delta_{jt-i} = 1$, for all j.

The expected market clearing price is then

(7.31) $$E(P_t) = B^{-1} \, A - B^{-1} \, \bar{\Gamma} \, G_t \, (L) \, E(P_t),$$

where $\bar{\Gamma} = E(\Gamma_t)$ has ijth element $\gamma_{ij} = \lambda_{ij} \, w_{ij}$, and $w_{i_j} = E(e_{ij})$.

Equation (7.31) defines a system of mth order difference equations, and if convergence occurs, has the particular solution

(7.32) $$\lim_{t \to \infty} E(P_t) = [I + B^{-1} \, \bar{\Gamma} \, G]^{-1} \, B^{-1} \, A,$$

where we have assumed that $G = \lim\limits_{t \to \infty} G_t$ exists.

Conditions for convergence can be developed as follows. Decomposing the G_t matrix in equation (7.31) and applying the lag operator gives the more explicit version:

$$EP_t = B^{-1} \, A - B^{-1} \, \bar{\Gamma} \, [G_{1t} \, EP_{t-1} + G_{2t} \, EP_{t-2}$$

$$+ \dots + G_{mt} EP_{t-m}],$$

where G_{it}, $i = 1$ to m, are $n \times n$ diagonal matrixes with jth diagonal element δ_{jt-1}. Necessary and sufficient conditions for equation (7.32) to hold are that the roots of the polynomial equation

$$\lim_{t \to \infty} [G_{1t} \, p^m + G_{2t} \, P^{m-1} + \dots + G_{mt} \, P] = 0,$$

where P^i is an $n \times 1$ vector with jth element $[p^i]$, are all within the unit circle. Sufficient conditions for convergence can, however, be obtained by noting that attainment of equation (7.32) is guaranteed if

$$\lim_{t \to \infty} (B^{-1} \bar{\Gamma} G_i)^t = 0, \quad i = 1 \text{ to } m,$$

but

$$\lim_{t \to \infty} (B^{-1} \bar{\Gamma} G_i)^{+t} = \lim_{t \to \infty} (H_i \Psi_i H_i^{-1})^t = \underline{0},$$

where

$$G_i = \lim_{t \to \infty} G_{it},$$

Ψ_i is the diagonal matrix formed by the characteristic roots of $B^{-1} \bar{\Gamma} G_i$, and H_i is the appropriate orthogonal transformation matrix.[11] This result implies that all the roots of the matrix $B^{-1} \bar{\Gamma} G_i$ must be real, nonnegative, and equal or less than 1, a result that generalizes the single product condition $(\lambda / b) < (1 / w)$.[12]

Turning now to the interaction price vector $(P^\star)$ for demands and anticipated supplies, we find that this occurs where:

$$\bar{\Gamma} G P^\star = A - BP^\star.$$

Solving, we obtain $P^\star = [I + B^{-1} \bar{\Gamma} G]^{-1} B^{-1} A$, which is the same as

$$\lim_{t \to \infty} E (P_t)$$

in equation (7.32).

References

Baumol, W. J. 1963. "An Expected Gain-Confidence Limit Criterion for Portfolio Selection." *Management Science*. Vol. 10 (October), pp. 174–82.

Bergendorff, H. G., Peter B. R. Hazell, and Pasquale L. Scandizzo. 1974. "On the Equilibrium of a Competitive Market When Production Is Risky." Washington, D.C.: World Bank, Development Research Center. Restricted circulation.

Charnes, A., and W. W. Cooper. 1959. "Chance Constrained Programming." *Management Science*. Vol. 6 (October), pp. 70–79.

Day, R. H.. 1963. "On Aggregating Linear Programming Models of Production." *Journal of Farm Economics*. Vol. 45 (November), pp. 797–813.

Dhrymes, P. J. 1970. *Econometrics*. New York: Harper and Row.

Duloy, John H., and Roger D. Norton. 1971. "Competitive and Noncompetitive

11. Note that B, $\bar{\Gamma}$, and G are assumed to be positive definite.

12. This result corresponds to the well-known stability requirement for the system of stochastic difference equations in equation (7.30). (See Dhrymes 1970, p. 519.)

Demand Structures in Linear Programming Models." Paper presented at the Summer Meeting of the Econometric Society, Boulder, Colorado, August 1971.

———. 1973. "CHAC, A Programming Model of Mexican Agriculture." In *Multi-Level Planning: Case Studies in Mexico*. Edited by Louis M. Goreux and Alan S. Manne. Amsterdam/New York: North-Holland/American Elsevier, pp. 291–337.

Fisher, R. A. 1920. "A Mathematical Examination of the Methods of Determining the Accuracy of an Observation by the Mean Error and by the Mean Square Error." *Royal Astronomical Society (Monthly Notes)*. Vol. 80, pp. 758–69.

Freund, R. J. 1956. "The Introduction of Risk into a Programming Model." *Econometrica*. Vol. 24 (July), pp. 253–63.

Hazell, Peter B. R. 1971a. "A Linear Alternative to Quadratic and Semivariance Programming for Farm Planning under Uncertainty." *American Journal of Agricultural Economics*. Vol. 53 (February), pp. 53–62.

———. 1971b. "A Linear Alternative to Quadratic and Semivariance Programming for Farm Planning under Uncertainty: Reply." *American Journal of Agricultural Economics*. Vol. 53 (November), pp. 664–65.

Hazell, Peter B. R. and Pasquale L. Scandizzo. 1973. "An Economic Analysis of Peasant Agriculture under Risk." Paper presented at the 15th International Conference of Agricultural Economists, San Paulo, Brazil, August 1973.

———. 1974. "Competitive Demand Structures under Risk in Agricultural Linear Programming Models." *American Journal of Agricultural Economics*. Vol. 56 (May), pp. 235–44.

Heady, Earl O., and Wilfred V. Candler. 1958. *Linear Programming Methods*. Ames: Iowa State University Press.

Hillier, F. S., and G. J. Lieberman. 1967. *Introduction to Operations Research*. San Francisco: Holden-Day.

Madansky, A. 1962. "Methods of Solution of Linear Programming under Uncertainty." *Operations Research*. Vol. 10, pp. 463–71.

Magnusson, G. 1969. *Production under Risk: A Theoretical Study*. Upsala: ACTA, Universitatis Upsaliensis.

Markowitz, H. M. 1959. *Portfolio Selection: Efficient Diversification of Investments*. New York: John Wiley and Sons.

Maruyama, Yoshihiro. 1972. "A Truncated Maximin Approach to Farm Planning under Uncertainty with Discrete Probability Distributions." *American Journal of Agricultural Economics*. Vol. 54 (May), pp. 192–200.

McFarquhar, A. M. M. 1961. "Rational Decision Making and Risk in Farm Planning." *Journal of Agricultural Economics*. Vol. 14 (December), pp. 552–63.

Officer, R. R., and A. N. Halter. 1968. "Utility Analysis in a Practical Setting." *American Journal of Agricultural Economics*. Vol. 50, pp. 257–77.

O'Mara, Gerald T. 1971. "A Decision-Theoretic View of the Microeconomics of Technique Diffusion in a Developing Country." Ph.D. dissertation. Stanford University.

Samuelson, Paul A. 1952. "Spatial Price Equilibrium and Linear Programming." *American Economic Review*. Vol. 42 (June), pp. 283–303.

Sengupta, J. K., and J. H. Portillo-Campbell. 1970. "A Fractile Approach to Linear Programming under Risk." *Management Science*. Vol. 16, pp. 298–308.

Stovall, J. G. 1966a. "Sources of Error in Aggregate Supply Estimates." *Journal of Farm Economics*. Vol. 48, pp. 477–79.

———. "Income Variation and Selection of Enterprises." *Journal of Farm Economics*. Vol. 48 (December), pp. 1575–79.

Takayama, T., and G. G. Judge. 1964. "Spatial Equilibrium and Quadratic Programming." *Journal of Farm Economics*. Vol. 46 (February), pp. 67–93.

———. 1971. *Spatial and Temporal Price and Allocation Models*. Amsterdam: North-Holland.

Thomson, K. J., and Peter B. R. Hazell. 1972. "Reliability of Using the Mean Absolute Deviation to Derive Efficient E, V Farm Plans." *American Journal of Agricultural Economics*. Vol. 54 (August), pp. 503–06.

Turnovsky, S. J. 1968. "Stochastic Stability of Short-Run Market Equilibrum under Variations in Supply." *Quarterly Journal of Economics*. Vol. 82 (November), pp. 666–81.

Zusman, Pinhas. 1969. "The Stability of Interregional Competition and the Programming Approach to the Analysis of Spatial Trade Equilibria." *Metroeconomica*. Vol. 21, fasc. 1 (January-April), pp. 45–57.

8

The Importance of Risk in Agricultural Planning Models

PETER B. R. HAZELL, ROGER D. NORTON,
MALATHI PARTHASARATHY, AND CARLOS POMAREDA

AGRICULTURAL PRODUCTION, particularly in developing countries, is generally a risky process, and considerable evidence exists to suggest that farmers behave in risk-averse ways.[1] Yet considerations of risk are rarely incorporated into regional or sectoral planning models; rather, farmers are assumed to behave in a risk-neutral, profit-maximizing way. Explicit representations of uncertain outcomes and farmers' attitudes toward them have appeared mostly in farm-level models. At the aggregate level, more indirect approaches have been used that amount to specifying "conservative" reactions to changes in the light of uncertainty; for example, the flexibility constraints of recursive programming (Day 1963).

Overview

On theoretical grounds, neglect of risk-averse behavior in agricultural planning models can be expected to lead to important overstatements of the output levels of risky enterprises (often reflected in overly specialized cropping patterns), hence also to overestimates of the value of important resources (for example, land and irrigation water). This chapter applies the theoretical framework of chapter 7 and reports results from two case studies in Mexico that were designed to measure the magnitudes of some of these biases; in the process, the chapter attempts to provide a quantification of risk aversion at aggregate farm levels. Other case studies using this methodology are found in chapters 12, 13, and 15 of this volume.

The two models used in this chapter are both regional linear programming models of the production of annual crops grown under irrigated conditions. They are static equilibrium models and, through incorpora-

---

1. See, in particular, the following case studies: Cancian (1973); Dillon and Anderson (1971); Francisco and Anderson (1972); Lin, Dean, and Moore (1974); Officer and Halter (1968); and O'Mara (1971, also chapter 9 of this book).

tion of linear demand functions, simulate competitive market equilibriums in which both prices and quantities are endogenous. The first model, ALPHA, comprises eight of the irrigation submodels of CHAC, and therefore it may be considered reasonably representative of the role of risk in CHAC. The second model, BETA, is a slightly revised and updated version of the model by Pomareda and Simmons (chapter 12).[2]

In the following we briefly review the risk specification of the models and show why ignoring risk-averse behavior in programming models can lead to biases in estimated crop outputs and the values of scarce resources. In the section "Description of the Models," we present the two models in some detail and discuss the methods used to estimate salient risk parameters. The section "Results of the Models" presents findings and demonstrates the numerical importance of incorporating risk-averse behavior into agricultural planning models. The final section contains our brief, concluding remarks.

Method of Incorporating Risk

The underlying behavioral assumption in our models is that farmers maximize expected net income less its standard deviation—an (E, σ) utility function—rather than expected profits. This assumption follows a tradition begun by Markowitz (1959) and Tobin (1958) regarding choice under uncertainty. More precisely, however, our assumption is a variation on the approach of Baumol (1963) who used an $(E, \phi\sigma)$ formulation that gives more reasonable answers than the straight (E, σ) approach in some cases. With Baumol's approach, the decisionmaker is assumed to establish subjectively a confidence limit and a floor on expected returns, to which the limit is applied. In this chapter, parametric programming techniques are used to derive values of the subjective parameter ϕ.

It is well known that a pure (E, σ) function, taken over a significant range of values, has some rather strange behavioral properties, and even the assumption of normally distributed outcomes will not rescue it from that fact. Nevertheless, it has been shown that, over a limited range of values, such a function can be viewed as an approximation (via truncated Taylor's series) to a polynomial function with all the desirable properties. More precisely, the first two terms of a Taylor's series yield an (E, V) utility function, and the (E, σ) efficiency frontier is a tangency to it at the equilibrium point.

2. Several versions of this model have been prepared over time at the request of the vegetable producers' association in northwest Mexico for use in their planting decisions.

Tsiang (1972, 1974) has provided a careful defense of the (E,σ) approach as a useful approximation in certain classes of choice, and it has proved to be helpful in improving predictions of behavior under risk at the farm level. This chapter builds on Tsiang's defense, with the purpose of showing one way in which it may be used in an aggregate rather than a micro model. One of Tsiang's conditions for (E,σ) analysis to be useful is that the risk be "small" relative to the total wealth of the risk taker. Although this condition may be met for the commercial, irrigated farms covered in our models, it may not be satisfied for the case of rainfed (*temporal*), subsistance farmers confronted with new technologies.

Given the underlying $(E,\phi\sigma)$ behavioral specification, the objective function of an aggregate linear programming model that simulates competitive market equilibria can be written as:

$$(8.1) \quad \max U = X'M (A - 0.5BMX) - C'X - \Phi(X'\,\Omega X)^{1/2},$$

where

X = A vector of aggregate crop levels

M = A diagonal matrix of average yields

C = A vector of cost coefficients

A,B = The coefficient matrixes of the linear demand structure $EP = A - BMX$, where EP is the expected price and B is assumed to be diagonal[3]

Φ = An appropriate aggregate of individual farm ϕ coefficients

Ω = An appropriate aggregate of individual farm covariance matrixes of activity revenues.

The derivation and justification of this objective function was provided by Hazell and Scandizzo in the preceding chapter. It is equivalent to the sum of expected values of producers' and consumers' surplus over all markets, and it gives the asymptotic values of expected quantities and market-clearing prices in equilibrium.

The usual assumption of profit-maximizing behavior is, of course, equivalent to setting Φ equal to zero in equation (8.1). Thus, the effect of ignoring risk-averse behavior depends on the properties of the term $\Phi(X'\,\Omega X)^{1/2}$, at least within the confines of our behavioral assumptions.

Let the constraint set of the aggregate model be denoted by

$$DX \le b,$$

where D is a matrix of technical coefficients, and b is a vector of resource supplies.

---

3. Diagonalization is obtained in the models by grouping commodities into demand independent groups (see chapter 3).

The Lagrangian of the model is then:

$$(8.2) \qquad L = X'M(A - 0.5BMX) - C'X$$

$$- \Phi(X'\Omega X)^{1/2} + v'(b - DX),$$

where v is a vector of dual values.

Now, from the necessary Kuhn-Tucker conditions, we obtain

$$(8.3) \qquad \frac{\partial L}{\partial x_j} = m_j(a_j - b_j m_j x_j) - c_j - \Phi(X'\Omega X)^{-1/2}$$

$$\sum_{i=1}^{n} \omega_{ij} x_i - \sum_{k=1}^{s} v_k d_{kj} \leq 0, \quad j = 1 \text{ to } n,$$

where lower case letters denote elements of the corresponding capital matrixes, n denotes the number of crops, and s denotes the number of constraints.

Complementary slackness conditions further require that, for all nonzero x_j in the solution, equation (8.3) must hold as a strict equality. Thus, we can rewrite equation (8.3) for any nonzero x_j as:

$$(8.4) \qquad EP_j = \frac{1}{m_j}\left[c_j + \sum_{k=1}^{s} v_k d_{kj} + \Phi(X'\Omega X)^{1/2} \sum_{i=1}^{n} \omega_{ij} x_i\right],$$

where we have used the fact that, since B is diagonal, then

$$EP_j = a_j - b_j m_j x_j.$$

In words, equation (8.4) states that, for each nonzero activity, the expected marginal cost per unit of output must be exactly equal to expected price. The expected marginal cost comprises expected own marginal cost, c_j, plus expected opportunity costs, $\Sigma_k v_k d_{kj}$, as reflected in the dual value of the resources used by that activity, plus a marginal risk factor, $\Phi(X'\Omega X)^{-1/2} \Sigma_i \omega_{ij} x_i$.

Now, had we made the usual assumption of risk neutrality, the marginal risk term in equation (8.4) would disappear (because $\Phi = 0$). Consequently, incorporating risk-averse behavior leads to different output levels in the model solution, and the direction of change from risk neutrality depends critically upon the sign of $\Sigma_{i=1}^{n} \omega_{ij} x_i$. Crops which have large variances in revenues or positively correlated revenues with most other crops, or both, will tend to have a positive marginal risk term, and this will lead to a lower output under risk-averse behavior. In contrast, crops that have negatively correlated revenues with most other crops will tend to have a negative marginal risk term, and hence their output will be increased under risk-averse behavior.

To show the effect of risk-averse behavior on the valuation of scarce resources, we can rearrange equation (8.4) as:

$$(8.5) \qquad v_s = \left[m_j\, E(P_j) - c_j - \Phi(X'\Omega X)^{-1/2} \sum_{i=1}^{n} \omega_{ij}\, x_i - \sum_{k=1}^{s-1} v_k\, d_{kj} \right] / d_{sj},$$

where, by suitable rearrangement over k, v_s can be the shadow price of any selected resource.

Clearly, the imputed value of the sth resource for the jth nonzero activity will be greater or smaller than its value under risk neutrality depending upon the sign of $\Sigma_i\, \omega_{ij}\, x_i$. But the outcome also depends on the valuations v_k, $k \neq s$. Equation (8.5) must, of course, hold for all nonzero activities, and so we have a simultaneous model in the v_k. The overall effect of risk-averse behavior on the value of v_s is not therefore apparent from equation (8.5). It depends on the total risk effect of the complete portfolio of farm crops.

We do know, however, that since $\Phi(X'\Omega X)^{1/2} \geq 0$ in the model objective function (8.1), then the value of the objective is smaller under risk-averse behavior than under risk neutrality. Euler's theorem then implies that the total valuation of the scarce resources must be smaller. This, however, still permits the possibility of some resources' increasing in value, providing that others are reduced by sufficiently large amounts. For those resources whose imputed value is reduced by the inclusion of risk (and very likely that means most, if not all, resources), how do we interpret the new valuation? To the extent that the model's structure faithfully reflects individual and market decisions, it can be said that an adequate treatment of risk normally reduces the price that farmers would be willing to pay for their production inputs.

Description of the Models

The two models to be described both deal with annual crop production in selected irrigation districts in Mexico.

The ALPHA model

The first model (hereafter called the ALPHA model) is in actuality part of CHAC and encompasses eight of the more than one hundred administrative districts of the Mexican Ministry of Water Resources (Secretaría de Recursos Hidráulicos, SRH). These selected districts are not contiguous but are scattered throughout the arid agricultural areas of Mexico; they are among the largest districts in their respective regions:

	District
Pacific Northwest	Culiacán, Comisión del Fuerte, Guasave, Río Mayo, Santo Domingo
North Central	Ciudad Delicias, La Laguna
Northeast	Bajo Río San Juan

Table 8-1. *Average District Cropping Patterns, 1967–68 to 1969–70, ALPHA Model*
(harvested hectares)

					District					Percentage of national production
Crop	El Fuerte	Culiacán	Río Mayo	Guasave	Cuidad Delicias	Bajo Río San Juan	Santo Domingo	La Laguna	Aggregate	
Dry alfalfa	1,988	—	2,144	—	6,510	1,190	285	5,498	16,425	34
Cotton	46,364	—	15,535	—	7,903	—	17,585	67,964	156,541	25
Green alfalfa	—	543	—	—	—	—	—	5,224	5,767	2
Rice	11,335	23,568	—	3,480	—	—	—	—	38,383	25
Sugarcane	12,706	24,172	—	3,737	—	—	1,098	—	36,878	12
Safflower	4,790	13,374	10,435	—	—	—	—	—	33,434	29
Barley	—	—	112	—	—	—	—	—	112	1
Chile	386	1,570	—	48	—	—	—	—	2,004	9
Beans	16,224	11,024	—	202	—	—	—	—	27,450	3
Chickpeas	561	938	—	271	—	—	—	—	1,770	1
Tomatoes	3,049	9,563	—	581	—	—	—	—	13,193	37
Sesame	3,010	2,815	8,390	144	—	—	—	—	14,359	4
Maize	10,792	4,302	4,071	2,420	10,053	54,269	1,038	6,213	93,158	2
Cantaloupe	231	387	—	722	—	—	—	—	1,340	4
Potatoes	1,320	—	—	—	—	—	—	—	1,320	5
Cucumbers	—	—	—	8	—	—	—	—	8	0
Watermelons	775	325	—	41	—	74	—	—	1,197	5
Sorghum	24,238	22,795	10,616	1,238	7,719	19,876	—	5,592	92,074	11
Soybeans	16,264	4,392	11,886	—	—	—	—	—	32,542	20
Wheat	23,561	3,057	29,969	5,742	29,668	1,048	11,738	16,150	120,933	16
Total	177,576	122,825	93,158	18,634	61,853	76,457	31,744	106,641	688,888	
Number of farms	16,484	6,224	9,185	2,984	10,710	4,480	647	48,341	99,055	
Available hectares per farm	10	12	8	6	4	16	47	2	5.8	

— Zero.

Taken together, the eight districts account for significant shares of the national production of cotton, tomatoes, dry alfalfa, rice, soybeans, and safflower (see table 8-1). They also produce a wide range of cereal crops and vegetables, and some sugarcane. Some double cropping is practiced in all the districts, particularly in the vegetable-growing areas. The average district cropping patterns for the crop years 1967–68 and 1969–70 are given in table 8-1; not covered is a small percentage of land devoted to crops that are not included in the models. Crop production is almost entirely dependent on irrigation in all eight districts, and any small areas of rainfed land have been excluded.

In total, the eight district models cover 99,000 farms, 5.8 hectares in average size, and a district breakdown is included in table 8-1. For modeling purposes, each district is treated as a single large farm. The farms are presumed to be sufficiently homogeneous so that this procedure is unlikely to lead to any serious problems of aggregation bias. The model activities provide for the production, in each district, of crops in table 8-1 grown by that district, each with a choice of three mechanization levels and two planting dates. A set of labor activities provides flexiblity in selecting seasonal combinations of family and hired day labor. Family labor is charged a reservation wage of half the hired day-labor rate, a value derived from CHAC (see chapter 2). Purchasing activities provide for the supplies of mules, machinery, and irrigation water. Seasonal constraints are imposed on land and labor, and an annual constraint is imposed on water supplies.[4] Technical coefficients and costs are taken at average levels for 1967–68 to 1969–70. The model constraints are also based on this period. Average yields are based on the six-year period from 1966–67 to 1971–72, and risk parameters were estimated from time-series data spanning the period from 1961–62 to 1970–71.

The district models are linked in block diagonal form and are integrated into an aggregate market structure, similar to that in CHAC. That is, the market comprises linear domestic demand functions of the form $EP = A - BMX$, and has import and export possibilities at fixed prices. For simplicity, and to approximate cross-elasticity relations in demand, the crops are classified into demand independent groups, and linear substitution is allowed between products within each group at rates fixed by base-year relative prices.[5] The definition and characteristics of these demand groups are summarized in table 8-2. The demand curves for each

---

4. We assume the input costs and the resource constraint values to be nonstochastic; this assumption can be relaxed, using one of several available techniques to handle stochastic constraints. See, for example, Charnes and Cooper (1959), Madansky (1962), and Maruyama (1972).

5. For a more detailed description, see chapter 3. Income effects are ignored in this procedure because this is a partial equilibrium model.

Table 8-2. *Characteristics of Demand Groups, ALPHA Model*

Demand group	Commodity	Base-period price (pesos per metric ton) Commodity	Group-price index[a]	Own-price elasticity
1	Sugarcane	70	70	−0.25
2	Tomatoes	1,150	1,150	−0.4
3	Chile	1,500	1,500	−0.2
4	Cotton fiber	5,770	5,770	−0.5
5	Dry alfalfa	400		
	Green alfalfa	100		
	Barley	930		
	Chickpeas	990	446	−0.3
	Maize	860		
	Sorghum	630		
6	Rice	1,220		
	Beans	1,830		
	Chickpeas	990	1,285	−0.3
	Potatoes	930		
7	Maize	860	817	−0.1
	Wheat	800		
8	Cantaloupe	680	741	−2.0
	Watermelons	780		
9	Safflower	1,550		
	Sesame	2,410		
	Cottonseed oil	830	1,164	−1.2
	Soybeans	1,600		
10	Cucumbers	590	590	−0.6

a. Group-price indexes are computed using base-year quantity weights.

group have the same price elasticities as in CHAC but are located at mean output levels appropriate for the eight district aggregates according to the procedure in chapter 11. Export and import constraints are also taken from CHAC and are prorated according to the ratio of output from the eight districts to national output for each product.

The BETA model

The second model (hereafter called the BETA model) was developed by Pomareda and Simmons (see chapter 12) and, for the purposes of this study, can be viewed as a model of the irrigation districts of Culiacán, Humaya, El Fuerte, and Guasave in the state of Sinaloa, with a simplified representation of competitive supplies from Guatemala. Taken together, these four districts accounted, in 1973–74, for about 90 percent of the national exports of tomatoes, green peppers, and cucumbers, and about 40 percent of the national exports of melons.

For modeling purposes, Culiacán and Humaya are grouped as a single region [elsewhere in this volume, they are grouped with San Lorenzo under the name "Culmaya"], as are El Fuerte and Guasave. This grouping permits satisfactory consideration of water transfers between district irrigation authorities. The two regions are then treated as single decision-making units.

Although all short-cycle crops in each region are included in the model, special emphasis was given to the modeling of activities representing vegetable production. Each vegetable is allowed several planting dates; and the yields for each planting date are disaggregated by months (the harvest period of a hectare of tomatoes planted in September, for example, may last as long as three and a half months). Yields are also disaggregated into exportable and nonexportable quality according to U.S. Department of Agriculture regulations on the quality of imports. Nonexportable qualities are channeled into domestic Mexican markets.

The principal resource constraints for each region are monthly supplies of land and labor and the annual supply of water. Labor requirements are specified by three categories: labor for cultivation, harvesting, and packing vegetables. Hiring activities provide for unlimited supplies of labor for each category of work, but at different wage rates. All input costs are at 1973–74 prices.

The market structure in the BETA model is rather more complicated than that in the ALPHA model, although prices again are made endogenous through the incorporation of linear demand schedules. The complexity arises because Mexican vegetable exports (produced mainly in Sinaloa) have significant price effects in the U.S. market and because they compete directly with Guatemalan exports. Consequently, to model export demands adequately, the U.S. and Guatemalan vegetable markets are incorporated directly into the model, along with the supply of melons produced competitively in the Apatzingan district of Mexico. A further source of complexity arises because U.S. prices are treated endogenously on a monthly basis.

Mexican domestic demands are treated according to CHAC, with the demand schedules and price elasticities given in table 8-3. Prices of vegetables other than tomatoes (peppers, cucumbers, and cantaloupes) are fixed with perfectly elastic demands. The supply of melons from Apatzingan is simply incorporated through four seasonal linear programming activities.

Vegetable supplies from Guatemala (peppers, cucumbers, cantaloupes, and honeydews) are incorporated through another linear programming submatrix, which simply contains a single production vector per crop. Domestic demands in Guatemala are assumed to be perfectly elastic at fixed prices.

The U.S. market is incorporated by means of monthly linear supply

Table 8-3. *Demand Functions for Tomatoes and Traditional Crops in Mexico, BETA Model*

Crop	Demand equation[a]	Direct-price elasticity
Tomatoes	$P = 2.993 - .00008372Q$	-0.5
Sesame	$P = 3.068 - .00011210Q$	-1.2
Cotton	$P = 3.276 - .00000537Q$	-0.5
Rice	$P = 2.960 - .00001536Q$	-0.3
Safflower	$P = 2.069 - .00000531Q$	-1.2
Beans	$P = 5.573 - .00008800Q$	-0.3
Chickpeas	$P = 3.448 - .00007409Q$	-0.3
Maize	$P = 1.126 - .00001938Q$	-0.2
Sorghum	$P = 1.185 - .00000208Q$	-0.3
Soybeans	$P = 2.334 - .00000423Q$	-1.2
Wheat	$P = 0.936 - .00000107Q$	-0.5

Note: Price elasticities were taken from CHAC. Mean prices and quantities were taken from SRH (1972).

a. Quantity (Q) in thousands of kilos; price (P) in pesos per kilo.

and demand functions (see table 8-4). It is assumed that supplies and demands are both independent through time, and the U.S. prices in other periods are fixed (Mexican exports do not compete in these months).

The model maximand is a simple generalization of the objective function of the CHAC kind, in that the producers' and consumers' surplus are summed by product, by season, and by country.[6] Risk is introduced only in the two regional models for Sinaloa, again using an $(E,\phi\sigma)$ utility formulation.

Estimation of risk parameters

To incorporate risk-averse behavior into the models in accordance with equation (8.1), it is necessary to have estimates of Φ and Ω for each irrigation district. In principle, Φ and Ω should be suitable weighted averages of the parameters of risk aversion and the covariance matrixes of revenue of individual farms. Such information is not, of course, available—even if it were, suitable aggregation procedures have not yet been developed. Consequently, in formulating the ALPHA and BETA models, more approximate and aggregate procedures were used.

ESTIMATION OF Ω MATRIXES. The only available data on revenue variations were time-series data on prices and yields at the level of irrigation districts. Using these data (thirteen years in the case of the ALPHA model,

---

6. For a mathematical description, see chapter 12 and Pomareda and Simmons (1977).

Table 8-4. *Supply and Demand Functions for Vegetables, BETA Model*

Product	Month	Supply	Demand
		Equation[a]	
Tomatoes	December	$P = .0000770Q$	$P = 7.241 - .000043755Q$
	January	$P = .0001052Q$	$P = 6.116 - .000036722Q$
	February	$P = .0001864Q$	$P = 4.719 - .000020254Q$
	March	$P = .0001705Q$	$P = 7.098 - .0000044305Q$
	April	$P = .0001278Q$	$P = 5.398 - .000019006Q$
	May	$P = .0000541Q$	$P = 8.248 - .000039815Q$
Peppers	December	$P = .0002715Q$	$P = 10.935 - .00046591Q$
	January	$P = .0003822Q$	$P = 11.041 - .00040690Q$
	February	$P = .0007818Q$	$P = 9.075 - .00026592Q$
	March	$P = .0006631Q$	$P = 10.835 - .00035203Q$
	April	$P = .0004780Q$	$P = 12.203 - .00042629Q$
Cucumbers	December	$P = .0003792Q$	$P = 4.046 - .00016589Q$
	January	$P = .0007304Q$	$P = 3.905 - .00011539Q$
	February	$P = .0010917Q$	$P = 3.968 - .00009907Q$
	March	$P = .0009442Q$	$P = 4.852 - .00015937Q$
	April	$P = .0002798Q$	$P = 3.562 - .00008637Q$
Cantaloupes	January	$P = .240380Q$	$P = 5.126 - .0004930Q$
	February	$P = .048850Q$	$P = 5.126 - .0004930Q$
	March	$P = .024020Q$	$P = 7.389 - .0001910Q$
	April	$P = .016450Q$	$P = 5.699 - .0000411Q$
	May	$P = .000110Q$	$P = 6.125 - .0000205Q$
Honeydews	January	n.a.	$P = 5.110 - .0013200Q$
	February	n.a.	$P = 5.110 - .0013200Q$
	March	n.a.	$P = 2.843 - .0002630Q$
	April	n.a.	$P = 3.170 - .0002240Q$

n.a. Not available.

Note: For supply, an elasticity of 1.00 is assumed.

a. Quantity (Q) in thousands of kilos; price (P) in pesos per kilo.

but only six years for the BETA model), covariance matrixes of crop revenues were calculated for each district modeled after detrending the original crop revenues by linear regressions.

There are two potential problems with these estimated Ω matrixes. First, the appropriate Ω matrixes should be aggregates of the covariance matrixes subjectively perceived by farmers at market equilibrium, and these may well differ from observed statistical relations. The estimated matrixes are therefore good to the extent that farmers' perceived values have converged to the observed statistical relations and that the time-series data covered an equilibrium period. The latter requirement may not be too unrealistic, in that few crops showed statistically significant or numerically important trends in their cropping areas during the time periods under consideration.

Second, the Ω matrixes should be direct aggregates of covariance matrixes of individual farms and should not incorporate any covariance relations that may exist between farms within a district. This is because it is required that:[7]

$$\Phi_h \left(X_h' \, \Omega_h \, X_h \right)^{1/2} = \sum_{k \in h} \Phi_k \left(x_k' \, \Omega_k \, x_k \right)^{1/2},$$

Where h denotes the hth irrigation district and k denotes the kth farm. The estimated Ω_h matrixes are therefore good to the extent that the Ω_k matrixes are the same for all farms in the hth district, and that the covariance between the revenues of any two crops i and j are the same between farms as within farms. That is,

$$\text{cov} \left(r_{ki}, \, r_{kj} \right) = \text{cov} \left(r_{\ell i}, \, r_{\ell j} \right) = \text{cov} \left(r_{\ell i}, \, r_{kj} \right),$$

where r denotes activity revenue and k, ℓ are farms.[8] Unfortunately, suitable cross-sectional samples of farm data could not be found to test these requirements.

ESTIMATION OF ϕ. In the absence of any empirical data from which to estimate the ϕ coefficients, the basic procedure followed was to search, through postoptimality techniques, values of ϕ that enabled the models to best describe a set of base-year prices. In both models we make the simplifying assumption that ϕ is the same for all irrigation districts. Various measures of "goodness of fit" were tried, but a simple average of the absolute values of the deviations in prices served as well as any (hereafter called the MAD, mean absolute deviation).

This method of estimating ϕ poses the obvious difficulty that it will pick up errors in model misspecification and data, and that there is no way

---

7. See chapter 7.

8. Under these assumptions, and letting R_j denote the average revenue of the jth crop observed at the district over all farms, and K the number of farms in the district, we obtain:

$$\text{cov} \left(R_i \, R_j \right) = \frac{1}{K^2} \sum_k \sum_\ell E(r_{ki} \, r_{\ell j}) - \frac{1}{K^2} \sum_k \sum_\ell E(r_{ki}) \, E(r_{\ell j})$$

$$= \frac{1}{K^2} \sum_k \sum_\ell \text{cov} \left(r_{ki}, \, r_{\ell j} \right)$$

$$= \frac{1}{K^2} \sum_k \sum_\ell \text{cov} \left(r_{ki}, \, r_{kj} \right)$$

$$= \frac{1}{K} \sum_k \text{cov} \left(r_{ki}, \, r_{kj} \right)$$

$$= \text{cov} \left(r_{ki}, \, r_{kj} \right), \text{ for all } k.$$

Table 8-5. *Price Solutions by Commodity Group for Different Values of* Φ, *ALPHA Model*
(pesos per metric ton)

Demand group	Values of Φ						Base-period index
	0	0.5	1.0	1.5	2.0	2.5	
1. Sugarcane	68	70	72	68	68	68	70
2. Tomatoes	321	703	988	1,200	1,494	1,772	1,150
3. Chile	1,031	1,125	1,124	1,352	1,330	1,547	1,500
4. Cotton fiber	5,770	5,770	5,770	5,770	5,770	5,770	5,770
5. Forage crops	499	475	445	449	444	443	446
6. Food crops	1,373	1,316	1,241	1,233	1,227	1,217	1,285
7. Cereals	1,044	1,043	992	970	958	989	817
8. Melons	446	416	420	468	476	416	741
9. Vegetable oils	1,014	1,004	1,022	1,058	1,133	1,200	1,164
10. Cucumbers	516	436	284	148	148	148	590
MAD[a]	219	175	153	123	146	172	

Note: Prices are reported as group indexes, using base-year quantity weights (these weights are discussed in chapter 3).

a. The MAD is the mean absolute deviation of the solution value from the base-period values.

of knowing how serious these errors might be other than to judge the "reasonableness" of the estimated parameter.

Results of the Models

The model Φ coefficients are a direct representation of risk-averse behavior at the aggregate farm level. Thus, solving the models[9] for different values of Φ provides direct information about the effects of different degrees of risk aversion on equilibrium prices and quantities and gives a basis for quantifying the actual value of Φ.

Quantification of risk aversion at the aggregate level

Pertinent results from the ALPHA model are presented in tables 8-5 and 8-6. Table 8-5 shows the effect of different Φ values on the domestic equilibrium prices for the commodity groups delineated in table 8-2. On

9. The models are solved with a linear programming algorithm using the linearization techniques of Duloy and Norton (chapter 3 of this volume) and Hazell (1971).

Table 8-6. *Quantities Produced for Domestic Market for Different* Φ *Values, ALPHA Model*
(metric tons)

Crop	Values of Φ						Base-year quantities
	0	0.5	1.0	1.5	2.0	2.5	
Dry alfalfa	137,323	139,120	140,916	140,916	140,916	140,916	179,019
Cotton	484,488	476,111	445,869	358,485	285,965	268,476	243,454
Green alfalfa	173,445	175,714	177,983	177,983	177,983	177,983	226,109
Rice	115,369	115,784	118,774	118,774	141,885	143,695	126,197
Sugarcane	2,659,859	2,659,859	2,659,859	2,659,859	2,659,859	2,659,859	2,627,020
Safflower	75,777	74,467	72,752	67,437	86,449	70,914	72,490
Barley	510	517	523	523	523	523	665
Chile	15,409	15,161	15,161	14,789	14,789	14,294	14,459
Beans	30,278	30,278	31,060	31,060	26,428	26,765	33,001
Chickpeas	1,239	1,264	1,272	1,272	1,250	1,251	1,585
Tomatoes	223,682	202,712	185,237	171,257	153,781	136,241	174,752
Sesame	9,643	9,475	9,258	8,581	8,161	7,662	9,224
Maize	194,989	200,993	198,093	163,852	163,851	163,406	210,801
Cantaloupe	9,966	10,589	10,589	9,967	9,966	10,589	6,935
Potatoes	45,065	45,065	46,228	46,228	21,734	22,010	27,139
Cucumbers	20	33	35	359	2,195	2,301	19
Watermelons	22,213	23,601	23,601	22,213	22,213	23,601	10,850
Sorghum	336,856	341,263	345,670	345,669	345,670	345,669	285,818
Soybeans	59,815	58,781	63,359	75,595	63,982	68,706	57,220
Wheat	326,064	320,453	328,585	369,565	369,566	367,959	343,979

the one hand, the prices of groups 2 (tomatoes), 3 (chile), and 9 (vegetable oils) all increase with Φ, indicating corresponding reductions in the quantities produced for the domestic market. On the other hand, the prices for groups 5 (forage crops), 6 (food crops), 7 (cereals), 8 (melons), and 10 (cucumbers) decrease as Φ increases, indicating that production of these crops for a domestic market increases as producers become more risk averse. The quantity effects are shown in detail in table 8-6 at the individual commodity level. The prices of groups 1 (sugarcane) and 4 (cotton fiber) show no response to the risk-aversion parameter.

These results confirm the ambiguities involved in predicting the effect of risk-averse behavior on the supplies of individual crops as discussed in the section "Method of Incorporating Risk," above. They also suggest a useful definition of riskiness in crop production that takes intercrop relations into account. High- (low-) risk crops can be defined as those in which production decreases (increases) as producers become more risk averse, whereas risk-neutral crops are those whose production is unaffected by Φ.

The last columns in tables 8-5 and 8-6 contains the base-year values (1967–68 to 1969–70 averages) of prices and quantities. By comparing the model solutions for different Φ values with these base-year values, we have a basis for selecting the "best-fitting" value of Φ.

Clearly, the solution corresponding to risk neutrality ($\Phi = 0$) is quite unsatisfactory. It predicts unrealistically high levels of production of cotton, tomatoes, cantaloupes, potatoes, watermelons, and sorghum, and particularly low prices for groups 2, 3, and 8. There is a definite improvement in both the price and quantity fits as Φ increases, but this deteriorates again as Φ approaches 2.5. In the latter solution, the quantities produced of tomatoes, maize, and potatoes become unrealistically low, whereas those for rice, cucumbers, sorghum, and wheat become too large.

In selecting a value of Φ, we have chosen to concentrate on the commodity group prices because the market structure of the model can only be expected to work best at the level of the demand group. The last row of table 8-5 reports the simple MAD's (mean absolute deviations) of the price fits and clearly demonstrates the superiority of the solution for $\Phi = 1.5$.

We conclude that introducing risk-averse behavior into the ALPHA model significantly improves its predictive power compared with the more usual assumption of risk neutrality (the price MAD is reduced by 44 percent when Φ is increased from 0 to 1.5), and that a reasonable measure of the risk parameter at the aggregate farm level may be about 1.5.[10]

Table 8-7 reports the comparable price solutions from the BETA model. The commodities were not grouped in this model (demand independence is assumed), and, because some prices are fixed exogenously to the model, these are not reported in table 8-7. Table 8-8 shows the effect of different Φ values on the cropping patterns in the state of Sinaloa. We reported production in metric tons for the APLHA model because there are important yield differences between districts in that model, but this is not a problem with the BETA model. The figures in hectares incorporate production for both the export and domestic markets, which is also an essential feature given the importance of vegetable exports in the BETA model.

The model shows considerable flexibility in its response to different Φ values, a feature reflected in the wide range of hectarage values for each crop. Tomatoes, peppers, cotton, safflower, and soybeans exhibit high-risk behavior in that their production falls rapidly as risk aversion increases, whereas wheat, sesame, and maize are clearly low-risk crops.

Comparison of the model solutions with base-period values (1973–74 actual figures in this case) again shows that introducing risk-averse behavior leads to significant improvements in the model's predictive powers as compared with risk neutrality. In this case, the best price fit occurs when $\Phi = 0.5$ (the price MAD is reduced by 52 percent as Φ is increased from 0 to 0.5).

---

10. Assuming normal distributions, this Φ value corresponds to a 6.7 percent confidence band, which is very close to the sort of risk levels accepted by statisticians!

Table 8-7. *Price Solutions for Different Values of* Φ, *BETA Model*
(pesos per kilo)

Crop	Values of Φ				Base-year values
	0	0.5	1.0	1.5	
Vegetables[a]					
Tomatoes	3.493	3.879	4.034	4.226	3.850
Peppers	4.002	4.747	4.759	4.978	4.852
Cucumbers	2.824	2.888	2.900	2.676	2.764
Other[b]					
Cotton	2.414	2.524	2.560	2.619	2.470
Safflower	1.371	1.526	1.658	1.839	1.600
Soybeans	1.485	1.490	1.534	1.622	1.450
Chickpeas	1.427	1.447	1.449	1.451	2.100
Rice	1.082	0.987	0.871	0.884	1.075
Beans	2.001	1.993	1.979	2.026	1.980
Wheat	0.825	0.775	0.743	0.725	0.800
Sorghum	0.655	0.637	0.589	0.616	0.625
Sesame	3.055	2.842	2.340	2.286	2.778
Maize	0.995	0.873	0.745	0.654	0.844
MAD	0.210	0.101	0.164	0.215	

a. Export prices; that is, average U.S. wholesale prices for winter season.
b. Domestic prices.

Effects on supply response behavior

Our results have already shown that the introduction of risk-averse behavior into the models leads to different levels of production for most crops compared with the assumption of risk neutrality. These results are, of course, estimates of single points on the product supply functions, but they amply demonstrate the biases inherent in ignoring risk-averse behavior in aggregate linear programming models. We turn now to an exploration of the broader effects of risk-averse behavior on both the slope and location of the domestic supply response functions for selected crops.

Given that most domestic prices are endogenous to our two models, it is not possible to derive directly the effects of price changes on domestic supplies. Rather, the location of the domestic demand curves must be shifted, and the model allowed to determine the new equilibrium values of both prices and the quantities supplied. Supply response functions derived in this way also allow for price and quantity adjustments in all other markets; they are not, therefore, the partial supply functions described in economic textbooks but must be considered as total supply response relations.

Table 8–8. *Cropping Area in Sinaloa for Different Values of* Φ, *BETA Model*
(hectares)

Crop	Values of Φ				Base-year quantities
	0	*0.5*	*1.0*	*1.5*	
Vegetables					
Staked tomatoes	21,713	15,117	11,996	8,595	14,200
Ground tomatoes	4,459	0	0	0	1,400
Green peppers	11,984	3,379	3,379	2,409	2,909
Cucumbers	2,107	1,869	1,697	2,531	3,643
Other					
Cotton	64,232	53,409	51,904	46,210	48,075
Safflower	98,976	72,983	52,751	27,187	81,471
Soybeans	104,685	103,267	102,005	89,111	117,827
Chickpeas	17,115	16,646	15,396	15,396	25,844
Rice	31,308	33,047	35,576	34,733	51,505
Beans	32,165	32,165	32,187	32,180	35,524
Wheat	28,685	48,776	54,142	68,697	49,989
Sorghum	51,376	55,266	57,005	55,717	45,798
Sesame	218	2,597	8,558	9,155	2,584
Maize	7,087	14,875	18,769	18,889	17,740

To simplify the presentation, we have selected two commodities from each model: commodity groups 2 (tomatoes) and 7 (cereals) from the ALPHA model, and sorghum and safflower from the BETA model. In each experiment, the demand curve for the relevant group or commodity was rotated to the right at discrete intervals of 5 percent each on the quantity axis while holding all other demand curves at their initial positions. The experiments were repeated for both risk neutrality and the best-fitting values of Φ. The supply response functions so derived are reported in figures 8–1 through 8–4. The points on these functions depict model solutions, and these are numbered so that 1 represents the base-period solution, 2 represents the solution with a 5 percent quantity shift in the relevant demand, 3 represents a 10 percent shift, and so on. Thus, points labeled with the same numbers in each figure correspond to solutions with different Φ values but with identical demand structures.

The two models show markedly different supply response behavior. The response relations obtained from the ALPHA model (figures 8–1 and 8–2) are all highly elastic, with supplies to the domestic market increasing by about the same percentage as the corresponding shifts in demand. Apparently, relative equilibrium prices in the ALPHA model are such that production of tomatoes and cereals is constrained by the volume of domestic demand rather than by the marginal profitability of these crops.

Figure 8-1. *Domestic Supply Response for Tomatoes, ALPHA Model*

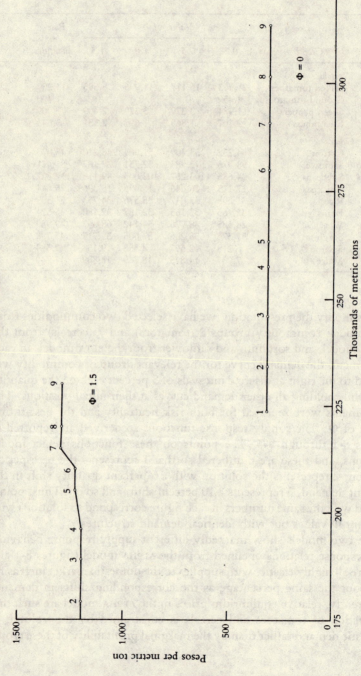

Pesos per metric ton

Thousands of metric tons

Φ = 1.5

Φ = 0

Note: Solutions are numbered as follows: 1, base period; 2, with 5 percent quantity shift in relevant demand; 3, with 10 percent quantity shift; 4, with 15 percent quantity shift; and so on.

242

Figure 8-2. *Domestic Supply Response for Cereals, ALPHA Model*

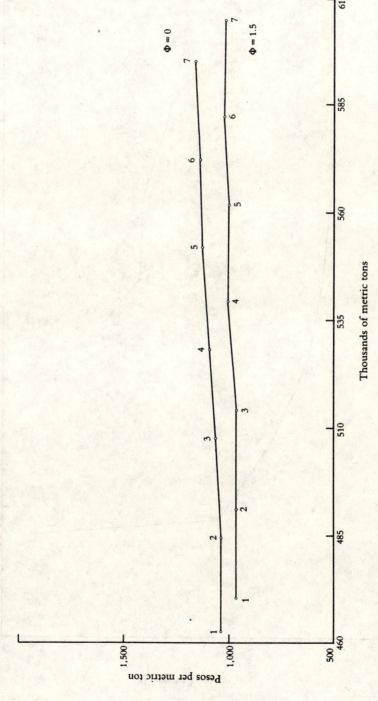

Φ = 0

Φ = 1.5

Pesos per metric ton

1,500

1,000

500

460   485   510   535   560   585   610

Thousands of metric tons

Note: See the legend to figure 8-1.

243

Figure 8-3. *Domestic Supply Response for Sorghum, BETA Model*

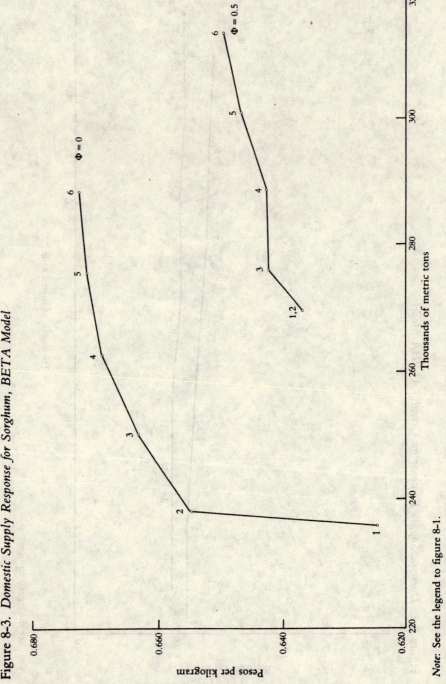

Note: See the legend to figure 8-1.

244

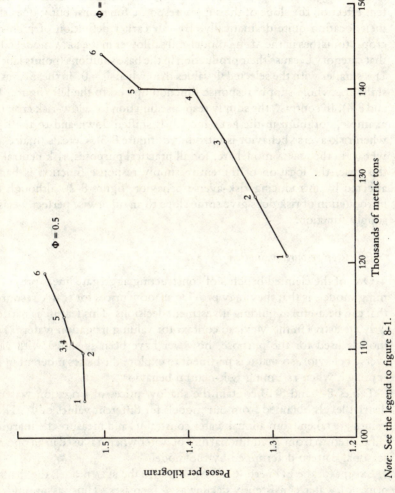

Figure 8-4. Domestic Supply Response for Safflower, BETA Model

Pesos per kilogram

Thousands of metric tons

$\Phi = 0$

$\Phi = 0.5$

Note: See the legend to figure 8-1.

245

The BETA model, in contrast, consistently yields more inelastic supply response functions. In this case, production increases less than proportionally with demand, causing prices to increase above their initial equilibrium values. Clearly, there is keener competition for resources from other crops in this model.

Generally, the risk parameter Φ does not have a pronounced or consistent effect on the slope of the supply response functions, but it does shift their location quite dramatically. By our earlier definition of high-risk crops, tomatoes in the ALPHA model and safflower in the BETA model fall in that category because their production in the base solutions (points labeled 1) is smaller with the selected Φ values than when $\Phi = 0$. In these cases, Φ shifts the whole supply response function up and to the left (figures 8-1 and 8-4). In contrast, the supply response function for a low-risk crop (for example, sorghum in the BETA model) is shifted down and to the right when risk-averse behavior is introduced (figure 8-3). Cereals (maize and wheat) in the ALPHA model are, for all practical purposes, risk neutral. In this case, the location of the entire supply response function is barely affected by introducing risk-averse behavior (figure 8-2), although the introduction of risk does give some slope to an otherwise perfectly elastic supply function.

Effects on resource valuation

One of the claimed benefits of constructing aggregate linear programming models is that they may provide shadow prices for scarce resources that can be useful in guiding investment decisions. This feature is particularly attractive in the Mexican context for valuing irrigation water.[11] Few models used for the purpose, however have been specified with risk-averse behavior, so that it is pertinent to explore the biases inherent in the approach when assuming risk-neutral behavior.

Tables 8-9 and 8-10 contain the shadow prices of irrigation water at district levels obtained from our models for different values of Φ. These values are taken from annual water constraints and measure the marginal annual return from an additional unit of water when it is used in an optimal seasonal pattern determined by the model.

Except for the El Fuerte-Guasave region in the BETA model, the shadow prices of water consistently decline as Φ increases. Thus, assuming risk neutrality in model specification will most likely lead to an upward bias in the marginal valuation of irrigation water. As we argued theoretically in the section "Method of Incorporating Risk," however, and as the El Fuerte-Guasave region demonstrates, such consistent results need not

11. See, for example, chapter 15.

Table 8-9. *Shadow Prices for Water*
with Different Values of Φ, *ALPHA Model*
(pesos per thousand cubic meters)

	Values of Φ					
District	0	0.5	1.0	1.5	2.0	2.5
El Fuerte	313	0	0	0	0	0
Culiacán	0	0	0	0	0	0
Río Mayo	1,516	1,248	1,123	1,033	851	845
Guasave	782	355	0	0	0	0
Ciudad Delicias	719	714	523	346	182	115
Bajo Río San Juan	842	598	172	0	0	0
Santo Domingo	2,380	1,934	1,672	1,285	943	656
La Laguna	816	620	487	478	418	352

hold. Assuming risk neutrality in the BETA model would actually lead to an underestimation of the value of water in the El Fuerte-Guasave region when compared with assuming the "best-fitting" value of Φ. This result occurs because the incorporation of risk happens to shift the optimal cropping pattern toward the more water-intensive crops.

Conclusions

This chapter has demonstrated that: (1) reasonable estimates of risk-aversion coefficients can be obtained at aggregate levels through programming techniques; (2) biases in estimates of supply response and resource valuation may be quite significant in planning models that ignore risk-averse behavior; (3) the descriptive preformance of agricultural planning models can be considerably improved by introducing risk-averse behavioral assumptions, even when such assumptions are based on the theoretically problematic (E, σ) utility function. The additional data re-

Table 8-10. *Shadow Prices for Water*
with Different Values of Φ, *BETA Model*
(pesos per thousand cubic meters)

	Values of Φ			
Region	0	0.5	1.0	1.5
Culiacán-Humaya	732	404	0	0
El Fuerte-Guasave	523	558	196	42

quirements for the incorporation of risk are time-series observations on prices and yields, by crop, for the relevant producing regions. Unfortunately, this requirement is least likely to be met for the more backward regions. The next two chapters explore techniques that may be more applicable for these areas, although more research is required to incorporate the approaches into mathematical programming models.

References

Bassoco, Luz María, Roger D. Norton, and José S. Silos. 1974. "Appraisal of Irrigation Projects and Related Policies and Investments." *Water Resources Research*. Vol. 10, pp. 1071–79.

Baumol, W. J. 1963. "An Expected Gain-Confidence Limit Criterion for Portfolio Selection." *Management Science*. Vol. 10, pp. 174–82.

Cancian, F. 1972. *Change and Uncertainty in a Peasant Economy*. Stanford, Calif.: Stanford University Press.

Charnes, A., and W. W. Cooper. 1959. "Chance Constrained Programming." *Management Science*. Vol. 6 (October), pp. 70–79.

Day, R. H. 1963. *Recursive Programming and Production Response*. Amsterdam: North-Holland.

Dillon, J. S., and Jock R. Anderson. 1971. "Allocative Efficiency, Traditional Agriculture, and Risk." *American Journal of Agricultural Economics*. Vol. 53, pp. 26–32.

Duloy, John H., and Roger D. Norton. 1973. "CHAC: A Programming Model of Mexican Agriculture." *Multi-Level Planning: Case Studies in Mexico*. Edited by Louis M. Goreux and Alan S. Manne. Amsterdam/New York: North-Holland/American Elsevier, pp. 291–337.

Francisco, E. M., and Jock R. Anderson. 1972. "Choice and Chance West of the Darling." *Australian Journal of Agricultural Economics*. Vol. 16, pp. 82–93.

Hazell, Peter B. R. 1971. "A Linear Alternative to Quadratic and Semi-Variance Programming for Farm Planning under Uncertainty." *American Journal of Agricultural Economics*. Vol. 53, pp. 53–62.

Lin, W., G. W. Dean, and C. V. Moore. 1974. "An Empirical Test of Utility vs. Profit Maximization in Agricultural Production." *American Journal of Agricultural Economics*. Vol. 56, pp. 497–08.

Madansky, A. 1962. "Methods of Solution of Linear Programming under Uncertainty." *Operations Research*. Vol. 10, pp. 463–71.

Markowitz, Harry M. 1959. *Portfolio Selection*. Cowles Foundation Monograph, no. 16. New York: Wiley.

Maruyama, Yoshihiro. 1972. "A Truncated Maximum Approach to Farm Planning under Uncertainty with Discrete Probability Distributions." *American Journal of Agricultural Economics*. Vol. 54 (May), pp. 192–200.

Officer, R. R., and A. N. Halter. 1968. "Utility Analysis in a Practical Setting." *American Journal of Agricultural Economics*. Vol. 50, pp. 257–77.

O'Mara, Gerald T. 1971. *A Decision-Theoretic View of the Microeconomics of Technique Diffusion in a Developing Country*. Ph.D. dissertation. Stanford: Stanford University.

Pomareda, Carlos, and Richard L. Simmons. 1977. "A Programming Model with Risk to Evaluate Mexican Rural Wage Policies." *Operational Research Quarterly*. Vol. 28, pp. 997–1011.

Secretaría de Recursos Hidráulicos (SRH). 1972. *Estadística agrícola del ciclo 1972–73*. Mexico City.

Tobin, J. 1958. "Liquidity Preference as Behavior towards Risk." *Review of Economic Studies*. Vol. 25, pp. 65–85.

Tsiang, S. C. 1972. "The Rationale of the Mean-Standard Deviation Analysis, Skewness Preference and the Demand for Money." *American Economic Review*. Vol. 62, pp. 354–71.

――――. 1974. "The Rationale of the Mean-Standard Deviation Analysis: Reply and Errata for Original Article." *American Economic Review*. Vol. 64, pp. 442–50.

9

The Microeconomics of Technique Adoption by Smallholding Mexican Farmers

GERALD O'MARA

A SIGNIFICANT AND GROWING BODY of evidence attests to the static efficiency of decisions concerning resource allocation made by small-holding farmers in developing countries.[1] These studies, however, are based on production or profit functions fitted to cross-section data and use aggregate value of farm production as a measure of output. Although these studies show that farmers tend to be efficient decisionmakers, when the production decision is analyzed at a less aggregate level, it has been found that profit maximization is not the sole motivation of producers. Studies of large- and medium-scale farmers in developed countries (for example, Freund 1956; McFarquhar 1961; Halter and Officer 1968; and Lin, Dean, and Moore 1974) have confirmed the superiority of an expected utility-maximization hypothesis in explaining farmers' choice of crop mix in an environment with price, technical, or weather risk. Of course, profit maximization is simply the special case of risk neutrality in the expected utility model, and these results suggest that even large farmers in developed countries tend to be risk averse.

There is much casual empirical evidence that attests to the prevalence of risk aversion among smallholding farmers in developing countries. Indeed, for a peasant farmer at the margin of subsistence to seek profit maximization would be irresponsible behavior if it exposed his family to a

Note: This chapter was adapted from my Ph.D. dissertation, written at Stanford University under the guidance of Alan Manne. Donald Keesing, Robert Masson, and Roger Norton provided helpful comments. The research on which the paper is based was supported by a grant from the Committee on International Studies, Stanford University. The extension department of the Mexican National School of Agriculture (Chapingo) assisted me by making personnel and facilities available when needed while I was in Mexico. I also benefited from discussions with, and suggestions from, Ing. Alberto Zuloaga, Ing. Abdo Magdub, Lic. Reuben Garcia, Dr. Gregorio Martínez, and Ing. Edilberto Nino Velásquez. I am grateful for the help and support of all.

1. See studies by Hopper (1965), Lau and Yotopoulos (1971; 1973), Massell (1967), Schultz (1964), Wellisz and others (1970), and Yotopoulos (1967).

significant probability of starvation or other dire consequences. If small-holders are risk averse, then it should be expected that their response to a new crop or technique being promoted by government extension agents will be cautious and skeptical. The new crop or technique presents un-known risks, and risk-averse farmers will want to accumulate sufficient information to assess these risks more precisely before jeopardizing their livelihood. Moreover, the conventional wisdom of traditional agriculture usually foresees any innovation as predestined to failure. Hence, farmers are likely to be unwilling to invest much in acquiring information about a new technique they presume to be of dubious merit.

Costs of information acquisition can be quite high for functionally illiterate peasant farmers, who are largely limited to oral communication and personal observation in acquiring information. Also, the information cost of a new technique is a set-up cost; that is, independent of scale. This indivisibility produces increasing returns to scale that favor farmers with large landholdings. Similar economies of scale exist for farm credit and often in the distribution of important purchased inputs. Such factors tend to inhibit adoption of new techniques by smallholding farmers, implying adverse effects on the rural income distribution from technical progress in agriculture. Hence, one concern of many writers over the consequences of the "green revolution" is that the resulting increase in inequality is likely to prove politically intolerable.[2]

One apparent solution to this problem is programs that combine low-cost dissemination of technical information to users (via oral communication and demonstration plots) with concurrent development of distribution systems for principal inputs and credit for small farmers. Such packages offset the inhibiting effects of economies of scale for information, credit, and purchased inputs on technique adoption by small farmers. Yet, even with such program packages, it is not clear that smallholders would respond readily. A conservative tradition reinforced by risk-averse responses to uncertainty about new techniques may restrain them. Evidence will be presented of smallholder response to one such program package.

Overview

The farmer's current stock of information is an important variable to include in a model of farmer behavior, at least if the objective is to explain the speed of adjustment to changes in the availability of farm techniques.

2. See Brown (1970), Falcon (1970), Frankel (1971), Ladejinsky (1970), and Wharton (1969).

To capture this aspect of his behavior, a dynamic model that employs concepts from statistical decision theory is used. The model applied in this chapter presupposes the axioms required for the validity of the expected utility theorem; for a discussion of these and proofs, see Arrow (1971, chapter 2). Note that the expected utility model is consistent with the mean-variance approach of the preceding two chapters if utility is quadratic or if the distribution of outcomes is normal.

This chapter is structured as follows: the following section sketches the theory; the next section gives an empirical test of the model for a program promoting a new maize technique to small farmers in Mexico; and the final section considers policy implications. The derivation of equation (9.7) is given in an appendix to the chapter.

Dynamic Microtheory of Choice of Technique

The intertemporal problem of choice of technique that a farmer faces in a dynamic world of imperfect information turns out to be remarkably similar to a naive, one-period choice (given some reasonable simplifying assumptions about technology and the environment). Since this similarity is not obvious, the reasons for it will be sketched in this section.

The farmer's varying stock of relevant information on payoffs is represented by a subjective probability distribution defined over the set of conceivable outcomes from farm investment activities. In an uncertain world, profit maximization is an acceptable decision criterion only for farmers who are risk neutral, and a more acceptable criterion is maximization of an indirect (or derived) utility of wealth function. That is, given reasonable assumptions with respect to the form of the utility function for lifetime consumption, an indirect (or derived) utility of wealth function $V_0(W_0)$ can be found, where

$$(9.1) \qquad V_0(W_0) = \max_{(C_0, \ldots, C_T)} U(C_0, \ldots, C_T),$$

with the function U representing the utility of consumption over the farmer's T-period lifetime, W_0 denoting initial wealth, and C_t consumption in period t.[3] This derived or behavioristic utility function is found by assuming that all future decisions are made optimally. Information about present gambles may be evaluated in a dynamic world by use of a derived utility of wealth function that satisfies the conditions of a one-period, von

---

3. Although it is not strictly necessary to the derivation of an indirect utility function, it is analytically convenient to make the Ramsey assumption of additivity and stationarity of intertemporal utility. The model of expected utility maximization used here does require, however, that preferences be independent of the state of the world.

Neumann–Morgenstern utility function. The function $V_0(W_0)$ is handled as a utility function, but it summarizes the intertemporal utility function as well as presently anticipated future wealth and production constraints.

Although the farmer's lifetime consumption–investment plan in general involves many farm investment alternatives, the introduction of n such activities would only serve to obscure the optimization problem of present interest; that is, his choice between a new and an existing technique. Hence, it is assumed that regional comparative advantage is such that a given crop rotation cycle is optimal, and each farmer completely specializes in this crop cycle. Thus, optimal crop activities are determined except for the choice between the new and existing techniques in the production of one of the crops in the cycle.

Let the new technique be denoted by the subscript 1 and the existing technique by the subscript 2. The relative return from a technique in period t is denoted by r_{it} ($i = 1, 2$) and is defined by

$$(9.2) \qquad r_{it} = p_t y_{it} / c_{it} \quad (i = 1, 2)$$

$$(9.3) \qquad c_{it} = \sum_{j=1}^{n} q_{jt} x_{ijt},$$

where p_t is the market price of output; y_{it} is the quantity of output (a random variable from the ith activity; x_{ijt} is the input of the jth factor of production to the ith activity; and q_{jt} is the market price of the jth factor. Output, y_{it}, is derived from a production function with a random component because of environmental or informational uncertainty, and for simplicity production is characterized by fixed proportions in factor inputs.[4] Product and factor prices are nonstochastic.[5]

The proportion of the farmer's total investment in period t that is allocated to the new technique, w_t, is given by

$$(9.4) \qquad w_t = c_{1t} / (W_t - C_t),$$

where W_t denotes the farmer's total wealth at the beginning of period t. It is assumed that all crop investments (that is, in a given technique) are for one period, and that a competitive market exists where capital services can be rented. Landholdings, however, are assumed to be fixed by existing tenure arrangements.

---

4. See O'Mara (1971, chapter 5.2) for a discussion of the more difficult variable proportions case. Note also that, given factor prices, optimal factor proportions are fixed even though factor proportions are freely variable. Thus, in intertemporal allocation problems, a fixed proportions assumption simply implies static factor price expectations.

5. The existence of a government-guaranteed price of 940 pesos per metric ton (summer 1970) for the area seemed to ensure that farmers would not regard the price of output as stochastic. Assessment of farmer beliefs with respect to price variation confirmed this supposition.

Given independence and stationarity in intertemporal utility of consumption (and neglecting boundary effects such as bequests), the farmer's intertemporal optimization can now be written:

$$(9.5) \qquad \max_{(C_t,\ w_t)} E_t \left\{ \sum_{t=0}^{T} \text{ß}^t U(C_t) \right\},$$

subject to the constraints

$$W_{t+1} = (W_t - C_t)\,[w_t r_{1t} + (1 - w_t)r_{2t}]$$

$$0 \leq C_t \leq W_t;\ 0 \leq w_t \leq 1,$$

where W_0 is given; $U(C_t)$ is the utility of consumption in period t; $F_t(r_{1t}, r_{2t})$ is the subjective joint probability distribution of the relative returns; E_t denotes the expectation operation with respect to the distribution F_t; and ß is a time-preference discount factor $(0 < \text{ß} < 1)$.

Starting at period 0, an initial consumption level C_0 and technique investment allocation w_0 are selected, given knowledge of W_0 but not W_1. One period later the values of the relative returns are known, determining W_1, and a new decision selecting C_1 and w_1 is made. Note that the optimization selecting C_1 and w_1 employs the subjective probability distribution of relative returns F_1 (r_{11}, r_{21}), which in general will be different from the distribution F_0 (r_{10}, r_{20}) employed in the period 0 optimization.

For mathematical simplicity, assume that the farmer's subjective probability distribution belongs to the class of distributions with a sufficient statistic, which implies that the subjective distribution is self-reproducing in the sense that the posterior has the same form as the prior distribution.

An example may help clarify this point. Suppose the farmer knows the variance but initially has a uniform prior distribution for the expected value of the yield; that is,

$$(9.6) \qquad g'(\mu) = \frac{1}{b-a} \quad (a \leq \mu \leq b)$$

$$= 0 \quad \text{(otherwise)},$$

whereas yields are actually distributed normally, $f_N\,(y|\mu, \sigma^2)$. Then, by Bayes' theorem, the posterior distribution g'' of the expected value μ is proportional to the product of the prior g' and the likelihood of a sample of n from the process; that is,

$$g''(\mu) = g'\,(\mu)\ \ell(y|\mu)$$

$$= \frac{1}{b-a} \exp\left(-\frac{1}{2\sigma^2} \sum_{i=1}^{n} (y_i - \mu)^2\right)(2\pi)^{-n/2}\,\sigma^{-n}$$

$$= K \exp\left(-\frac{n}{2\sigma^2}\,(\mu - \bar{y})^2\right),$$

where K incorporates all terms not involving μ, and $\bar{y}$ is the mean of the first-period observations. Hence, the posterior distribution (that is, the prior distribution of period 2) is normal with mean $\bar{y}$ and standard deviation $\sigma/\sqrt{n}$. Repeated application of Bayes' theorem demonstrates that all subsequent prior distributions will also be normal. In fact, the transformation equations for the mean and variance of subsequent priors are easily derived (see Pratt, Raiffa, and Schlaifer 1965) and are given by

$$m_{t+1} = \frac{(1/v_t)m_t + (n_t/\sigma^2)\bar{y}_t}{(1/v_t) + (n_t/\sigma^2)}$$

$$\frac{1}{v_{t+1}} = \frac{1}{v_t} + \frac{n_t}{\sigma^2},$$

where m_t and v_t are the mean and the variance, respectively, of the prior distribution of the expected yield in period t; σ^2 is the known process variance; $\bar{y}_t$ is the mean of observed yields in period t; and n_t is the number of observations in period t. Thus, we see for this normal case that, given a lack of strong prior beliefs, the data generated have imposed a self-reproducing form.

The assumption that the prior distribution is self-reproducing may be written:

(9.7) $$g(\theta_{t+1}) = T(\theta_t, \hat{\theta}_t),$$

where $\hat{\theta}_t$ is a sufficient statistic of fixed dimensionality that transforms the prior distribution of the parameter θ_t into a posterior distribution of the same form, and θ_t is the parameter of the joint distribution of relative returns, $F_t(r_{1t}, r_{2t})$. For a mathematical derivation of equation (9.7), see the appendix to this chapter [equations (9.20a–g)]. The assumption of a self-reproducing prior is only for mathematical convenience and is not essential to the model, since revised probabilities can always be computed numerically. Nor is it implied that peasant farmers are conversant in algebra and calculus. But if complicated, third–order differential equations are needed to describe the flight behavior of the birds in the field, certainly elementary calculus may be used to describe the decisionmaking processes of farmers.

The probability density function corresponding to F_t is

(9.8) $$f_t(r_{1t}, r_{2t}) = \int_R f_t(r_{1t}, r_{2t}|\theta_t)g(\theta_t)d\theta_t,$$

which defines subjective beliefs with respect to relative returns from the two techniques (and R is the set over which θ_t is defined). In view of equation (9.7), we can also write

(9.9) $$f_t(r_{1t}, r_{2t}) = f_t(r_{1t}, r_{2t}|\hat{\theta}_{t-1}, \hat{\theta}_{t-2}, \ldots, \hat{\theta}_0, \theta_0),$$

expressing the dependence of probabilistic beliefs on previous experience, observation, and initial beliefs.

The problem as stated in equation (9.5) can be solved by dynamic programming methods. The backward recursion, which decomposes the τ period problem into τ one-period problems commences with the final decision:

$$(9.10) \qquad V_{\tau-1}(W_{\tau-1}) = \max_{(C_{\tau-1},\ w_{\tau-1})} U(C_{\tau-1}) + \beta E_{\tau-1}U(C_\tau).$$

Although $V_{\tau-1}$ is written as a function of $W_{\tau-1}$, we must remember that it also depends on $\theta_{\tau-1}$, which enters through the expectation operation. Optimization yields the Kuhn-Tucker inequalities:

$$(9.11) \quad U'(C_{\tau-1}) \leq E_{\tau-1}U'(W_\tau)\ [W_{\tau-1}r_{1,\ \tau-1} + (1-w_{\tau-1})r_{2,\ \tau-1}]$$

$$(9.12) \quad E_{\tau-1}U'(W_\tau)r_{1,\ \tau-1} \leq E_{\tau-1}U'(W_\tau)r_{2,\ \tau-1},$$

which utilize the fact that

$$C_\tau = W_\tau = (W_{\tau-1} - C_{\tau-1})\ [w_{\tau-1}r_{1,\ \tau-1} + (1-w_{\tau-1})r_{2,\ \tau-1}].$$

Since U is strictly concave by assumption, $V_{\tau-1}$ is concave, and the Kuhn-Tucker conditions are sufficient. The optimality conditions can be solved for the optimal decisions $C^*_{\tau-1}$ and $w^*_{\tau-1}$, and proceeding recursively, the following system is derived:

$$(9.13) \qquad V_{\tau-t}(W_{\tau-t}) = \max_{(C_{\tau-t},\ w_{\tau-t})} U(C_{\tau-t})$$

$$+ \beta E_{\tau-t}V_{\tau-t+1}(W_{\tau-t+1}) \quad (t=1,\ \ldots,\ \tau).$$

As with the first stage of the backward recursion, optimization at each of the τ stages yields optimal decisions:

$$(9.14) \qquad C^*_{\tau-t} = C_{\tau-t}(W_{\tau-t},\ \hat{\theta}_{\tau-t-1},\ \ldots,\ \hat{\theta}_0,\ \theta_0)$$

$$(9.15) \qquad w^*_{\tau-t} = w_{\tau-t}(W_{\tau-t},\ \hat{\theta}_{\tau-t-1},\ \ldots,\ \hat{\theta}_0,\ \theta_0).$$

The solution for the case of learning by doing (and observing), although formally similar to solutions of stochastic problems in dynamic programming with stationary probabilities (see Samuelson 1969), is actually quite different in character. The computation of optimal decision rules for all τ periods implicates all future prior distributions, all of which are unknown when the optimization is done at period 0. This lack of stationarity implies that the optimal decisions for periods 1 and T are subject to significant change as new information alters the farmer's subjective probabilities. Hence, the optimality of the solution for period 0 depends critically on the assumption that all investments are for one period, and the farmer will want to reevaluate the optimal decision rules at the start of each period if new information has altered his subjective probabilities.

Thus, we find that our solution may be characterized by a problem of expected utility maximization that looks like a one-period problem at period 0. The farmer simultaneously selects C_0 and w_0 just as if he had a utility function V_1 for income from the outcome of his investment decision.[6]

Empirical Application of Theory to Technique Adoption

The optimal decisions at each stage of the T-period horizon depend upon the farmer's preferences and beliefs, yet very little is known about the form of the utility functions and subjective probability distributions that are used to represent these preferences and beliefs. Both utilities and probabilities, however, are capable of measurement.[7] Therefore, an empirical test requires measurement of both beliefs and preferences as well as a context consistent with the special assumptions made in developing the model.

Such conditions were found in the Programa de Altos Rendimientos (PAR), a program of the Mexican National School of Agriculture to improve techniques of maize and alfalfa production in the area around its location in Chapingo, Mexico.[8] The improved technique of maize cultiva-

6. This argument for the optimality of a myopic policy based on the individual's perception of a dynamic situation in which subjective probabilities are changing, perhaps rapidly, should be carefully distinguished from proofs establishing the optimality of a myopic policy given stationary probabilities. For discussions of this latter case, which involves conditions on the individual's utility function, see Mossin (1968) and Samuelson (1969). An implicit assumption of the argument for the case of nonstationary probabilities is that costs of active information seeking (via experiment, observation, or both outside the immediate environment) are so large relative to passive information seeking (via assimilation of whatever observations are generated in the ambient environment) that a policy of passive information seeking is optimal.

7. Arrow (1971, p. 37) notes that "Both the von Neumann-Morgenstern utility theory and the present extension to a priori probabilities are ideally capable of refutation. Both the utilities and the probabilities can be discovered by suitable formulations of choices in simple situations; then behavior in more complicated situations can be checked against that predicted by the theory with the numerical data supplied by study of simple cases."

8. PAR is limited to a few *municipios* (the Mexican equivalent of a U. S. county) east and northeast of Mexico City in the vicinity of Chapingo. Smallholding farmers predominate in this area; over 75 percent of the cultivable land is in holdings of less than 5 hectares, and over 75 percent of these smallholdings are under the *ejido* system. An ejido is a communal land

(Note continues on following page.)

tion that PAR offered area farmers consisted of a package that specified: (1) a hybrid variety; (2) two fertilizer applications precisely defined by types and quantities of fertilizer and dates and method of application; (3) plant density per hectare; (4) critical periods for water application; and (5) cultivation practice and pest control. An extension agent was assigned to each participant for periodic consultation and inspection of crop progress. The agent also handled the details of the associated credit transaction (which was given in the form of physical inputs) and arranged for the timely delivery of principal inputs. Farmers could adopt the new technique without participating in the program because purchased inputs and credit could be obtained elsewhere, but, nonparticipant adopters would not receive the services of an extension agent. The technique was designed for, and participation limited to, farmers operating irrigated land.[9] The improvement was such that PAR agents were able to claim an expected yield of 6 metric tons per hectare in contrast to an expected yield of 2.5 metric tons per hectare from preprogram best practice. Given these estimates, a rough calculation indicates the one-period return on estimated incremental investment (all one-period inputs) is about 250 percent.

Investigation disclosed that over 80 percent of cropland in the program area was committed to either maize or alfalfa, and this cropping pattern had been stable for at least ten years. In general, farmers felt that rotation between maize and alfalfa was desirable, and a farmer specializing in one or the other would be planning a rotational change in his crop pattern. Thus, the maize-alfalfa cycle of the Chapingo area approximates the assumption of regional comparative advantage in specialization that was used in developing the model. Moreover, the dominance in the area of the *ejido* tenure system, which prohibits land sale or lease, implies that the assumption of fixity in landholdings is also approximated. Note also that fixed proportions in production and constancy of landholdings imply that an allocation of land is equivalent to an allocation of total crop investment.

---

tenure system fostered by the Mexican agrarian reform law. The farmers belong to a communal organization, the ejido, which controls the land (transferred to it by the agrarian reform) and assigns the individual farmers the right to work a specified parcel(s) for their lifetime (or as long as they continue farming in the community). They are legally proscribed from transferring or leasing these rights.

9. Limitation of the program to irrigated farming minimizes risk from uncontrollable environmental variation. Similarly, the location in the vicinity of Mexico City allows most farm families to derive a significant proportion of total family income from off-farm sources. Obviously, these factors tend to reduce either the absolute or relative risk from a trial of the PAR technique and, hence, tend to ensure program success. Note, however, that the farmer's residual risk is largely because of lack of information (that is, uncertainty) with respect to yield outcomes from the new technique. Thus, PAR is, in effect, a controlled experiment with respect to the effects of Bayesian uncertainty (that is, learning by doing, observing, or both).

As we have seen, the period-0 optimization of the farmer's lifetime plan is indistinguishable from a one-period optimization that treats the derived utility of wealth function V_1 as a utility of consumption function from income from the period-0 investment. The dynamic programming formulation of this decision is written as follows:

$$(9.16) \qquad V_0\left(W_0\right) = \max_{(C_0,\ w_0)} U(C_0) + \beta E_0 V_1\left(W_1\right).$$

It is clear that this optimization involves the simultaneous selection of a consumption level and an investment allocation between the two techniques. But, if the farmer's utility of consumption function is isoelastic in marginal utility (that is, an x percent change in consumption implies a kx percent change in the marginal utility of consumption, where k is a constant number), then the optimization decomposes into a choice of C_0 and an independent selection of w_0 that allocates $W_0 - C_0$ between the two techniques.[10] Therefore, assuming that farmers' preferences are representable in isoelastic form (and there is some empirical evidence supporting this assumption),[11] the Chapingo-area farmer's optimization between the two maize techniques can be written:

$$(9.17) \qquad \max_{\substack{(w) \\ 0 \le w \le 1}} EV\Big[\big\{wr_1 + (1-w)r_2\big\}A\Big] =$$

$$= \max_{\substack{(w) \\ 0 \le w \le 1}} \iint V\Big[\big\{wr_1 + (1-w)r_2\big\}A\Big]f(r_1,\ r_2)dr_1 dr_2,$$

where A is the amount of maize land operated (in hectares); w is the proportion of A planted to the PAR technique; r_i is the relative return from

10. For a proof of this proposition, see Samuelson (1969). The class of functions that are isoelastic in marginal utility have the general form

$$U(c) = d + b\,C^{1-a} \quad (a>0).$$

For this class, the elasticity of marginal utility is given by

$$e_{u'} = -\left(U''/U'\right)C = a.$$

Since it can be shown that, for this class of utility of consumption functions, the derived or indirect utility of wealth function belongs to the same class (see Samuelson 1969), the relative risk-aversion function of Arrow (1971) and Pratt (1964) also has the constant value a, and the absolute risk-aversion function defined by Arrow and Pratt is given by a/w; that is, the individual exhibits decreasing (in wealth) absolute risk-aversion and constant relative risk aversion.

11. If U is isoelastic in form, then V is too (see Samuelson 1969). Regression analysis of the utility of wealth assessments obtained from survey interviews showed that most of these were representable in isoelastic form (see O'Mara 1971, chapter 7). Although Samuelson did not show that isoelastic V implies isoelastic U, it is clear from his argument that the converse proposition also holds except for a possible additive constant (which is meaningless for von Neumann-Morgenstern utility functions).

the ith technique ($i = 1, 2$); and the functions V and f are representations of risk preferences and beliefs obtained through empirical assessment.

As noted above, under our assumptions a land allocation is equivalent to a crop investment allocation. Now A in equation (9.17) denotes maize land operated, which may not be the same as total land operated if the farmer divides his land into concurrent maize and alfalfa hectarage and rotates by switching land use. The availability of credit and technical assistance from PAR, however, eliminates other resources as constraints in an optimization over maize land.

Survey methods

The model of equation (9.17) was empirically tested by utilizing a sample survey (summer 1970) of participant and nonparticipant maize farmers (operating irrigated land) in the program area. The first step was a special census of water users to specify the population for sampling. From this census, a list of program participants, and a table of random numbers, two samples of fifty were drawn—participants and nonparticipants.

The basis of the assessment of each respondent's joint probability distribution of returns from the two maize techniques was the determination of marginal and conditional distributions by the method of fractiles.[12] Questions were phrased so as to state the fractiles in terms of betting odds: for example, "what do you believe is the number of kilograms (or tons, *cargas*, and so forth) of maize that would be exceeded by the yield from method x three out of every four times you tried it?" In practice, it was necessary to limit assessment to five fractiles.[13]

The definition of the "other" technique proved troublesome because it varied significantly among farmers. It was finally decided to let each farmer provide his own definition: the other technique was the one currently used by nonparticipants and by participants before joining PAR. The PAR technique, of course, was defined by the program management.

The basis of the estimation of farmers' risk preferences was the von Neumann-Morgenstern (1953) theorem that, if a decisionmaker is indif-

12. This method has also been called the technique of CDF-fractiles. See Winkler (1967) for a description and comparison with other methods of assessing beliefs.

13. To define a joint distribution, without specifying its functional form, would require determining a number of conditional distributions. If a functional form is assumed, then assessment need only meet the lesser information requirements of parameter estimation. The choice of a functional form, however, is bound to be arbitrary—especially if evidence is lacking to justify the choice. The solution adopted was to examine the marginal distributions obtained by the method of fractiles for goodness of fit, with respect to several broad and flexible families of distributions—for example, normal and log normal. To determine the covariance of yields from the PAR and other techniques, an assessment with respect to PAR yield conditional on a specified value for the other yield was obtained.

ferent toward a certain return C and a gamble G, then, given the von Neumann-Morgenstern assumptions with respect to preferences, $U(C) = EU(G)$, where the expectation operation is with respect to the distribution of G. The method of assessment of utility employed in the field work in Mexico is what Fishburn (1967) has called the "probabilistic midpoint" method and what Halter and Officer (1968) have named the "modified von Neumann-Morgenstern" method. Briefly, the method seeks to find a sure return x that is indifferent to a fifty-fifty gamble resulting in either y or z. Hence, from the von Neumann-Morgenstern theorem, we have

$$U(x) = \frac{1}{2} \left(U(y) + U(z) \right).$$

Since U is only unique up to a positive linear transformation, an origin and scale may be selected arbitrarily, thus defining $U(y)$ and $U(z)$ for an initial gamble. Once an X is found such that the individual is indifferent between this sure quantity and the fifty-fifty gamble on y or z, $U(x)$ is determined from the right-hand side of the equation. Points of indifference were estimated by incrementing returns in units of 100 pesos until the individual switched his preference, at which point it was assumed indifference was approximated.

To bridge cultural differences, interviewers were employed in the field work. The interview proper usually took one and one-half to two hours and was conducted at the farmer's residence. The distribution of interviews was as follows:

	Sample from participant list	Sample from nonparticipant list
Sample size	50	50
Number dropped	1	2
Interviews obtained	44	40
Interviews (percentage of effective sample size)	90	83

Since the parcel was actually worked by a participant in two cases of the nonparticipant sample, the final totals were forty-six participants and thirty-eight nonparticipants.

Assessments of probabilistic beliefs

A crucial element of this study was the assessments of farmers' beliefs about the probabilities of returns from the two maize techniques. This information, when combined with estimates of their derived utility of wealth functions, may be used to test whether the farmers act according to their beliefs and preferences as measured. If they do so, then these measures may be used to gain additional insights into the development process. One important insight yielded by this work is that farmers view the new technique in a more favorable light (as measured by changes in their subjective probability distributions) as they gain more experience with it. This insight is important for understanding the dynamics of technique diffusion.

ANALYSIS OF NONPARTICIPANTS' BELIEFS. This group of thirty-eight farmers separates into a group of fourteen who were unable or unwilling to express beliefs with respect to the PAR technique and a group of twenty-four that did. Examination of these groups' beliefs with respect to their own "other" technique reveals a striking difference, as may be seen in the first and fourth rows of table 9-1: the composite distributions for the two groups are so clearly separated that the tails do not even overlap.

Data of the former Mexican Ministry of Water Resources (Secretaría de Recursos Hidráulicos, SRH) show that the average yield was 2.56 metric tons per hectare on all irrigated maize land in SRH districts during the 1968–69 crop cycle. PAR managers estimated that a farmer using preprogram best practice on irrigated land in the Chapingo area would have an average yield of 2 to 3 metric tons per hectare. In contrast, the group of

Table 9-1. *Assessed Beliefs of Nonparticipants in the Programa de Altos Rendimientos (PAR), Chapingo, Mexico, 1970*
(maize yields in metric tons per hectare)

	Mean value of unconditional assessment (by fractile)									
	Other technique					PAR recommendations				
Group (number)	0.00	0.25	0.50	0.75	1.00	0.00	0.25	0.50	0.75	1.00
1. Farmers giving no assessment of PAR (14)	0.69	1.00	1.24	1.53	1.83	—	—	—	—	—
2. Low assessors in first group (10)	0.27	0.46	0.60	0.76	0.94	—	—	—	—	—
3. High assessors in first group (4)	1.74	2.24	2.69	3.26	4.06	—	—	—	—	—
4. Farmers giving an assessment of PAR (24)	2.17	2.67	3.12	3.71	4.25	3.51	3.83	4.20	4.76	5.35
5. Fourth group less four removals (20)	2.39	2.83	3.30	3.75	4.45	3.32	3.68	4.02	4.57	5.17
6. Low assessors in fifth group (10)	0.96	1.26	1.48	1.77	2.19	2.11	2.59	2.78	3.08	3.29
7. High assessors in sixth group (10)	3.83	4.40	4.83	5.91	6.70	4.53	5.01	5.58	6.38	7.05

— Not applicable.

non-PAR assessors had a mean estimate of the median yield of only 1.24 metric tons per hectare.

Yields at this level would imply a very low standard of living on a 1–2-hectare plot if there were no other source of income. Of the eighty-four respondent farmers, however, fifty-four, or 64 percent, had other employment that contributed substantially to family income and involved a significant proportion of their time. In an additional fifteen cases where full-time farming was reported as an occupation, there was another person (or persons) with significant outside income in the household. Most of the farmers reporting other employment had a full-time job elsewhere and could farm their holdings only by significant purchases of agricultural labor and other services. Summary statistics with respect to income (both farm and off-farm), wealth, land operated, age, formal education, family size, and so forth are given in table 9-5, below.

Furthermore, other natural breaks are evident within these groups. Application of the Wilcoxon rank–sum test (one-sided) to the fourteen assessments of the non-PAR assessors indicates a group of four that have distributions significantly different from the other ten at the 5 percent level.[14] The mean values for the fractile assessments of these two groups are given in the second and third rows of table 9-1 and by curves A and B of figure 9-1. The mean of the median assessments for the four high assessors is 2.7 metric tons per hectare.

Why should these farmers within the mean of the median for high assessment, whose assessments indicate an ability to obtain a return consistent with efficient resource management, lack the information needed to assess the program recommendations? Old age and its attendant shortening of the planning horizon may be responsible in the case of two of the four, whose ages were 70 and 79 years. In the case of a third, who had full-time industrial employment, a tour of his property with his son made it clear that the bad experience of a neighbor had caused the farmer to form a negative opinion of the program. It seems likely that he chose to suppress this opinion in favor of a "no-information" response. In the

---

14. It is possible to view the answer by a respondent with respect to any fractile as a random drawing from a subset of the set of perceived possible outcomes. Thus, the answer given for the median would be a random selection from some neighborhood about the median. An assessment on this view is a stratified sample of observations. Measures of location should not vary significantly between completely random samples and those chosen by this method of stratification. Hence, if a nonparametric test sensitive to location, such as the Wilcoxon rank–sum test (see Bradley 1968), is applied to the data, the efficiency of the test should not be impaired. In fact, this sampling procedure increases the probability of drawing observations from the tails and hence reduces the probability of a type-1 error. Our null hypothesis is, of course, that the samples being compared (that is, different assessments) were drawn from the same distribution.

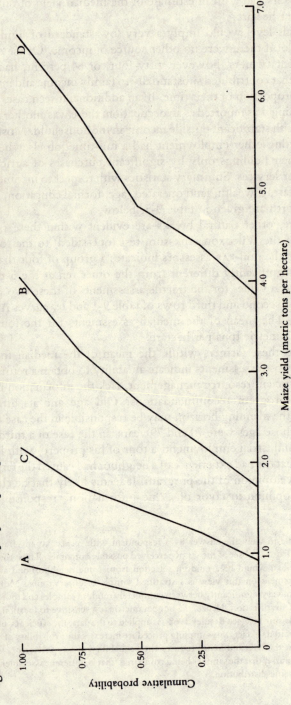

Figure 9-1. Assessed Beliefs of Nonparticipants with Respect to Self-defined "Other Technique"

Note: PAR = Programa de Altos Rendimientos [high-yield program]; A = low "other" assessors (non-PAR-assessing group); B = high "other" assessors (non-PAR-assessing group); C = low PAR assessors; D = high PAR assessors. Curves were constructed by plotting mean fractile assessments and connecting adjacent points with straight lines.

fourth case, supplementary evidence does not provide any direct insight, although this farmer defined an "other" technique and expressed beliefs about it that indicated some diffusion of information about the improved technique. For the group of ten low assessors, the mean value of the median is 0.6 metric tons per hectare, and all mean fractile values are less than 1 metric ton (see table 9-2). The almost negligible proportion of family income obtained from farming by this quite technically backward group (see table 9-2) suggests that these men are probably best regarded as nonfarmers. To a considerable degree, the decision of these men to function as entrepreneurs rather than to opt for the less demanding role of *rentier* is because of the peculiarities of land tenure under the ejido system.

Of the twenty-four nonparticipants who gave assessments of the PAR technique, four had to be dropped because of circumstances in the interview that made their assessments of the two techniques incomparable. Mean fractile assessments of the remaining twenty are given in the fifth row of table 9-1. Again, a significant intragroup variation can be seen. This group was decomposed into two groups of ten farmers, each of whose members have assessments (of their own "other" techniques) that are significantly different from the assessments of all of the members of the other group in a one-sided Wilcoxon rank-sum test at the 5 percent level. Mean fractile assessments for these two groups are given in the sixth and seventh rows of table 9-1 and by curves C and D of figure 9-1.

The mean assessments of their own technique by the group of ten low assessors indicate a level of technique about ten years behind the mean for Mexican farmers with irrigated land. (Of course, it is possible that this group has exceptionally poor land, but the soils and topography around Chapingo are sufficiently homogeneous to make this a remote possibility, at least with respect to irrigated land.) This same group also gave PAR assessments at about the level that farmers using preprogram best practice should get from irrigated land. When a one-sided Wilcoxon rank-sum test is used to compare the assessments of the two techniques for each member of this group, in all but two cases the test rejects the hypothesis of equality at the 5 percent level.

The mean fractile assessments of their own technique by the group of ten high assessors are actually higher than the mean fractile assessments of the PAR technique by participants with one to three years of program experience. When assessments of the two techniques for each member of the group are compared in a one-sided Wilcoxon rank-sum test, the hypothesis of equality is rejected (at the 5 percent level) in only three cases. In fact, the two assessments are identical in several instances, and comparison of the definition of their own technique with program recommendations shows few significant differences in almost all cases. In all cases the high-PAR-assessing nonparticipants also reported maize yields (for the

Table 9-2. Income, Wealth, and Selected Characteristics of Respondent Farmers, 1970 (group means)

Group	Number	Income (thousands of current pesos)			Wealth (thousands of current pesos)		Head of household				
		Total (1969)	Farm (1969)	Off-farm (1969)	Material (mid-1970)	Material and human[a] (mid-1970)	Age (years)	Formal education (years)	Full-time farmer (percent)	Family size (persons)	Land operated (hectares)
Total (all farmers less outliers)[b]	77	25.5	9.6	15.9	60.2	197.0	46.7	3.3	31.2	7.7	2.38
Nonparticipants	34	23.0	6.5	16.5	41.1	173.7	44.9	3.4	26.5	7.9	1.90
No information about PAR											
Low assessors of other technique	9	14.9	0.4	14.5	19.9	124.7	42.9	3.0	11.1	6.9	0.83
High assessors of other technique	4	20.3	3.8	16.5	36.8	163.4	59.5	0.8	75.0	6.3	1.88
PAR assessors[c]											
Low assessors of PAR	7	15.6	3.8	11.8	33.0	127.7	39.0	4.0	14.3	7.9	1.65
High assessors of PAR	10	33.3	14.8	18.5	68.7	240.6	41.4	3.7	20.0	10.0	2.87
Participants[d]	43	27.5	12.0	15.5	75.4	215.5	48.1	3.3	34.9	7.5	2.75
First-year	19	21.2	6.5	14.7	39.1	158.7	47.6	3.1	36.8	6.7	2.15
Second-year	11	21.9	8.6	13.3	54.4	169.1	42.5	4.1	9.1	8.1	2.26
Third-year	7	50.0	27.1	22.9	229.8	455.6	52.7	3.6	57.1	9.7	5.54
Fourth-year	5	30.9	17.7	13.2	48.2	205.8	52.0	2.2	40.0	6.4	2.40
Adopters[e]	53	28.6	12.5	16.1	74.1	220.2	46.8	3.3	32.1	8.0	2.78
Nonadopters	24	17.4	3.0	14.4	29.6	145.8	46.3	3.3	29.2	7.0	1.52

a. Includes capitalized future labor and entrepreneurial services.
b. Group means were calculated after removal of seven outliers whose answers to questions about income and wealth were significantly biased.
c. Four nonparticipants who assessed PAR and the other technique on inconsistent bases are excluded.
d. One participant who did not tell when he joined the program is excluded from the subtotals.
e. Includes participants and high-PAR-assessing nonparticipants.

previous year) that were comparable to the yields reported by PAR participants. Clearly, *the improved technique of the PAR recommendations has effectively diffused to the members of this nonparticipant group.* This finding furnishes a basis for an estimate of total nonparticipant adoptions and, hence, total adoptions.

To estimate the extent to which the PAR technique had diffused among nonparticipant farmers (with irrigated land), the population of nonparticipant farmers was partitioned into adopters and nonadopters. Now, a random sample drawn from this population will yield a sample proportion, $\bar{p}$, of adopters that will have a binomial sampling distribution. If the high-PAR assessors are identified as the group of adopters in the nonparticipant sample, and the evidence from the interviews makes it clear this is a reasonable identification, then a basis for statistical inference is obtained with respect to the proportion of the population, p, that are adopters. Since the sample size might range from thirty-eight to fifty depending on the assumptions made, interval estimates were computed using both the upper and lower bounds for n.[15] The resulting estimates for $n = 50$ and $n = 38$ were $0.09 \leq p \leq 0.31$ and $0.12 \leq p \leq 0.40$, respectively.[16] Applying the midpoints of these intervals to an estimated total population of 2,500 nonparticipant farmers (with irrigated land) planting maize in 1970, it would appear that 500 to 650 of these farmers planted maize using the PAR technique in 1970. If the 1970 program participants are added to this group, the estimate of the total number of farmers using the PAR technique in 1970 is 918 to 1,068, out of an estimated total of 2,918 farmers in both groups. Thus, it appears that the proportion of users is probably between 31.5 and 36.6 percent. Even if the lower-bound value of 9 percent diffusion among nonparticipants is used, the estimate of overall adopters is 643, or 22.0 percent of the total.

It is assumed that this rapid process of diffusion can be represented by

---

15. A sample of fifty was drawn originally, but two of those drawn had to be discarded. From the effective sample of forty-eight, forty interviews were obtained. Two of these, however, turned out to be participants, reducing the effective number of nonparticipant interviews to thirty-eight.

16. In either case, sample size is sufficiently large to employ the normal approximation to the binomial, and the interval estimate will have the form

$$\bar{p} \pm z_{\alpha/2} \, s_{\bar{p}},$$

where z indicates the value of the standard normal variate that leaves $\alpha/2$ probability density in the right-hand tail; $1 - \alpha$ is the confidence coefficient; and $s_{\bar{p}}$ is an estimator of the standard deviation of $\bar{p}$ derived from sample statistics. The definition of this estimator is

$$s_{\bar{p}} = \sqrt{\frac{\bar{p}(1-\bar{p})}{n-1}} \sqrt{\frac{N-n}{n-1}},$$

where n is the sample size and N is the population size.

a logistic curve, as Griliches (1969) was able to do for the pattern of adoption of hybrid corn in the United States, then it is possible to derive some rough estimates of the rate of acceptance or slope parameter of the logistic for comparison with the rates of acceptance Griliches found in the United States. The logistic can be written as follows:

$$p_t = \frac{p^*}{1 + e^{-(a+bt)}},$$

where p_t is the proportion of users at time t; p^* is the ultimate proportion of users; and a and b are parameters. It is easily seen that

$$\ln\left[\frac{p_t}{p^* - p_t}\right] = a + bt.$$

Given the evident comparative advantage of the maize-alfalfa cycle in the Chapingo area, and the significant differential in profitability of the PAR technique, it seems reasonable to assume that the ceiling proportion p^* is unity. There remains the question of the proportion of users in the total population at the start of the PAR program in 1967. Fragmentary evidence suggests that this proportion was not greater than 5 percent and probably was around 1 percent. The origin of the time scale can be selected arbitrarily, so let $t = 0$ be set at 1967. Then we have the estimates shown in table 9–3. Of course, these calculations involve large assumptions and are in the nature of very rough estimates. They do, however, provide estimates of the slope (or rate of acceptance) parameter b that provide a basis for comparison with the estimates of b that Griliches found for the diffusion of hybrid corn in the United States. Griliches' estimates of b for crop-reporting districts in the corn-belt states of Iowa, Illinois, Indiana, and Minnesota ranged from 0.64 to 1.36. Elsewhere, Griliches' estimates of b were lower. They ranged from 0.41 to 0.90 in the prairie states of Kansas and Nebraska, and from 0.32 to 0.57 in the southern states of Alabama and

Table 9-3. *Estimates of Parameters of Logistic Function for Role Acceptance of the Recommended Techniques*

Proportion of initial users	Proportion of users in period 3	Estimate of origin, parameter a	Estimate of rate of acceptance, parameter b
0.01	0.220	−4.595	1.11
0.01	0.315	−4.595	1.27
0.01	0.366	−4.595	1.35
0.05	0.220	−2.945	0.56
0.05	0.315	−2.945	0.72
0.05	0.366	−2.945	0.80

Arkansas. Moreover, Griliches' estimates were based on statistics of percent of *total corn acreage* planted with hybrid seed, while the Chapingo area
estimates are based on estimates of the *proportion of farmers* planting maize
according to the PAR technique. The estimates from the Chapingo area
would undoubtedly be larger proportions if translated into area planted.

It is not intended that the estimates above be taken as anything other
than very rough calculations. The rather fragmentary evidence does indicate, however, a significant rate of diffusion, probably comparable to that
which occurred in the United States corn belt with the adoption of hybrid
varieties.

ANALYSIS OF PARTICIPANTS' BELIEFS. These assessments closely conformed to prior expectations—that is, in almost all cases the PAR technique
was believed to produce significantly better yields most of the time, as
would be expected from the logic of revealed preference. Possibly more
important to an understanding of the diffusion of agricultural technology
is the finding that PAR assessments show uniformly higher yields as
experience with the program increases.

In thirty-six cases, the hypothesis that the two techniques have the same
yield distribution was rejected at the 1 percent level (using a one-sided
Wilcoxon rank-sum test), whereas rejection occurred in forty-two out of
the forty-six cases at the 10 percent level. Since objective evidence supports claims for significantly improved expected yields from use of the PAR
technique and it seems that the entire yield distribution is shifted upward
more or less, this result is in accordance with the theory of learning by
doing (and observing) embodied in the model.[17] In most instances, assessments of the two techniques do not even overlap, and this separation is
reflected in the mean fractile assessments for all participants given in the
first row of table 9-4.

PAR had been in existence for four years (in 1970) and the sample of
participants was partitioned into groups that had been in the program for
one, two, three, and four years. At the time of the interviews, first-year
participants did not yet know the outcome of their initial trial. All other
participants had harvested one, two, or three crops using the PAR technique. To the extent that accumulation of information depends on actual

---

17. Program records indicated mean yields for participants of over 6 metric tons per
hectare in 1968 and 1969. The assessments of program participants do not, however, indicate
yields quite this high. Also, reported yields for the preceding year by second-, third-, and
fourth-year participants suggested a mean yield of between 4½ and 5 metric tons per hectare
from the PAR technique. There were enough cases of quite high yields (7–9 metric tons per
hectare), however, to suggest that, given enough time, mean yields might reach a level of 6
metric tons or better.

Table 9-4. *Assessed Beliefs of Participants in PAR*
(maize yields in metric tons per hectare)

	Mean value of unconditional assessment (by fractile)									
	Other technique					PAR recommendation				
Group (number)	0.00	0.25	0.50	0.75	1.00	0.00	0.25	0.50	0.75	1.00
1. All participants[a] (46)	0.79	1.16	1.46	1.83	2.30	2.97	3.58	4.13	4.66	5.22
2. Four-year participants (4)	0.49	0.90	1.34	1.75	2.17	3.90	4.60	5.43	6.00	6.48
3. Three-year participants (7)	0.91	1.46	1.91	2.28	2.99	3.57	4.26	4.75	5.29	5.84
4. Two-year participants (12)	0.69	1.00	1.28	1.68	2.13	2.69	3.43	4.04	4.65	5.49
5. One-year participants (21)	0.85	1.20	1.44	1.71	2.23	2.76	3.30	3.78	4.23	4.68

a. One participant who did not say how long he had been in PAR and one four-year participant giving an incomplete assessment were excluded from the subtotals.

program experience, systematic differences in beliefs should be observable among the four groups. Such variation is shown by the mean fractile assessments of PAR yields by experience groups (the second through fifth rows of table 9-4 and curves P1, P2, P3, and P4 of figure 9-2). *The perceived distribution of PAR yields becomes more optimistic with greater experience.* For almost any cumulative subjective probability, the average minimum yield shifts upward—for example, the mean value of the median increases monotonically from 3.78 to 5.43 metric tons per hectare as experience increases from one to four years.

The null hypothesis of static beliefs clearly involves arbitrary and improbable assumptions about farmer behavior. If beliefs are stationary, the choice of the "pioneer" fourth-year participants is easily explainable—but the choices of first-, second-, and third-year participants are not. Why, having rejected the PAR technique in the initial year, should they choose to adopt it in later years if their beliefs are unchanged? Moreover, why should those farmers whose beliefs can be ranked second most favorable choose precisely the second program year for adoption? A similar question can be raised with respect to the timing of choices by first- and second-year participants.

Beliefs about maize yields may also be related to income, wealth, farm and off-farm earnings, age, education, size of household, and the amount of land operated. Table 9-5 presents data on these characteristics by adoption class.[18] (For a breakdown by assessment class, see table 9-2.) Differences between adopters and nonadopters are not significant (at the 5 percent level) with respect to age, formal education, size of household, proportion of full-time farmers, and off-farm income. Differences in total income, farm income, wealth (both categories), and land operated, however, are significant.[19] The higher mean incomes of adopters are almost entirely because of farm incomes more than four times as great as those of nonadopters. This is because adopters operate almost twice as much land and own over twice as much material capital—including livestock, trees, and farm buildings and equipment. In accordance with the greater material wealth of the adopters, greater absolute returns accrue to agricultural innovation and thus induce more intensive information-search activity and faster response to technical progress. As already noted, that information acquisition is a set-up cost should tend to create a greater responsiveness to technical change by larger farmers. This greater responsiveness for larger farmers seems to hold even for comparisons between quite small farmers.

Assessment of risk preferences

Out of the eighty-four completed interviews, usable assessments of risk preferences were obtained in seventy-two. A graphic approximation of the utility function for each farmer was obtained by connecting the nine points of each assessment in piecewise linear segments. Six of these graphic approximations are presented in figures 9-3 through 9-5. Inspection of these graphs reveals significant variation in form among them. Thus, case 44 appears classically concave; cases 6 and 32, when smoothed, would have alternately concave and convex sections that might represent segments of a Friedman-Savage (1948) utility function; case 22 shows a discontinuous change in otherwise almost constant marginal utilities at 300 pesos; and case 46 is approximately linear for gains. It should be noted that the apparent Friedman-Savage functions may actually reflect the individual's perception of an imperfect capital market and not true risk preferences (see Hakansson 1970; and Masson 1972), since the indirect

18. For a description of the methodology used in preparing the estimates of income and wealth, see O'Mara (1971, appendixes B–F).

19. Tests for significance were as follows. Confidence intervals (at the 95 percent level) were estimated for the population mean of each characteristic for the adopter group. The endpoint of the confidence interval closest to the value of the sample mean for nonadopters was then selected as a conservative estimate of the population mean, and the null hypothesis—that nonadopters came from the same population as the adopters—was tested in a one-sided *t* test. Results (at the 5 percent significance level) were as stated.

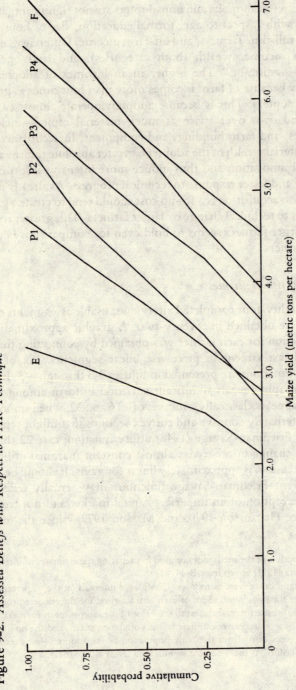

Figure 9-2. Assessed Beliefs with Respect to PAR Technique

Note: E = low-PAR-assessing nonparticipants; F = high-PAR-assessing nonparticipants; P1 = first-year participants; P2 = second-year participants; P3 = third-year participants; P4 = fourth-year participants. Curves were constructed by plotting mean fractile assessments and connecting adjacent points with straight lines.

Table 9-5. *Income, Wealth, and Selected Characteristics*
of Adopters and Nonadopters
(group means)

Item	Adopters[a]	Non-adopters[b]	Significant difference at 5 percent level
Income			
Total family, 1969			
(thousands of current pesos)	28.6	17.4	Yes
Farm, 1969			
(thousands of current pesos)	12.5	3.0	Yes
Off-farm, 1969			
(thousands of current pesos)	16.1	14.4	No
Wealth			
Material, mid-1970			
(thousands of current pesos)	74.1	29.6	Yes
Material and human, mid-1970			
(thousands of current pesos)	220.2	145.8	Yes
Land operated (hectares)	2.78	1.52	Yes
Household			
Size (number of persons)	8.0	7.0	No
Head of			
Age (years)	46.8	46.3	No
Formal education (years)	3.3	3.3	No
Full-time farmer (percent)	32.1	29.2	No

Note: Means exclude seven outliers who gave biased answers to income and wealth questions.

a. Includes participants and nonparticipant adopters (that is, high-PAR-assessing nonparticipants). Sample size is fifty-three.

b. Excludes nonparticipant adopters from nonparticipant group. Sample size is twenty-four.

utility of wealth function is a reduced-form concept that cannot discriminate between pure risk preferences and environmentally induced distortions of preferences. Moreover, these functions are unique only up to a positive linear transformation, and each has a different origin (that is, the farmer's existing wealth). This characteristic has led theorists to devise measures of risk aversion that are invariant to linear transformation; for example, the absolute risk aversion function, $A(w) = -U''(w)/U'(w)$. In general, the form of the assessments of risk preference does not seem to be systematically related to farmer wealth. The freedom to shift location and expand or compress the scale of individual functions facilitates fitting to a common functional form. Yet when the assessments are reduced to a measurement invariant to linear transformation, the hypothesis of a common utility function is rejected.[20]

20. See O'Mara (1971, chapter 7) for a discussion of these tests of the risk preference assessments.

Figure 9-3. *Piecewise Linear Approximations
of Utility of Wealth Assessments, Cases 6 and 22*

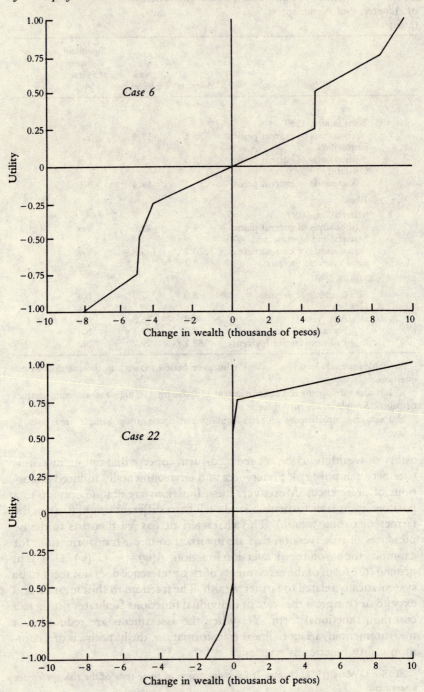

Figure 9-4. *Piecewise Linear Approximations of Utility of Wealth Assessments, Cases 32 and 44*

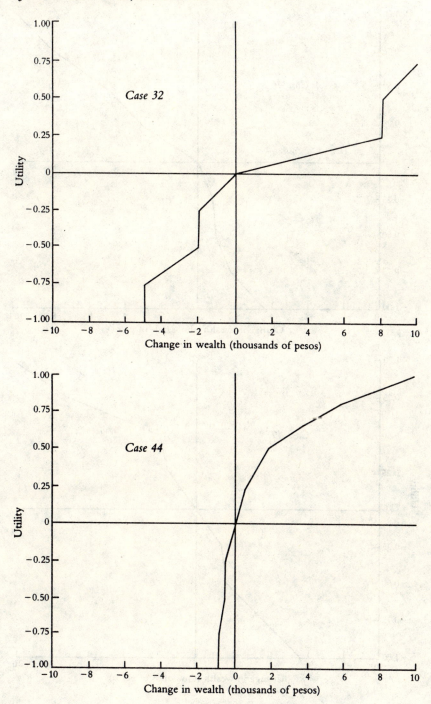

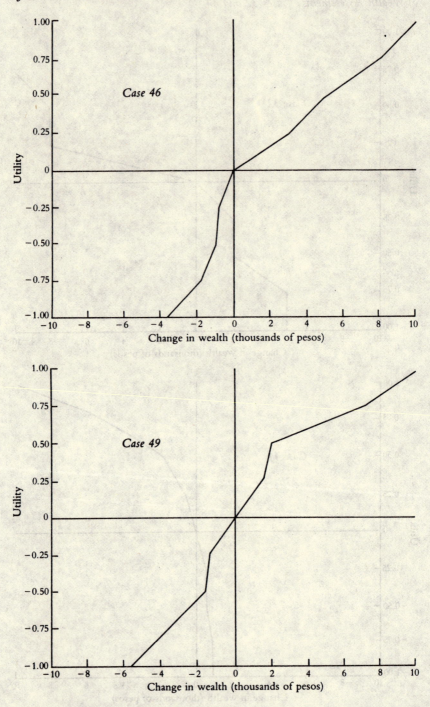

Figure 9-5. *Piecewise Linear Approximations of Utility of Wealth Assessments, Cases 46 and 49*

Transformation of yield assessments into return assessments

The model is formulated in money returns, whereas the survey assessments of probabilistic beliefs are in yields. The problem of transforming yield assessments into money return assessments is complicated by the tendency of harvest costs to vary with the level of yield. But estimates of the value of by-product fodder produced with the grain, which also varies with the yield in grain, were approximately equal to estimates of harvest costs. These revenues canceled with harvest costs, and thus both items were left out of the calculations transforming yields into returns. The remaining factor inputs do not vary with yield;[21] and with prices and technology given, returns could be derived from yields by

$$(9.18) \qquad \pi = py - \sum_{j=1}^{n} q_j x_j = py - c,$$

where π denotes the money return; p is the wholesale price of maize (per metric ton); y is the yield of grain (in metric tons); x_j is the input (in units) of factor j; q_j is the market price (per unit) of factor j; and c is total costs.

Quantities of factor inputs for the "other" technique were determined for each farmer from his personal definition. Program recommendations plus estimates of water use, cultivation practice, and so forth were used to define a representative input vector for the PAR technique.

There remains the question of what are the appropriate prices to use in this transformation. To use market prices as measures of the opportunity cost of inputs requires the potential marketability of on-farm factors at these prices. An abundance of off-farm employment opportunities, the high proportion of total family income derived from off-farm sources, and the dependence of most farmers on purchased inputs (extending, in many cases, to purchased tractor services for plowing) all suggest that opportunity costs may be appropriately valued at market prices.

A priori it was not completely clear that farmers in the Chapingo area regarded the price of output as truly fixed despite a guaranteed price for

21. In general, farmers will tend to vary the quantity of factor inputs in the interval between planting and harvest in accordance with their revised beliefs about yield on the current crop and in a manner that optimizes over a multi-period horizon (given opportunity costs of factor inputs). Thus, if yield prospects are seriously compromised by adverse circumstances, a farmer may forgo a second fertilizer application, preferring to hold the fertilizer for a subsequent crop season when its productivity will be much higher. Given irrigation, however, the environment around Chapingo offers very little in the way of risks that are capable of seriously affecting yield prospects. One has to envision quite special circumstances to have this result—for example, failure of a pump on the well bringing irrigation water at a crucial time or virtual destruction of the crop by a hail storm. Hence, in practice nonharvest costs do not vary with the level of yield.

sales to government warehouses, but an interview assessment of beliefs about maize prices indicated that treating the maize price as given was a justifiable approximation.

Calculation of theoretically optimal decisions

Evaluation of the marginal distributions of money returns from the two techniques for each respondent showed that, in fifty-five out of sixty-six cases, the farmer's evaluation of one technique is such that it would be preferred under any circumstances, independently of risk preferences. (Since only sixty-six farmers gave comparable assessments of yields from the two techniques, only sixty-six sets of derived assessments of returns were obtained.) In these fifty-five cases, the return distribution for the dominant technique is such that the probability of a return equal to or less than any specified money value is at least as small as the corresponding probability for the other technique and is smaller for some money value.[22] That is, if F is the dominant technique and G is the dominated technique, then

$$(9.19) \qquad \begin{aligned} F(\pi) &\leq G(\pi) \text{ for all } \pi, \text{ and} \\ F(\pi) &< G(\pi) \text{ for some } \pi. \end{aligned}$$

These conditions may be represented graphically as in figure 9-6. Given the conditions in equation (9.19), it follows from some theorems of Hadar and Russell (1969) and Hanoch and Levy (1969) on efficient choice under risk that the distribution F dominates the distribution G in the first degree; that is, for nondecreasing utility functions bounded for finite arguments.[23] This result implies that the dominant distribution will be strictly preferred to the dominated one for any nondecreasing utility function. These results are summarized in the following comparison of derived beliefs with respect to returns:

	PAR technique dominates other technique	Other technique dominates PAR technique	Neither technique dominates
Participants	37	2	7
Nonparticipants	8	8	4
Total	45	10	11

---

22. Strictly speaking, this ordering only applies to the five fractiles of the returns distributions for which values have been assessed. If it is assumed that the corresponding probability density functions f and g are continuous nonzero functions (so that F and G are strictly increasing functions) and that the observed ordering applies to all fractiles of the assessed distributions, the necessary and sufficient conditions for the dominance result are obtained.

23. See the discussion by Jock R. Anderson (chapter 10 of this book) for a derivation and discussion of stochastic dominance theorems of first, second, and third degree.

Figure 9-6. *Return Distributions for Dominant and Dominated Techniques*

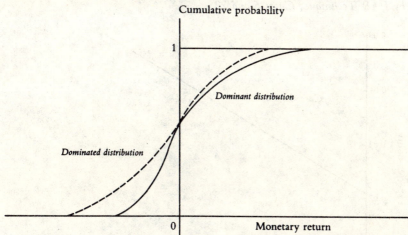

Cumulative probability

Dominant distribution

Dominated distribution

1

0   Monetary return

Clearly, the eight nonparticipants for whom the PAR technique dominates the other technique, and the two participants for whom the "other"
technique dominates, have made decisions contrary to those predicted by
the model, although, with respect to the latter group there is evidence that
suggests the contradiction is more apparent than real.[24] The eight nonparticipants who chose a dominated technique are harder to evaluate. A
number of interpretations are possible, including the hypothesis that some
farmers are not utility maximizers. These cases notwithstanding, the
important finding is that forty-five (and perhaps forty-seven) cases out of
fifty-five conform to the prediction of the model.

In the eleven cases where neither technique dominated, calculation of an
optimal decision is possible for only eight, since three of these farmers did
not give utility of wealth assessments. In two of the three cases, the
following condition holds:

$$\int_{-\infty}^{\pi} F(x)dx \leq \int_{-\infty}^{\pi} G(x)dx, \text{ for all } \pi,$$

24. One of these farmers was employed at the Mexican National School of Agriculture
and had managed to obtain program credit despite not having irrigated land. He had
substituted a more drought-resistant variety of maize in the definition of his own "other"
technique. He made it quite clear that he was using his own technique and not that of the
program. In the other case, the interview was conducted in the presence of the head of the
local ejido, which gave the interview an unintended social content. His answers to interview
questions were subject to bantering comments, and it became quite clear that the farmer, a
first-year participant, was taking great pains not to display great expectations with respect to
his trial of the PAR technique.

Figure 9-7. *Variation of Expected Utility with Proportion Planted by PAR Technique, Cases 6 and 22*

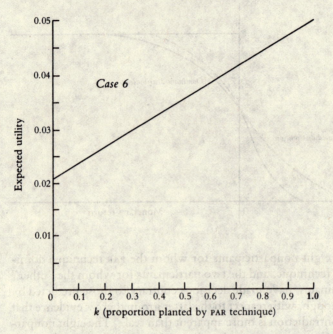

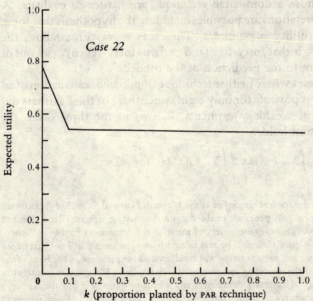

Note: Boldface numbers on the horizontal scales denote actual decisions.

where the distribution F denotes the PAR technique and distribution G the other technique. That is, the area under the PAR curve never exceeds the area under the curve for the other technique in the graphs of the cumulative probability functions, and strict inequality holds for at least one return. Thus, the conditions are met for stochastic dominance of the PAR technique in the second degree; that is, for risk-averse (strictly concave) utility functions.[25] Neither technique, however, dominates in any degree for the remaining case. Actual decisions were an equal division of maize land between the two techniques in the undominated and one of the dominated cases, whereas the farmer of the other dominated case planted all of his maize land by the PAR technique. Each of these decisions is conceivably consistent with a theoretically optimal decision (since the diversifying farmer with second-degree dominance for the PAR technique may not have had a risk-averse utility function).

Of the eight remaining cases, calculation is unnecessary in two.[26] For the six cases that are left, maximum expected utility was calculated using the utility of wealth function obtained in the interviews.[27] Graphic approximations of the utility functions of these six farmers were given in figures 9-3 through 9-5. Graphic plots of the expected utility calculations are given in figures 9-7 through 9-9. A divergence of 0.1 between actual and theoretically optimal decisions can be explained by transaction costs, which the model neglects.

The results of the expected utility calculations are summarized in table 9-6. As the table shows, there is close agreement between actual and theoretically optimal decisions in four of the six cases. The divergence in one of the remaining cases is probably because of an extraneous factor not fully reflected in the optimization. In this case, the farmer was a first year participant who was farming cooperatively with his brother, a second-

25. The reader is again referred to the study of Jock R. Anderson (chapter 10 of this book) for a derivation and discussion of the theorems on second-degree stochastic dominance, which were developed by Hadar and Russell (1969) and Hanoch and Levy (1969).

26. In one of these cases, the other technique was defined identically with the PAR technique, and the choice between them was null. For the other case, all possible returns to both techniques were negative and the optimal decision was to stop cropping maize. Of course, it is conceivable that returns to the alfalfa portion of a crop-rotation cycle would be sufficiently large to more than offset negative returns in the maize portion, and operation of the cycle even with maize losses could be optimal. This interpretation, however, strains credibility both with respect to farming in the Chapingo area in general and this farmer in particular. For this farmer, the theoretically optimal—which has been structured to neglect decisions of a purely sampling nature—is to stop cropping maize.

27. Calculation of expected utilities for these six cases was done on the computer using a Gauss-Legendre quadrature rule. In these calculations, V was a piecewise linear approximation of the utility of wealth assessment and the joint probability distribution of returns was a bivariate distribution fitted to the assessments given in the interviews. A complete description of these calculations is given in O'Mara (1971, chapter 4).

Figure 9-8. *Variation of Expected Utility with Proportion Planted by PAR Technique, Cases 32 and 44*

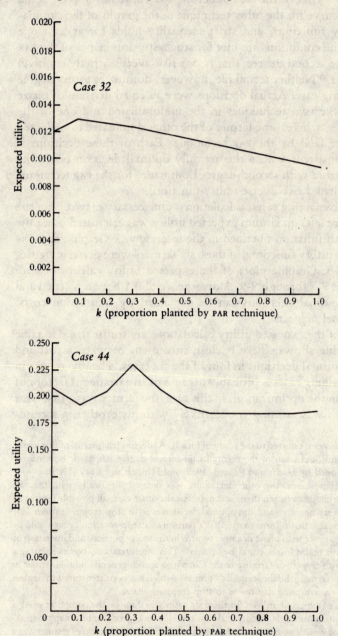

Note: Boldface numbers on the horizontal scales denote actual decisions.

year participant. It was clear that the farmer was not as sanguine with respect to the returns from the PAR technique as his more experienced brother. If he were not constrained by the need to reach an agreement with his brother in respect to their joint operations, he probably would have chosen a proportion closer to the 0.3 indicated as his theoretical optimum.

Implications for Policy

The PAR technique is clearly superior. For this reason, information flows can be expected to yield higher assessments of the PAR technique and a rapid rate of adoption. This seems to have been the case. Rough calculations indicate a rate of diffusion comparable to the rapid diffusion of hybrid corn in the U.S. corn belt. Furthermore, we find patterns of beliefs indicating that the information flows produce more optimistic assessments of the new technique as familiarity with it increases. This finding is the basis for a microeconomic explanation of technique adoption as a time-dependent process.

It is remarkable that first-degree stochastic dominance should prevail in fifty-five out of sixty-six cases. The evidence suggests, however, that this result is largely because of the character of the PAR innovation. Given an expected return on incremental investment of several hundred percent and what appears to be an upward shift of the entire yield distribution, it is scarcely surprising that dominance should prevail. It might be expected that nonparticipant adopters would not show stochastic dominance in their beliefs, since the other and PAR techniques ought to be equivalent. Six of the eight cases where the other technique dominates the PAR technique, however, were nonparticipant adopters. In each such case, the farmer indicated a modification that significantly reduced input costs—for example, substitution of heavy manuring for chemical fertilizer where a farmer has a dairy herd—without indicating an equivalent reduction in beliefs with respect to yield variation.

From another project, some evidence suggests that a stochastic dominance condition may be an important criterion for distinguishing innovations that will meet with widespread acceptance among small farmers (see chapter 10). This project promotes a similar high-yield innovation (that is, calculation based on conservative estimates of representative costs and mean yield indicates a return of 115 percent on one-period incremental inputs) for maize production in an adjacent valley of the Mexican Central Plateau but is designed for rainfed land.[28] An early evaluation of this

28. See International Maize and Wheat Improvement Center (1970, p. 92) for a representative crop budget and calculation of mean return.

Figure 9-9. *Variation of Expected Utility with Proportion Planted by PAR Technique, Cases 46 and 49*

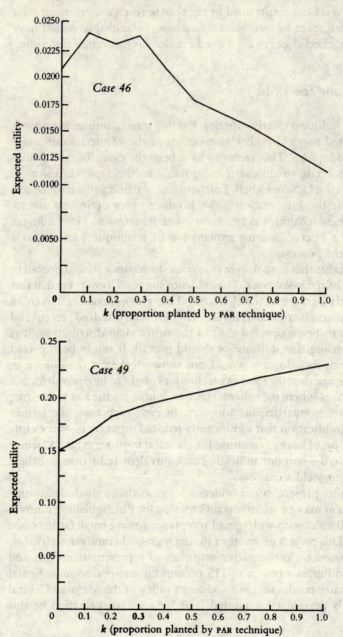

Note: Boldface numbers on the horizontal scales denote actual decisions.

Table 9-6. *Proportion of Maize Land Planted by the PAR Technique, 1970*

Group (number)	Actual decicion	Theoretical decision	Group (number)	Actual decision	Theoretical decision
Participants (3)	1.0	1.0	Nonparticipants (3)	0.0	0.0
	0.8	0.3		0.0	0.1
	0.3	1.0		0.0	0.1

program by Winkelmann (1973) suggests that acceptance of this innovation has been disappointing compared with acceptance of other agricultural innovations offering similar increases in expected return. Winkelmann also notes that discussions with farmers in the project area have disclosed an awareness of possible outcomes from using the innovation that are distinctly worse than the worst outcome using existing cropping technique—that is, the innovation is not stochastically dominant.[29] Even where an innovation does not meet the dominance condition, there may still be significant net social benefits from widespread adoption; hence, sponsors of a nondominating innovation may want to include in their effort a program of crop insurance that effectively makes the innovation a dominating one.

Conclusions

The evaluation of the evidence presented in this chapter depends in part on how it integrates with the body of more or less accepted results on the economic behavior of small farmers. Thus, the decision-theoretic interpretation of the results from this study is reinforced by the growing body of evidence that small farmers make decisions concerning resource allocation that are statically efficient[30] and by the more sketchy evidence that smallholders are dynamically efficient when not handicapped by significant economies of scale from costs of a set-up nature. Thus, Lau and Yotopoulos (1971) found that Indian smallholders were technically more efficient by a factor of 20 percent. The data set they used, however, was for a period before the rapid technical change induced by the introduction of

29. Where the improved technique has a worse-case return that is less than the worse-case return of the existing technique, it will not dominate in any degree. See the discussion by Jock R. Anderson (chapter 10 of this book) regarding the importance of the left-hand tail of the distribution of returns for dominance criteria.

30. See studies of Hopper (1965), Lau and Yotopoulos (1971; 1973), Massell (1967), Schultz (1964), Wellisz and others (1970), and Yotopoulos (1967).

the Mexican dwarf wheats and the "miracle" rice varieties developed by the International Rice Research Institute.

In summary, the evidence of this study is generally consistent with the implications of a decision-theoretic model; and in view of accepted results concerning smallholding farmers, confidence in the validity of this interpretation is reinforced. This may have significant implications for agricultural development policies. In particular, this suggests the feasibility of program packages that extend the benefits of technical progress to small farmers. If this seems a modest result, the reader is reminded of the pessimism of Mexican policymakers about smallholder responsiveness (see Solís 1971, section II). The effects of risk-averse behavior on acceptance, however, must be considered if the innovation is not stochastically dominant as the example of Plan Puebla demonstrates. Clearly, investigation of smallholder acceptance of nondominating innovations is an important area for further research. The results of this chapter suggest that subjective probability distributions are effective and useful devices for summarizing the state of technical information of a decisionmaker—at least in highly structured decisionmaking situations.

Appendix. Derivation of Equation (9.7)

In this appendix, we derive equation (9.7), that is,

$$g(\theta_{t+1}) = T(\theta_t, \hat{\theta}_t),$$

where $\hat{\theta}_t$ is a sufficient statistic of fixed dimensionality and θ_t is the parameter of the joint distribution of relative returns for period t.

If $\hat{\theta}$ is sufficient, then the likelihood can be factored:

(9.20a) $$\ell(x|\theta) = k(\hat{\theta}(x)|\theta)p(x),$$

where the factor k, called the kernel, is the only one involving θ.

From Bayes' theorem, we have

(9.20b) $$g''(\theta) = g'(\theta)\ell(x|\theta)n(x),$$

where prime and double prime denote prior and posterior distributions, respectively, and $n(x)$ is a normalizing constant that insures that

(9.20c) $$\int g''(\theta)d\theta = 1.$$

Hence,

(9.20d) $$g''(\theta) = g'(\theta)k(\hat{\theta}(x)|\theta)c(x),$$

where the normalizing constant $c(x) = p(x)\,n(x)$ can be determined from equation (9.20c). If $\hat{\theta}$ is a sufficient statistic of fixed dimension, say m, then Raiffa and Schlaifer (1961, chapter 3) have established:

THEOREM. *Let* $\hat{\theta}^1 = \hat{\theta} \ (x_1, x_2, \ldots, x_p)$ *and* $\hat{\theta}^2 = \theta(x_{p+1}, \ldots, x_n)$. *Then, it is possible to find a binary operation* (⋆) *such that*

$$\hat{\theta}^1 \star \hat{\theta}^2 = \hat{\theta}^\star = (\hat{\theta}_1^\star, \hat{\theta}_2^\star, \ldots, \hat{\theta}_m^\star)$$

has the properties:

$$\ell(x_1, x_2, \ldots, x_n | \theta) \propto k(\hat{\theta}^\star | \theta)$$

$$k(\hat{\theta}^\star | \theta) \propto k(\hat{\theta}^1 | \theta) \ k(\hat{\theta}^2 | \theta).$$

If $g'(\theta)$ is defined by

(9.20e) $$g'(\theta) = f(\theta | \hat{\hat{\theta}}) = k(\hat{\hat{\theta}} | \theta) \ p(\hat{\hat{\theta}}),$$

where $p(\hat{\hat{\theta}})$ is a normalizing constant and $\hat{\hat{\theta}}$ is the parameter of $g'(\theta)$. The kernel function k is now considered a function on parameter space rather than sample space, and the function f is called a natural conjugate of the kernel function k. Hence,

(9.20f) $$g''(\theta) \propto k(\hat{\hat{\theta}} | \theta) \ k(\hat{\theta} | \theta) \propto k(\hat{\hat{\theta}} \star \hat{\theta} | \theta).$$

That is, the kernel of the prior density combines with the sample kernel in exactly the same way that two sample kernels combine. Hence, if $g'(\theta)$ is so defined,

(9.20g) $$g''(\theta) = g'(\theta)k(\hat{\theta} | \theta)c(x)$$
$$= k(\hat{\hat{\theta}} | \theta)k(\hat{\theta} | \theta)p(\hat{\hat{\theta}})c(x)$$
$$= k(\hat{\hat{\theta}} \star \hat{\theta} | \theta)C(\hat{\hat{\theta}}, x) = T(\theta, \hat{\theta}),$$

since the constant $C(\hat{\hat{\theta}}, x) = p(\hat{\hat{\theta}})c(x)$ is easily determined from equation (9.20c).

References

Arrow, Kenneth J. 1971. *Essays in the Theory of Risk Bearing*. Chicago: Markham.

Bradley, James V. 1968. *Distribution-Free Statistical Tests*. Englewood Cliffs, N.J.: Prentice-Hall.

Brown, Lester R. 1970. *Seeds of Change: The Green Revolution and Development in the 1970's*. New York: Praeger.

Falcon, Walter P. 1970. "The Green Revolution: Generations of Problems." *American Journal of Agricultural Economics*. Vol. 52 (December), pp. 698–710.

Fishburn, Peter C. 1967. "Methods of Estimating Additive Utilities." *Management Science*. Vol. 13 (March), pp. 435–53.

Frankel, Francine R. 1971. *India's Green Revolution: Economic Gains and Political Costs*. Princeton, N. J.: Princeton University Press.

Freund, R. J. 1956. "The Introduction of Risk into a Programming Model." *Econometrica*. Vol. 24 (July), pp. 253–63.

Friedman, M., and L. J. Savage. 1948. "The Utility Analysis of Choices Involving Risk." *Journal of Political Economy*. Vol. 56 (August), pp. 279–304.

Griliches, Zvi. 1969. "Hybrid Corn: An Exploration in the Economics of Technological Change." *Econometrica*. Vol. 25 (October 1957), pp. 501–52. Reprinted in *Readings in the Economics of Agriculture*. Edited by Karl Fox and D. Gale Johnson. Homewood, Ill.: Irwin.

Hadar, J., and W. R. Russell. 1969. "Rules for Ordering Uncertain Prospects." *American Economic Review*. Vol. 59 (March), pp. 25–34.

Hakansson, N. 1970. "Friedman-Savage Utility Functions Consistent with Risk Aversion." *Quarterly Journal of Economics*. Vol. 84 (August), pp. 472–87.

Halter, A. N., and R. R. Officer. 1968. "Utility Analysis in a Practical Setting." *American Journal of Agricultural Economics*. Vol. 50 (May), pp. 257–77.

Hanoch, G., and H. Levy. 1969. "The Efficiency Analysis of Choices Involving Risk." *Review of Economic Studies*. Vol. 36 (July), pp. 335–46.

Hopper, W. D. 1965. "Allocational Efficiency in a Traditional Indian Agriculture." *Journal of Farm Economics*. Vol. 47 (August), pp. 611–24.

International Maize and Wheat Improvement Center. (Centro Internacional para el Mejoramiento de Maíz y Trigo, CIMMYT). 1970. *The Puebla Project 1967–69*. Mexico City.

Ladejinsky, Wolf. 1970. "Ironies of India's Green Revolution." *Foreign Affairs*. Vol. 48 (July), pp. 758–68.

Lau, Lawrence J., and Pan A. Yotopoulos. 1971. "A Test for Relative Efficiency and an Application to Indian Agriculture." *American Economic Review*. Vol. 61 (March), pp. 94–109.

———. 1973. "A Test for Relative Economic Efficiency: Some Further Results." *American Economic Review*. Vol. 63 (March), pp. 214–23.

Lin, G., G. W. Dean, and C. V. Moore. 1974. "An Empirical Test of Utility vs. Profit Maximization in Agricultural Production." *American Journal of Agricultural Economics*. Vol. 56 (August), pp. 497–508.

Masson, R. T. 1972. "The Creation of Risk Aversion by Imperfect Capital Markets." *American Economic Review*. Vol. 54 (March), pp. 77–86.

Massell, B. F., 1967. "Farm Management in Peasant Agriculture: An Empirical Study." *Food Research Institute Studies*. Vol. 7, no. 2, pp. 205–15.

McFarquhar, A. M. M. 1961. "Rational Decision Making and Risk in Farm Planning." *Journal of Agricultural Economics*. Vol. 14 (December), pp. 552–63.

Mossin, J. 1968. "Optimal Multiperiod Portfolio Policies." *Journal of Business*. Vol. 41 (April), pp. 215–29.

O'Mara, Gerald T. 1971. "A Decision-Theoretic View of the Microeconomics of Technique Diffusion in a Developing Country." Ph.d. dissertation. Stanford University.

Pratt, J. W. 1964. "Risk Aversion in the Small and in the Large." *Econometrica*. Vol. 32 (January-April), pp. 122–36.

Pratt, J., H. Raiffa, and R. Schlaifer. 1965. *Introduction to Statistical Decision Theory*. New York: McGraw-Hill.

Raiffa, H., and R. Schlaifer. 1961. *Applied Statistical Decision Theory*. Boston: Division of Research, Graduate School of Business Administration, Harvard University.

Samuelson, Paul A. 1969. "Lifetime Portfolio Selection by Dynamic Stochastic Programming." *Review of Economics and Statistics*. Vol. 51 (August), pp. 239–46.

Schultz, Theodore W. 1964. *Transforming Traditional Agriculture*. New Haven: Yale University Press.

Solís, L. 1971. "Mexican Economic Policy in the Post-War Period: The Views of Mexican Economists." *American Economic Review*. Vol. 61 (June, Supplement), pp. 2–67.

von Neumann, J., and O. Morgenstern. 1953. *Theory of Games and Economic Behavior*. 3d ed. Princeton, N. J.: Princeton University Press.

Wellisz, S., B. Munk, T. P. Mayhew, and C. Hemmer. 1970. "Resource Allocation in Traditional Agriculture: A Study of Andhra Pradesh." *Journal of Political Economy*. Vol. 78 (July-August), pp. 655–84.

Wharton, Clifton R., Jr. 1969. "The Green Revolution: Cornucopia or Pandora's Box?" *Foreign Affairs*. Vol. 47 (April), pp. 464–76.

Winkelmann, Donald L. 1973. "Factors Inhibiting Farmer Participation in Plan Puebla." *LTC Newsletter*. Land Tenure Center, University of Wisconsin Madison, no. 39 (January-March), pp. 1–5.

Winkler, Robert L. 1967. "The Assessment of Prior Distributions in Bayesian Analysis." *Journal of the American Statistical Association*. Vol. 62 (September), pp. 776–800.

Yotopoulos, Pan A. 1967. *Allocative Efficiency in Economic Development: A Cross Section Analysis of Epirus Farming*. Athens: Center of Planning and Economic Research.

10

Reviewing Agricultural Technologies When Farmers' Degrees of Risk Aversion Are Unknown

JOCK R. ANDERSON

RISK IS NOW WIDELY RECOGNIZED as a critical factor in nearly all farming activities and, especially, as an important factor in the adoption of new technology by farmers, particularly those in traditional agricultures.

The analytical framework of Bernoullian decision theory exists for incorporating consideration of risk in planning. As outlined by Dillon (1971), however, it is a personalized structure that emphasizes the individual's preferences for risk and his individual feelings of uncertainty and perception of risk. Even such decision-theoretic methods have so far found very restricted practical application because of the difficulties and costs involved in eliciting farmers' subjective probabilities and in encoding their preferences in utility functions. To date, including the study by O'Mara (the preceding chapter of this book), preferences have probably been elicited from something less than 400 farmers. How, then, can anything useful be said about risky planning for the remaining millions of farmers? Chapters 7 and 8 of this volume approach the problem by making assumptions at the individual preference level and by applying the resultant structure over aggregates of farmers.

An alternative approach, which holds some promise and which does not require assumptions about farmers' preferences, is that of ordering risky prospects according to stochastic dominance rules. These rules were discovered more or less independently by Fishburn (1964), Hadar and Russell (1969), Hammond (1974), Hanoch and Levy (1969), and Quirk and Saposnik (1962). Detailed exposition of the rules is available in Anderson (1974b) and Anderson, Dillon, and Hardaker (1977); for brevity, only a brief description is offered herein. Most of the nonagricultural work on stochastic dominance has had as its rationalization the evident superiority

Note: This chapter is a condensed version of Anderson (1974b). Permission of the original publisher to use material here, with some editorial changes, is gratefully acknowledged. Most of the work was done at the Centro Internacional para el Mejoramiento de Maíz y Trigo (International Maize and Wheat Improvement Center, CIMMYT), El Batán, Mexico, where the author especially enjoyed the encouragement of Don Winkelmann.

of dominance orderings to orderings of uncertain quantities based only on moments, notably the mean and variance. This is not regarded here as a crucial advantage. The prime purpose is to explore how far one can go in reviewing new agricultural technologies in the absence of specific assumptions about the algebraic form of farmers' preference functions.

The empirical applications described below all relate to problems in the interpretation of agricultural research results. It is believed that, whenever research is addressed to the development of new crop varieties and practices that are intended for adoption by risk-averse farmers, the principles of stochastic efficiency are pertinent and, indeed, offer an important method of filtering out inefficient technological packages (that is, packages that would not be preferred and adopted by those averse to risk) so that they are not extended to the farming community.

Overview of Stochastic Dominance Rules

Consider two probability density functions (PDF) $f(x)$ and $g(x)$ for the random variable x, which does not take values outside the range $[a, b]$ (that is, outside $[a, b]$, $f(x)$ and $g(x)$ are everywhere zero). Assuming x to vary continuously over its range so that the PDF are continuous, "less-than" cumulative distribution functions (CDF), the CDF can be defined as:

(10.1)
$$F_1(R) = \int_a^R f(x)dx, \text{ and}$$
$$G_1(R) = \int_a^R g(x)dx,$$

so that R varies continuously on the interval $[a, b]$. The procedure of accumulating areas under $f(x)$ to define $F_1(R)$ can be applied to $F_1(R)$, that is,

(10.2)
$$F_2(R) = \int_a^R F_1(x)dx, \text{ and}$$
$$G_2(R) = \int_a^R G_1(x)dx.$$

Analogously, define

(10.3)
$$F_3(R) = \int_a^R F_2(x)dx, \text{ and}$$
$$G_3(R) = \int_a^R G_2(x)dx.$$

These functions are illustrated in figure 10-1.

We will be concerned with choice between alternatives described by a single uncertain quantity, x. A decisionmaker's preferences for x are encoded in a utility function $U(x)$ that is defined for all x in $[a, b]$. Several increasingly restrictive assumptions are introduced concerning the prefer-

Figure 10-1. *Schematic Probability and Derived Stochastic Dominance Functions*

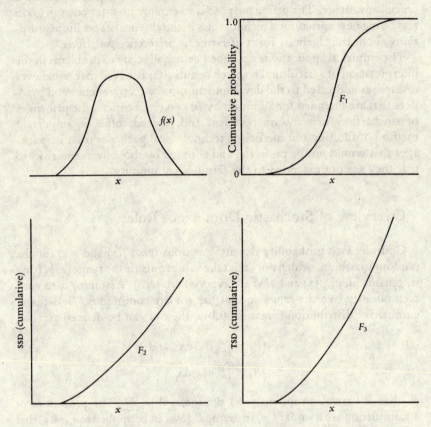

Note: SSD, second-degree stochastic dominance; TSD, third-degree stochastic dominance.

ence function. These involve the first three derivatives of $U(x)$, of which the ith is written $U_i(x)$, which are also defined for all x in $[\mathfrak{a}, b]$. Under risk, utility maximization implies maximizing expected utility.

The distribution $f(x)$ dominates $g(x)$ by *first-degree stochastic dominance* (FSD) if, and only if, $F_1(R) \leqslant G_1(R)$ for all R in $[a, b]$ with strict inequality for at least one value of R. This initial case presumes only that decision-makers prefer more of x to less. This implies that the function $U(x)$ is monotonically increasing between a and b or, equivalently, $U_1(x) > 0$.

This is eminently reasonable for the objectives of profits, gross margins, net farm produce, and the like that farmers are usually thought to be interested in. The rule (and those that follow) works in two ways, of which only one is of central concern in this chapter. If $f(x)$ dominates $g(x)$ in the sense of FSD, then $f(x)$ is preferred to $g(x)$ by all expected utility maximizers with strictly increasing utility functions. The second and presently less important converse result is that, if $f(x)$ is preferred to $g(x)$, (that is, has higher expected utility) for all utility functions with $U_1(x) > 0$, then $f(x)$ dominates $g(x)$ in the sense of FSD.

In graphic terms, a dominant (less-than) CDF lies nowhere to the left of a dominated CDF. This means that pairs of CDF that cross at least once reveal that FSD is not present.

The FSD results hold for discrete distributions, too. If x takes only the finite values in ascending order x_i, $i = 1, \ldots, n$, with respective probabilities $f(x_i)$, then the corresponding cumulative mass function is defined as

$$(10.4) \qquad F_1(x_r) = \sum_{i=1}^{r} f(x_i), \; r = 1, \ldots, n.$$

The FSD ordering rule for the discrete case is then: $f(x_i)$ dominates $g(x_i)$ by FSD if, and only if, $F_1(x_i) \leqslant G_1(x_i)$ for all x_i's with strict inequality for at least one value.

Unfortunately, whereas the FSD is based on a completely innocuous behavioral assumption, the majority of empirical CDF's associated with alternative technologies do cross, and so the FSD rule does not allow too many "inefficient" technologies to be identified in this way. It is useful then to introduce an additional assumption to "tighten-up" the ordering procedure. The assumption of aversion to risk seems a reasonable one to make. At least the conventional wisdom is that farmers generally, and small farmers in particular, are technically risk averse. That is, their utility functions are strictly concave; that is, $U_2(x) < 0$. This assumption then leads to a second ordering rule.

The distribution $f(x)$ dominates $g(x)$ by *second-degree stochastic dominance* (SSD) if, and only if, $F_2(R) \leqslant G_2(R)$ for all possible R with strict inequality for at least one value of R.

Figure 10-2 depicts the SSD graphically, where f is dominant if the F_2 curve lies nowhere to the left of the G_2 curve. Intuitive interpretation of this rule is not easy in terms of the F_2-type curves but is simplified by observing the corresponding CDF curves of figure 10-2. A necessary condition for f to be dominant in the sense of SSD is that the area labeled A is not less than the area labeled B.

Analogous to the FSD case, if $f(x)$ dominates $g(x)$ in the sense of SSD, then $f(x)$ is preferred by all farmers who are averse to risk and who maximize expected utility because $f(x)$ will always have higher expected utility. In this way, then, inefficient or undesirable distributions (technologies) can be identified and discarded.

Figure 10-2. *SSD Where Cumulative Distribution Functions (CDF) Cross Twice (Area A > Area B)*

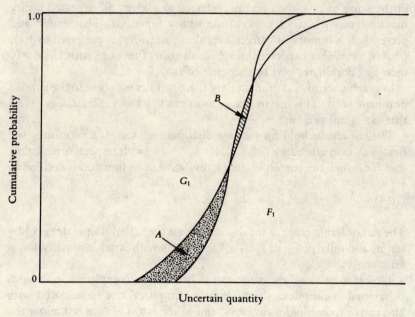

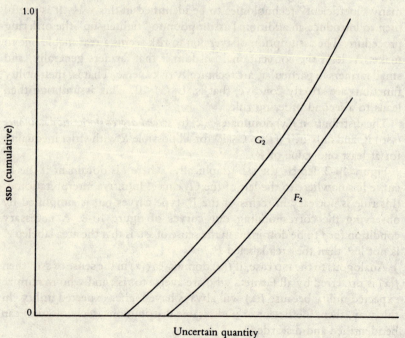

Hadar and Russell (1969) have also proved equivalent results for the discrete case. This can be stated, in following the earlier notation, by defining $\Delta x_i = x_i - x_{i-1}$, x_n as the highest value taken by x, and defining the analogue of $F_2(R)$ as

$$(10.5) \qquad F_2(x_r) = \sum_{i=2}^{r} F_1(x_{i-1})\, \Delta x_i, \; r = 2, \ldots n, \text{ and}$$

$$F_1(x_1) = 0.$$

Then $f(x_i)$ is preferred to $g(x_i)$ by risk averse farmers if $F_2(x_r) \leqslant G_2(x_r)$ for all $r \leqslant n$, and the strict inequality holds at least once.

SSD can be thought of as potentially ordering some prospects that are not orderable under FSD. Although the set of stochastically efficient alternatives under SSD will usually be smaller than under FSD, there is no guarantee that the set will be small, and to make further progress on narrowing down the efficient set it is necessary to make more restrictive assumptions about the nature of preferences. Of course, the limit to such activity is to define a particular preference function that will inexorably lead to the identification of a unique efficient (utility-maximizing) prospect. There is, however, one more fairly reasonable assumption that can be introduced to narrow down the utility-nonspecific efficient set; namely, a constraint on the third derivative.

The final assumption introduced is that the third derivative is positive, $U_3(x) > 0$, as well as $U_1(x) > 0$ and $U_2(x) < 0$. This additional restriction is not so strong in intuitive appeal as the former but is likely to characterize the preferences of many decisionmakers, including small farmers. The restriction is implied by the requirement that decisionmakers become decreasingly averse to risk as they become more wealthy. It also usually implies that owners of the utility functions prefer positive skewness in distributions of x to negative skewness.

The distribution $f(x)$ dominates $g(x)$ by *third-degree stochastic dominance* (TSD) if, and only if, $F_3(R) \leqslant G_3(R)$ for all R in $[a, b]$ with strict inequality for at least one value of R, and if $F_2(b) \leqslant G_2(b)$. Whitmore (1970) found that, if $F_3(x) \leqslant G_3(x)$ for all x in $[a, b]$, $U_1(x) > 0$, $U_2(x) < 0$ and $U_3(x) > 0$, then $f(x)$ is at least as preferred as $g(x)$. The condition that $F_2(b) \leqslant G_2(b)$ is equivalent to the requirement that the mean of f is not less than the mean of g and so is a necessary condition for TSD and SSD.

The discrete version of the TSD rule has been inferred by Porter, Wart, and Ferguson (1973) and requires the cumulative function

$$(10.6) \qquad F_3(x_r) = 1/2 \sum_{i=2}^{r} [F_2(x_i) + F_2(x_{i-1})]\, \Delta x_i,$$

$$r = 2, \ldots, n, \; F_3(x_1) = 0.$$

Then $f(x_i)$ is preferred to $g(x_i)$ by all risk averse farmers with the third derivative of their utility function positive if $F_3(x_r) \leqslant G_3(x_r)$ and the strict inequality holds at least once, and if the mean of f, E_f, is not less than the mean of g, E_g.

Rules analogous to FSD, SSD, and TSD have proposed, based on further assumptions about the higher derivatives of the utility function. These are not discussed here because the additional utility assumptions are difficult to rationalize and because they seem unlikely to add significantly to the discriminating power of the rules already discussed. Indeed, even the TSD extension appears to be only of very marginal usefulness because it seldom is able to reduce the size of sets efficient according to the SSD rule.

One pays a price for unrestrictive generality in a choice criterion. In all the stochastic dominance criteria the "price" seems to be the important emphasis placed on the lower tails of the distributions compared. A review of the criteria reveals that a necessary condition for FSD, SSD, and TSD is that the lower bound of a dominant distribution not be less than that of an unpreferred distribution. As an empirical matter, this places inordinate emphasis on estimation of (lower) extreme values of uncertain quantities. One can easily envisage preferring one distribution to another in spite of the fact that the preferred has some small probability in its relative leftward lower tail. When people talk of risk in farming, however, it is usually the prospect of falling into the lower tails of probability distributions of yields, prices, profits, or sustenance consumption that they have in mind. It thus seems appropriate to focus attention on these tails.

A second necessary condition for FSD, SSD, and TSD is that the mean of a dominant distribution cannot be less than the mean of an unpreferred distribution. It might appear at first glance that these two necessary conditions (on means and lower bounds) provide an expedient first approach in computations for reviewing stochastic dominance. But, since they are necessary and not sufficient conditions, they can in general only identify pairs of distributions that cannot be separated through the dominance criteria. General implementation of the rules requires identification (and then elimination) of distributions that are dominated, and for this review process there is no alternative to comprehensive pairwise comparisons based on the rules elaborated above. The exception, as elaborated by Hammond (1974), occurs in the case where the CDF's are known to intersect only once.

Applications with Discrete Probability Distributions

The principles of stochastic dominance are applicable to many decision problems in agriculture, provided that probability distributions can be usefully specified and that inherent computational problems can be overcome. Probability specification is clearly of fundamental importance. It is a broad topic and is treated by length by Anderson, Dillon, and Hardaker (1977).

A review of several risky prospects using the dominance rules involves pairwise comparisons among the prospects while progressively eliminating from further comparison those prospects (actions) that are revealed as being dominated at any degree, commencing with degree one (FSD). Conceptually, distributions of any type can be compared, but the simplification adopted here is to confine comparisons to those among distributions of the same category. The category considered in this section is that of discrete distributions. Such analysis of stochastic efficiency is the simplest from a computational point of view. Against this simplicity must be balanced a recognition that the assumption that a random variable is discrete is usually a rather simplistic interpretation of a probabilistic situation that is properly continuous. Because of analysts' frequent resort to the assumption, however, the case is important and deserves careful attention.

For any two probability mass functions $f(x_i)$ and $g(x_i)$, compatibly defined in terms of an appropriate uncertain quantity x_i analysis of stochastic efficiency proceeds straightforwardly by first listing all the combined values taken by x in ascending order such that if $i < j$ then $x_i < x_j$. If two or more have the same numerical value, each value is considered to be distinct and the rank allocated to ties is lowest for those ties associated with the distribution with the nonzero probability for the lowest value of x or, where this is also tied, with the potentially dominated distribution. With the x_i listed, the $f(x_i)$ and $g(x_i)$ can then be written out, and the cumulative functions $F_1(x_i)$, $F_2(x_i)$, and $F_3(x_i)$ and those corresponding for $g(x_i)$ readily computed.

Two empirical examples are now considered. The first treats the case of discrete distributions of unequal probability elements. In the second, the distributions are assumed to have equal probability elements corresponding to the (assumed equal) relative sample frequencies. Thus, if there are N_f distinct observations on prospect f, then $F(x_i) = 1/N_f$; and, in comparing two probability functions $f(x_i)$ and $g(x_i)$, there is a total of $N = N_f + N_g$ distinct observations, and the relative frequencies of $g(x)$ are $g(x_i) = 1/N_g$ or $g(x_i) = 0$. That is, if the ith of the merged observations belongs to prospect f, then $f(x_i) = 1/N_f$ and $g(x_i) = 0$. The assumptions of equal probability elements and equal numbers of observations also simplify the FSD review procedure because it is only necessary to show that, for f to dominate g, the ranked outcomes for f are never less than the corresponding ranked outcomes for g.

Control of a maize insect pest

Yield distributions of crops under different regimes of chemical controls are doubtless continuous. Agronomists and farmers, however, often think and talk about such distributions as if they were discrete (that is, have only a finite, usually small, number of states). The specific problem considered under this assumption is the control of whorlworm (fall army worm) on maize at Poza Rica, Mexico. This pest is often a serious problem for farmers in this and many other areas. Quite satisfactory control measures have been evolved, even though they are not used as widely as would appear to be socially desirable.

Two treatments are compared here. The simplest is a seed treatment with an appropriate insecticide in wettable powder form that is available at about 120 pesos per hectare for purchased ingredients. Since farm labor can readily perform this task during slack periods, labor costs are ignored. The second treatment consists of the seed treatment plus a foliar application, in the form of granules of another appropriate insecticide, early in the growth of the crop. The cost of the granules and labor is about 150 pesos per hectare—so that the total cost of the second ("both") treatment is about 270 pesos per hectare. More sophisticated treatments, such as deciding on foliar applications on the basis of observed levels of insect infestation, are presently ignored.

Three discrete states are assumed; namely, low, medium, and high degrees of infestation prevailing. No infestation was thought not to be a possibility. With these assumptions, the basic data assumed for this example are presented in table 10-1. The yields and probabilities are agronomists' estimates of typical situations prevailing on farms in the area with appropriate fertilization. Valuing maize at 1.2 pesos per kilogram and subtracting the respective treatment costs converts the yields of table 10-1 to the gross margins of table 10-2.

Table 10-1. *Technical Data for the Insect Control Problem*

Degree of infestation	Probability	Treatment yields (kilograms per hectare)		
		No treatment	Treated seed	Treated seed and foliar application ("both")
Low	0.15	4,500	4,600	4,700
Medium	0.70	2,200	4,000	4,650
High	0.15	600	2,500	4,600

Table 10-2. *Economic Data for the Insect Control Problem*

Degree of infestation	Probability	Treatment returns (pesos per hectare)		
		No treatment	Treated seed	Treated seed and foliar application ("both")
Low	0.15	5,400	5,400	5,370
Medium	0.70	2,640	4,680	5,310
High	0.15	720	2,880	5,250
Expected returns[a]		2,106	4,086	5,328

a. Returns weighted by respective probabilities. The "both" treatment has the highest expected money value and accordingly would be preferred by farmers who are indifferent to risk.

From the data displayed in table 10-2, one simple form of dominance is immediately apparent: namely, treated seed is dominant over no treatment since, no matter how much infestation occurs, the payoff from seed treatment exceeds or equals that from no treatment. This means that "no treatment" need not be considered further in this analysis.

The decision problem is now reduced to exploring whether the dominance rules permit the identification of either the use of treated seed or both treated seed and foliar application as being stochastically efficient. The analysis proceeds in table 10-3 by first ranking the discrete payoffs in the two treatments to be considered and then defining the cumulative mass and SSD functions. Since only the "both" treatment is efficient in the sense of SSD, analysis can cease at that point: otherwise the TSD functions would be computed as in equation (10.6) and these values compared in the hope of identifying one of the treatments as being inefficient. Thus, as one might anticipate when there is an effective and fairly cheap measure available for control of an important pest, risk-averse farmers should adopt the safest treatment irrespective of their own particular attitudes to risk.

Selection of wheat varieties

Several methods have been used for identifying crop varieties that have wide environmental adaptability. The basic data for such work are usually obtained from nursery trials conducted in diverse environments, sometimes across many countries, such as in the collaborative International Spring Wheat Yield Nursery (ISWYN) administered by the International Maize and Wheat Improvement Center (CIMMYT 1972). Without critical

Table 10-3. *Stochastic Dominance Review for the Insect Control Problem*

	Financial return from treatment (pesos per hectare)					
x_i	2,880	4,680	5,250	5,310	5,370	5,400
$f(x_i)$						
Treated seed	0.15	0.70	—	—	—	0.15
"Both"	—	—	0.15	0.70	0.15	—
$F_1(x_i)$						
Treated seed	0.15	0.85	0.85	0.85	0.85	1.0
"Both"	0	0	0.15	0.85	1.00	1.0
Δx_i	—	1,800	570.00	60.00	60.00	30.0
$F_2(x_i)$						
Treated seed	0	270	754.50	805.50	856.50	882.0
"Both"	0	0	0	9.00	60.00	90.0

— Not applicable.

Note: The "both" (both treated seed and foliar application of insecticide) SSD function is less than the "treated seed" function at each value of x, so "both" is efficient in the sense of SSD.

review, it is clear that, in the absence of specifically and carefully elaborated criteria, there can be no one perfect method of appraisal. The present example examines the question of adaptability from the point of view of risk aversion and stochastic efficiency—a point of view believed to be relevant if the ultimate purpose of identifying widely adapted varieties is to make them available for adoption by farmers who generally are averse to risk. As Finlay and Wilkinson (1963) have observed earlier: "Plant breeders are inclined to ignore the results obtained in low-yielding environments, eg. drought years, on the basis that the yields are too low and are therefore not very useful for sorting out the differences between selections. This is a serious error, because high-yielding selections under favourable conditions may show relatively greater failure under adverse conditions."

The notions of stochastic dominance and efficiency seem to provide a useful framework for posing the essentially empirical question of how different selections perform in diverse risky environments. This analysis is straightforward, as is demonstrated below, with some important provisos—most important of which is that it makes good sense to speak of a world (or regional) probability distribution of wheat yields. Such a concept seems implicit and inherent in the conduct of international yield nurseries and in the comparison of means from such nurseries. A second proviso, which amounts to a presumption, is that the selection of sites, cooperators, fields, and growing conditions is representative of the relevant domain of production.

A third proviso is of importance for the logical application of the principles of stochastic dominance: that yield provides a reasonable surrogate for the argument of the implicit utility function. This assumption, which involves ignoring likely varying production costs, is unavoidable in processing international nursery data, since each trial is in general grown under differing regimes of irrigation (where practiced), tillage, fertilizers, and weed and pest control that are most difficult to cost.

Attention is now concentrated on the data from the Sixth ISWYN (CIMMYT 1972). In this nursery, forty-nine varieties were compared in trials at sixty locations in thirty-seven countries during 1969–70. For each variety, each trial observation is regarded as a distinct component of the discrete sample probability function of that variety. The pairwise comparison of forty-nine discrete actions involves up to $(49)(48)/2 = 1,176$ FSD comparisons at each of up to $(60)(2) = 120$ values of the uncertain quantity yield. Such a computational burden can be faced with equanimity only with the aid of an electronic digital computer. To this end, a program was prepared to undertake analysis of stochastic efficiency in this case of discrete distributions from samples of equal size. This program worked on a fully defined yield matrix for varieties and states (sites) and lists the varieties that are stochastically efficient of degrees 1, 2, and 3.

The results of applications to two sets of the Sixth ISWYN yield data are summarized in table 10–4. The results can be appraised from two viewpoints: first, the empirical identification of the numbered varieties into categories of stochastic efficiency, information that would seemingly be useful to plant breeders; second, there is the methodological question of the relation between stochastic efficiency and rank according to mean yield of each variety. Not surprisingly, there is a close relation, especially between first-degree stochastic efficiency (FSE) and rank. The FSE set includes most of the top-ranked varieties, and perhaps the most useful aspect of this identification is in pinning down a cutoff point of mean yields that is less than arbitrary. It should also be emphasized that risk-efficient varieties in the second-degree stochastically efficient (SSE) and third-degree stochastically efficient (TSE) sets are selected from the FSE set.

These applications offer little scope for generalizations about the likely general composition of the SSE and TSE sets, but it seems reasonable to risk one; namely, that as the environmental scope of analysis becomes more restrictive (for example, as in the nonirrigated cases reported in table 10-4), the greater is the chance that only the very highly mean-ranked varieties will be SSE or TSE. The implication of this shaky generalization is that, providing breeders limit the environmental scope for selecting broadly adapted varieties in some way, then by focusing on mean yield they will most probably be selecting varieties that are also stochastically efficient for risk-averse growers. To summarize, the analysis using no-

Table 10-4. Results of Stochastic Dominance Analysis of Yield Data from the Sixth International Spring Wheat Yield Improvement Nursery (ISWYN)

Sites (type and number)		First degree									Second degree			Third degree		
Stochastically efficient varieties of wheat and their mean ranks[a]																
All sites (60)	Variety	25	47	33	15	34	11	23	30	31	25	47	34	25	47	34
	Rank	1	2	3	4	5	6	7	8	9	1	2	5	1	2	5
	Variety	45	1	38	42	44	20	40	18	13	14	21		14	21	
	Rank	10	11	12	13	14	15	16	17	18	27	34		27	34	
	Variety	17	36	24	41	29	26	14	21	16						
	Rank	19	21	22	23	24	26	27	34	35						
Nonirrigated sites not affected by severe biological limitations and for which rainfall was reported (22)	Variety	15	47	25	1	33	34	30	23	44	15	47		15	47	
	Rank	1	2	3	4	5	6	7	8	9	1	2		1	2	
	Variety	31	40	42	11	13	41	22	16	45						
	Rank	10	11	12	13	14	15	16	17	18						
	Variety	17	26	24	36	18	43	21	28							
	Rank	19	20	21	24	25										

— Not applicable.

a. Ranks of varietal means calculated within the specified sites.

b. Risk-efficient varieties variously include: 14, Giza 155 (Egypt); 15, Siete Cerros 66 (Mexico); 25, LR-P4160[3] (E) (Pakistan); 34, Tabari 66 (Mexico); 45, CIANO 5 (Mexico); 47, Sonalika (India).

tions of stochastic efficiency also seems to provide useful information, but it may be most applicable at a later stage of screening materials that are relatively more similar in their adaptation and yield characteristics.

Applications with Continuous Probability Distributions

The probability distributions that emerge from analyses of data or from judgments are often asymmetric or otherwise irregular, so that a search for a convenient theoretical distribution that fits adequately may be tedious or simply too costly. One pragmatic alternative is simply to describe a cumulative distribution function, CDF, by a smooth hand-sketched curve and in turn approximate this curve by a number of segments of simple algebraic form.

The simplest form consists of linear segments, and for several reasons this is the alternative adopted here. A linear-segmented CDF corresponds to a rectangular histogram probability density function, PDF, and, if sufficient segments are taken, this is bound to be an adequate approximation. Apart from conceptual simplicity, the advantage of specifying a CDF in linear segments is the relative simplicity afforded to the integrations required to specify the SSD and TSD functions and to the solution of the simultaneous equations required in comparisons done on a computer.

There are many ways of arranging linear approximation of an arbitrary CDF. The alternative adopted here is to assume that each of a total number of segments spans an equal interval of cumulative probability. The examples presented here are based on the additional simplifying assumption that each distribution has the same number of segments. This assumption results in some simplification of computation and programming. The algebra of this case is outlined by Anderson (1974*b*) and a computer program for implementing the method is reported in Anderson, Dillon, and Hardaker (1977).

Adoption decision concerning
a new technology package for maize

The next example with rather crude distributional representation based on $NS = 4$ illustrates graphically the nature of the computations. No generality should be attached to the empirical aspects of this example, which is not typical of the technological situation for maize in Mexico. For more representative examples indicating the stochastic efficiency of improved technologies, see O'Mara (chapter 9 of this book) and Villa Issa (1974). The example concerns a hypothetical choice between a maize technology in current use and a new technology based on improved varieties and more intensive use of fertilizer, seed, and irrigation. The new technology here is that recommended in the Programa de Altos Rendimientos (PAR) in the Texcoco area. All data come from the study by O'Mara (chapter 9 of this volume and 1971) and the particular subjective yield distributions are those elicited in case 49 of his farmer interviews. The use made here of these data is simplistic in several ways, including the simplification of couching the adoption question in an "all or nothing" farm in which, in reality, partial adoption (that is, on part of the maize area of a farm) is clearly an important possibility. This simplification, however, permits us to ignore the dependence (included in O'Mara's work) between yield distributions under the two technologies.

The data are a farmer's subjective estimates of several fractiles[1] of grain yield distributions (in metric tons per hectare) under each technological alternative and are presented in the following tabulation:

	Fractile				
	0	0.25	0.50	0.75	1.00
Present technology	0.75	1.25	1.5	2.50	3.0
New technology	1.00	2.00	3.5	4.75	6.0

The bimodal character of these histograms makes them atypical of yield distributions of field crops, and a more appropriate analysis would first submit these to a smoothing process and would probably result in smooth unimodal distributions.

Economic analysis of such data first involves bringing them to a common basis with due account of the costs and returns in each technology. Following O'Mara (1971, pp. 97, 100, 268ff.), this is done by computing returns per hectare as: grain yield times the government's guaranteed grain price of 940 pesos (1970 pesos) per metric ton less variable costs excluding harvest costs, land rent, and fixed labor costs. This simplification is based on the assumption that returns from fodder and costs of harvesting grain and fodder (returns and costs both vary with grain yield) are equal. The variable costs budgeted for the present and new technologies were 866 and 1,770 pesos per hectare, respectively. The fractile data for yields were then transformed to fractiles for net returns (in pesos per hectare) and are reported in the following tabulation (they are also sketched in figure 10-3):

	Fractile				
	0	0.25	0.50	0.75	1.00
Present technology	−161	309	544	1,484	1,954
New technology	−830	110	1,520	2,695	3,870

1. A "point b fractile," $f \cdot b$, is that value of a random variable x such that $P(\alpha \leqslant f \cdot b) = \cdot b$.

Figure 10-3. *CDF (FSD Functions) for Profits under Two Maize Technologies*

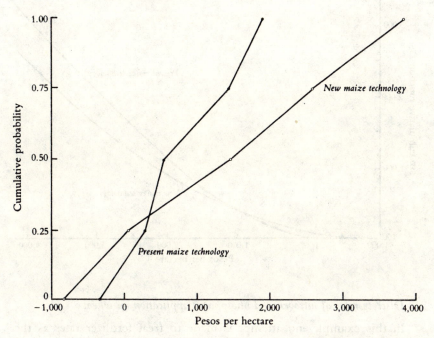

Note: FSD, first-degree stochastic dominance.

Since the CDF intersect, there is clearly no FSD. Viewing the CDF of figure 10-3 and recalling the discussion of figure 10-2, it should also be clear that this case will not reveal SSD either. This is confirmed by observing the intersection of the spliced quadratic SSD cumulatives in figure 10-4.

Intuition would also lead one to guess that this particular case, which reveals a pessimistic emphasis of unfavorable outcomes on the new technology, will also not reveal TSD. This too is confirmed by observing the intersection of the (spliced cubic) TSD cumulatives in figure 10-5.

This means that, given the farmer's subjective probabilities and the arbitrary representation accorded them here, it is not possible to say which he would prefer without knowing something more of his specific attitudes to risk beyond the assumptions involved in the SSD and TSD ordering rules. The virtue of this example is to make explicit the pertinent review procedures that, even from this small exemplification, will be sensed to be computationally burdensome when many (say twenty) linear segments are used to approximate a CDF and (especially) where many distributions are to be reviewed. Such an example is explored next.

Figure 10-4. *SSD Functions for Profits under Two Maize Technologies*

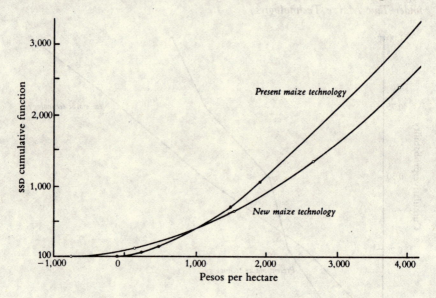

Discrete rates of nitrogen and phosphorus application on wheat

In this example, no attempt is made to treat fertilizer rates as the continuous variables that (up to limits of machinery calibration) they are. The example builds on material described more fully by Anderson (1973; 1974*a*). The starting point here is to use the thirty-six probability distributions of nonirrigated wheat yield estimated for each of the design points of a 6 × 6 complete factorial. The treatments are for *N* approximately 0, 22.5, 44.9, 67.4, 89.8, 112.2 (or 0, 22, 45, 67, 90, 112 kilograms per hectare) and for *P* approximately 0, 9, 18, 27, 36, 45 kilograms per hectare.

The estimation of the distributions, which reflect between-year variability, was based on sparse data, and some examples of these are shown in figure 10-6 in the linear-segmented CDF form in which all are subsequently described in this exemplification. Each distribution is described by twenty linear segments of equal height. The yield distributions were transformed to return distributions by the linear expression $R_{ijk} = p_y Y_{ijk} - p_n N_i - p_p P_j$, where *R* denotes return, *Y* denotes yield, *N* denotes nitrogen applied, and *P* denotes elemental phosphorus applied (all per unit area); p_y, p_n, and p_p are the respective unit prices of *Y*; and *N*, and *P*, and the subscripts *ijk* denote, respectively, the *i*th level of *N*, the *j*th level of *P*, and the *k*th fractile. Prices assumed are as previously reported. Note that fixed costs are not included in this expression because they have no influence on the determination of stochastic efficiency.

Figure 10-5. *TSD Functions for Profits under Two Maize Technologies*

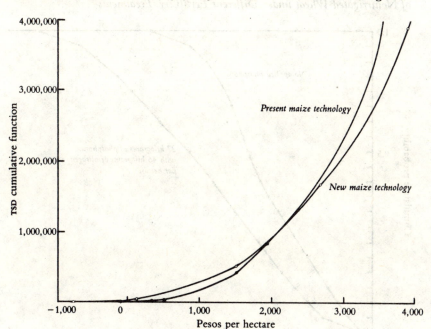

The outlined procedure for reviewing stochastic dominance with arbitrary continuous distributions was applied to these thirty-six discrete actions through the implemented computer program. The results are most easily viewed in table 10-5, in which rows and columns are defined by rates of N and P, and the table entries refer to the degree of stochastic efficiency, where zero denotes being dominated in the sense of FSD. FSE combinations of fertilizers are indicated by an integer ≥ 1. FSE combinations become candidates for SSD review, and those that are not dominated in this sense (the SSE set) would be indicated by an integer ≥ 2 and in turn become candidates for TSD review. Those not dominated in the sense of TSD are TSE and are indicated by the integer 3. In this case the SSE and TSE sets are identical so that no "2" entries appear in table 10-5. The SSE and TSE combinations are referred to as "risk-efficient."

These results indicate a fairly consistent pattern wherein a necessary condition for stochastic efficiency of any order is a reasonable dose of phosphorus. Nitrogen application is indicated as being a rather risky proposal, since most of the risk-efficient combinations involve zero levels of N and the highest risk-efficient rate of N is 45 kilograms per hectare in this nonirrigated situation.

Figure 10-6. *Examples of Linear-segmented CDF for Yields of Nonirrigated Wheat under Different Fertilizer Treatments*

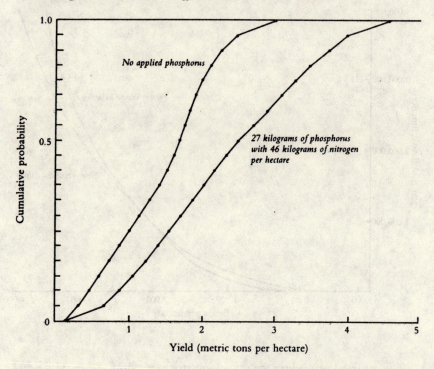

Yield (metric tons per hectare)

Within the risk-efficient set, choice of fertilizer rate properly depends upon the risk preferences of individual farmers. Given the consistent pattern of risk-efficient rates, it seems reasonable to interpolate within the set. In this particular example, as in others from the field of portfolio analysis, it does turn out that the risk-efficient set corresponds very closely with the mean-variance efficient set.[2]

Rice production packages

Procedures for normal distributions are now introduced by reference to an example from Roumasset (1973). It concerns four alternative packages of rice technology under irrigated conditions in the Philippines (Bicol region). Techniques M_1, M_2, and M_3, involve progressively more inten-

---

2. A risky prospect is mean-variance efficient if it is not dominated by another in this sense. Distribution f dominates g in the mean-variance sense if mean $f \geqslant$ mean g and variance $f \leqslant$ variance g, with at least one of the strict inequalities holding.

Table 10-5. *Degree of Stochastic Efficiency of Specified Combinations of Nitrogen and Phosphorus Fertilizer*
(kilograms per hectare)

	Phosphorus					
Nitrogen	0	9	18	27	36	45
	Degree of stochastic efficiency					
0	0	1	3	3	3	3
22	0	0	0	3	3	1
45	0	0	1	3[a]	1	1
67	0	0	0	1	1	1
90	0	0	0	1	1	1
112	0	0	0	0	1	1

a. The discrete combination with greatest mean return.

sive use of fertilizer, other chemicals, and labor on improved variety IR-5, whereas technique T involves traditional practices applied to local varieties. Mean profits are budgeted in Roumasset (1973, table 3-2, p. 54) and are 960, 1,055, 1,135, and 510 pesos per hectare, respectively. The introduced assumption of normality, and the estimation of standard deviations, are much more arbitrary when designed to equilibrate the profit distributions at two standard deviations below each mean. The standard deviations suggested are, respectively, 320, 400, 480, and 128 pesos per hectare.

Since all four distributions are normal, this devolves to a mean–variance (equivalently, mean–standard–deviation) analysis. Comparison among these distributions, however, reveals that for no pair does the condition hold that simultaneously a mean is greater and a standard deviation lesser. Thus, all four distributions belong to the SSE set.

One of the difficulties associated with the assumption of normality is that the range is necessarily plus or minus infinity. Clearly, this is not a very realistic assumption for most empirical phenomena in agriculture. Presumably, however, the frequent resort to the normal distribution must imply that it does fit empirical distributions tolerably well over much of their relevant domains. Thus, perhaps in risk analysis we should "go along" with the assumption of normality but somehow ignore the extreme tails. This is the rationale behind the following suggestion for a modified definition of FSD for normal distributions.

Consider two normal distributions with means μ_i, μ_j and standard deviations σ_i, σ_j, respectively. Assume that $\mu_i > \mu_j$ and that $\sigma_i \neq \sigma_j$ so that the CDF's will intersect once at $x^\star$, which lies $z^\star$ standard deviations from each mean and at a cumulative probability $P^\star$. The important case is where $\sigma_i > \sigma_j$, $z^\star < 0$ and $P^\star < 0.5$, but for completeness we look also to the case of upper-tail intersections where $\sigma_i < \sigma_j$ and $P^\star > 0.5$. For the lower-tail intersection, $z^\star = (\mu_i - x^\star)/\sigma_i = (\mu_j - x^\star)/\sigma_j$, from which it follows that $C^\star = 1/z^\star = (\sigma_i - \sigma_j)/(\mu_i - \mu_j)$ and, analogously, for intersections in both tails, $C^\star = |\sigma_i - \sigma_j|/(\mu_i - \mu_j)$. Now, if intersections in the extreme tails of normal distributions can in some pragmatic sense be ignored, then pragmatic FSD orderings can be based on calculated values of $C^\star$. The procedure is as follows: first, define a critical left-tail probability, $P^{\star\star}$, below which CDF crossovers can be ignored; second, from tables of the cumulative standard normal distribution, look up $z^{\star\star}$ and find $C^{\star\star} = 1/z^{\star\star}$; third, compute $C^\star$; and fourth, if $\mu_i > \mu_j$ and $C^\star \leqslant C^{\star\star}$, distribution i pragmatically dominates j in the sense of approximate FSD.

Although choice of any critical probability is necessarily arbitrary, the suggestion offered here is that a value of 0.01 seems a reasonable guideline, bearing in mind the typical adequacy of the normal assumption in empirical applications. This is the critical probability used in applying this approximate FSD procedure to the data for alternative rice technology. The calculations are reported in table 10-6.

According to the criteria developed and the critical levels reported, only one technique, T, is dominated in the sense of approximate FSD—for with $P^{\star\star} = 0.01$ and $C^{\star\star} = 0.429$, since $320 > 128$ and $0.426 < 0.429$, then M_1 dominates T. Similarly, if the critical probability is raised to 0.05, T is also dominated by M_2 and M_3. For critical probabilities up to 0.1, however, all of the three improved-variety techniques achieve FSE.

Methodological Extensions

The approaches outlined above can be extended to deal with other problems in the assessment of agricultural technologies.

One important class of decisions includes what could be termed "continuous-action" problems, where the decision variables are intrinsically continuous (such as fertilizer levels). Anderson (1974*b*) has illustrated the

Table 10-6. *Approximate First-degree Stochastic Dominance (FSD)*
Analysis of Four Normal Distributions

Rice production technique	Data	Ratio
M_1M_2	I(320–400)I/(960–1 055)	0.842
M_1M_3	I(320–480)I/(960–1 135)	0.914
$M_1 T$	I(320–128)I/(960–510)	0.426
M_2M_2	I(400–480)I/(1 055–1 135)	1.000
$M_2 T$	I(400–128)I/(1 055–510)	0.499
$M_3 T$	I(480–128)I/(1 135–510)	0.563

application of stochastic dominance rules to such problems, specified variously by arbitrary or theoretical (normal and beta) continuous probability distributions. That article also examines the question of the effect of uncertain product prices on fertilizer rates in the context of stochastic efficiency.

Another important consideration in reviewing new technologies concerns their integration into a whole-farm plan. As illustrated by Anderson (1975), it is also possible to include the concepts of stochastic dominance in whole-farm planning models of the risk-programming type. A discussion of these and other possible uses of stochastic efficiency, however, is beyond the scope of this chapter.

Implications for Policy

Increasingly over recent years, lip service has been paid to the notion that risk is an important aspect of agricultural technology. Although this recognition is valuable in itself, a machinery that deals analytically with risk in the absence of knowledge of farmers' individual attitudes to risk has not hitherto been exploited in agricultural research.

There appear to be several implications in such machinery for workers in agricultural research and extension. Rather than focus only on estimation of treatment means, agricultural research should explore and estimate whole probability distributions to complement conventional "average-oriented" research if risk-averse consumers are to be well served. This implies that risk-oriented research will be generally more demanding and more expensive than average-oriented research. This apparently is the price one must pay for work that is potentially relevant.

More particularly, appraisal of stochastic efficiency in the absence of knowledge of individuals' attitudes of risk demands pinning down the lower tails of probability distributions. This estimational task suggests that agricultural innovations need to be evaluated and reported under the bad, as well as the typical or average, environmental conditions that potential adopters face (for example, with respect to moisture stress, disease exposure, nutrient suboptimization, and so forth). Risk-oriented research should deliberately span an appropriate range of both environments and environmental conditions, which will usually mean replication over space and time. There seems to be much unexploited scope for formal documentation of research agronomists' considerable experience of, and largely unpublished knowledge of, the tails (especially the left tails) of relevant probability distributions.

Extension of technological advice in risky agricultures will probably be more effective if due recognition is given to the effect of risk and the importance of technologically induced risk. Such extension could be simplified by grouping farmers according to the worst environmental conditions they face, and its success could be increased by promoting practices that are tailored to be stochastically efficient (at least to the second degree) for the identified groups.

In judging extension efforts, recognition that a recommendation efficient in terms of average profit may not be risk efficient should temper appraisal of programs.

Perceptive practitioners of the arts of agricultural research and extension inevitably develop a keen intuition for the importance of risk in most agricultural production. Their formal training, however, has usually done little or nothing to equip them with an analytic apparatus with which to deal directly with this aspect of their work. Clearly, educational programs should do more to sensitize prospective practitioners to the effect of risk in farming, and consequently in research and extension. Particular attention needs to be drawn to the fact that espoused experimental methodology is addressed to estimating only average effects, responses, and the like and, accordingly, is only directly applicable to risk-indifferent farmer-users.

The principal pertinent policy instruments in dealing with risk have been crop insurance schemes and minimal price supports for agricultural products. Our discussion of the ordering of stochastic dominance places a new emphasis on such schemes. By stressing values in the lower tails of distributions of yields and prices rather than on values near the averages, relatively low premiums may still exert significant adjustments to actions of farmers.

More specifically, for example, a crop "insurance" scheme that effectively truncates yield distributions below a crossover point in the lower tails of two varietal distributions makes the variety that yields higher on average stochastically dominant in the sense of FSD. Ensuant adoption by farmers of the new variety, now with FSE, would doubtless be in the national interest. Typically, a recommended technological practice will dominate (have FSD) traditional practices if it can be "insured" to the extent that, under really poor eventualities, farmers are not disadvantaged by adoption.

Conclusions

Although well-known procedures are available for formulating recommendations destined for farmers concerned solely with (average) profits, this is not the case for farmers concerned with risks as well as with average profits. Farmers are generally influenced in their decisions by aspects of profit distributions other than the averages in a way that can be described as risk averse. This is not to say that farmers do not take risks—they must do so continually—but rather that they can take into account variability of profits as well as the average. A second, widely agreed upon belief is that farming generally is risky and that new technology in particular is perceived as risky. This study then is predicated on the notions that research, extension, and new technology will be more effective and successful if proper account is taken of risk and that it is impossible to account individually for the attitudes to risk of the millions of farmers who feed the world.

A critical question, and one to which the answer has not been obvious, is whether proper account of risk can be taken in research and extension. Perhaps this study has contributed to an affirmative answer. The principles of stochastic dominance permit orderings of risky prospects that are as complete as is theoretically possible without knowing more—much more—about farmers' attitudes to risk than will ever be possible.

References

Anderson, Jock R. 1973. "Sparse Data, Climatic Variability, and Yield Uncertainty in Response Analysis." *American Journal of Agricultural Economics.* Vol. 55, pp. 77–82.

_____. 1974*a*. "Sparse Data, Estimational Reliability, and Risk-Efficient Decisions." *American Journal of Agricultural Economics.* Vol. 56, pp. 564–72.

_____. 1974*b*. "Risk Efficiency in the Interpretation of Agricultural Production Research." *Review of Marketing and Agricultural Economics.* Vol. 42 (September 1974), pp. 131–84.

_____. 1975. "Programming for Efficient Planning against Nonnormal Risk." *Australian Journal of Agricultural Economics.* Vol. 19 (August), pp. 94–107.

Anderson, Jock R., J. L. Dillon, and J. B. Hardaker. 1977. *Agricultural Decision Analysis.* Ames: Iowa State University Press.

CIMMYT (Centro Internacional para el Mejoramiento de Maíz y Trigo, International Maize and Wheat Improvement Center). 1972. *Results of the Sixth International Spring Wheat Yield Nursery.* CIMMYT Research Bulletin, no. 23. Mexico City.

Dillon, J. L. 1971. "An Expository Review of Bernoullian Decision Theory." *Review of Marketing and Agricultural Economics.* Vol. 39 (March), pp. 3–80.

Finlay, K. W., and G. M. Wilkinson. 1963. "The Analysis of Adaptation in a Plant Breeding Programme." *Australian Journal of Agricultural Research.* Vol. 14 (November), pp. 742–54.

Fishburn, Peter C. 1964. *Decision and Value Theory.* New York: Wiley.

Hadar, J., and W. R. Russell. 1969. "Rules for Ordering Uncertain Prospects." *American Economic Review.* Vol. 59 (March), pp. 25–34.

Hammond, J. S., III. 1974. "Simplifying the Choice between Uncertain Prospects Where Preference Is Nonlinear." *Management Science*. Vol. 20 (March), pp. 1047–72.

Hanoch, G., and H. Levy. 1969. "The Efficiency Analysis of Choices Involving Risk." *Review of Economic Studies*. Vol. 36 (July), pp. 335–46.

O'Mara, Gerald T. 1971. "A Decision-Theoretic View of the Microeconomics of Technique Diffusion in a Developing Country." Ph.D. dissertation. Stanford University.

Porter, R. B., J. R. Wart, and D. L. Ferguson. 1973. "Efficient Algorithms for Conducting Stochastic Dominance Tests of Large Numbers of Portfolios." *Journal of Financial and Quantitative Analysis*. Vol. 8, pp. 71–81.

Quirk, J. P., and R. Saposnik. 1962. "Admissibility and Measurable Utility Functions." *Review of Economic Studies*. Vol. 29 (February), pp. 140–46.

Roumasset, J. 1973. "Risk and Choice of Technique for Peasant Agriculture: The Case of Philippine Rice Farmers." Ph.D. dissertation. University of Wisconsin.

Villa Issa, Luis A. 1974. "Adopción de tecnología neuva en zonas de temporal: El efecto del factor de incertidumbre." M.A. thesis. Colegio de Postgraduados, Escuela Nacional de Agricultura, Chapingo, Mexico.

Whitmore, G. A. 1970. "Third-Degree Stochastic Dominance." *American Economic Review*. Vol. 60 (June), pp. 457–59.

Part Three

Regional Programming Models

11

A Regional Agricultural Programming Model for Mexico's Pacific Northwest

Gary P. Kutcher

THE POLICY USES OF A MODEL that possesses spatial distinctions can be divided into two categories, the micro and the macro. Micro refers to an experiment that affects (directly) only one of the spatially distinguished parts of the model; macro refers to an experiment that affects all such parts of the model. An example of a micro experiment with CHAC would be the introduction of an investment choice, such as the selection of an irrigation project, in a particular submodel. An example of a macro experiment would be the placing of a foreign exchange premium on agricultural exports, or the subsidizing of fertilizer.

Overview

The objectives of this chapter are to determine to what extent a model of one region can serve as a substitute for a full sectoral model in both of these types of uses. Consider first the micro use. If any possible externalities of a local project or local policy change can be ignored, then recourse to a full sectoral model clearly is not necessary. If, however, the policy intervention occasions changes in national prices, can the regional model adequately account for these in appraising the effects? Although the regional model may be able to assess the affects of, say, price changes on the region's production and resource use, to what extent can it describe the reaction of the rest of the sector?

Now consider a macro experiment. In this case, the regional model might serve as a model for the entire sector. If the region is représentative of the sector in its responses to administered "shocks," then the regional model suffices. But suppose the region is not representative of the sector. It may be more or less responsive to shocks than the sector as a whole. Can assumptions about the behavior of the sector relative to that of the region

Note: This chapter is derived from my Ph.D. dissertation (Kutcher 1972). I am grateful to Clopper Almon, John H. Duloy, and Roger Norton for useful discussions of this study.

be made so that at least directions of change in critical variables can be predicted? This is the kind of question that is addressed in the present chapter (and in chapter 16 as well).

A model of a single agricultural region in Mexico is presented, and investigations are made concerning the appropriate structure of such a model, including alternative assumptions about underlying market structure. Demand functions for the region's output are constructed under various assumptions about the relation of the region's behavior to that of the rest of the sector. Finally, numerical experiments of both a micro and a macro nature are conducted. The region selected for analysis is Mexico's Pacific Northwest coast.

Although this chapter primarily is a methodological exercise in regional analysis and the design of planning models, a semblance of realism has been attempted. One of the goals is to construct a model that can closely simulate the activity of a particular region and that can be used in policy planning decisions without recourse to the full CHAC model.

The Regional Approach to Development Planning: The Pacific Northwest

Of the several regions included in CHAC, the Pacific Northwest of Mexico is the most appropriate for the present purposes, for several reasons. Most other regions in CHAC are not generally contiguous areas but are defined according to farm size, climatic characteristics, or some combination thereof. The agricultural land of the Northwest region, however, is contiguous, and the region is represented in CHAC primarily by four submodels for irrigated districts based on geographical districts that have been defined for administrative purposes by the former Mexican Ministry of Water Resources (Secretaría de Recursos Hidráulicos, SRH). A fifth "district," the regional residual, accounts for the rest of the producing area in the Northwest. Because these district submodels are based more closely on SRH data, they possess more detail and are probably more reliable than the other submodels. The regional model constructed on the basis of these five CHAC submodels is called PACIFICO.

Prior to the 1940s, the Northwest states of Baja California, Sonora, and Sinaloa were virtual deserts, barely able to support a sparse population. Almost all of the agricultural development in this region therefore occurred through irrigation, and its resultant crop yields are among the highest in Mexico. Transformation of the region into Mexico's major commercial agricultural zone has occurred while large areas elsewhere in the country exhibited slow agricultural development and remained largely traditional. Given the presence of several rivers and aquifers in the Pacific

Northwest, there was a much greater scope for the expansion of cultivation through irrigation than in other areas. And because of the sparse population and relatively level land, encouragement was given to the use of larger plots and more mechanized cultivation techniques. This region also has been favored in the allocation of public investment in water resources and other social overhead capital.

One reason for this allocational policy has been a desire to provide incentives for agricultural workers to migrate from Mexico's densely populated Central Plateau and South.

Because the Pacific Northwest region is more advanced agriculturally than the rest of the country, the results of the regional model analysis may be suggestive not only for this region but also for other areas of the country as they reach a more advanced stage of development.

The Model

R. G. D. Allen, in a demonstration model, showed how both demand and supply conditions could be incorporated into a programming model so as to achieve market-clearing equilibrium (1965, pp. 685–91). His model was divided into two types of activities, "technology" and "consumption tastes." The technology part comprises production activities that use resources and contribute positive outputs to commodity balance equations. The tastes section is derived from a utility function and transforms the outputs into the objective function, which maximizes utility of a representative consumer subject to the production restrictions. The structure of CHAC, though arrived at independently, largely follows this pattern. The technology side of CHAC is represented by the production submodels. A set of selling activities, which measures the area under the demand curves, yields "utility." As in Allen's model, production is required to equal consumption plus exports by means of commodity balances.

There are several advantages to this approach besides the ability to represent both sides of the market. With the utility-generating activities separated from the production activities, various forms of utility functions can be used without changing the production side (which is likely to be the largest component). If downward-sloping demand functions are used as in CHAC, the positive contributions to the utility function can be accounted for in the pricing activities, and the familiar "profit" coefficients on production, which otherwise must represent constant unit returns, may be dispensed with. This kind of general structure will be adopted for PACIFICO. Given that the production side will be similar to that of CHAC, it is possible to construct and validate this part of the model independently.

Table 11-1. Schematic Structure of PACIFICO

Row	Selling activities	Factor supply activities	District production activities					RHS
			Yaqui (F)	Culmaya (I)[a]	Río Colorado (A)	El Fuerte (D)	Residual (U)	
Objective function	+	−						
Commodity balances			+	+	+	+	+	= 0
Factor balances and constraints		−	+	+	+	+	+	= 0
Constraints								
Yaqui			+					≤ B_F
Culmaya[a]				+				≤ B_I
Río Colorado					+			≤ B_A
El Fuerte						+		≤ B_D
Residual							+	≤ B_U

Note: + (or −) indicates that the submatrix contains a set of positive (or negative) entries. The letters below the district names are the identifying symbols used in the computer version of the model. RHS indicates right-hand side.

a. "Culmaya" stands for Culiacán, Humaya, and San Lorenzo.

A good deal of flexibility will then be available for the "tastes" part of the model.

Table 11-1 shows a schematic tableau of the general form of the model. The model is static, representing one year's flow of inputs and outputs. Many of the inputs are dated by month. On the production side, cropping activities will be defined for each of the five district submodels: Yaqui, Culmaya,[1] Río Colorado, El Fuerte, and the Residual. These activities use both region-wide inputs and district-specific resources and contribute positive outputs to the commodity balances. A set of selling activities, which have positive objective function coefficients and negative entries in the commodity balances, registers the utility of this production. Also shown in the tableau is a set of factor-supplying activities for the region-wide inputs. Although constraints and prices of these inputs can be accounted for directly in the production activities, I follow the CHAC procedure of separating them for two reasons. First, if only one activity charges the cost of a given input rather than have that cost included in the production activities, one can vary the price of this input by changing only one number. Second, one may make the prices of some inputs endogenous; that is, effectively insert upward-sloping supply functions. This can only be accomplished if the supply for such a factor is separated from the demand.

The downward-sloping demand functions are incorporated in the model according to the methods of chapter 3. The competitive, rather than monopolistic, version of the demand structures is selected initially. In effect, it is assumed that each farmer in the region is a price taker but that the region as a whole, because of aggregation effects, is not. Variations on the competitive assumption are explored later in the chapter. (See also chapters 12 and 16 for discussions of the appropriate market form.)

Market structures and alternative maximands

The separation of technology and tastes permits a great deal of flexibility in the selection of maximands. At one extreme, the set of selling activities shown in table 11-1 could be deleted and a cost-minimization model with fixed production quotas constructed. But other types of maximands more relevant to the market structure are of greater interest. For example, if one considers the Northwest region to be a price-taking region, unit selling activities would be formed which may sell unlimited amounts of the commodity at the given market price.[2] This price-taking assumption is undoubtedly not valid for the Northwest, however, and

1. This name stands for the districts Culiacán, Humaya, and San Lorenzo.
2. This structure is typical of many agricultural models; for example, see Gotsch (1971).

downward-sloping demand functions are used for PACIFICO instead. (As much as 80 percent of Mexico's production of certain crops takes place in the Northwest.) There are about thirty-five short-cycle crops that are produced in significant quantities in Mexico. Of these, the fifteen included in PACIFICO accounted for over 98 percent of the value of the Northwest's production in the base year, 1968.[3] All outputs in the model are valued at the average 1968 farm-gate prices for the Northwest.

Specified inputs include irrigated land, farm labor, hired (day) labor, irrigation water, tractors, draft animals, seeds, chemicals, and short-term credit. Of these, labor and water are the most interesting for different reasons. On the one hand, unemployment and underemployment are severe problems in Mexican agriculture, and a goal of the Northwest's development is presumably to provide increased employment opportunities. On the other hand, the availability of irrigated cultivable land is considered a primary constraint to the expansion of agricultural output. As explained later in this chapter, the programming formulation is used to provide marginal evaluations of these inputs, which can then be of use in cost-benefit project analysis and policy experiments.

The district production submodels

Of the three types of activities that PACIFICO comprises (production, factor supply, and output demand), production activities are the most numerous. In most versions of the model, they comprise between 70 and 80 percent of the total number of activities. Because the production activities yield the implicit supply functions, they are of primary importance in obtaining accurate simulations of output and input patterns.

The technical coefficients in PACIFICO's production activities are derived from those in CHAC.[4] Modifications to those coefficients for PACIFICO were of the nature of refinement, elimination of redundancies, and exclusion of detail that would have been superficial for the present purposes. As in the CHAC study, the immobility of certain factors of production in agriculture is handled by segregating production by geographical district. Each of these districts has its own set of production opportunities and its own endowment of immobile factors, and each can be considered as a distinct submodel.

A district submodel comprises two types of activities: activities supplying or charging (or both) the cost of the district-level resources, and cropping activities. The cropping activities are defined as cultivation of 1

---

3. Because cotton has a joint product of fiber and seed, there are sixteen outputs in the model.

4. For a discussion of the derivation of the CHAC production coefficients, see chapter 4.

hectare of land in a particular crop by one of several techniques. All crops that were produced in nonnegligible amounts in a district in the base period are represented by at least two techniques in the model. The variations in the techniques arise from two sources: up to three different degrees of mechanization are possible, and, in many cases, two different planting dates are included.

Of the three degrees of mechanization, *fully mechanized* is the most prevalent and requires use of tractors for planting, harvesting, and some cultivation activities. *Partially mechanized* techniques substitute draft animals and labor for some of the operations. For example, a partially mechanized technique may utilize tractors for plowing operations and animal power for all others. *Nonmechanized* techniques utilize animal power wherever possible. These last techniques are employed almost exclusively in the traditional, dryland (*temporal*) areas of Mexico, where the land is not sufficiently level for tractors and farmers are generally not able to afford the hiring of tractors. The implicit capital-labor ratios in the various techniques vary according to the crop, but production isoquants can be traced out from the technical coefficient data included in the model. CHAC includes as alternatives all three variations in the degree of mechanization in its representation of the Northwest, but, except in a few cases, unrealistically high capital-labor price ratios would be required to induce the model to use nonmechanized techniques. Such techniques therefore were not included in the model so that its size could be reduced.

Because virtually all the Northwest's production takes place on irrigated land and is thus relatively independent of seasonal rainfall, farmers have some flexibility in choosing planting dates. This flexibility is captured by including, for most crops, additional activities representing a planting date that differs by one month from that in the basic activities. This generally entails only shifting the monthly requirements of land and labor forward by one month, but in some cases it also creates different requirements of irrigation water.

Although most of the Northwest is irrigated by means of gravity-fed canals, some of the districts are located over an aquifer, which permits part of the production to be irrigated by tubewells. This additional water source gives rise to another variation in cropping activities. Separate production activities for land irrigated by wells are therefore included, which differ from the ones for canal-irrigated land only in the water constraints.

The cropping activities have entries in the commodity balances representing the average yields (in metric tons per hectare) in the particular district and use two types of resources: those classified as region–wide and those that are district specific. The region–wide resources are tractors, draft animals (mules), and day labor. It is assumed that these inputs (except day labor) are available in perfectly elastic supply and are fully mobile within the region; casual observation shows that markets for these factors are well developed in the Northwest. Rows for seeds, fertilizers and insecticides (grouped together as chemicals), and agricultural credit are also included and may be considered as region-wide resources as well. The rows for these inputs work against unbounded region-wide input supply activities, and so it is possible to change an input's price by changing only one number in the model. (Alternatively, the costs of these inputs could be entered directly into the objective function coefficients in the cropping activities, and the rows for them eliminated—but then, for a solution investigating the effects of input prices, it would be necessary to change hundreds of numbers.)

The inputs constrained at the district level are land, water, and farmers' labor. These inputs cannot be transferred among the districts in the region in the short run. The use of water is constrained on an annual basis to be less than the average annual increment to the reservoir (less evaporation), and it is charged the administered SRH price for the district in the base year. Land availability is constrained on a monthly basis because of the seasonal nature of many crops. Finally, the cropping activities also have entries in monthly labor balances. The basic unit for labor in the model is man-days, but, in the interest of better scaling, figures in man-days have been divided by 10.[5]

The treatment of labor

As in CHAC, labor is classified as either farm labor or day (hired) labor, but the two types are assumed to have equal productivity in field labor tasks. The definition of farm labor includes farmers who own their land (or are attached to it through some legal means, such as the *ejidal* system) and the number of full-time equivalent workers in the household. The distinction between farm labor and day labor is made because the opportunity cost of farmers is considered to be lower than that of day laborers in the short-run (farmers are less mobile in the short run).[6] The opportunity cost of farmers' labor (sometimes termed the "reservation wage rate") is taken to be 10 pesos per day, about two-fifths of the wage rate for day labor. This is charged to the objective function WELFAR and is included in the calculation of farmers' real income.

The monthly supply curves for day laborers are assumed to be of the form shown in figure 11-1. In the figure, W_m is the nonpeak wage rate

---

5. As many readers are aware, cumulative round-off error (as well as computational time) in the procedure for linear programming solution is minimized if the numbers in the linear programming matrix are more nearly uniform in magnitude.

6. See chapter 15 for an extensive discussion of the pricing of farm labor.

Figure 11-1. *Supply Curve for Hired Labor*

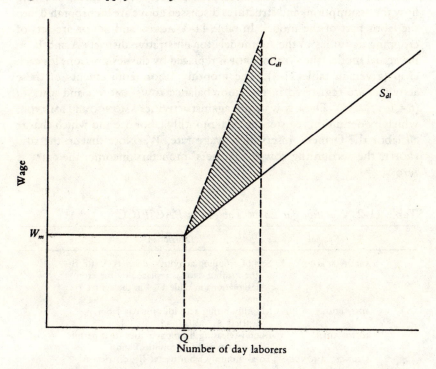

Note: W_m = nonpeak wage rate observed during most of the year; $\bar{Q}$ = number of day laborers in the region; S_{dl} = supply of day laborers; C_{dl} = marginal cost of day labor.

observed most of the year, and $\bar{Q}$ is the number of day laborers in the region. In most months, the demand for day labor will not exceed $\bar{Q}$, and the supply curve is taken to be perfectly elastic up to that point. In peak demand months, however, the supply of laborers in the region may be exhausted, and a wage higher than W_m must be paid to attract workers from other regions and other professions. Thus the upward-sloping portion of the supply curve $S_{d\ell}$ begins at $\bar{Q}$.

Because this upward-sloping portion of the supply curve of day labor reflects the opportunity cost of employing the labor in the region in agriculture, the area under $S_{d\ell}$ is charged to the objective function WELFAR. In effect, relative to wage and employment outcomes, the region is assumed to behave as a perfectly discriminating monopsonist in hiring labor—the equivalent of a competitive situation. But in the definitions of income (both real and monetary), it is assumed that each worker is paid the same daily rate. Thus the area under the marginal cost curve $C_{d\ell}$ is considered in the income accounting rows.

Table 11-2 identifies the notation used in table 11-3, a tableau showing how the assumptions and structures discussed above are incorporated into the labor part of the model. In table 11-3, ACROP and BCROP are sets of cropping activities in the submodels for illustrative districts A and B; in the actual model, the symbol CROP is replaced by the designations for each crop (given in table 11-4). The monthly labor requirements of these activities are registered in the labor balance rows AMONC*t* and BMONC*t* ($t = 1, \ldots, 12$). These rows work against activities AMONCC and BMONCC, which by definition register the total monthly labor use and which charge *all* labor the farmers' reservation wage rate, W_f. Note that in the row PROFIT, the accounting row for farmers' monetary income, the entry is zero.

Table 11-2. *Notation for Labor Tableau in PACIFICO*

Symbol	Description
ACROP, BCROP	Sets of cropping activities, districts A and B (the symbol CROP is replaced by the crop designations in table 11-4 in the actual model)
ADLt, BDLt	Monthly hiring activities for day labor, districts A and B ($t=1, \ldots, 12$)
RDLHi, RDLHj	Bound-releasing hiring activities (that permit hiring beyond $\bar{Q}$), months i and j
AMONCC, BMONCC	Total monthly labor use at W_f, districts A and B
CDL	Charging activity (the difference between W_m and W_f)
WELFAR	Objective function for opportunity cost of farmers' labor
PROFIT	Accounting of farmers' monetary income
CCCi	Convex combination constraint, month i
RCDL	Total district hiring activities
RDLBt	Regional monthly day-labor balances ($t=1, \ldots, 12$)
AMONCA, BMONCA	Monthly labor constraints bounded by number of farm laborers in districts A and B
AMONCt, BMONCt	Monthly labor requirements of district cropping activities ($t=1, \ldots, 12$)
S	Supply of day labor
W_f	Farmers' reservation wage rate
W_d	Peak demand (premium) wage rate
W_m	Nonpeak demand wage rate
C	Marginal cost of day labor
$\bar{Q}$	Number of laborers in the region
L_t	Number of day laborers available in time period t
A_f, B_f	Number of farm laborers, districts A and B

The monthly labor constraints AMONCt and BMONCt are bounded by the number of farm laborers in each district, A_f and B_f. If these are exhausted in any one month, the constraints may be released by operating the hiring activities for day labor, ADLt and BDLt ($t = 1, \ldots, 12$). Row RCDL totals the activity levels of the district hiring activities and works against a charging activity, CDL. CDL charges the difference, $\bar{W}_d$, between the nonpeak demand wage rate, W_m, and the farmers' reservation wage rate, W_f, to WELFAR (recall that all labor has been charged at W_f). In the income accounting row, PROFIT, CDL charges the full wage rate.

The monthly district-level hiring activities are bounded by $\bar{Q}$, the number of laborers in the region, through the regional day labor balances, RDLBt ($t = 1, \ldots, 12$). Now suppose that the demand for day labor exceeds $\bar{Q}$ in months i and j. Activities RDLHi and RDLHj serve to release the bounds, as they permit hiring beyond $\bar{Q}$, up to the incremental amounts L_i and L_j. These sets of hiring activities represent the upward-sloping portion of the supply curve in figure 11-1 as a set of steps, releasing the bounds by increments at progressively higher costs. The objective function entries account for the area below the supply curve and above the basic market wage W_m, and the entries in the PROFIT row account for the area under the marginal cost curve and above W_m. These implicit steps have been transformed according to a method described in Kutcher (1972, chapter 3), so that a convex combination constraint, CCC$_i$, replaces the constraints on the steps or segments.

A summary of the above structure is as follows: farmers in a given district will be "employed" first, and their reservation wage rate is charged to the objective function measuring social welfare. When the stock of farm labor is exhausted, a given district may hire day labor at the nonpeak wage rate. If the sum of all districts' hiring of day labor in any month exceeds the number of agricultural day laborers in the region, further labor may be drawn from other sources at progressively higher wage rates.

Validation of the Production Side of PACIFICO

To be sure of the meaning of the results, it is desirable to test the production side of the model independently of the demand side. For, if this "technology" part (in the terminology of Allen 1965) does not adequately represent the production choice set of the Northwest, experiments with the "tastes" side of the model will have little meaning.

Nugent (1970) has explored the possibility of validation tests of programming models. In his work with a multisectoral, multitemporal

Table 11-3. Labor Tableau in PACIFICO

Row	ACROP	ADLt	BCROP	BDLt	RDLHi	RDLHj	AMONCC	BMONCC	CDL	RHS
WELFAR	+	+	+	+	$-S_1^i \cdots -S_n^i$	$-S_1^j \cdots -S_n^j$	$-W_f$	$-W_f$	$-W_d$	≤ 1
PROFIT	+	+	+	+	$-C_1^i \cdots -C_n^i$	$-C_1^j \cdots -C_n^j$	0	0	$-W_m$	≤ 1
CCCi	+	+	+	+	$1 \cdots 1$	$1 \cdots 1$				$= 0$
RCDL									-1	
RDLBt	+	−	+	−	$-\tilde{Q}_1^i \cdots -\tilde{Q}_n^i$	$-\tilde{Q}_1^j \cdots -\tilde{Q}_n^j$				$\leq L_t$
AMONCA	+	+					-1			$= 0$
AMONCt	+	+								$\leq A_f$
BMONCA			+	+				-1		$= 0$
BMONCt			+	+						$\leq B_f$

Note: + (or −) indicates that the submatrix contains a set of positive (or negative) entries.

model of the Greek economy, he discerns three broad reasons why a model may not perfectly simulate the actual economy: there may be errors of specification in the model's constraint set; the underlying market structure may be incorrectly represented numerically in the model; and a programming model optimizes a particular objective function, whereas the real world may not optimize or may optimize a combination of several micro objective functions.

Since one of the goals of this chapter is to produce a model that can closely simulate the agricultural economic activity of the Northwest region of Mexico, these causes of distortion are relevant even though it is a single-period model of a subsector that is considered. There may well be misspecifications of the production side; perhaps the market structure does include monopolistic elements so that our demand specification is not the correct one; and it may be that no single objective function can describe the motivations of the economic agents of the region. This last possibility must largely be ignored simply because of the inability of an optimization framework to handle a multiplicity of objective functions.[7] But tests for the first two distortions can be derived even if we consider only the production side of the model. I propose two tests. The first is a "weak" test of the production capacity of the model. If the constraints and production possibilities have been correctly specified, then the model can be expected at least to duplicate the production levels observed in the base period. In fact, the model should be expected to show some excess capacity, in recognition that the real world undoubtedly has some deviations from the norm of perfect competition (for example, some form of supply control may be operating).

A second, somewhat stronger test also has been devised that, to a certain extent, can validate the assumptions on the market structure. Here the approach is to use PACIFICO for estimating the marginal costs of production at the observed base-period production levels. These costs are then compared to the base-period market prices. This test will reveal whether the production activities contain errors of specification, or whether farmers in the aggregate in fact do not equate marginal costs and prices, through discrepancies between the estimated marginal cost and the observed price. Simple tests of correlation are employed to measure any discrepancies.

A capacity test

First, the production capacity of the model is tested by attempting to force the model to produce at least as much of every output as the region

---

7. To some extent, the problem of multiple objectives can be handled in the constraints. See the discussion below on tomatoes, the market for which does not strictly conform to the assumptions made in this chapter regarding competitive markets.

produced in the base period. Many objective functions would suffice for the test, but, since the aim is to determine if any excess capacity is present (and how much), extra production should be profitable; the objective is therefore defined to be maximization of producers' profits at fixed prices. A set of fixed-price selling activities is added to the basic production model, each selling one ton of the output in question at the base price. These activities simply consist of a −1 in the commodity balance and the output's price (in model units) in the objective function. The amounts sold via the selling activities are constrained to be greater than or equal to base-year quantities. If any excess capacity does exist, production of the most profitable crop beyond the required minimum will take place. Finally, a set of artificial activities is added to ensure feasibility. These artificial activities "produce" one unit of each output at an arbitrarily high cost.[8]

The solution to this problem showed that the model could reproduce the base-period production levels of all commodities. And it did, in fact, possess a small degree of excess capacity that was used to produce more alfalfa and tomatoes than required. If outputs are measured in their base-period prices, then this test reveals 4.2 percent of excess capacity.

A test of the competitive market assumption

At the outset of this study, it was assumed that the market structure represented by PACIFICO is competitive in the sense that equilibrium prices will be equal to marginal costs of production. Because this assumption is crucial to any further development of the model, a means of testing it is given in the following.

The commodity balances of the capacity test are modified by deleting the selling activities and requiring production of each output to be at least as great as in the base period. The objective function is then redefined to be minimization of the costs of producing these output levels. Thus the shadow prices on the minimum output constraints are the costs of producing an extra unit of each crop (the desired marginal cost concept). These costs can then be compared with the base-period market prices. Table 11-4 shows the results of this test.

With the exception of dry alfalfa, green chile, and tomatoes, all the calculated marginal costs are reasonably close to the base prices. The correlation coefficient for price and marginal cost for all sixteen crops is 0.903. We consider this sufficient to validate the assumption of a competitive market structure. Only if the marginal costs were generally far below the market prices could such a test indicate that the market has monopolis-

8. These artificials are similar to those used by Heady, Randhawa, and Skold (1969, p. 376).

Table 11-4. *Marginal Cost Test for PACIFICO*
(1968 pesos)

Crop (symbol and name)[a]	Base-year (1968) price	PACIFICO marginal cost[b]	Minimum cost[b]
ALA (dry alfalfa)	354	551 (D)	329 (D)
ALV (green alfalfa)	126	107 (I)	73 (U)
ALG (raw cotton)	2,500	2,246 (F, A, D, U)	2,031 (F)
ARO (rice)	1,134	1,324 (I)	887 (U)
AZU (sugarcane)	68	65 (I, D)	25 (D)
CAR (safflower)	1,541	1,680 (F)	866 (U)
CEG (barley	1,014	981 (A)	577 (U)
CHV (green chile)	1,413	592 (U)	355 (U)
FRI (beans)	1,834	1,590 (D)	774 (D)
JIT (tomatoes)	1,150	278 (I)	135 (D)
JON (sesame)	2,407	2,516 (I)	1,734 (D)
MAI (maize)	861	929 (F)	491 (F)
SOR (sorghum)	633	625 (F, I)	353 (F)
SOY (soybeans)	1,600	1,436 (F, D)	731 (F)
TRI (wheat)	800	738 (F, A, I, D)	640 (F)

a. The symbols are derived from the Spanish names of the crops.

b. The letters after the cost figures indicate the producing districts: F, Yaqui; I, Culmaya; A, Río Colorado; D, El Fuerte; U, residual.

tic elements (since demand functions for agricultural outputs are presumably highly price inelastic). The exceptions, however, require further investigation.

Dry alfalfa has a marginal cost significantly higher than the price. Assuming that the technical coefficients used here are correct, the following explanation can be given. The input requirements of a close substitute, green alfalfa (both are used principally as animal feed), are very similar. The major difference is that dry alfalfa is allowed to dry in the field about two months longer before collection. Since the Northwest has a longer and more flexible growing season than other regions, the opportunity cost of the extra time is significant. Furthermore, the transport costs of this crop from other nonirrigated regions are small, suggesting that perhaps the Northwest should not be producing the crop.

An inquiry into the characteristics of tomato production revealed two reasons that its measured marginal cost is far below the market price.[9] First, the production of this crop requires extensive management skills that are possessed by only a few farmers in the region, and these skills are not assigned a price in the model. Second, a large part of the Northwest's tomato production is exported to the United States, which imposes import limitations on this commodity. These limitations are enforced at

9. These explanations were suggested by Luz María Bassoco.

the farm level through the SRH water allocations by crop, and they help maintain the price at levels above marginal costs. In an attempt to account for both the water restrictions and the management constraints, we added bounds on tomato production in each district that permitted an expansion in the number of hectares cultivated in this crop of only 20 percent above the base-year behavior.

Green chile also had a marginal cost significantly below the price, and was produced solely in the residual district. A closer look at the underlying data revealed that only one component district of the residual submodel was producing significant quantities of the crop in the base year. This district (Río Mayo) apparently has a strong comparative advantage in this crop that should not be extended to the rest of the residual submodel. To reduce this aggregation bias, a constraint was added that permitted the implicit Río Mayo district to expand its cultivation of green chile by no more than threefold.

These additional constraints for tomatoes and green chile remedied the more obvious discrepencies and permitted relatively close simulation of the base-period production levels and prices under further solutions of the model.

As an aside, this version of the model was also solved with the right-hand-side (RHS) values of the output constraints set to one unit (1,000 metric tons). That is, the costs of producing 1,000 metric tons of each output were minimized. This solution is instructive on two counts. First, the shadow prices on the output constraints are then the minimum costs of production for the region as a whole. These costs are shown in the right-most column of table 11-4 under the heading "Minimum cost." As such, each represents the intercept of the region-wide supply curve. That these minimum costs are substantially lower than the marginal costs at the base-period output levels is a confirmation that the implicit supply curves are indeed upward-sloping, despite the fact that all inputs except farmers' labor and land are available in infinitely elastic supply. This, of course, is due to the inherent competition among crops and technologies for those scarce resources. Second, the district that produces the single unit can be considered to have an absolute advantage in production of this output in situations where no resource is binding. (The symbol for the district producing the single unit is given in parentheses beside the minimum cost in table 11-4.)

Validation of the full model

The results of the validation tests described above suggest that the full price-endogenous model with the additional constraints should closely simulate the output levels and prices of the region in the base year. For the full model, correlation tests on output levels are made, as in the case of the marginal cost test. We are also interested in the patterns of input use. Although no data are available on capital-labor ratios, for example, it is known that mechanized techniques are used almost exclusively in the Northwest, so that informal validation can be conducted by the incidence of mechanization.

Alternative Demand Structures

Although the Northwest region is capable of producing quantities of several crops sufficient to affect national prices, a price-taker version of the model is instructive on several counts.

A price-taker version

Little and Mirrlees (1969) have argued that in cost-benefit analyses the outputs involved in a project decision should be valued at their international prices. This would imply that the region should be considered as a price taker, with the linear programming selling activities formed so that the outputs face perfectly elastic demand curves. A solution of the model under this assumption is compared with one in which the output prices are endogenous in the next subsection. Also, a price-taker solution will indicate at what output levels marginal costs are equal to the base-period prices. If the Northwest is indeed perfectly competitive and the model simulates the region perfectly, these output levels will be identical to those observed in the base period. But small discrepancies, either in the actual market structure or in the model, are likely to cause wide divergences from the observed output levels because, given the inherent competition for the fixed resources and the wide flexibility in production choices, the supply functions can be expected to be relatively elastic.

The price-taker version of PACIFICO is similar to the capacity-test version, except that output levels are not constrained. The set of fixed-price selling activities is reinserted in the model, and the prices used were those observed in the base period. Since these selling activities represent perfectly elastic demand functions, the maximand is simply producers' surplus (profits). In this measure of producers' surplus, farmers' reservation wages are charged against the objective function, so that we have a measure of farmers' net profit above and beyond their own-wage income. Tables 11-5 and 11-6 report the solution to this version of the model.

Although the pattern of input use is about what could be expected a priori (97 percent of the cultivated land used fully mechanized techniques), this version's simulation of the base-period output levels was very poor.

Table 11-5. *Solution of Price-taker Version
of PACIFICO, Aggregate Variables*

Variable	Solution
Ten millions of pesos	
Producers' surplus (maximand)	233.98
Income	
Producers	266.86
Labor	38.94
Regional[a]	317.59
Thousands of man-years (or equivalent)	
Employment	
Farmers	128.26
Day labor	56.72
Total	184.98
Tractor use	13.76
Mule use	0.18

a. Regional income includes farmers' monetary income, day-labor income, and tractor operators' wages.

Very large differences between the actual and computed output levels occurred, despite the relative closeness of marginal costs and base prices as determined by the marginal cost test. Yet producers' surplus in this solution was only about 13 percent greater than in the capacity-test version, which implies that the supply curves have relatively small slopes and that the differences in profitability among crops is relatively small.

Other model builders who have employed the price-taker structure (for example, Gotsch (1971) have experienced similar problems with non-realistic, "optimal" output levels. A common means of forcing a programming model to yield more realistic output levels is to add a set of upper and lower bounds (flexibility constraints) on the production of certain outputs. This approach, however, can lead to difficulties in the interpretation of the dual solution, and it precludes analysis of the determinants of the true optimal production levels.

A price-endogenous version

The price-taker assumptions above are inadequate for a version that allows for interaction between regional production levels and national price levels. Downward-sloping demand functions confronting the region are introduced in this subsection, and it is shown how these may be incorporated into the model. Throughout, intraregional transport costs are ignored. There are three sources of demand for the Northwest's output: consumption in the region, exports to the rest of Mexico, and exports to the rest of the world. Data are not available on the destination of

Table 11-6. *Solution of Price-taker Version
of PACIFICO, Crop Variables*

Variable[a]	Base price (pesos per metric ton)	Output (thousands of metric tons)[b]		Variable[a]	Base price (pesos per metric ton)	Output (thousands of metric tons)[b]	
		Actual	Solution			Actual	Solution
ALA	354	153.08	0	CHV	1,413	21.45	24.95
ALV	126	60.57	3,895.07	FRI	1,834	20.62	141.30
ALG	5,767	290.50	393.03	JIT	1,150	268.94	339.08
SAL	831	539.50	685.98	JON	2,407	21.98	0
ARO	1,134	109.20	0	MAI	861	125.11	0
AZU	68	3,689.29	1,988.26	SOR	633	288.08	132.97
CAR	1,544	42.17	0	SOY	1,600	232.73	203.26
CEG	1,014	12.29	0	TRI	800	966.16	785.00

a. Crop symbols are identified in table 11-4 (except for SAL, cottonseed).
b. The correlation coefficient for output levels was 0.380.

the Northwest's output, so these sources of demand cannot be distinguished. The consideration of a total demand function confronting the region—instead of considerations of the three sources separately—should not introduce problems, since the following solution deals only with f.o.b. prices for the Northwest.[10] These prices are generally lower than those prevailing in the national market centers; much of the observed difference in price is probably attributable to transport costs.

To introduce price-responsive demand, the selling activities for crops in previous versions are replaced by a set of segmented demand functions for each crop. To do this in an efficient manner, the CHAC procedure is followed (see chapter 3). When this is done, the maximand changes in nature; under fixed prices it was maximization of profits, and now it becomes maximization of consumer and producer surplus. As discussed in Kutcher (1972, chapter 3), this maximand may be viewed as a device for ensuring that the model replicates the price-taking behavior of individual farmers under the assumption that producers as a group face downward-sloping demand functions. One of the chief concerns of this chapter is what degree of slope in the demand functions is appropriate for regional, as opposed to sector-wide (national), models.

To initiate this line of inquiry, a price-endogenous version of PACIFICO is presented under the assumption that the price elasticities of demand facing regional producers are equal in value to the national elasticities. The implications of this assumption, and other alternative assumptions, are

10. If separate demand functions were known, they could be separately specified in the model; the procedure adopted here therefore involves no loss of generality.

Table 11-7. *Solution of Basic Price-endogenous Version*
of PACIFICO, Aggregate Variables

Variable	Solution
Ten millions of pesos	
WELFAR (maximand)	802.08
Surplus	
Consumers'	633.34
Producers'	168.74
Income	
Producers'	205.26
Labor	24.51
Regional[a]	240.21
Thousands of man-years (or equivalent)	
Employment	
Farmers	143.12
Day labor	35.35
Total	178.47
Tractor use	13.40
Mule use	1.71

a. See table 11-5 for composition.

discussed below. In specifications of the demand functions at the regional level, the assumptions are passed through the point in price-quantity space that corresponds to observed base-year values.

The solution to this version of the model is shown in tables 11-7 and 11-8. The close correspondence of the solution was the base-year actual values is indicated by the high correlation coefficients. If only the output levels had shown a close correlation, this could have been attributed to the low price elasticities assumed for the outputs, and hence the relatively constrained output choice set in the model (imagine a very steep demand curve passing through the point representing the base-period output and price). But since the dual prices also exhibit almost perfect correlation (0.99), we can be satisfied that this simulation validates the model fairly well.

Two types of prices, the dual and the primal, are listed in the tables. As pointed out in chapter 3, the two types of prices are equal, except for approximation errors arising from the segmentation of the demand functions. The primal prices were tabulated ex post by noting the price in the last segment of the demand function employed. The dual prices are the shadow prices on the commodity balances. In cases where the discrepancy is judged to be too great, the solutions could be repeated under a fine segmentation of the demand functions. Henceforth in this chapter, the dual prices will be reported as the prices.

Table 11-8. *Solution of Price-endogenous Version of PACIFICO, Crop Variables*

	Output (thousands of metric tons)[b]		Price (pesos per metric ton)		
Variable[a]	Actual	Solution	Actual	Primal	Dual
ALA	153.08	130.12	354	531	552
ALV	60.57	63.60	126	109	108
ALG	290.50	314.71	5,767	4,966	4,799
SAL	539.50	544.85	831	808	808
ARO	109.20	100.10	1,134	1,355	1,331
AZU	3,689.29	3,638.05	68	70	66
CAR	42.17	39.83	1,544	1,673	1,670
CEG	12.29	12.50	1,014	986	984
CHV	21.45	22.32	1,413	1,138	1,138
FRI	20.62	21.77	1,834	1,681	1,593
JIT	268.94	339.08	1,150	767	767
JON	21.98	20.76	2,407	2,608	2,509
MAI	125.11	123.72	861	933	929
SOR	288.08	288.08	633	651	629
SOY	232.73	258.59	1,600	1,467	1,448
TRI	966.16	976.90	800	778	741

a. Crop symbols are identified in table 11-4.
b. The correlation coefficient for both output levels and dual prices was 0.99.

That a relatively high proportion of the objective function represents consumer surplus can be explained by the low price elasticities. These magnitudes are important, however, only when policy experiments are conducted and changes in the welfare of the various parties concerned are measured. Yet the relatively low income share of landless day laborers is of significance. Given that the reservation wage rate for family farm labor has been set at 10 pesos a day, whereas the minimum wage rate for day labor is 26 pesos, the model obviously is biased toward maximum use of farm labor. Family farm labor was exhausted in every district in at least one month. This reservation wage would appear realistic, though, in light of actual labor market conditions. There is virtually no alternative employment for the Northwest farmer except migration to a distant urban center. Furthermore, more than half the farm labor force is family labor. The short-run opportunity cost of labor for the farmers' families is undoubtedly extremely low. Also, the relatively high level of farm profits, over and above the reservation wage, would provide a disincentive to Northwest farmers' migration.

The model's monthly hiring pattern for day labor shows extreme variation, as is shown in table 11-9. Only in June was the regional labor

Table 11-9. *Seasonal Employment of Day Labor in PACIFICO, Basic Price-endogenous Version*

Month	Labor hired (thousands)	Month	Labor hired (thousands)
January	20.09	July	84.13
February	20.59	August	47.25
March	7.84	September	5.17
April	18.45	October	16.61
May	48.82	November	21.70
June	106.02	December	27.52

force fully employed. In that month the upward portion of the labor supply curve was used so that the wage rate in that month rose about 1 peso daily, to about 27 pesos. Only in four months was the employment rate for the region's 100,000 landless laborers near or above 50 percent.[11] Also, the pattern of input use appears close to the assumed true pattern. Although the ratio of tractor-mule use is down from the solution of the price-taker version, 97 percent of the land was cultivated by fully mechanized techniques in both solutions.

Alternative demand structures and regional-sectoral linkage

The demand functions employed in the initial price-endogenous version of PACIFICO were scaled-down national demand functions. Figure 11-2 shows alternative demand functions for a typical commodity. The national demand function is D, and D_1 is taken as the demand function confronting the region. D_1 was passed through the observed regional point (P^o, Q_1^o), assuming that the elasticity of the function at $P = P^o$ is the same as that of the national function. Since both D and D_1 are linear and have the same intercept, it is implicitly assumed that the Northwest's share of national output will be constant for all P. This formulation thus considers the region to be a representative sample of the entire sector: any shift in the Northwest's implicit supply function will apply to that of the rest of the sector proportionately, and any change in quantity supplied because of a price change will be duplicated proportionately by the rest of the sector. In other words, using this demand function is equivalent to assuming that the elasticities of supply are identical in the region and the rest of the sector.

In figure 11-2, D_1' would be the demand curve for the region under consideration if the supply response elsewhere in the sector were zero,

11. Informal discussions with Mexican economists indicate that this pattern of labor hiring is realistic for the Northwest.

Figure 11-2. *Alternative Demand Functions for the Region*

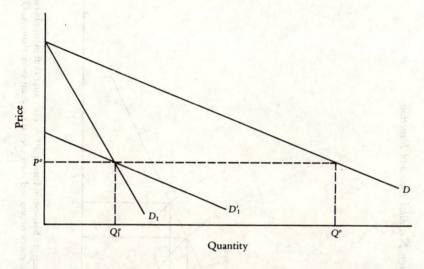

Note: P = price; Q = quantity; D = national demand function; D_1 = demand function confronting the region, assuming that the elasticity is the same as the national demand elasticity at the point $P = P^o$; D'_1 = regional demand curve if supply response elsewhere in sector is zero; P^o, Q^o_1 = observed regional point.

even when production and prices change in this region. That is, the supply functions of the other regions are assumed to be perfectly inelastic and not affected by any shift in this region's supply function. Note that D'_1 has the same *slope* as the national function.

For validation tests of the model, the form of the demand function is of no consequence. Any downward-sloping demand curve passing through the point (P^o, Q^o_1) will allow calculation of a divergence between price and marginal cost and a consequent failure of the model to simulate the base-period production levels. But for most intended uses of the model, the "correct" demand function is likely to lie between D_1 and D'_1. The elasticity of supply in the Northwest is probably significantly higher than that in the rest of the sector since the average farm size is much larger and modern techniques are more readily available.[12]

Thus, it is useful to derive alternative demand functions for the model under various assumptions about the degree to which a shift in the region's supply function applies to that of the rest of the sector and about the

---

12. *Editors' note:* Although this statement is likely to be true for any specific crop produced in the Northwest, it is unlikely to be true for aggregate production in the Northwest. See chapter 5 for a discussion of this issue in reference to irrigated versus nonirrigated areas.

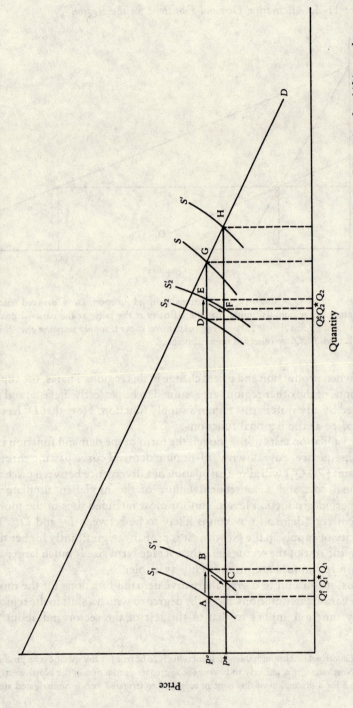

Figure 11-3. *Supply and Price Response in Two Regions Facing a Common National Demand Function*

Note: P = price; Q = quantity; S = supply; D = national demand function; P^o = observed base-period price; $P^\star$ = price after shift in supply curve; Q_i^o = base-period output level for region i (1 or 2) at P^o; Q_i = region i's output level if shock occurs and there are no resultant price changes; $Q_i^\star$ = equilibrium output level in region i after shock and price responses.

340

elasticity of supply of the rest of the sector relative to that of the region. Note that the primary interest is the *total* demand function confronting producers in the region, not the region's demand for its own output. Thus, transport costs of any output not consumed locally are ignored.

In the derivation of the general demand function, the following notation is employed: subscript 1 refers to the Northwest region, and subscript 2 refers to the rest of the sector (region 2). In addition,

Q_i^o = The observed base-period output level, region i, at base period price P^o

Q_i = The output level in region i if some shock (shift in supply curve) occurred in the absence of any resultant price changes

Q_i^* = Final equilibrium output level in region i after both the shock and the price responses have been taken into account.

The elasticity of supply for region i, at $P = P^o$ can be written

(11.1)
$$\gamma_i = \frac{\partial Q_i}{\partial P} \cdot \frac{P^o}{Q_i^o}.$$

The slope of the supply curve for region i at $P = P^o$ is

(11.2)
$$s_i = \frac{\gamma_i \, Q_i^o}{P^o},$$

and national output is

(11.3)
$$Q = Q_1 + Q_2.$$

The national demand function is assumed to be given exogenously and is linear:

(11.4)
$$P = a + bQ.$$

Figure 11-3 presents the problem graphically for the case of one product. Point A represents region 1's output at P^o before the shock. The output level at point B reflects the shift in the supply curve from S_1 to S_1'. This shift also applies to region 2 (to possibly a nonproportionate extent) such that S_2 shifts to S_2'. These shifts cause the sector supply curve to become S', which leads to a fall in price to $P^\star$. The equilibrium production levels for the two regions are then Q_1^* and Q_2^* at points C and F, respectively. Thus we wish to derive the function

(11.5)
$$P = f(Q_1),$$

which passes through points A and C and, for convenience, is assumed to be linear. This function will depend, of course, not only on the supply functions S_1 and S_1', but also on S_2 and S_2'.

Because we are dealing with PACIFICO alone and have no means of determining regions 2's behavior through a simulation, the following assumptions are made:

ASSUMPTION 1. For a shift in S_1 of ΔQ_1 at price P^o,

$$(A.1) \qquad \frac{\Delta Q_2}{Q_2^o} = \theta \frac{\Delta Q_1}{Q_1^o},$$

where $Q_1 = AB$ and $Q_2 = DE$ in figure 11-3 and θ is a multiplicative factor. Note that, if the shift does not apply to region 2, $\theta = 0$; if the shift in S_2 is proportionate, $\theta = 1$.

ASSUMPTION 2. At the points under consideration, the slopes (s_i) of the supply curves are related by $s_2 = \alpha s_1$, or

$$(A.2) \qquad \frac{\Delta Q_2 - \Delta Q_2^{\star}}{\Delta P} = \alpha \frac{\Delta Q_1 - \Delta Q_1^{\star}}{\Delta P},$$

where $\Delta Q_i = Q_i - Q_i^o$ and $\Delta Q_i^{\star} = Q_i^{\star} - Q_o$.

It is also assumed that s_1 is the same at points A, B, and C, and S_2 is the same at points D, E, and F. (This assumption can be relaxed.) Note that the s_1 and α can be derived from programming estimates of supply elasticities if desired.

These assumptions made, analysis can proceed. First, the slope (b_1) is derived for the desired function $P = f(Q_1)$. From the national demand function,

$$(11.6) \qquad \Delta P = b(\Delta Q_1^{\star} + \Delta Q_2^{\star});$$

and, substituting from the relation of slopes,

$$(11.7) \qquad \Delta P = b(\Delta Q_1^{\star} + \Delta Q_2 - \alpha s_1 \Delta P).$$

Substituting from the shift relation,

$$\Delta P = b[\Delta Q_1^{\star} - \alpha s_1 \Delta P + \theta Q_2^o (\frac{\Delta Q_1}{Q_1^o})];$$

$$\Delta P = b[\Delta Q_1^{\star} - \alpha s_1 \Delta P + \theta Q_2^o (\frac{\Delta Q_1^{\star} + s_1 \Delta P}{Q_1^o})];$$

$$\Delta P = b \Delta Q_1^{\star} - b \alpha s_1 \Delta P + \frac{b \theta Q_2^o \Delta Q_1^{\star}}{Q_1^o} + \frac{b \theta Q_2^o s_1 \Delta P}{Q_1^o};$$

$$\Delta P + b \alpha s_1 \Delta P - \frac{b \theta Q_2^o s_1 \Delta P}{Q_1^o} = \Delta Q_1^{\star} (b + \frac{b \theta Q_2^o}{Q_1^o});$$

$$\Delta P(1 + b \alpha s_1 - \frac{b \theta Q_2^o s_1}{Q_1^o}) = \Delta Q_1^{\star} (b + \frac{b \theta Q_2}{Q_1^o});$$

and therefore

$$(11.8) \qquad b_1 = \frac{\Delta P}{\Delta Q_1^{\star}} = \frac{b(1 + \dfrac{\theta Q_2^o}{Q_1^o})}{1 + bs_1 (\alpha - \dfrac{\theta Q_2^o}{Q_1^o})}.$$

The function linearized about (P^o, Q^o), the intercept is

$$(11.9) \qquad a_1 = p^o - b_1 Q_1^o.$$

Table 11-10 shows how the intercept and slope of a typical regional demand function vary as the shift parameter (θ) and the elasticity relation parameter (α) vary. The figures in this table are those for cotton fiber (ALG), which is one of the more important crops in PACIFICO. The numerical behavior of the parameters of the demand functions for this example can be considered as roughly representative of the other outputs in PACIFICO. The elasticity of supply of $+1.18$ used in the example was calculated from parametric solutions of the model (Kutcher 1972).

The most significant results from the experiments shown in the table concern the insensitivity of the function to the elasticity relation parameter, α. Although the parameters in the table were calculated to an accuracy of 10^{-7}, none of them change as α is varied from unity to 3. This result is most encouraging, since the determination of the elasticity relation parameter is highly arbitrary at best.

Table 11-10. *Sample Parametric Values of a Regional Demand Function*

θ	α	a_1	b_1
0	1.0	1.2225977	−0.0022234
0	1.5	1.2225977	−0.0022234
0	2.0	1.2225977	−0.0022234
0	2.5	1.2225977	−0.0022234
0	3.0	1.2225977	−0.0022234
0.5	1.0	1.4763204	−0.0030968
0.5	1.5	1.4763204	−0.0030968
0.5	2.0	1.4763204	−0.0030968
0.5	2.5	1.4763204	−0.0030968
0.5	3.0	1.4763204	−0.0030968
1.0	1.0	1.7300721	−0.0039703
1.0	1.5	1.7300721	−0.0039703
1.0	2.0	1.7300721	−0.0039703
1.0	2.5	1.7300721	−0.0039703
1.0	3.0	1.7300721	−0.0039703

Note: θ, shift parameter for regional demand function; α, elasticity relation parameter; a, demand intercept parameter; b demand slope parameter. This table is based on the following data: $\eta = -0.5$, $P^o = 0.5767$, $Q^o = 290.5$, $Q_2^o/Q_1^o = 0.7857$, $\gamma_1 = 1.18$, $s_1 = 0.00168$, and $b = -0.0022234$, where η is the national demand elasticity at (P^o, Q^o) and γ_1 is region 1's supply elasticity at (P^o, Q^o).

Table 11-11. *Effects of a Cotton Subsidy in PACIFICO*

Variable	Change (pesos)	Variable	Change (pesos)
		Foreign exchange earnings	56,000
Producers' income	141,000	Consumers' surplus	−50,000
Day-labor income	15,000	Cost of subsidy	91,000

The numbers in table 11-11 were subsequently recalculated with a much lower supply elasticity, but the new values did not differ significantly from the ones reported in the table. Thus, the determination of θ (which also is quite arbitrary) is the only remaining problem in specifying "general" demand functions for the region.

Policy Experiments

Now that PACIFICO has been subjected to validation tests, this section demonstrates a few of the uses to which such models can be put. The experiments include subsidies on irrigation water and cotton inputs and reductions in the short-term interest rate. Following these, experiments are made to examine the derived demand for labor in agriculture in the Northwest.

Subsidies for irrigation

Because virtually all agricultural production in the Northwest region requires irrigation water as an input, this resource provides an important policy tool. In all the PACIFICO solutions, the constraints on gravity-fed irrigation water have been binding, but the pump water constraints generally have not been. The structure of the model permits variations in the administered price of water charged farmers. An upward change can represent water taxes and a downward one subsidies.

For this numerical experiment, the water price was decreased by 20 percent because such a policy may be an efficient way of both raising farm income and stimulating output. For the experiment, the general price-endogenous version of PACIFICO was used (θ = 0), assuming that the water subsidy would apply only to the Northwest region. The solution showed only a small increase in the use of pump water in Río Colorado (13 percent), although the constraints on gravity-fed water, of course, remained binding. Output, consumers' surplus, farm income, and day labor employment and earnings also increased, but not substantially. In monetary terms, the subsidy cost 65,000 pesos, farm income increased by

33,000 pesos, and day-labor wage income increased by 6,000 pesos. The slightly higher total output led to a monetary increase in consumers' surplus of about 33,000 pesos. The overall benefit-cost ratio is about 1.0 when total net benefits are measured by the sum of Marshallian surpluses, but one suspects that further investment in water resources would yield much higher returns.[13]

Subsidies for cotton inputs

Cotton fiber is one of the few exportable Mexican crops that is not subject to import quotas in the consuming countries. Because of this, and the presence of relatively good climate and soil conditions for cotton in the northern regions, policymakers view cotton as an important commodity for earning foreign exchange. Thus, there is a good deal of interest in ways to stimulate cotton production and exports. (There is an important secondary benefit connected with increased production of cotton fiber: the joint product of fiber, cottonseed, is an oilseed that can be an important addition to the often nutrient-deficient diets of rural residents in developing countries.)

One means of stimulating cotton production is to subsidize its inputs. For such an experiment, farmers were charged in PACIFICO only 80 percent of the base-period prices of cottonseed, fertilizer, and insecticides. We assumed that this policy would apply to the entire country, so θ, the shift parameter, was set at 0.5, since the Northwest produced about half the country's output of cotton in the base year. An export activity for cotton fiber also was added to the model for this experiment. This export activity permits an unlimited quantity to be sold on the international market at a price slightly below the equilibrium price in the basic case. Thus, there were no "exports" in previous PACIFICO solutions (except the substitution in demand case).

In the numerical experiment, the subsidy caused cotton production to increase by only about 4 percent. This increase, however, led to increases in farm income of 7 percent and day labor wages of 6 percent. The fiber export activity did come into play, and some foreign exchange earnings were generated. Because some resources were diverted from other commodities to produce the exported cotton, consumers' surplus fell by about 1 percent. Some of the costs and benefits of the subsidy in monetary terms are shown below in table 11-11.

13. *Editors' note:* The difference from 1.0 is only because of approximation errors introduced by the segmentation of the demand functions. It can be shown that this measure of the benefit-cost ratio will *always* be equal to 1.0 for input subsidies when producers are assumed to be profit maximizers.

Given the problems of handling international exports in a model of a single region, we can draw conclusions about only the direction of the changes indicated by the experiment. The results, however, do indicate that the subsidization of cotton inputs can be an effective means of raising farm incomes and stimulating additional employment. Foreign exchange earnings would also increase by this policy, but consumers would be somewhat worse off.[14]

Subsidies for short-term interest rates

All of the cropping activities in PACIFICO require the use of short-term credit in varying amounts. It is assumed that the payback period of this credit is one year, although many cropping activities require less than twelve months. A short-term annual interest rate of 12 percent is charged to the objective function and income rows. Another policy instrument that could be used to increase farm income and output is the subsidization of this short-term interest rate. For experiments with PACIFICO involving this policy tool, a value of 0.3 is assumed for the relative shift parameter, θ, because it is estimated that agricultural production in the rest of the country uses about this proportion of purchased inputs relative to the Northwest. (In other regions, family farm labor and draft animals are still used almost exclusively; use of the more sophisticated inputs such as fertilizers and insecticides is nonexistent in some areas.)

The experiments involved reducing the interest rate to 10 percent and then 8 percent. At the 10 percent rate, consumers' surplus and day labor employment increased only marginally, but farm income rose by 39,000 pesos. The cost of the subsidy was 25,000 pesos. On the basis of this experiment, the credit subsidy appears to be a highly effective way of raising farm income. But when the interest rate was further reduced to 8 percent, the results were somewhat different. Again, consumers' surplus and day labor employment were affected only slightly, but the cost of this subsidy was slightly more than the increase in farm income (51,000 versus 49,000 pesos). The implication is that, if such a policy were to be undertaken, the subsidy rate should be kept small for maximum efficiency of the policy.

Wage variations and labor demand

In an attempt to raise subsistence income levels, Mexico, like many other countries, has initiated minimum wage laws. Although these laws

---

14. *Editors' note*: If total net benefits are measured by the sum of Marshallian surpluses, then the benefit-cost ratio turns out to be 1.0 again; see footnote 13.

are not tightly enforced in most of the country, they are, for the most part, operative in the Northwest, where the larger average farm size and relatively higher level of organizational sophistication allows better enforcement. For this reason, the minimum wage rate (26 pesos per day) has been used in all of the versions of PACIFICO.

One of the objectives of the development of the Northwest region, however, has been to draw excess labor from the densely populated Central and Southern regions of the country. For this objective, the programs have been somewhat less than successful. The PACIFICO solutions show that, in general, full employment of regional landless laborers occurs only in one month of the calendar year. The question arises concerning the influence of the minimum wage policy on the demand for labor. If the model could show that the demand for labor would significantly increase if the wage rate were reduced, then removing the minimum could be a useful policy instrument for promoting employment. The tradeoff is the classic one in a surplus labor situation: maintaining the income levels of those who are already employed versus increasing total employment. The model can be used to quantify the tradeoff. A priori, one would expect that there is much room for labor substitution in the Northwest: the apparently strong preference for mechanized techniques does not appear consistent with the vast numbers of unemployed and underemployed in Mexico.

Using the basic price-endogenous version of PACIFICO ($\theta = 1.0$), the derived demand for hired labor was estimated by varying the wage rate. It was assumed that day labor is available in infinitely elastic supply. Table 11-12 shows the results of these experiments for wage rates varying from the current minimum of 26 pesos per day down to 18 pesos per day. In all solutions, the reservation wage rate for farmers was held constant at 10 pesos per day.

The results of these experiments require some elaboration. Consider first the variation in total annual-equivalent day-labor employment. This variable increased monotonically as the wage rate was reduced (figure 11-4). In the neighborhood of the current wage, the labor demand function is highly inelastic. But as the wage is reduced, the function becomes less inelastic until, in the last segments, the arc elasticity is above 1.0. A large increase in the capital-labor price ratio apparently is required before significant labor substitution can take place, implying that the limits of capital-for-labor substitution are approached with the current wage, and perhaps even higher wage rates would not significantly decrease employment of day laborers. Note also the variations in land use by degree of mechanization: a reduction in the wage rate to 20 pesos per day is required before any significant shift away from fully mechanized techniques is induced.

Table 11-12. *PACIFICO Results under Variations in Day-labor Wage*

Concept	Day-labor wage				
	18 pesos	*20 pesos*	*22 pesos*	*24 pesos*	*26 pesos*
WELFAR[a]	810.56	808.20	805.99	804.03	802.08
Consumers' surplus[a]	647.93	645.49	638.67	635.95	633.31
Producers' surplus[a]	162.63	162.71	167.32	168.08	168.77
Producers' income[a]	198.43	199.27	203.85	204.57	205.55
Wage payments (day labor)[a]	22.16	22.24	21.73	23.31	24.26
Day-labor income (pesos) per man-year)	2,212.50	2,237.50	2,175.00	2,325.00	2,425.00
Total income per man-year[a]	230.86	231.89	236.02	238.28	240.25
Day labor employment[b]	46.64	42.32	37.42	36.79	35.34
Farmer employment[b]	135.59	138.50	138.37	138.23	139.34
Machinery use[b]	13.19	13.19	13.40	13.36	13.39
Mule use[b]	4.12	4.24	1.68	1.74	1.71
Distribution of land by technique (percent)					
Mechanized	87	90	93	93	94
Partially mechanized	9	6	6	6	5
Nonmechanized	4	4	1	1	1
Monthly labor hire (thousands of man-months)					
January	18.70	19.16	19.31	19.34	20.09
February	31.31	23.95	23.92	23.29	20.60
March	14.44	6.49	7.28	7.22	7.85
April	18.58	18.06	19.06	18.72	18.46
May	68.13	64.78	51.83	51.46	48.80
June	142.41	131.34	111.75	109.62	106.00
July	114.72	105.95	89.02	87.52	84.12
August	53.44	49.16	50.65	49.82	47.23
September	14.14	10.46	7.49	6.93	5.16
October	28.54	24.25	19.25	18.22	16.60
November	29.40	27.60	22.90	22.58	21.69
December	25.78	26.60	26.64	26.70	27.54
Output price index	0.976	0.980	0.991	0.994	1.00

a. Ten millions of pesos.
b. Thousands of man-years.

The monthly pattern of day-labor hiring also shown in table 11-12 for the various wage rates. Even with the lowest wage, the demand for labor exceeds the region's supply in only two months. It is highly unlikely that a decision to migrate to the region would be made if a farmer could expect only two months' employment. In quantitative terms, the income-

Figure 11-4. *Day-labor Demand Curve Calculated by PACIFICO*

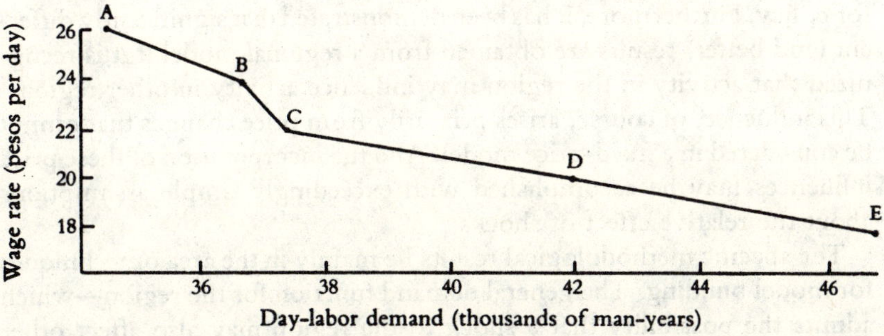

Note: Arc elasticities are: AC, 0.34; CE, 1.09; AE, 0.76.

employment tradeoff comes out as follows: a 9 percent reduction in annual incomes of the current set of day laborers leads to a 13 percent increase in day-labor employment. In the peak month, the employment increase is 34 percent. These experiments show some employment responsiveness to downward wage flexibility. In practice, such a policy would be implemented by raising the nominal minimum wage at a rate slower than the rate of price inflation.

The maximand, WELFAR, was not particularly sensitive to the wage variations. But the redistribution between producers' and consumers' surpluses was significant. As expected, consumers gain slightly when the cost of the labor input falls. But producers appear to fare better, both in profit income (producers' surplus) and total monetary income, at the higher wage rates. This surprising result was simply because of the output effect: on the basis of a Fisherian production index, output fell by a bit over 1 percent between daily wage rates of 18 and 26 pesos. Since most of the demand functions are inelastic, this decline in output resulted in a rise in prices sufficiently large that producers were better off even though they were paying higher wages.

Conclusions

The possible conclusions to this chapter may be classified as (1) generally methodological, (2) specifically methodological, and (3) specifically relevant to policy.

The general methodological conclusions are straightforward. It has been shown that it is possible (and relatively simple) to construct a programming model of a single region that can provide useful suggestions for policy. Furthermore, it has been demonstrated that significantly different (and better) results are obtained from a regional model if it is recognized that activity in the region may influence activity in other regions. This influence, of course, arises primarily from price changes that cannot be considered in a fixed-price model. And the incorporation of these price influences may be accomplished with exceedingly simple assumptions about the relative effect of shocks.

The specific methodological results lie mainly in the area of techniques for model building. The general demand function for the region—which admits the possibility that a shock to the region may also affect other regions and allows for price responses both in the region and elsewhere—could have wider applicability. Some other specific advantages of the PACIFICO (and CHAC) formulation are the following. First, the downward-sloping demand functions have eliminated the need for the (always arbitrary) flexibility constraints that impose upper and lower bounds on production levels by crop. Second, the disaggregation of production technologies for each crop permits an analysis of the factor substitution that arises not only from shifts in product mix but also from changes in factor intensity in the production of each crop. Third, the wage differentiation by farmers and nonfarmers allows some recognition of differing degrees of labor mobility in the short run.

A model such as PACIFICO is primarily useful for static "impact" experiments concerning the effects of hypothetical policies, such as those illustrated in the latter part of this chapter. To address it to questions of specific investment projects raises problems that are the subjects of chapters 15 and 16 in this volume.

References

Allen, R. G. D. 1965. *Mathematical Economics*. London: Macmillan.

Gotsch, Carl H. 1971. "A Programming Approach to Some Agricultural Policy Problems in West Pakistan." In *Studies in Development Planning*. Edited by Hollis B. Chenery. Cambridge, Mass.: Harvard University Press.

Heady, Earl O., Narindar S. Randhawa, and Melvin D. Skold. 1969. "Programming Models for the Planning of the Agricultural Sector." In *The Theory and Design of Economic Development*. Edited by Irma Adelman and Erik Thorbecke. Baltimore, Md.: Johns Hopkins University Press.

Kutcher, Gary P. 1972. "Agricultural Planning at the Regional Level: A Programming Model of Mexico's Pacific Northwest." Ph.D. dissertation. University of Maryland.

Little, Ian M. D., and James A. Mirrlees. 1969. *Manual of Industrial Project Selection in Developing Countries*. Vol. 2. *Social Cost-Benefit Analysis*. Paris: Development Centre of the Organisation for Economic Co-operation and Development.

Nugent, Jeffrey B. 1970. "Linear Programming Models for National Planning: Demonstration of a Testing Procedure." *Econometrica*. Vol. 38, no. 6 (November).

12

A Risk Programming Model for Mexican Vegetable Exports

Carlos Pomareda and Richard L. Simmons

As in the preceding chapter, the region examined in this chapter is Mexico's Pacific Northwest; specifically, the state of Sinaloa. Whereas chapter 11 addressed generic issues of regional models, the focus here is on one group of farmers—those who grow winter vegetable crops for export.

Overview

The model developed in this chapter is of aggregate producer behavior in Mexico's winter vegetable-exporting regions; it is used to evaluate the effects of changes in economic factors on the optimal timing and quantity of tomato, bell pepper, and cucumber exports. The model takes monthly net import demand functions in the United States and Canada as given and uses linear programming to generate static industry equilibriums under a range of alternative specifications concerning risk, competitive supply structure, and wage rates. It also takes into account sales to the Mexican domestic market by the same groups of producers. Since vegetable prices are endogenous to the model, the solutions simultaneously generate equilibrium prices for the U.S. and Mexican markets and the Mexican production response. The supply responses refer to the three export vegetables and ten competing crops.

Note: We wish to thank José S. Silos for generous use of the staff resources of the Coordinating Commission for the Agricultural Sector (Comisión Coordinadora del Sector Agropecuario, COCOSA) in supporting this study. We are also grateful to Luz María Bassoco, Roberto Castro, Gustavo Grey, and James Seagraves for help in implementing the model, and to Peter Hazell and Roger Norton for assistance on the conceptual framework. The cooperation from the Unión Nacional de Productores de Hortalizas (UNPH) and Confederación de Asociaciones Agrícolas del Estado de Sinaloa (CAADES) in data collection is also appreciated. This chapter is based on work financed under U.S. Agency for International Development contract number AID-CSD-3632.

Alternative Specifications

Specifications for risk, market structure, and input prices are described in this section.

Risk

Export vegetables are notoriously risky and risk should therefore be included in any behavioral model purporting to analyze producers' decisions. Following the procedures of chapters 7 and 8, in this study risk is treated as a cost to be subtracted from revenues in the objective function. The cost attributable to risk may be described by the term:

$$- \phi (X' \Omega X)^{1/2},$$

where ϕ is a risk-aversion coefficient, X is a vector of activity levels, and Ω is a variance-covariance matrix of gross activity returns.

Conceptually, ϕ is an aggregation of the risk-aversion coefficients of individual micro units. No attempt is made to estimate ϕ values for either the micro units or the corresponding aggregate coefficient. Instead, alternative levels of ϕ are used to determine the sensitivity of the optimal solution as ϕ varies and to determine the value of ϕ, which yields solutions most closely corresponding to real-world situations. (Chapter 8 reports a more exhaustive analysis of ϕ on the basis of an updated version of this chapter's model.)

Market structure

By virtue of national and state producer organizations, producers of export vegetables in Mexico have developed a partial framework for a system of supply management. Coordination of planting programs for individual producers was instituted for the first time in the 1973–74 season, after several years of reliance on quality controls, shipping holidays, and informal coordination. The actual degree of monopolistic power exercised both before and after the 1973–74 season is a matter for question. This study compares optimal shipments under a purely competitive structure, implying price equals marginal cost for each activity, as well as under the monopolistic assumption of marginal revenue equaling marginal cost. A comparison of the solutions under the two alternative specifications with actual 1972–73 plantings is used as an informal test of the hypothesis that Mexican producers acted competitively prior to the enactment of the 1973–74 controls. Implications regarding future production trends as monopolistic controls become more effective are discussed. Also, the effects of hypothetical relaxation of U.S. trade barriers are explored by means of the model.

Input prices

One of Mexico's production advantages has been relatively low costs of farm labor. Recent emphasis by the government on increasing rural living standards has brought about a doubling of farm wage rates from 1968 to 1974, and further increases are likely. With use of current technologies, labor for the vegetables included in this study constitutes about 40 percent of total production costs. Hence, wage rates are important in determining future production trends. The wage rate is entered at alternative levels, ranging from 36 to 70 pesos per day, to determine the effect on resultant solutions. (The actual 1972–73 wage was 36 pesos per day.)

Technology Base

Two regions in the state of Sinaloa, Culiacán and Fuerte Sur, are analyzed. Together these two regions in 1971–72 supplied 90, 88, and 80 percent of Mexico's tomato, bell pepper, and cucumber exports, respectively. Although the two regions are approximately 100 miles apart and have different climates, each can produce a wide variety of intensive, irrigated crops on a year-round basis. Each region is treated as a single, aggregate decision unit, implying homogeneity in resource quality and a relative absence of restrictions in resource combination in the individual micro units.

Livestock and perennial crop activities are omitted, leaving thirty-three short-cycle crop activities in Culiacán and thirty-nine in Fuerte Sur. The principal resource restrictions for each region are monthly land and water supplies and an annual water restriction. Other input supplies are assumed to be perfectly elastic at existing market prices.

Plantings of tomatoes, bell peppers, and cucumbers are entered on a monthly basis during the winter season, allowing monthly changes in production conditions for the three export vegetables to interact with changes in monthly demand so that an optimal program of plantings over the season can be determined.

Input requirements reflect only a single technology, which in all cases is a machine-oriented, high-technology production with substantial use of fertilizers and pesticides. The harvesting of cotton and vegetables, and part of the weeding, is done by hand in the regions.

Since the study focuses primarily on three export vegetables, the technical data for these crops are substantially more detailed than for the traditional crops. Planting dates for vegetables affect total yield over the picking season, and these factors are taken into account according to experimental data. Tomatoes grown in Culiacán are staked, whereas in Fuerte Sur tomatoes are grown both staked and unstaked (as ground tomatoes). Ground tomatoes, besides being unstaked, use less fertilizer, are picked less frequently, and yield about one-third as much per hectare as staked tomatoes. According to the producers, the attractiveness of ground tomatoes lies in the lower levels of investment and risk.

Structure of the Model

The model draws heavily on previous work by Duloy and Norton (chapters 2 and 3 of this book), Hazell (1971), and Hazell and Scandizzo (chapter 7 of this book).

The competitive objective function may be written as follows:

$$(12.1) \quad \max \Pi = X' W (A - 0.5\, BWX) - C' W - \phi(X'\, \Omega\, X)^{1/2},$$

where X is a vector of aggregate cropping activity levels; W is a diagonal matrix of average yields; C is a vector of cost coefficients; A and B are, respectively, the coefficient vector and matrix B of the linear demand structure $P = A - BWX$, where market quantities (Q) equal WX (all individual demand functions are assumed independent), and Γ is a covariance matrix of activity revenues.

The term $X' W(A - 0.5\, BWX)$ is the sum of areas under the product demand functions, and $C'X + \phi(X'\Omega X)^{1/2}$ is total variable production costs or, equivalently, the sum of areas under the product supply functions. Consequently, the difference between these two terms is the expected sum of producers' and consumers' surpluses over all markets, the maximization of which yields the competitive market solution.

For the monopolistic case, the objective function may be modified as follows:

$$(12.2) \quad \max \Pi^\star = X' W (A - BWX) - C'X - \phi(X'\Omega X)^{1/2}.$$

That this formulation yields a solution that equalizes marginal revenues with marginal costs for each product can be seen by partially differentiating the objective function with respect to $X' W = Q$ under the given resource constraints and by noting the following Kuhn–Tucker condition:

$$(12.3) \quad X' WA - 2BWX - \frac{\partial}{\partial Y}\{C'X - \phi(X'\Omega X)^{1/2}\} - \lambda D \le 0,$$

where λ is the set of dual values and D is the set of resource constraints. Note that the term $X'WA - 2BWX$ is the marginal revenue that is equated to marginal cost in the solution for the monopolistic case.

These objective functions are nonlinear, but they can be linearized according to the procedure of chapter 3 by letting V_j be the area under the demand curve from 0 to Q_j for the jth product. Then

$$(12.4) \qquad Q'(A - 0.5BQ) = \sum_{j=1}^{n} V_j,$$

where Q is the vector of quantities sold and V_j is a quadratic, concave function of Q_j. It can be linearly approximated to any degree of accuracy by choosing a number of segments on the V_j functions; assigning upper bounds to each segment; letting a single value, V_{ij}^*, approximate V_j over the segment; and finally by replacing that part of the programming problem involving max $Q'(A - 0.5BQ)$ with

$$(12.5) \qquad \max \sum_{j} \sum_{i} d_{ij} V_{ij}^*,$$

subject to

$$(12.6) \qquad X_j' W_j - \sum_{i} d_{ij} Q_{ij} \geq 0 \quad \text{(all } j\text{) and}$$

$$(12.7) \qquad \sum_{i} d_{ij} \leq 1 \quad \text{(all } j\text{)},$$

where Q_{ij} is the quantity of product j sold at the upper end of segment i.

Since equation (12.6) is the market-clearing commodity balance requirement, which is necessary in any case, the Duloy-Norton procedure adds only one additional row per product but permits inclusion of as many V_{ij} activities as desired for accuracy in approximation.

The measure of variation in gross returns to activities used in this study is the mean absolute deviation (MAD), a method first proposed for farm-level analysis by Hazell (1971) and later adapted to market situations by Hazell and Scandizzo (chapter 7 of this book).

Let r_{jt} denote the gross revenue from cropping activity j in historical year t ($t = 1, \ldots, T$), and let $\bar{r}_j = 1/T\Sigma_t r_{jt}$. Then, an estimate of the variance of the set of activities that relies on the MAD is:

$$(12.8) \qquad \text{est}\,(X'\Omega X) = \Delta \left\{ \frac{1}{T} \sum_{t} \left| \sum_{j} (r_{jt} - \bar{r}_j) X_j \right| \right\}^2,$$

where $\Delta = T\pi/\sqrt{2(T-1)}$ is a correction factor to convert the square of the MAD's to an estimate of the population variance. The measure used in this study is not the estimate of the variance, but rather Sir Ronald Fisher's estimator of the standard deviation; that is,

$$(12.9) \qquad S = \sqrt{\Delta/T} \left\{ \sum_t \left| \sum_j \left(r_{jt} - \bar{r}_j \right) X_j \right| \right\},$$

which is not dependent on assumed normality of the population distribution. This formulation can be entered in the objective function according to the procedure of chapter 7 by defining new variables,

$$(12.10) \qquad Z_t \geq 0 \quad \text{(all relevant historical values of } t\text{)},$$

and by forming the problem

$$(12.11) \qquad \max \Gamma = \sum_j E(Y_j) - \phi \hat{S},$$

such that

$$(12.12) \qquad \sum_j \left(r_{jt} - \bar{r}_j \right) X_j + Z_t \geq 0 \quad \text{(all } t\text{), and}$$

$$(12.13) \qquad \sum_t Z_t - \hat{S} \frac{T}{2\sqrt{\Delta}} = 0.$$

The Z_t variables then measure the negative deviations in total revenue from the mean for the activity revenue outcomes, and therefore $2 \sum_t Z_t$ is the sum of total deviations.

The linear programming tableau for the model is given in table 12-1.

Estimation of Product Demand

Monthly demand equations were estimated for tomatoes, bell peppers, and cucumbers for the winter season. The winter season was defined as December through May for tomatoes and December through April for bell peppers and cucumbers.

The statistical model was a single equation of the least-squares, pooled-data type that uses dummy variables to allow for changes in intercepts and in coefficients of explanatory variables. Each month was defined as a "class," and a dummy variable was used to shift the demand relations between classes. To facilitate subtraction of marketing costs, the form of the demand functions was assumed to be linear. Tests of hypotheses using the error sum of squares and the F statistic were accomplished to determine if monthly differences in slopes or intercepts (or both) were statistically significant. Tests indicated that monthly differences in slopes and intercepts were significant for all three products (Castro and Simmons 1974).

The statistical model was as follows:

Table 12-1. Linear Programming Tableau for Vegetable Exports Model

Constraints	Production activities — Culiacán $X_1\ X_2 \cdots X_n$	Production activities — Fuerte Sur $X_1\ X_2 \cdots X_m$	Selling activities (r sets of activities to linearize areas under demand functions) $V_{11}\cdots V_{1k}\ \ V_{21}\cdots V_{2k}\ \cdots\ V_{r1}\cdots V_{rk}$	T negative deviation counters (Culiacán)	T negative deviation counters (Fuerte Sur)	Z_1 (Culiacán)	Z_2 (Fuerte Sur)	RHS
Resource constraints (Culiacán)	C'_{jm}	0	0	0	0	0	0	$\leq b_j^A$
Resource constraints (Fuerte Sur)	0	C_{jm}	0	0	0	0	0	$\leq b_j^B$
r Commodity balance constraints	$V_1\ V_2 \cdots W_n$	$W_1\ W_2 \cdots W_m$	$-g_{11}\cdots -g_{1k}\ \ -g_{21}\cdots -g_{2k}\ \cdots\ -g_{r1}\cdots -g_{rk}$	0	0	0	0	≥ 0
r Convex combination constraints	0	0	$1\cdots 1\ \ 1\cdots 1\ \cdots\ 1\cdots 1$	0	0	0	0	≤ 1
T Gross revenue deviation constraints (Culiacán)	$r_{11}-\bar r_1\ \ r_{21}-\bar r_2 \cdots r_{m1}-\bar r_m$ $r_{12}-\bar r_1\ \ r_{22}-\bar r_2 \cdots r_{m2}-\bar r_m$ $\vdots$ $r_{1r}-\bar r_1\ \ r_{2r}-\bar r_2 \cdots r_{nr}-\bar r_n$	0	0	$1\ \,1\ \cdots\ 1$ (diagonal)	0	0	0	≥ 0
T Gross revenue deviation constraints (Fuerte Sur)	0	$r_{11}-\bar r_1\ \ r_{21}-\bar r_2 \cdots r_{m1}-\bar r_m$ $r_{12}-\bar r_1\ \ r_{22}-\bar r_2 \cdots r_{m2}-\bar r_m$ $\vdots$ $r_{1r}-\bar r_1\ \ r_{2r}-\bar r_2 \cdots r_{nr}-\bar r_n$	0	0	$1\ \,1\ \cdots\ 1$ (diagonal)	0	0	≥ 0
Z_1 Identity (Culiacán)	0	0	0	$2\ 2\cdots 2$	0	$-\left(\dfrac{\sqrt{A}}{T}\right)^{-1}$	0	$= 0$
Z_2 Identity (Fuerte Sur)	0	0	0	0	$2\ 2\cdots 2$	0	$-\left(\dfrac{\sqrt{A}}{T}\right)^{-1}$	$= 0$
Objective function	$-C'_1\ -C'_2 \cdots -C'_n$	$-C_1\ -C_2 \cdots -C_m$	$W_{11}\ W_{21}\ W_{12}\ W_{22}\ \cdots\ W_{1k}\ W_{2k}\ \cdots\ W_{r1}\ \cdots\ W_{rk}$	0	0	0	0	$\to$ max

$$(12.14) \qquad P_{it} = \alpha_o + \sum_{j=1}^{5} \alpha_j D_{jit} + \beta_o X_{it}$$

$$+ \sum_{j=1}^{5} \beta_j S_{jit} + \gamma_o I_{it} + \varepsilon_{it},$$

where

ij = Class 1, 2, 3, 4, or 5 ($i, j = 1$ for December . . . $i, j = 5$ for April)

t = Crop years 1, 2, . . . n

P_{it} = Monthly average price for the vegetable in question, in cents per pound for ith month of year t

X_{it} = Monthly shipments of the vegetable in question to U.S. and Canadian markets, in pounds per capita in ith month of year t

I_t = Disposable income in dollars per capita

ε_{it} = Random disturbances with zero mean and constant variance

D_{jit} = Intercept-shifting variables with $D_{jit} = 1$ when $i = j$ and 0 when $i \neq j$

S_j = Slope coefficient–changing variables with $S_{jit} = D_{jit} X_{it} = X_{it}$ when $i = j$ and 0 when $i \neq j$.

Quantities shipped and income were assumed to be predetermined.

Demand equations for bell peppers and cucumbers used 1972 prices from the Florida Department of Agriculture and Consumer Services for Florida shipping points, and the demand equations for tomatoes used prices for Nogales, Arizona, for the appropriate product classifications (U.S. Department of Agriculture 1963–73). The use of retail or wholesale prices would have involved the estimation and subtraction of a complex system of commission, brokerage, and shipping charges to derive the on-farm demand.

The statistical estimates of the monthly demand functions were converted to net import demand functions by subtracting estimated shipments from Florida and other production areas (average of 1970–72) and substituting the 1972–73 per capita income into the equation.

From the net import demand functions were subtracted the sales commission of 12 percent, the U.S. tariff, and transport costs from Culiacán to Nogales. Finally, the demand functions at the Culiacán level were converted to pesos per kilogram for use in the model. The resulting demand functions and price elasticities for export vegetables are given in table 12–2.

The Mexican domestic market demand functions for tomatoes and other crops were estimated by assuming values for direct price elasticities and passing the demand equation through the 1972 price-quantity equilibrium points. Time-series quantity data were not available to estimate least-squares equations. The demand for tomatoes in Mexico was assumed equal in all of the months included. Mexican demand functions for bell peppers and cucumbers were omitted from the model because

Table 12-2. *Estimated Demand Functions and Elasticities for Export Vegetables, f.o.b. Culiacán, Mexico*

Product	Month	Demand equation[a]	Price elasticity	
			At 1972–73 quantity (1)	At mean of observation (2)
Tomatoes				
(United States	Dec.	$P = 3.351 - 0.000043755Q$	−18.4	−0.75
and Canada)	Jan.	$P = 3.184 - 0.000036722Q$	− 1.4	−0.75
	Feb.	$P = 2.856 - 0.000020254Q$	− 1.5	−1.39
	Mar.	$P = 4.309 - 0.000044305Q$	− 0.3	−0.61
	Apr.	$P = 3.229 - 0.000019006Q$	− 1.6	−1.46
	May	$P = 3.533 - 0.000039815Q$	− 1.1	−1.26
Bell peppers	Dec.	$P = 1.905 - 0.00046591Q$	− 2.3	−0.52
	Jan.	$P = 4.270 - 0.00040690Q$	− 0.3	−0.50
	Feb.	$P = 5.092 - 0.00026592Q$	− 1.1	−0.92
	Mar.	$P = 5.968 - 0.00035203Q$	− 0.8	−0.71
	Apr.	$P = 5.896 - 0.00042629Q$	− 1.9	−0.67
Cucumbers	Dec.	$P = 1.752 - 0.00016589Q$	[b]	−0.57
	Jan.	$P = 2.419 - 0.00011539Q$	− 0.6	−0.83
	Feb.	$P = 2.636 - 0.00009907Q$	− 1.7	−0.94
	Mar.	$P = 3.121 - 0.00015937Q$	− 0.5	−0.86
	Apr.	$P = 1.591 - 0.00008637Q$	− 1.3	−0.83

Note: Column 2 refers to Nogales, Arizona, prices, whereas column 1 refers to Nogales prices *minus* border crossing charges, transport, and all Culiacán to Nogales charges; that is, at the level of the Culiacán packing plant.

a. P = price in pesos per kilogram; Q = quantity in metric tons.

b. The abnormally large quantity in December 1972–73 gave a zero price on the estimated net farm demand curve.

these products are not produced in these regions in significant quantities for Mexican consumption. Demand functions for traditional crops and for tomatoes in Mexico are listed in table 12-3.

The model allows the Mexican market to absorb tomatoes of nonexportable quality as well as the transfer of tomatoes of exportable quality according to the principles of market allocation.

Production Data

Most of the input–output data for cropping activities were taken from unpublished budgets prepared by the Confederación de Asociaciones Agrícolas del Estado de Sinaloa (CAADES) and were verified in part by several informal field visits. The yield distribution of staked tomatoes by

Table 12-3. *Demand Functions for Tomatoes
and Traditional Crops in Mexico*

Crop	Demand function	Direct-price elasticity
Tomatoes	$P = 2.993 - 0.00008372Q$	-0.5
Sesame	$P = 3.068 - 0.00011210Q$	-1.2
Cotton	$P = 3.276 - 0.00000537Q$	-0.5
Rice	$P = 2.960 - 0.00001536Q$	-0.3
Safflower	$P = 2.069 - 0.00000531Q$	-1.2
Beans	$P = 5.573 - 0.00008800Q$	-0.3
Chickpeas	$P = 3.448 - 0.00007409Q$	-0.3
Corn	$P = 1.126 - 0.00001938Q$	-0.2
Sorghum	$P = 1.185 - 0.00000208Q$	-0.3
Soybeans	$P = 2.334 - 0.00000423Q$	-1.2
Wheat	$P = 0.936 - 0.00000107Q$	-0.5

months over the harvest season (table 12-4) was estimated from experimental data published by López and Chan (1973).

Variation in gross revenues per hectare for all crops over six cropping years was taken from published CAADES bulletins (1967–72). Table 12-5 indicates the large variation in gross revenues per hectare of export vegetables compared with traditional crops.

Estimates of water requirements and availabilities in the two regions were obtained from the state's Division of Water Resources. Good measurements of water requirements by months for specific crops are scarce, and the water constraints are considered the weakest part of the data base. Optimal hectarages for vegetables, however, are not greatly affected by inaccuracies in the water constraints.

Solutions I: Risk and Market Structure

The first set of solutions used the 1972–73 daily wage rate of 36 pesos, the objective function corresponding to the competitive case, and risk-aversion levels of 0, 0.5, 1.0, and 1.5. Optimal hectarages for each solution were compared with actual hectarages planted in 1972–73 to determine the level of ϕ that yielded the solution most closely corresponding to the actual 1972–73 plantings. The results are presented in table 12-6. The risk-aversion level of $\phi = 1.0$ appears to give the closest correspondence with the actual situation. Optimal hectarages of tomatoes, cotton, soybeans, safflower, and sesame are clearly closer to actual values with

Table 12-4. *Yield of Vegetables According to Planting Date* (kilograms per hectare)

Crop and zone	Market	Month of planting	November	December	January	February	March	April	May	June	July	Total
Tomatoes (Culiacán)	Export	Sep.	1,001	9,181	6,342							16,524
		Oct.		511	4,061	9,177	3,227					16,996
		Nov.			521	4,130	9,289	3,269				17,199
		Dec.				847	6,743	5,908	3,374			16,877
		Jan.					833	6,657	5,813	3,322		16,625
		Feb.						1,120	6,745	5,620	a	13,485
Tomatoes (Culiacán)	Mexican	Sep.	670	3,934	2,718							7,322
		Oct.		219	1,749	3,933	1,383					7,284
		Nov.			219	1,770	3,931	1,401				7,371
		Dec.				363	2,892	2,952	1,446			7,633
		Jan.					357	2,853	2,492	1,424		7,126
		Feb.						480	2,890	2,409	3,900	9,679
Tomatoes (Fuerte Sur)	Export	Oct.			3,441	7,571	2,753					13,675
		Nov.				3,532	7,700	2,826				14,058
		Dec.					4,347	7,245	2,898			14,490
		Jan.						4,238	7,063	2,826		14,127
		Feb.							4,818	6,121	a	10,939
Tomatoes (Fuerte Sur)	Mexican	Oct.			1,475	3,245	1,180					5,900
		Nov.				1,513	3,330	1,212				6,055
		Dec.					1,863	3,105	1,242			6,210
		Jan.						1,816	3,027	1,212		6,055
		Feb.							2,065	3,095	3,566	8,726

Product	Type	Month								Total
Ground tomatoes (Fuerte Sur)	Export	Oct.		5,225						5,225
		Nov.			5,362					5,362
		Dec.				5,500				5,500
		Jan.					5,362			5,362
		Feb.						5,225		5,225
Ground tomatoes (Fuerte Sur)	Mexican	Oct.		2,230						2,230
		Nov.			2,298					2,298
		Dec.				2,397				2,397
		Jan.					2,298			2,298
		Feb.						2,230		2,230
Bell peppers (Culiacán)	Export	Sep.	1,900	6,175	1,425					9,500
		Oct.		2,000	6,500	1,500				10,000
		Nov.			1,500	6,500	2,000			10,000
		Dec.				1,425	6,175	1,900		9,500
Bell peppers (Fuerte Sur)	Export	Oct.		1,520	4,940	1,140				7,600
		Nov.			1,600	5,200	1,200			8,000
		Dec.				1,200	5,200	1,600		8,000
		Jan.					1,144	4,900	1,500	7,544
Cucumbers (Culiacán)	Export	Oct.		11,650						11,650
		Nov.			11,600					11,700
		Dec.				11,550				11,550
		Jan.					11,250			11,250
		Feb.						11,050		11,050
Cucumbers (Fuerte Sur)	Export	Oct.		4,640						4,660
		Nov.			4,660					4,900
		Dec.				5,380				5,380
		Jan.					5,140			5,140

a. Because there are no exports in July, that month's production is sold in the domestic market.

Table 12-5. *Variation in Gross Revenues, Total Costs
and Input Requirements per Hectare for Cropping Activities*

Crop	$\Sigma_{t=1}^{6}\|r_{jt}-\bar{r}_j\|$		Total cost[a]		Annual water requirements (both regions; ten thousands of cubic meters)	Labor requirements (both regions, man-days)
	Culiacán (pesos)	Fuerte Sur (pesos)	Culiacán (pesos)	Fuerte Sur (pesos)		
Tomatoes	99,204	68,405	41,175	36,236	0.960	133.0[b]
Bell peppers	134,178	92,878	22,500	21,020	0.960	132.0[b]
Ground tomatoes	n.a.	26,396	n.a.	13,893	0.710	28.2[b]
Cucumbers	58,731	33,792	16,418	10,045	0.470	19.7[b]
Sesame						
Spring	997	889	2,112	2,064	0.820	15.8
Summer	997	889	2,112	2,112	0.650	15.8
Rice	2,916	2,834	2,227	3,235	1.950	9.7
Safflower	3.662	3.293	1.678	1.704	0.700	3.9
Kidney beans	2,389	2,114	2,160	2,170	0.890	14.1
Chickpeas	3,796	3,274	2,586	2,614	0.990	10.7
Maize						
Summer	997	724	2,052	2,212	0.910	18.4
Winter	997	724	2,052	2,212	0.840	18.4
Sorghum	3,220	1,456	2,574	2,629	0.790	9.5
Soybeans	3,174	3,537	2,110	2,159	0.960	5.4
Wheat	2,009	1,922	2,510	2,540	0.880	9.6
Cotton	n.a.	3,830	n.a.	6,022	n.a.	72.2

n.a. Not available.
a. Wage rate of 36 pesos per day.
b. Preharvest labor only.

$\phi = 1.0$, and the correspondence for the other crops is at least as good as with other levels of ϕ. Although empirical experience useful in judging acceptable levels of risk-aversion coefficients is not extensive, it is interesting that recent work for other parts of Mexico (see chapter 8) lends support to the credibility of ϕ levels of 1.0 and 2.0.

When the objective function corresponding to the monopolistic case for export vegetables was used, solutions were also generated for the same levels of ϕ and the same wage rate as for the competitive case just described. These solutions are presented in table 12-7. For all levels of ϕ, optimal hectarages of the three vegetables are unrealistically low when compared with actual 1972–73 plantings. It is therefore concluded that vegetable producers were operating in the context of a competitive environment in 1972–73. The monopolistic solution indicates the possible future trend in planted hectarage if, in fact, recent controls enacted by the vegetable producers have created significant monopolistic characteristics.

The competitive case with a risk-aversion level of 1.0 is used in the remainder of the analysis.

Table 12-6. *Comparison of Actual 1972–73 Plantings with Optimal Solutions for the Competitive Case*

Crop	Optimal planted area (hectares) by level of risk aversion				1972–73 actual (hectares)
	$\phi = 0$	$\phi = 0.5$	$\phi = 1.0$	$\phi = 1.5$	
Tomatoes	19,239	15,709	13,356	9,217	16,382
Bell peppers	3,633	3,332	3,128	2,440	4,869
Cucumbers	3,177	2,803	3,447	3,552	5,614
Sesame	0	3,194	4,386	4,983	4,883
Rice	33,047	34,733	33,890	33,047	33,047
Safflower	96,986	65,196	43,219	29,885	51,837
Beans	33,555	33,555	32,166	30,350	47,192
Chickpeas	13,352	14,980	13,730	12,897	25,580
Maize	14,875	18,769	18,769	18,769	21,503
Sorghum	55,717	54,749	46,552	47,495	68,608
Soybeans	104,037	104,170	79,478	79,478	75,048
Wheat	26,552	39,588	26,034	10,479	45,620
Cotton	57,024	46,210	39,001	31,792	37,056
Average percentage of absolute deviation	38.9	24.4	21.8	30.2	

Note: The daily wage rate used was the 1972–73 rate of 36 pesos.

Solutions II: Effects of Trade Barriers

The effects of a parallel shift of the export demand function on vegetable exports and prices and on the allocation between the export and national markets was examined. Such a shift might occur because of changes in either tariffs, Florida supplies, or (possibly) per capita consumer incomes.

The specific demand shift simulated was a removal of all tariffs on tomatoes, peppers, and cucumbers.[1] The consequences of this change in reference to allocation between the export and domestic markets and resultant revenues to producers are seen in table 12-8. The quantity exported increases by 27.5 percent and the weighted average export price increases 10 percent, increasing total export revenues by 37.5 percent. The increase in quantity produced also automatically increases the quantity available for the domestic market, since 30 percent of total field production is considered to be of nonexportable quality. The domestic price therefore decreases by about 13 percent. Net producer revenues above

1. The tariff rates are variable by month. For tomatoes, they vary from US$0.015 to US$0.021 per pound.

Table 12-7. *Comparison of Actual 1972–73 Plantings with Optimal Solutions for the Monopolistic Case*

Crop	Optimal planted area (hectares) by level of risk aversion			1972–73 actual (hectares)
	$\phi = 0$	$\phi = 0.5$	$\phi = 1.0$	
Tomatoes	9,644	7,653	6,720	16,382
Bell peppers	1,921	2,047	1,936	4,869
Cucumbers	1,572	1,628	1,679	5,614
Sesame	810	0	4,386	4,883
Rice	33,890	33,047	33,890	33,047
Safflower	99,575	64,982	31,962	51,837
Beans	34,018	32,165	32,200	47,192
Chickpeas	13,352	13,730	13,314	25,580
Maize	14,875	18,769	18,769	21,503
Sorghum	55,717	55,717	50,855	68,608
Soybeans	111,460	104,243	75,614	75,048
Wheat	26,447	39,588	25,034	45,620
Cotton	57,024	24,583	35,397	37,056

Note: The daily wage rate used was the 1972–73 rate of 36 pesos.

costs increase by an estimated 69 percent for the combined markets. Although the tariff elimination leads to some substitution of tomatoes for other crops, there is a net increase in planted area and in employment. Total wages paid increase by 17 percent with the tariff elimination.

Any other demand shift in the export or domestic markets could be analyzed in the same way. Since simultaneous results in both markets affect the revenue outcome, the model is particularly suited to handle questions of this type.

Solutions III: Allocations over Seasons and Markets

The distribution of exports over the season is affected by monthly changes in demand, production costs, and the normal pattern of competitive shipments from Florida and Caribbean countries. These three factors are taken into account in the model. Table 12-9 compares the model's simulated monthly distribution of exported tomatoes, peppers, and cucumbers over the winter season with the 1970–73 average actual pattern of shipments. The table indicates that tomato producers should attempt to increase December shipments and reduce shipments in February and March. The actual and optimal patterns of tomato shipments are illustrated in figure 12-1. By using the monthly demand functions to estimate the expected prices for the optimal and actual quantities shipped each

Table 12–8. *Comparison of Marketed Quantities, Prices,
and Revenues in Export and Domestic Markets
for Tomatoes, with and without Tariffs*

Quantity, price, and revenue	With tariff	Without tariff
Tomatoes		
Quantity exported (metric tons)	186,568	237,931
Average export price (pesos per kilogram at packing plant)	2,400	2,640
Total export revenue (thousands of pesos)	462,140	635,080
Quantity sold domestically (metric tons)	79,994	96,478
Average domestic price (pesos per kilogram at packing plant)	1,738	1,514
Total domestic revenue (thousands of pesos)	139,068	146,514
Total revenue, both markets (thousands of pesos)	608,654	774,148
Total costs (thousands of pesos)	494,352	580,979
Net revenue to producers (thousands of pesos)	114,302	193,169
All crops		
Employment (thousands of man-years)	13,426	15,126
Total area planted (hectares)	400,945	436,734
Total wages paid (thousands of pesos)	526,165	616,558

month, it was estimated that net revenue from tomatoes could be increased by 10 percent by adopting the optimal shipment pattern.

After comparison of the normal shipment pattern for bell peppers with the optimal pattern, it can be concluded that shipments of bell peppers should be reduced in December and January and increased in March and April, for net revenue could be increased by about 6 percent by doing so. Cucumber shipments should be reduced in December and increased in February and March.

Solutions IV: Allocation between Export and National Markets

Although tomatoes are grown primarily for the export market, about 30 percent of the total field production is sold in Mexico. The Mexican

Table 12-9. *Optimal Monthly Distribution of Exports of Tomatoes, Peppers, and Cucumbers Compared with Actual 1972–73 Shipments*
(metric tons)

Month	Tomatoes		Bell peppers		Cucumbers	
	Optimal	Actual	Optimal	Actual	Optimal	Actual
Dec.	20,217	3,959	0	1,227	2,725	14,029
Jan.	24,528	39,228	1,811	8,155	9,698	12,841
Feb.	27,451	57,044	7,491	11,100	13,485	9,914
Mar.	38,737	72,972	8,147	9,530	11,426	12,998
Apr.	47,180	65,842	7,506	4,735	2,211	7,917
May	28,457	42,645	0	0	0	0
Total	186,570	281,690	24,955	34,747	39,545	57,699
Net farm income (thousands of pesos)[a]	112,136	92,588	12,405	6,862	1,421	−49,293[b]
Wage bill (thousands of pesos)[a]	199,084	198,689	28,698	33,159	17,724	24,263
Employment (thousands of man-days)[a]	4,033	4,025	651	693	309	423

Note: $\phi = 1.0$; the daily wage rate was the 1972–73 rate of 36 pesos. "Actual" net farm income was calculated using actual volumes sold together with the model's input-output coefficients and estimated demand prices.

a. Only for these three crops.

b. 1972–73 was a year of abnormally high cucumber production, with the consequence that the Nogales, Arizona, border price was driven down to one-third its normal level. It indeed appears likely that farmers suffered losses on cucumbers in that year.

market is normally used for tomatoes of nonexportable quality and occasionally for diversion of exportable supplies when the U.S. market becomes temporarily oversupplied. In practice, the allocation of the crop between the two markets has been tentative and informal, and it is of interest to evaluate possibilities for more positive allocation according to established maximization criteria.

In the absence of supply restrictions or quality differentials, the prices in two competitive markets tend to be equalized (net of handling costs) by the decisions of individual producers to ship to the market yielding the highest price. The monopolistic case is similar, except that marginal revenues instead of prices are equalized in the two markets. The present model simulates this process by including a transfer activity to divert exportable supplies from the export market to the domestic market.

In the first set of solutions (table 12-10), the potentially exportable portion of total field production was entered at about 70 percent, in accordance with the recent practice of applying grading statements.

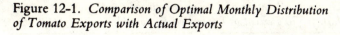

Figure 12-1. *Comparison of Optimal Monthly Distribution of Tomato Exports with Actual Exports*

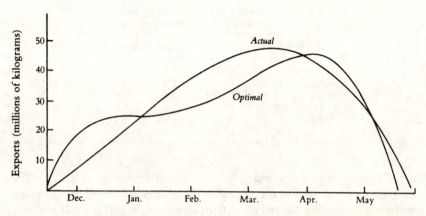

Under these conditions the model indicated that no transfers from the export market to the Mexican market were economical. The normal quantity of nonexportable tomatoes was sufficient in each month to keep the Mexican prices below U.S. prices and thus to prevent an optimal allocation.

To investigate the effects of a possible relaxation of qualitative export restrictions or a possible technological breakthrough that would increase the proportion of exportable fruit, the basic input data were changed to reflect a proportion of potentially exportable quality of 80 percent of total field production (table 12-11). In this case, the model shows that tomatoes of exportable quality would be transferred to the domestic market in December, January, February, and May, and prices would be equalized in the two markets according to principles of market allocation. In March and April, however, nonexportable supplies would be sufficient to keep the Mexican price below the export price, and no transfers would occur. Of course, additional work in estimating Mexican demand for tomatoes would make this kind of allocation more precise.

Solutions v: Increased Rural Wages

As a final numerical experiment, the effects of increased minimum wages for farm labor on optimal plantings of export vegetables and other crops were estimated. These comparisons are given in table 12-12. An increase in the wage rate causes sharp decreases in optimal plantings of the

370   POMAREDA AND SIMMONS

Table 12-10. *Optimal Allocation of Tomatoes between the Export and Mexican Markets, 70 Percent of Production Exportable*

	Quantities sold (metric tons)		Net farm price (pesos per kilogram)	
Month	Export	Domestic	Export	Domestic
Dec.	20,217	8,518	2.5	2.3
Jan.	24,528	10,273	2.3	2.1
Feb.	27,451	11,744	2.3	2.0
Mar.	38,737	16,700	2.6	1.6
Apr.	47,180	20,426	2.3	1.3
May	28,457	12,197	2.3	2.0

three export vegetables and other relative labor-intensive crops. The average arc elasticities of the vegetable hectarages in response to increases in the wage rate turn out to be -1.0 for tomatoes, -0.67 for bell peppers, and -0.71 for cucumbers. A policy of encouraging rapid increases in rural wages would be likely to have a substantial effect on vegetable exports.

In fact, the area planted (and hence production) declines for almost all crops; an exception is soybeans, a crop that is quite low in labor intensity. As a consequence, employment in the region declines as wages rise, but not enough to prevent the total wage bill from increasing.

Conclusions

By using demand functions and risk factors for cropping activities in the mathematical programming model for Culiacán and Fuerte Sur, acceptable representations of actual aggregate behavior for the base year 1972–73 were obtained.

Table 12-11. *Optimal Allocation of Tomatoes between the Export and Mexican Markets, 80 Percent of Production Exportable*

	Quantities sold (metric tons)		Net farm price (pesos per kilogram)	
Month	Export	Domestic	Export	Domestic
Dec.	20,878	5,888	2.4	2.4
Jan.	24,073	8,277	2.3	2.3
Feb.	27,451	8,277	2.3	2.3
Mar.	43,088	10,665	2.6	2.1
Apr.	54,141	13,535	2.3	1.8
May	30,968	8,277	2.4	2.4

From solutions for both the competitive and monopolistic cases, it was concluded that producer behavior in the base year corresponded more closely to the competitive case. Risk-aversion coefficients (ϕ) of 0, 0.5, 1.0 and 1.5 were tested, and the level of risk aversion corresponding to $\phi = 1.0$ appeared to function best in obtaining solutions most closely corresponding to actual plantings in 1972–73.

With the competitive objective function and $\phi = 1.0$, the model was then used to evaluate possible effects on tomato, bell pepper, and cucumber exports and related variables of changes in wage rates, changes in the percentage of total production that is of exportable quality, and shifts in export demand.

It was found that, given present technologies, an increase of 10 percent in the minimum daily wage would decrease exports by 10 percent for tomatoes, 6.7 percent for bell peppers, and 7.1 percent for cucumbers. Employment would also decrease, but the total value of regional wage payments would rise.

The monthly distribution of vegetable exports over the season was found to be important for maximizing total net revenue to producers. Tomato producers could increase net revenue by 10 percent by increasing exports in December and reducing them in February and March.

The effects of a possible shift in export demand were analyzed for the resulting revenues in both the export and domestic markets. It was found, for example, that the removal of tariffs on the three vegetables would increase the weighted average farm-gate price for exports by 10 percent, reduce the domestic price by 13 percent, increase exports by 27.5 percent, and increase the combined net revenues from the two markets by 69 percent. It would also increase employment and wages paid for field labor. In other words, because of the character of export and domestic tomatoes as joint products, all pertinent groups in Mexican society would benefit from the tariff removal: consumers, export traders, agricultural landowners, and rural landless laborers.

Table 12-12. Area Planted (hectares), Level of Employment (thousands of man-days), and Wage Bill (thousands of pesos) for Three Wage Rates (W)

Crop	W = 36 pesos			W = 50 pesos			W = 70 pesos			Man-days per hectare
	Area	Employment	Wage bill	Area	Employment	Wage bill	Area	Employment	Wage bill	
Tomatoes	13,356			9,891			6,642			
Cultivation	—	1,776	64,037	—	1,315	65,775	—	883	61,837	133.0
Harvesting	—	1,536	83,047	—	1,137	85,310	—	763	80,115	115.0
Packing	—	721	52,000	—	534	53,411	—	359	50,703	54.0
Peppers	3,128			2,366			2,022			
Cultivation	—	413	14,885	—	312	15,616	—	267	18,683	132.0
Harvesting	—	185	9,979	—	139	10,469	—	119	12,526	59.0
Packing	—	53	3,834	—	40	4,022	—	24	4,812	17.0
Cucumbers	3,447			2,439			2,171			
Cultivation	—	68	2,448	—	48	2,402	—	43	2,994	19.7
Harvesting	—	117	6,329	—	83	6,219	—	73	7,750	34.0
Packing	—	124	8,947	—	88	8,780	—	78	10,942	36.0
Watermelons*	320			320			320			
Cultivation	—	4	143	—	4	198	—	4	280	12.4
Harvesting	—	2	124	—	2	150	—	2	210	7.2
Packing	—	2	115	—	2	200	—	2	280	5.0
Melons*	1,403			1,403			1,403			
Cultivation	—	33	1,118	—	33	1,648	—	33	2,310	23.5
Harvesting	—	90	4,885	—	90	6,750	—	90	9,450	64.4
Packing	—	28	2,003	—	27	2,700	—	27	3,780	19.8

String beans*									
Cultivation	1,016	58	2,077	1,016	58	2,900	1,016	58	4,060	56.7
Harvesting	—	138	7,488	—	138	10,350	—	138	14,490	136.3
Packing	—	16	1,143	—	16	1,600	—	16	2,240	15.6
Squash*									
Cultivation	1,523	82	2,965	1,523	82	4,100	1,523	82	5,740	54.0
Harvesting	—	209	11,331	—	209	15,675	—	209	21,945	137.6
Packing	—	23	1,691	—	23	2,300	—	23	3,220	15.4
Eggplant*									
Cultivation	750	40	1,460	750	40	2,000	750	40	2,800	54.0
Harvesting	—	63	3,406	—	63	4,725	—	63	6,615	84.0
Packing	—	27	1,947	—	27	2,700	—	27	3,780	36.0
Sesame	4,386	69	2,498	810	13	640	—	312	21,866	15.8
Rice	33,890	329	11,851	33,047	320	16,028	32,204	157	10,986	9.7
Safflower	43,219	168	6,076	43,206	168	8,425	40,242	428	29,967	3.9
Beans	32,166	453	16,350	30,356	428	21,401	30,362	125	8,724	14.1
Chickpeas	13,730	147	5,296	12,897	138	6,900	11,648	59	4,112	10.7
Maize	18,769	345	12,450	10,982	202	10,100	3,193	395	27,663	18.4
Sorghum	46,552	442	15,943	45,993	437	21,864	41,599	461	32,271	9.5
Soybeans	79,478	429	15,472	80,980	437	21,864	85,374	90	6,350	5.4
Wheat	25,034	240	8,664	10,479	101	5,030	9,449	—	—	9.6
Cotton	39,001	2,816	101,512	13,769	994	71,775	—	—	—	72.2
Sugarcane	39,777	2,180	78,581	39,777	2,180	109,000	39,777	2,180	152,600	54.8
Total	400,945	13,426	562,095	342,004	9,928	603,027	309,695	7,630	626,101	

— Not applicable.

Note: Area of these crops (*) were fixed in the solution. **V**egetable harvest wage: 1.5W = 54.07 (W=36); 1.5W = 75 (W=50); 1.5W = 105 (W=70). Vegetable packing wage: 2W = 72.10 (W=36); 2.0W = 100 (W=50); 2W = 140 (W=70).

References

Castro, Roberto, and Richard L. Simmons. 1974. *The Demand for Green Peppers, Cucumbers, and Cantaloupes in the Winter Season.* Economics Research Report, no. 27. Raleigh: North Carolina State University.

Confederación de Asociaciones Agrícolas del Estado de Sinaloa (CAADES). 1967–72. *Análisis de la situación agrícola de Sinaloa.* Boletín bimestral. Various issues.

Hazell, Peter B. R. 1971. "A Linear Alternative to Quadratic and Semi-Variance Programming for Farm Planning under Uncertainty." *American Journal of Agricultural Economics.* Vol. 53, pp. 53–62.

López, Fidel, and José Luis Chan. 1973. "Efecto de la densidad de población y métodos de poda sobre el rendimiento y calidad del tomate en Espaldera." Paper presented at the 21st Congress of the American Society of Horticultural Science, Tropical Region, San José, Costa Rica, July 15–21, 1973.

Unión Nacional de Productores de Hortalizas (UNPH). 1973. *Informe de la II Convención Anual y XIII Asamblea General Ordinaria, Cuernavaca, March 4–6, 1973.*

U.S. Department of Agriculture. 1963–73. *Fresh Fruit and Vegetable Prices.* Various issues. Washington, D.C.: U.S. Government Printing Office.

13

Machinery-Labor Substitution in Mexico's Central Plateau

Hunt Howell

Issues of input substitution and employment creation in agriculture are addressed in this chapter. The perspective is from an irrigated area in the Central Plateau of Mexico, where plot sizes are small and there is substantial underemployment of farmers. The vehicle of analysis is a process-analysis model that is similar in many respects to CHAC but that contains more fully elaborated structures for depicting factor supply behavior.

Overview

Policies of changing the wage rate and the effective cost of farm machinery are considered with this model, called TOLLAN.[1] The model simulates the effects of changing factor prices, levels of factor use, production, and product prices. In another experiment, TOLLAN is used to present the effects of subsidizing the production of labor-intensive crops. A distinction among three sizes of landholdings permits a rudimentary analysis of the consequences of these policies for the rural income distribution also, but the focus of the study throughout is on the possibilities for employment creation in agriculture. As discussed in chapter 5, the model was also used to generate product supply functions that are pertinent to analysis of pricing policy.

Note: This chapter is based on my Ph.D. dissertation written at the University of Pennsylvania under Gerard Adams. Roger Norton, then of the World Bank, provided invaluable guidance throughout the project. I wish to thank both of them for their comments, suggestions, and encouragement. I also am indebted to the Centro Científico IBM de América Latina (IBM de México) and the Latin American Teaching Fellowship Program (Tufts University) for providing funds and facilities with which to carry out the research. Sra. Luz María Bassoco and Dr. José Silos, then of Comisión Coordinadora del Sector Agropecuario (COCOSA), generously provided assistance in collecting and organizing the data as well as improving the institutional realism of the model.

1. *Tollán* is the Toltec name for the ancient capital of the Toltec Indian civilization, which flourished at the site of Tula during the period 600–900 A.D.

The area under study is the Tula irrigation district in the state of Hidalgo, located approximately fifty miles northwest of Mexico City. Water is supplied to all farmers in the district through a reticulation network linked to three small dams. Systematic irrigated agriculture in the area dates back to 1904, although there is some evidence of water-control measures in Toltec times. Agricultural activity during a twelve-month period in a district is simulated by the model, and the parameters are derived from historical data averaged over the three-year period 1966–67 through 1968–69. During this period there were an average of 21,608 farmers cultivating 37,950 hectares (97,737 acres), with an average farm size of 1.76 hectares (4.34 acres). The principal crops grown in the district were maize (34 percent of the total land sown), alfalfa (30 percent), and wheat (19 percent). Almost all the rest of the land was allocated among sixteen other crops.

Both mechanized and nonmechanized field operations are observed in the area, although machinery use is confined, by and large, to 60 horsepower tractors used for plowing and harvesting. Some farmers own their machinery and others rent it. Mules and bullocks are the other source of traction power, but in TOLLAN the use of draft animals is expressed in equivalent amounts of mule time. On the larger farms, hired labor is employed in peak seasons. Some of these laborers come from the ranks of farmers with very small plots and others from the group of landless laborers who migrate temporarily to Mexico City (three to four hours by local bus) for work in the off-season.

Features of the TOLLAN Model

In TOLLAN, agricultural operations are represented in the form of a large set of alternative production activities in a linear programming model. Only crop-related activities are included; livestock raising is excluded from the specification because it is an incidental part of the region's economy. The procedure adopted for measurement of factor substitution is to simulate numerically the new equilibrium position, in the entire production set, that would be induced by factor price changes. This gives, then, as a by-product, the factor response surfaces. The resulting measurement of factor substitutions includes the effects of both changes in cropping patterns (product mixes) and changes in technology. Elasticities of factor substitution and other scalar measures are calculated ex post from the results of the solutions.

To carry this out, the model is structured to simulate the response of the producers and the regional market to changes in factor prices. In this respect, TOLLAN is similar to CHAC. Since TOLLAN represents only one producing area, however, it was possible to build into it greater detail on the factor markets than CHAC contains. As in CHAC, product and factor markets are linked through the production technology set, and the objective function is designed so that its maximization drives the model to the competitive equilibrium in prices and quantities on all markets. In addition, this equilibrium takes into account the farmers' desire to avoid risky crop combinations (farm plans). The mean-absolute-deviation (MAD) approach of Hazell and Scandizzo is used here.[2] The detailed treatment of factor and product markets and the inclusion of risk yield a model of 1,635 variables, 472 equations, and approximately 49,300 nonzero coefficients.[3] Since the model is essentially similar to CHAC (see chapters 2, 4, and 5), the following paragraphs focus on a few salient aspects of TOLLAN, especially those in which it differs from CHAC.

Because the geographical coverage is limited to a small area, the question of appropriate price elasticities of crop demand must be addressed. Should they be the same as those used at the national level in CHAC, or should they be local elasticities; that is, national values divided by TOLLAN's share of national production for each crop? The answer depends on the probable supply response of the rest of the country to the exogenous shocks used in policy simulation with TOLLAN (chapter 11 of this book). On the one hand, change in the interest or wage rate is probably going to affect all areas similarly; thus, their supply responses will be comparable to the response in the Tula irrigation district. In this case, market shares would remain constant, and the relevant price elasticities of demand for Tula's crops would thus be the national values.[4]

On the other hand, local price elasticities of crop demand would be appropriate in Tula if the supply response in the rest of the country were slight in relation to a change in one of the parameters in TOLLAN. Such a case would arise if changes were specific to the irrigation district; for example, an increase in the endowment of irrigated land. Because the policy instruments examined here—namely, wages and interest rates—affect other producing areas also, national price elasticities of crop demand are used.[5]

For the nineteen crops included in TOLLAN, there are 1,059 alternative production activities; all are done under irrigation and all use chemical fertilizer. One of the sources of this diversity is the three-way breakdown

2. A complete discussion of this technique is contained in chapters 7 and 8.

3. Despite the size of the model, it was readily constructed, modified, and its results put into an easily interpretable format by means of a series of computer programs. These are described in chapter 19.

4. As noted in chapter 2, these values are quite low, ranging from -0.10 to -0.60, with the majority in the -0.20 to -0.30 range.

5. Chapter 11 contains a rather thorough analysis of the appropriate concept of demand elasticities for use in a local model. The concept used here corresponds to the case of $\theta = 1$ in that chapter.

of farm size by the following criteria: 0–5 hectares (small), 5–10 hectares (medium), and more than 10 hectares (large). Essentially identical sets of potential production activities are repeated for all farm sizes, although, of course, the differing factor endowments lead to selection of differing cropping mixes in the model solutions. A second source of variation in the production set is the choice of planting dates; TOLLAN contains up to two planting dates per crop, with associated monthly schedules of other inputs up to harvest time. This follows the specifications of Bassoco and Rendón (chapter 4 of this book) for incorporating the calendars of agricultural activities.

A third, and final, source of diversity in production in TOLLAN is the variation in the possible degrees of intensity of mechanization in the cultivation operations. A rather complete specification of this area is obviously important to the measurement of capital-labor substitution, and in this respect TOLLAN is quite a bit more detailed than CHAC. In TOLLAN, cultivation may be carried out under as many as nine different degrees of mechanization per crop: fully mechanized, completely un-mechanized, and seven intermediate degrees. As in CHAC, the specification is crop specific, so that, for example, "full mechanization" may include mechanized harvesting for the grains but not for some vegetables. In consultation with Mexican specialists,[6] the nine alternative degrees of mechanization were worked out for each crop, and then simple tests were applied, introducing prices, to find out which degrees would be economically inefficient (that is, dominated by a convex combination of other degrees). Inefficient degrees were then eliminated from the model as being unlikely to be observed in actuality. These alternative methods of production are represented by a total of 353 potential production activities for each of the three farm sizes.

Thus, the model effectively contains—for each crop, farm size, and planting date—a piecewise linear isoquant with respect to machinery and labor, with up to nine linear segments. Each such isoquant may be thought of as pertaining to a single, micro-level production function, and the envelope of all such isoquants for Tula may be thought of as corresponding to the district's aggregate production function.

Treatment of Labor and Machinery in TOLLAN

Seasonal variation in factor demand is captured in TOLLAN, as it is in CHAC, by means of monthly balance equations. Labor, machinery, draft

6. I would like to express my appreciation to Luz María Bassoco, Teresa Rendón, and José Silos for generously spending countless hours with me working out the mechanization alternatives.

animals, and water are treated in this way. Demand for and supply of land are registered on a semimonthly basis, and the markets for the other three inputs—seed, credit, and chemicals—are represented on an annual basis. Thus there is only one balance equation for each of these inputs per farm-size block in the matrix of technical coefficients. As in CHAC, the monthly labor balance equations are instrumental in determining the level of farmer self-hiring versus the level of day-labor hiring in each month. A similar mechanism is incorporated in the model for determining the fraction of total machine services required that will be supplied by owned machines and the fraction that will be supplied by rented machines. These two factor markets are described in more detail in the following paragraphs.

Labor

Labor demand, as created by the cropping activities, is for unskilled field-labor service over a twelve-month period.[7] This demand is on a monthly basis so that seasonal variations in demand that are characteristic of agriculture may be captured. As will be demonstrated, the composition of labor supply is determined in part by these seasonal fluctuations in demand. Tasks performed by workers fall into two categories. One involves using hand implements to do jobs that need no traction force other than the worker himself. Canal cleaning is such an activity in this model. The second category of jobs involves walking behind a mule team for such tasks as plowing. This second category is subject to mechanization.

Unskilled labor is supplied from two sources: self-employed farm family labor and hired day labor. Workers from either source are equally qualified to carry out any of the tasks in the cropping activities. Day labor may be hired on the basis of monthly contracts and is available in unlimited quantities at an exogenously specified wage rate. Farmers' self-employment is limited in each of the three farm sizes to the number of farmers in that class. They are paid a reservation wage equal to one-half of the day-labor wage rate.[8] Farmers working on medium and large farms must hire themselves on the basis of yearly contracts, whereas those on small farms may hire themselves for only a month at a time. In the model

---

7. Demand for skilled machinery operators is treated as a joint demand with machinery. Because these operators constitute less than 5 percent of the total agricultural labor force, their supply is not considered to be a bottleneck in the CHAC model. The CHAC assumption of infinite elasticity of supply at a fixed wage is used here; thus, tractor operators are included in the costs of machine services and are not treated in the activities pertaining to the labor market. For further detail, see chapter 2.

8. The justification of a reservation wage rate equal to one-half the day-labor rate is contained in chapter 2.

the labor market for medium and large farms is represented by a matrix of sixteen variables and sixteen equations; for small farms the size of the matrix is twenty-eight variables and twenty-seven equations.

The supposition that day labor is available in unlimited quantities is based on the small size of the Tula irrigation district in comparison with the agricultural area surrounding it. Also, the abundance of landless agricultural labor in Mexico supports this hypothesis. This treatment of day labor differs from that of the CHAC model. Because CHAC is national in coverage, interregional labor transfers are explicitly considered.

The base wage at which day labor may be hired is 15.80 pesos per eight-hour day, or 4,171.20 pesos per year (assuming 264 workdays per year). This wage rate was the prevailing rate in Hidalgo during the 1966-69 base-period.

The maximum number of farm family laborers available for work is limited to 1.2 times the number of farmers (heads of households) registered in each farm size. The additional 20 percent represents an estimation of the economically active family members, other than the head of household, in adult equivalents. The total number of economically active family members in the Tula area by farm size, based on this ratio, is as follows:

	Number
Small	24,960
Medium	656
Large	313
Total	25,929

Machinery

The market for services of farm machines is specified in considerably more detail in the present model than it is in CHAC. In TOLLAN the machinery market is represented by a tableau of fourteen variables and seventeen equations for each farm size plus several accounting rows and columns common to all farm sizes. A distinction is made between owned and rented machines. As part of the model's solution process, the optimal level of use of each of the two forms of machinery is computed. The way this is accomplished may be described by outlining the set of decisions a farmer must make as represented in the model. Of course, all farm decisions are interrelated (and in the model they are solved simultaneously), but for the sake of exposition a causal chain of events may be depicted as follows:

Basing his decision on current prices, past experience of uncertainty associated with price and yield variations, and land and water availability, the typical farmer in a given farm size decides on which crops to grow.

Having decided on his cropping pattern, he chooses among the several alternative techniques of production available to him. He has a choice of up to nine techniques for each crop. Each technique represents a different combination of labor, machinery, and mules used in producing a constant output per unit of land of the specified crop. He chooses these techniques in accordance with relative factor prices and, in the case of family labor (which is priced at a reservation wage below market-clearing level), in accordance with physical factor availability.

Input requirements are considered on a monthly basis. Once the farmer has decided on his cropping pattern and the techniques he will use in producing these crops, he has a determinate monthly demand for each of the factor inputs.[9] Machinery is supplied either from rental of tractor services or from owned tractors. Tractors may be rented each month in unlimited quantities at a fixed rate.[10] In addition, or as an alternative, owned tractors may be used to meet monthly demand. The cost to a farmer of renting a tractor will be greater than the variable cost of operating his own tractor *at a comparable utilization rate.* He must, however, pay fixed costs on his tractor regardless of use. Thus, he must take into account seasonal variability in demand for tractor services in making the choice whether to rent or buy. The economically rational farmer will maintain a stock of owned machines sufficient to meet his base-load demand and will rent extra machines during the months of peak load. If the base-load level changes, the endogenously determined stock of owned machines adjusts accordingly. The model, however, makes no provision for the dynamics of this adjustment process. Problems of the fixity of assets and the depletion of stocks are not considered.

The Model's Base Solution

An important objective in building this model was to obtain reasonable simulated values for several variables when all parameters were fixed at their historical base-period values. The variables for which historical data also exist are area sown in each crop, total area sown, and water use. A comparison of the simulated values for the crop variables from the model's base solution and the historical values for these variables is contained in table 13-1.

9. In fact, the cropping pattern is influenced by factor costs in the simultaneous determination of optimal input and output levels.

10. This fixed rate accounts for expenditures on fuel, maintenance, amortization of the loan taken out to purchase the tractor, the tractor driver's salary, and a profit margin for the tractor owner. The resulting rate incorporated into the model is 240.00 pesos per eight-hour shift.

Table 13-1. *Comparison of Historical Cropping Pattern and Water Use in Tula with Values Obtained from the Base Solution of TOLLAN*

Crop	Area sown (hectares)			Difference as percentage of historical value
	Historical value[a]	Model solution	Differ- ence	
Garlic	157	−158	1	0.6
Dried alfalfa	11,419	12,253	834	7.3
Green alfalfa	2,048	1,708	−340	−16.6
Oats	0	0	0	—
Forage oats	1,896	2,015	119	6.3
Onions	22	24	2	9.1
Barley	1,362	1,336	−26	−1.9
Forage barley	197	149	−48	−24.4
Dried chile	0	1	1	—
Green chile	982	1,463	481	49.0
Beans	1,225	1,426	201	16.4
Chickpeas	4	4	0	0
Lima beans	37	24	−13	−35.0
Tomatoes	1,980	2,017	37	1.9
Maize	15,287	15,603	316	2.1
Forage maize	0	0	0	—
Potatoes	14	15	1	7.1
Cucumbers	15	18	3	20.0
Wheat	8,807	7,916	−891	−10.1
Total area	45,452	46,132	680	1.5
Water use (millions of cubic meters)	900	851.8	−48.2	−5.4

— Not applicable.

a. *Source*: Secretaría de Recursos Hidráulicos (SRH, 1966–69). The historical data presented are averages for this three-cycle period.

Unfortunately, published data are not available on the level of agricultural employment; nevertheless, it is instructive to analyze the base solution's employment characteristics. The base solution presents a picture of considerable unemployment among farmers and their families[11] in the Tula irrigation district. Furthermore, all of this unemployment occurs in the small-farms (0–5 hectares per farm). Farmers on the medium (5–10 hectares) and large (10 hectares and more) farms experience no permanent unemployment and no seasonal unemployment. These data are presented in table 13-2. Row 7 of this table contains the yearly average unemployment rate—which is 59 percent for the small farmers, zero for the medium and large farms, and 56 percent for the district as a whole. This yearly rate

11. The term "farmers" will henceforth be understood to include not only the head of the farm household but the economically active family members as well.

Table 13-2. *Farmer Employment in the Base Solution of TOLLAN*
(man-years per year unless otherwise noted)

| Hiring and unemployment | Farm size | | | |
	Small	Medium	Large	All
1. Farmer self-hiring[a]	10,320	656	313	11,289
2. Maximum monthly farmer employment[b]	15,654	—	—	—
3. Minimum monthly farmer unemployment[c]	—	0	0	0
4. Total farmer endowment[d]	24,960	656	313	25,929
5. Permanent farmer unemployment[e]	9,306	0	0	9,306
6. Yearly average unemployment[f]	14,640	0	0	14,640
7. Yearly average unemployment rate[g]	59	0	0	56
8. Land-to-farmer endowment (hectares per man)[h]	1.10	6.01	20.80	1.46
9. Day-labor hire	0	217	432	650
10. Total labor use[i]	10,320	873	745	11,939
11. Labor-to-land ratio (man-years per hectare-year)[j]	0.417	0.268	0.261	0.388

— Not applicable.

Note: The source for all data, except those appearing in line 4, was the base solution of TOLLAN, with all exogenously determined parameters fixed at base-period historical values.

a. In the case of small farmers, the datum is the average level of employment over the twelve-month period according to the monthly fluctuations in farmer self-hiring. In the case of medium and large farmers, the data represent the number of man-years of self-employed farmers on a yearly contract basis.

b. The number of small farmers hired in the month of peak demand for their services in units of man-months. (This row is inapplicable to medium and large farms.)

c. The number of man-months spent idle by medium and large farmers in the month of maximum demand for labor services on the medium and large farms, respectively. (This row is inapplicable to small farms.)

d. *Source*: SRH, unpublished information on number of farmers by farm size. Those data were adjusted by the author to include economically active family members.

e. Row 4 minus row 2 in the case of small farms. Row 4 minus row 3 in the case of medium and large farms.

f. Row 4 minus row 1.

g. Row 6 as a percentage of row 4.

h. Land data are from table 13-3, row 13, divided by row 4 of this table.

i. Row 9 plus row 1 (minus $\frac{1}{12}$ of row 3 for medium and large farms).

j. Row 10 divided by row 14 of table 13-3.

is averaged over seasonal fluctuations in demand for labor services. This 59 percent unemployment rate on small farms coincides closely with other estimates of unemployment for small farms and ejidos[12] for the country as a whole. The Centro de Investigaciones Agrarias derived a 60 percent

12. The average size of an ejido is well within the 0–5 hectare range. Therefore, ejidos are treated here as part of the "small farm" size category.

Figure 13-1. *Seasonal Pattern of Labor Use on Small Farms, Base Solution*

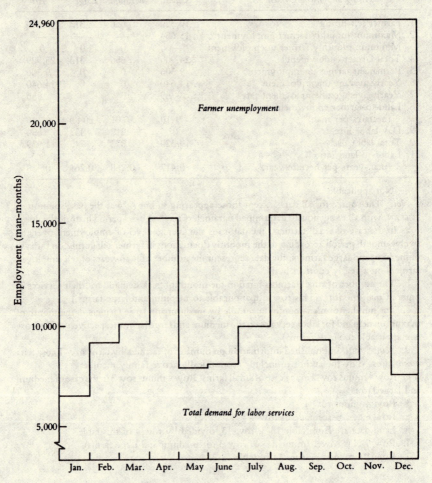

Note: The top horizontal line represents the total availability of labor from farmers on small farms.

unemployment rate on these farms for 1960.[13] Because the Center's estimate differs both in methodology and in coverage from that of this study, the coincidence of the two estimates should be interpreted as indicating the reasonableness of the present estimate but not as corroborating its exact value.

Though small farms have the least land per farmer, taken together they have most of the land (72 percent of the total) and most of the farmers (96 percent of the total), so conditions on these small farms strongly influence the overall employment situation in the district. There is only 1.1 hectare of land per full-time family laborer on small farms, whereas the corresponding numbers for medium and large farms, respectively, are 6.0 and 20.8 hectares.

The effect of these relative endowments of land in each farm size is manifest also in the type of labor employed. Row 9 of table 13-2 indicates the total number of man-years of hired day labor employed by each farm size. Day labor, which is twice as expensive as farmer labor, is not hired at all on small farms. Medium and large farms, however, employ day labor in large amounts. Day labor accounts for 25 percent of all labor use on medium farms and 58 percent on large farms.[14] This result follows from the observation that, in the month of peak labor demand on small farms, there is work for only 63 percent of the farmers,[15] whereas during every month of the year the supply of medium and large farmers is exhausted. In months of peak demand for labor services on medium and large farms, day labor therefore must be hired.

Figures 13-1 through 13-3 illustrate these monthly fluctuations in labor demand the way in which they are met from farmer and day-labor sources. Demand for labor in each farm size is created by the cropping activities, and labor requirements for each crop vary from month to month. Therefore, total monthly demand for labor will also vary, although not as extremely as the demand created by any one crop. The cropping pattern in the base solution creates peak demand for labor on small farms in August, on medium and large farms in July. Other months of strong demand are April and November. The extent of seasonal unemployment and permanent unemployment on small farms is readily apparent in figure 13-1. This contrasts sharply with figures 13-2 and 13-3, in

13. See Centro de Investigaciones Agrarias (1970). The unemployment rate is 1 minus the quotient of total man-years required divided by total man-years available according to the 1960 census.

14. This figure is derived by taking row 9 of table 13-2 as a percentage of row 10 in the table.

15. This figure is derived by taking row 2 of table 13-2 as a percentage of row 4. See also figure 13-1.

Figure 13-2. *Monthly Labor Use on Medium-size Farms, Base Solution*

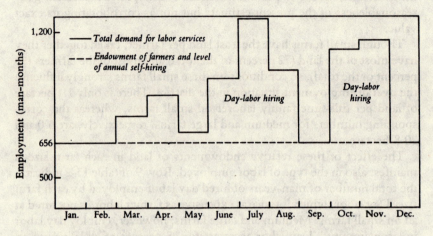

Note: The top horizontal line represents the total availability of labor from farmers on medium-size farms.

Figure 13-3. *Monthly Labor Use on Large Farms, Base Solution*

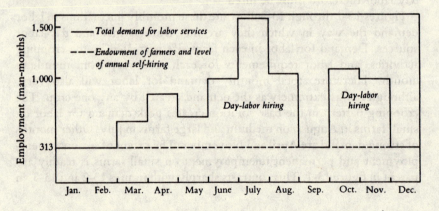

Note: The top horizontal line represents total availability of labor from farmers on large farms.

which all the farmers are employed continuously and a large day-labor component is also used.

Income Distribution in TOLLAN

A crude income distribution may be derived in TOLLAN from the calculated income levels of farmers in each farm size. (Chapter 5 of this book reports a comparable exercise for the entire sector.) It would have been desirable to stratify the farmers so that the three groups contained more nearly equal proportions as suggested, for example, by Ahluwalia (1974), but the available data on land distribution forced the particular stratification used here. Farm income levels from TOLLAN are presented in tables 13-3 and 13-4.

The disparity in mean incomes by farm size is readily apparent. Because small farmers constitute the vast majority (96 percent) of all farmers, their per capita income is close to the average. Meanwhile, farmers on medium and large farms are earning from two-and-one-half times the average per capita income, depending on which measure is used. On the one hand, only the small farmers receive implicit rent payments for their land because it is only on these small farms that land is fully utilized in some fortnights and thereby has a positive shadow price.[16] On the other hand, farmers on medium and large farms are earning shadow wages because they are fully employed throughout the year. (See row 8 of table 13-3.) Thus, labor income alone suggests a more skewed distribution of income than do the data on total income received by farmers.

It is difficult to extend these findings to day labor because the number of day laborers in the district and their alternative sources of income are unknown. Yet, it can be assumed that the supply of available day labor is just sufficient to meet the peak labor requirement of 1,872 men in July (see figures 13-2 and 13-3) and if it also can be assumed that these workers have no alternative source of income, then the conclusion that the average income per day laborer is 1,447 pesos per year can be made. This income level places the day laborer in a position superior to farmers on holdings of 0–5 hectares yet distinctly worse off than the medium or large farmers. In reality, some of the small farmers work part-time as day laborers on other farms, so this interpretation is difficult to maintain.

Given that the composition and size of this day-labor population is so ambiguous at the level of the irrigation district, an alternative working hypothesis is that the small farmers provide all the day-labor services

16. See row 10 ("Other rents accruing to farmers") of table 13-3. Irrigation water was not a binding constraint in this solution, so these rents pertain exclusively to land.

Table 13-3. *Employment, Income, and Land Use by Farm Size for the Base Solution of TOLLAN*

Employment, wages and income, and land use	Farm size			
	Small	Medium	Large	All
Employment (man-years)				
1. Farmers[a]	10,320	656	313	11,289
2. Unemployment rate[b]	59	0	0	56
3. Day-labor hire[c]	0	217	432	650
4. Total employment[d]	10,320	873	745	11,939
Wages and income (thousands of pesos)				
Wage payments				
5. Day-labor wage bill	0	907	1,803	2,710
6. Farmer reservation wage bill	21,524	1,368	653	23,545
7. Total wage bill	21,524	2,275	2,455	26,254
Farmer income				
8. Farmer shadow wage bill[e]	0	918	439	1,357
9. Total farmer labor income[f]	21,524	2,286	1,092	24,902
10. Other rents accruing to farmers[g]	12,015	0	0	12,015
11. Total farmer income[h]	33,539	2,286	1,092	36,917
12. Total district income[i]	33,539	3,193	2,895	39,627
Land use				
13. Land endowment (hectares)	27,497	3,944	6,509	37,950
14. Hectare-years cultivated	24,652	3,262	2,854	30,767
15. Utilization rate (percent)	90	83	44	81

Source: Base solution of TOLLAN.

a. See row 1, table 13-2.

b. See row 7, table 13-2.

c. See row 9, table 13-2.

d. See row 10, table 13-2.

e. The farmer shadow wage bill is the portion of profit (economic rent) accruing to his labor, above and beyond his implicit own-wage payment (reservation wage).

f. Total farmer labor income is defined as the reservation wage bill plus the shadow wage bill.

g. These are the portion of profits attributable to rents earned on fully utilized land and water.

h. Row 9 plus row 10.

i. Row 5 plus row 11.

required by the medium and large farms. Even during the month of peak demand for day workers, there are more than enough unemployed small farmers to carry out these tasks. (See figures 13-1 through 13-3.) If in fact all day labor is supplied by small farmers, then the two become one labor class, and the earnings of day laborers may be added to those of the small farmers to give these farmers a total annual income of 1,454 pesos per farmer. Thus, even if small farmers were to be the only source of day

Table 13-4. *Income Distribution over the Three Classes of Farm Size*

	Farm size			
	Small	Medium	Large	Average
Labor income per farmer				
(pesos per year)[a]	862	3,485	3,488	960
Percent of average	86	363	363	
Total farm income per farmer				
(pesos per year)[b]	1,344	3,485	3,488	1,424
Percent of average	94	245	245	
Percentage of all farmers				
represented by farmers				
in this size class[c]	96	3	1	

Source: Tables 13-2 and 13-3.

a. Total labor income (row 9 of table 13-3) divided by the total number of farmers (row 4 of table 13-2). Note that "Total farmer endowment" in table 13-2 includes the head of the farm plus economically active members of the household (as discussed in the section "Treatment of Labor and Machinery in TOLLAN").

b. Total farmer income (row 11 of table 13-3) divided by the total number of farmers.

c. Number of farmers in each group divided by the total number of farms in the district.

labor, their augmented income would still be substantially below the income of medium and large farmers.

Derived Demand for Labor

To determine the characteristics of the demand for farm labor as represented by the model, optimal solutions were computed under alternative assumptions about the wage rate. The model was solved for forty-one values of the wage rate ranging from 0 percent to 200 percent of the historical base-period rate.[17] All other parameters in the model were maintained at their base-period values while the wage rate was increased in 5 percent steps from 0 percent to 200 percent of the base rate. This procedure is equivalent to shifting the labor supply curve along the curve for implicit derived demand for labor, with the result that the latter is traced out explicitly.[18]

17. The base-period wage rate for day labor is 15.80 pesos per eight-hour day, or 4,171.20 pesos per year (assuming 264 work days per year).

18. Strictly speaking, the locus of equilibrium points in labor's price and quantity space is a factor response rather than a factor demand curve because other prices are not held constant. Each point on the curve represents a general equilibrium solution to the model in which other endogenously determined prices and quantities are free to vary. Specifically, no index of output was held constant in these wage-rate sweeps.

Figure 13-4. *Overall Employment Response Function: Total Gainfully Employed Labor on All Farms as a Function of Labor Unit Cost*

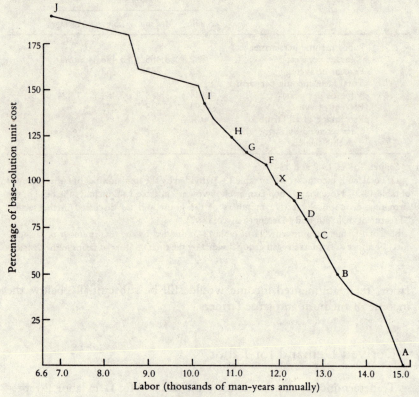

Note: The base-solution unit cost was 2,313 pesos per man-year.

Figure 13-4 is a graph of labor use plotted against the unit cost of labor.[19] Labor use is measured in thousands of man-years of gainful employment. In other words, only the time actually worked by farmers and day laborers is measured. Seasonally employed, self-hired farmers on the medium and large farms are not counted in the months when they are not employed. Point X on the curve represents the quantity and unit cost of labor in the base-period solution. The curve is essentially linear in the region around

19. The unit cost is a weighted average of the day-labor wage rate and the farmers' reservation wage rate; the weights are the proportion of each group in the total labor force. In addition, the farmers' "shadow wage" earned when they are fully utilized is included in the computation of unit cost. This "shadow wage" is equivalent to that portion of their profits attributable to their scarcity.

Table 13-5. *Arc Wage Elasticities of Derived Demand
for All Labor on All Farms*

Arc end points	Associated percentage of base wage	Elasticity
Group 1: Arcs of increasing length centered on base wage		
E–F	90–110	−0.28
D–G	80–120	−0.31
C–H	70–130	−0.30
A–J	50–200	−0.38
Group 2: Arcs extending upward from base wage		
X–F	100–110	−0.26
X–G	100–120	−0.38
X–H	100–130	−0.40
X–I	100–200	−0.42
X–J		−0.91
Group 3: Arcs extending downward from base wage		
X–E	100–90	−0.29
X–D	100–80	−0.26
X–C	100–70	−0.22
X–B	100–50	−0.17
X–A	100–0	−0.11

Source: Data from general-equilibrium wage-rate sweep (national product price elasticities) in model's solution. The letters A through J are keyed to the points in figure 13-4.

the base-period point. The nonlinearities at the extreme points may be
because of the increasing bias inherent in the model as it is forced to
simulate situations in which parameters are changed to values differing
greatly from their base-period values.

Table 13-5 contains arc wage elasticities of derived demand for labor.
Three groups of such elasticities are presented: the first represents arcs of
increasing length centered about the base wage and quantity point X; the
second group depicts arcs extending upward from that point; and the third
group corresponds to arcs extending downward from point X. The letters
alone the curve in figure 13-4 indicate the end points of the arcs listed in
table 13-5. All elasticities are computed at the midpoint of the arc.

The values for the elasticities of the arcs centered about the base unit cost
and quantity are essentially uniform, with values between −0.27 and
−0.30. The exception is the arc connecting the two extreme points that
has a value of −0.38. Given that structural rigidities in the model bias
simulations of extreme points, more confidence should be placed in the
−0.27 to −0.30 figures as indicative values about the base point.

The values of the arcs extending upward and downward demonstrate
the essential linearity of the curve: elasticities at higher points on the curve
are greater than those at lower points.

The relatively low values for the wage elasticities of demand for labor are in part a reflection of the low price elasticities of demand for crops that are incorporated in the model. As discussed above, national product price elasticities are used that have a mean value of approximately -0.40. Thus, shifts in the crop supply schedules brought about by variations in wage rates result in relatively small changes in the quantity of crops demanded. There is little opportunity at the level of the irrigation district for changes in crop composition toward (or away from) labor-intensive crops as wages fall (or rise).[20]

From the policymaker's perspective, perhaps the most significant aspect of the labor response curve is that it demonstrates the difficulty of significantly affecting employment by small adjustments in the wage rate. All wage elasticities are less than unity, and those in the neighborhood of the base price and quantity are quite low, ranging from -0.20 to -0.40.[21]

The total employment response function shown in figure 13–4 and table 13–5 may be decomposed into sources of variations. In TOLLAN, total employment is determined by two factors: the labor per hectare used to grow each crop and the number of hectares planted in each crop. A change in the first factor indicates a change exclusively in cultivation technique. A change in the second factor indicates a change in crop composition, a change in total area sown, or both.

To quantify these components of change in total labor use, the following identity between index numbers is used:

$$(13.1) \qquad \underbrace{\left[\frac{\sum\limits_{i=1}^{n} L_{it} A_{it}}{\sum\limits_{i=1}^{n} L_{io} A_{io}} \right]}_{[\text{LHS}]} \equiv \underbrace{\left[\frac{\sum\limits_{i=1}^{n} L_{io} A_{it}}{\sum\limits_{i=1}^{n} L_{io} A_{io}} \cdot \frac{\sum\limits_{i=1}^{n} A_{io}}{\sum\limits_{i=1}^{n} A_{it}} \right]}_{[\text{RHS1}]}$$

$$\cdot \underbrace{\left[\frac{\sum\limits_{i=1}^{n} L_{it} A_{it}}{\sum\limits_{i=1}^{n} L_{io} A_{it}} \right]}_{[\text{RHS2}]} \cdot \underbrace{\left[\frac{\sum\limits_{i=1}^{n} A_{it}}{\sum\limits_{i=1}^{n} A_{io}} \right]}_{[\text{RHS3}]},$$

where

L_{it} = The amount of labor per hectare necessary to grow crop i in solution t (L_{io} is the requirement for the base solution)

---

20. If these same variations in wage rate are carried out under conditions of local crop price elasticities, the wage elasticities approximately double in value. See Howell (1974, pp. 230–39, especially table 5–7).

21. The effect on income distribution of a change in wages is discussed in the section "Implications of the Elasticities . . . ," below, in which it is shown that raising wages may actually worsen the distribution in these circumstances.

A_{it} = The number of hectares sown in crop i in solution t (A_{io} is the area in the base solution)

n = Number of crops grown.

The terms in brackets are defined as follows:

LHS = An index of change in total labor use

RHS1 = A Laspeyres index of change in labor use because of changes in crop composition (deflated for changes in total area)

RHS2 = A Paasche index of change in labor use because of changes in technique

RHS3 = An index of change in total area.

Effects of changes in total area sown and changes in crop composition are separated by considering the latter deflated by the former and then considering the former separately—namely, RHS1 and then RHS3. The indexes could have been defined equivalently in other ways, but this form is convenient for computational purposes. Values for overall index and its components as defined in the equation-identity (RHS1) are presented in table 13-6. In addition, a Laspeyres crop production index is presented so that effects on output of variations in the wage rate may be observed. Values of the indexes were computed for each of the three farm sizes as well as for the district as a whole for each of five wage rate levels other than the base wage.

Looking first at the index values for the district as a whole, denoted by "All" in the farm-size stub of table 13-6, we see that changes in labor use are entirely because of changes in total area and changes in technique. Essentially, none of the change is because of changes in crop composition once that figure is deflated for changes in total area. This is reasonable given the low elasticities of product demand, as discussed above. The crop production index rises as wage rates are lowered and falls as the rates are raised, demonstrating the effect on output of decreasing and increasing costs in an environment of price-responsive demand. Because there is essentially no change in crop composition, the area index follows the production index almost exactly.

An interesting difference in the behavior of the Paasche index for change in technique may be observed when comparing its behavior in response to lowered wages with that when wages are raised. The technique index increases in response to a wage reduction less than it decreases in response to a comparable wage increase.[22] This may be explained by determining approximately where, on the irrigation district's envelope isoquant, pro-

22. It should be observed that in this case the Paasche index for change in technique essentially collapses to a Laspeyres index because of the negligible variation in crop composition; that is, $A_{io} = kA_{it}$, where k reflects changes in total area sown in crop i and is constant for all i. Thus, the Paasche index values may be compared with each other in the manner that Laspeyres indexes are.

Table 13-6. *Sources of Change in Total Labor Use by Farm Size*

Farm size	Arc end point and associated percentage of base wage					
	A (0)	B (50)	X (100)	F (110)	I (150)	J (200)
Index of total labor use						
Small	1.085	1.074	1.000	0.972	0.883	0.553
Medium	1.563	1.114	1.000	1.014	0.794	0.726
Large	3.031	1.652	1.000	0.973	0.603	0.480
Total	1.242	1.113	1.000	0.975	0.859	0.561
Laspeyres index of change in labor use because of changes in crop composition (deflated for changes in total area)						
Small	0.941	0.931	1.000	1.001	0.952	0.936
Medium	1.077	0.896	1.000	1.022	1.093	1.138
Large	1.390	0.950	1.000	0.989	1.295	1.355
Total	1.009	1.011	1.000	1.020	1.020	1.016
Paasche index of changes in labor use because of changes in techniques						
Small	1.163	1.177	1.000	0.973	0.907	0.581
Medium	1.554	1.304	1.000	0.992	0.709	0.627
Large	1.469	1.347	1.000	0.985	0.620	0.487
Total	1.181	1.086	1.000	0.995	0.849	0.561
Total area index						
Small	0.992	0.981	1.000	1.000	1.023	1.016
Medium	0.993	0.953	1.000	1.000	1.025	1.018
Large	1.484	1.291	1.000	0.999	0.752	0.727
Total	1.042	1.014	1.000	1.000	0.992	0.983
Laspeyres crop production index (using historical base-period prices)						
Small	0.913	1.020	1.000	1.000	0.959	0.920
Medium	1.098	0.783	1.000	1.015	1.196	1.310
Large	2.471	1.253	1.000	0.974	1.120	1.303
Total	1.035	1.014	1.000	1.000	0.991	0.981

Source: Data from general-equilibrium wage-rate sweep (national product price elasticities) in model's solution. The letters A through J are keyed to the points in figure 13-4.

duction is taking place in the base solution. In other words, how much labor substitution is possible when wage rates are lowered? (When wage rates are raised?) The answer is that there is relatively little opportunity for increased labor use when wage rates are lowered as compared with opportunities for increased machinery use (and decreased labor use) as wages are raised.

From the Paasche technique index broken down by farm size, it may be seen that small farms are responsible for this asymmetry in technique-change flexibility when comparing wage increases with decreases. Small farms are operating at the labor-intensive end of their crop isoquants at the base wage rate, and there is therefore little opportunity for switching to techniques that use even more labor. When wages are raised, however, employment drops off relatively easily given the broad range of machine-intensive techniques available to the small farmer.

Medium and large farms, in contrast, show approximately equal absolute changes in labor use in response to a given absolute change in the wage rate. As could be expected, they are much more mechanized at the base wage rate given their relative scarcity of cheap, farm family labor. Thus, as wage rates are lowered they have more opportunity for switching technique. As rates are raised, they are able to reduce labor use through technique changes with comparable ease because several crops, notably maize, are cultivated according to intermediate techniques in the base solution. For the district as a whole, the employment effects are dominated by the effects in the small farm group.

Isoquants and General Equilibrium Substitution

Traditionally input substitution is depicted in terms of isoquants. The Hicks-Robinson elasticity of substitution provides a measure of the freedom with which one input may be substituted for another.[23] In this chapter, measurements of machinery-labor substitution are possible in a context quite different from the traditional production function framework. Rather than two factors of production and one output, the present model considers eight inputs and nineteen outputs. Thus, the implicit isoquants are segmented linear surfaces in eight dimensions of space rather than smooth lines in two dimensions. The segmented nature of the isoquants follows from the limited and discrete number of substitution possibilities inherent in a process-analysis model. To measure machinery-labor substitution in the present context, several modifications of the traditional approach are necessary. (Chapter 5 of this volume addresses these same issues at the sectoral level.)

The first problem to be considered is how an isoquant should be defined in a multiproduct environment. Second, there is the question of whether an isoquant is really the path of factor substitution of greatest interest to the policy analyst. It is possible to trace out a set of points in input space each of which represents a general equilibrium position replicated by the model in which output and price levels of each of the nineteen crops are free to achieve their equilibrium levels. The resulting general equilibrium

23. See Hicks (1963) and Robinson (1969).

substitution locus (GESL) of one factor for another is perhaps of greater interest than an isoquant. A third question is how the elasticity of substitution should be measured when the isoquants (or GESL) are not smooth; that is, not twice differentiable. These and related issues are discussed in the following paragraphs.

Experiments with the model to determine the degree of substitutability between factors, specifically labor and machinery, are based on a particular characteristic of the model. For the given set of factor prices, technical coefficients of production, physical constraints, and product demand schedules, the optimal solution of TOLLAN represents a minimum-cost combination of inputs. This is because of the behavioral assumption built into the objective function of producers' profit maximization under conditions of perfect competition. Thus, the set of levels of input use associated with each optimal solution represent points on an isoquant. They are the least-cost combinations of factors of production for the given level of output. In essence, a point of tangency between the isocost plane and the isoquant surface has been found. By varying the price of one of the inputs, the slope of the isocost hyperplane is changed. A new solution to the model yields another point on the same isoquant if output is artificially held constant. If output is not held constant, the "point of tangency"[24] most probably will be on an isoquant for a different output level. By repeatedly changing the price of one input, a section of an isoquant or, in the case of unrestricted output, of a GESL, may be traced out.

Before presenting results, we must define what is being held constant in the case of tracing out an isoquant. A Laspeyres quantity index of the nineteen crops produced was the quantity held constant, at its base-solution level, while input prices varied. The price weights used are the three-year averages of the historical prices recorded in the irrigation district during the study period. This definition of an isoquant permits individual crop quantities to vary, thus allowing for changes in input use because of changes in crop composition.[25]

In this chapter, machinery-labor substitution possibilities are studied under four different conditions. These arise from varying the wage rate and from varying a component of machinery cost, namely the interest rate. The wage and interest rates are assigned different values under the conditions of no restriction on output and output that is restricted to a constant value. The four circumstances for machinery-labor variation are therefore:

24. It is possible that the isocost plane touches the isoquant surface on one of the latter's linear facets. In this case, the levels of input use are indeterminant within a specified range.

25. This index is identical to the "crop production index" for all farms that is used in table 13-6. In that application, the accounting variable in the model representing the index value is unrestricted. In the isoquant exercise it is "point bounded" at its base-solution level.

Figure 13-5. *Machinery-Labor Substitution on All Farms as Induced by Variations in the Wage Rate*

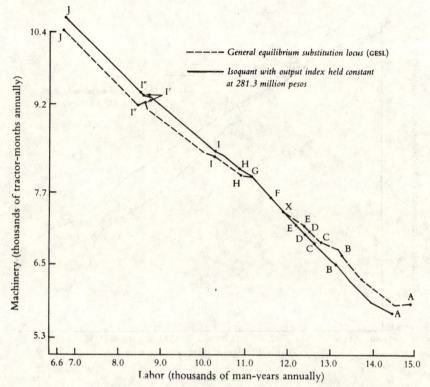

- Variation of wages allowing the model to achieve a general equilibrium for each wage-rate value
- Variation of wages but restricting an index of output to a constant level
- Variation of the interest rate allowing the model to achieve a general equilibrium for each value of interest (the long- and short-term rate are varied at the same rate)
- Variation of the interest rate but restricting an index of output of a constant level.

Figure 13-5 demonstrates TOLLAN's reactions to labor-machinery substitution as induced by variations in the wage rate, and figure 13-6 demonstrates this substitution as induced by interest rate variation.[26] Both the

26. In each case, only the wage or interest rate is altered. Other prices are either left at their base-solution, exogenous values or are left to be determined endogenously by the model, as in the base solution.

Figure 13-6. *Machinery-Labor Substitution on All Farms as Induced by Variations in the Interest Rate*

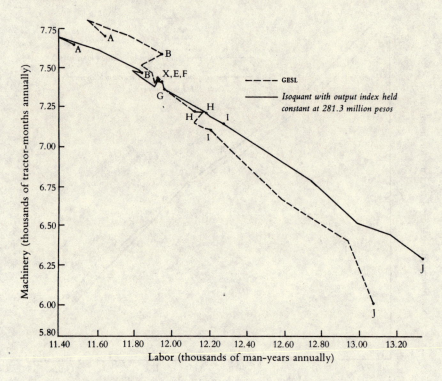

"isoquant" and the GESL traced out by the wage-rate sweep are shown in the figures. In figure 13-5, they both are downward sloping throughout except for a small segment of the GESL in its upper portion. The curves are less well behaved, however, in the second-order conditions. Neoclassical theory stipulates that the isoquant should be convex to the origin, although no such requirement is made of the GESL. The isoquant in figure 13-5 is not convex, and it is even more nonconvex in figure 13-6. In general, these kinks and nonconvexities may be attributed to characteristics of the model with which the substitution loci were derived. To the extent that this form of process-analysis model does a better job than the standard neoclassical model in capturing the complexities of observed production relations, these substitution paths may be better representations of actuality than are the smooth, convex isoquants posited by neoclassical theory. (Again, see chapter 5 for a related discussion.)

It is useful to quantify the degree of substitutibility of machinery for labor in these results, and, naturally, some measure of the elasticity of factor substitution comes to mind. In a world of discontinuous substitution paths and more than two inputs, however, there is a problem of definition of the elasticity. In the next few paragraphs, a definition of the elasticity is presented, and numerical results are subsequently reported.

The standard definition of the elasticity of factor substitution is essentially that given by Hicks (1963) and Robinson (1969). At a point along a smooth isoquant of a production function of the type $x = f(a, b)$, the elasticity is defined as[27]

$$e = \frac{\dfrac{a}{b} \, d\left|\dfrac{b}{a}\right|}{\dfrac{1}{r} \, dr},$$

where a and b are the two factors of production and $r = (\partial x / \partial a) / (\partial x / \partial b)$—that is, the marginal rate of factor substitution (MRS). In a perfectly competitive equilibrium, it is known that the MRS is equal to the factor price ratio.

In the current case this definition is inapplicable because elasticities cannot be computed at all points because of the piecewise-linear, rather than smooth, curvilinear isoquants implicit in a process-analysis model. The MRS is undefined at the "kink" between two segments. Rather than point elasticities, arc elasticities are computed here. A second reason the standard definition of elasticity of substitution is not strictly relevant is that it assumes only two factors of production. As observed in the preceding section, there are six other factors of production in TOLLAN besides machinery and labor. These factors are not held constant in the solutions, and their variation undoubtedly affects the values of the elasticities reported. Hicks (1963, p. 381) has shown that, in the case of multiple inputs, the elasticity of substitution will be higher when the other inputs are free to vary than when they are bounded. Although the circumstances of this chapter do not conform completely to those Hicks referred to, his conclusion is generally valid here also.

The definition of the elasticity of substitution that is used here is basically that of an arc elasticity computed at its midpoint. Specifically, it is:

$$e = \frac{(k_1 - k_2) / (k_1 + k_2)}{(\text{MRS}_1 - \text{MRS}_2) / (\text{MRS}_1 + \text{MRS}_2)},$$

27. This notation follows the presentation in Allen (1962).

where k_i is the quotient of machinery services used in solution $i \div$ labor services used in solution i, and MRS$_i$ is the quotient of labor's unit cost in solution $i \div$ machinery's unit cost in solution i.

Because the marginal rate of substitution is undefined at the segment end points, the device of using the inverse of the factor cost ratio is employed. This is based on the assumption that, if the factor substitution loci were smooth, the slope of a tangent to the curves at the equilibrium machinery-labor ratio would be equal to the factor price ratio.[28] Since these elasticities of substitution are not derived from a neoclassically specified model, the reader may object to their being called "elasticities of factor substitution" in the Hicks-Robinson sense. The objection may be particularly strong to applying the term to measures of substitution along the GESL because output is not held constant in this case. The view taken here is more agnostic. These values are measures of the percentage change in the machinery-labor ratio divided by the percentage change in the labor-cost-to-machine-cost ratio. They are called elasticities of factor substitution for want of a better term and because they are fashioned after the Hicks-Robinson elasticities.

Tables 13-7 and 13-8 contain elasticities of factor substitution over selected arcs of the isoquant and GESL as derived from the wage-rate and interest-rate sweeps, respectively. In each table, three groups of elasticities are presented. The first group contains the values for elasticities of substitution over arcs of increasing length that are centered on the base-period, machinery-labor ratio. The second and third groups contain values for arcs extending upward and downward, respectively, from the base ratio.

Two features of the elasticities reported in tables 13-7 and 13-8 stand out. The elasticities are far from constant, and they are generally less than unity. Furthermore, variation among the elasticities within each group is greater for the interest-rate sweep than for the wage-rate sweep. For instance, the isoquant elasticities by segment in group 1 vary from 0.072 to 0.987 in the case of the interest-rate sweep, but only from 0.504 to 0.663 in the case of the wage-rate sweep. Contrasts between the values in groups 2 and 3 are even more striking. Between points X and D (group 2) in the "interest-rate" isoquant there is even a negatively valued elasticity.

The kinks in the factor substitution paths, which are most prevalent in the interest-rate sweep, are the cause of the variations in the values for the elasticities pertaining to this sweep. A small change in the cost of machine-

---

28. This assumes that "entrepreneurs are able to change their position on the [production] function freely, that they wish to maximize profits, and that the prices of inputs are competitively determined" (O'Herlihy 1972, p. 264). These assumptions are substantially met in the model.

Table 13-7. *Arc Elasticities of Factor Substitution Derived from the Wage-rate Sweep*

Arc end points	Associated percentage of base wage	Elasticity Isoquant	Elasticity GESL
Group 1: Arc of increasing length centered on the base machinery-labor ratio			
E–F	90–110	0.562	0.574
D–G	80–120	0.663	0.645
C–H	70–130	0.541	0.545
B–I	50–150	0.504	0.506
A–J	0–200	0.591	0.593
Group 2: Arcs extending upward from the base machinery-labor ratio			
X–F	100–110	0.546	0.566
X–G	100–120	0.782	0.782
X–H	100–130	0.721	0.733
X–I	100–150	0.728	0.742
X–J	100–200	1.361	1.386
Group 3: Arcs extending downward from the base machinery-labor ratio			
X–E	100–90	0.559	0.580
X–D	100–80	0.512	0.533
X–C	100–70	0.406	0.404
X–B	100–50	0.344	0.340
X–A	100–0	0.216	0.220

Source: Data from general–equilibrium and isoquant wage-rate sweeps (national product price elasticities) in model's solution The letters A through J are keyed to the points in figure 13-5.

ry may have a greater effect on the use of other inputs than it does on machinery. The model is composed of many distinct production activities and factor supply activities. If the changes one makes in the exogenous parameters are small enough, one begins to observe marginal effects that make sense only in relation to the myriad of activities incorporated in the model. When short ranges are considered, these indirect adjustments may overpower the expected basic relationship of substituting cheap for costly inputs. Over longer ranges, these "anomalies" are submerged by the basic substitution relation.

The elasticities reported in tables 13-7 and 13-8 are sensitive to these indirect effects because they are not computed with all other conditions being equal. When these indirect effects dominate the direct adjustments of factor use and cause other inputs to be used in altered proportions, the elasticities of substitution will be altered accordingly. Table 13-8 (the interest-rate sweep) contains elasticities derived under such conditions,

Table 13-8. *Arc Elasticities of Factor Substitution Derived from the Interest-rate Sweep*

Arc end points	Associated interest rate (percent)	Elasticity Isoquant	GESL
Group 1: Arcs of increasing length centered on the base machinery-labor ratio			
E–F	11–13	0.072	0.028
D–G	10–14	0.143	0.066
C–H	9–15	0.987	0.426
B–I	5–20	0.550	0.365
A–J	0–40	0.858	0.754
Group 2: Arcs extending upward from the base machinery-labor ratio			
X–E	12–11	0.003	0.016
X–D	12–10	−0.096	0.000
X–C	12–9	0.410	0.110
X–B	12–5	0.269	0.137
X–A	12–0	0.570	0.323
Group 3: Arcs extending downward from the base machinery-labor ratio			
X–F	12–13	0.136	0.123
X–G	12–14	0.344	0.338
X–H	12–15	1.547	1.560
X–I	12–20	0.780	0.783
X–J	12–40	0.976	1.037

Source: Data from general-equilibrium and isoquant interest-rate sweeps (national product price elasticities) in model's solution. The letters A through J are keyed to points in figure 13-6.

while table 13-7 (the wage-rate sweep) displays elasticities for which these conditions are relatively unimportant because of the longer arcs over which these latter elasticities are measured. For purposes of policy analysis, the elasticities from the wage-rate sweep are more useful than those from the interest-rate sweep because they are not plagued by these technical problems. Yet, if one is interested in small changes in the interest rate, and one believes that the model's specification is faithful to the real world, then the elasticities from the interest-rate sweep are relevant. It is hard to defend the position that the model is so well specified that every small anomaly in the model's behavior occurs in exactly the same way in the real world. Because of this, more credence should be placed in the elasticities from the wage-rate sweep.

The present model of Tula is closely linked to the CHAC model, as previously demonstrated. The TOLLAN elasticities of machinery-labor sub-stitution, however, are substantially lower than those of the CHAC model. CHAC yields values uniformly above unity, and some as high as 5.9 depending on the arc (see chapter 5). The estimates of the machinery-labor elasticity of substitution made with TOLLAN are in general below unity, ranging from 0.344 to 0.782.[29] The reason for the CHAC–TOLLAN difference in results is to be found in the difference in geographical coverage of the two models. CHAC is national in scope, whereas TOLLAN includes only 0.26 percent of the total area planted in short-cycle crops in Mexico. There are more possibilities for machinery-labor substitution in CHAC than in TOLLAN, including substitution via interregional shifts in cropping patterns, since factor intensities for a given crop vary over regions. Thus, a dimension of machinery-labor substitution that exists in CHAC is largely absent in the TOLLAN model. This absence, which is largely justified by the geographic homogeneity of TOLLAN, results in fewer possibilities for substitution and, therefore, lower elasticities of substitution in TOLLAN than in CHAC.

Another difference between CHAC and TOLLAN, which may also be attributed to their differences in geographical coverage, is the international trade feature of the former. The CHAC model permits export of some crops when the domestic price falls below the international f.o.b. price. Once a crop is sold on the international market, its demand is infinitely elastic at that price until the volume sold abroad reaches an export quota if the crop is so regulated. Thus, in the case of internationally traded crops there is much greater latitude for changes in crop composition in response to changes in relative costs of production. In TOLLAN no such possibility of export at fixed prices is considered. Also, fewer crops are included in TOLLAN than in CHAC. Changes in crop composition are limited to nineteen crops in TOLLAN, whereas there are thirty-three crops included in CHAC. Thus, changes in the composition of national crop production in CHAC are achieved relatively freely in comparison with changes in crop composition in TOLLAN. As is demonstrated above, changes in crop composition are an important source of changes in factor demand.

In summary, the increased possibilities for technical substitution by means of interregional changes in crop production in CHAC, plus its more flexible demand structure, are the principal explanations for the difference in the values of the elasticity of machinery-labor substitution estimated from the CHAC model as compared with those from the TOLLAN model.

---

29. The extreme values for arcs $X–A$ and $X–J$ in groups 2 and 3, respectively, in table 13-7 fall outside this range. These are considered special cases and are relevant only to extreme variations in wage rates.

Implications of the Elasticities for Farm Wage Policy and Income Redistribution

Elasticities of capital–labor substitution have a direct bearing on farm wage policy. If the elasticity is greater than unity, then any government policy that increases farm wages will be of doubtful benefit. The capital-labor ratio would by definition increase in percentage by an amount greater than the percentage increase in the factor price ratios. This could be accomplished by increasing capital services or by decreasing labor services or by partially changing both. Assuming that capital services cannot be increased, that there are limited possibilities for selling the increased output (that is, the crop demand curves are strongly inelastic), or that both assumptions are operative, then a rise is wages would imply a comparable or greater percentage decline in employment. But, if the elasticity of capital–labor substitution is low, say around 0.5, increases in wage rates would not have such negative effects on employment. A 10 percent increase in wages would result in an approximately 5 percent decrease in employment (assuming constant capital services) and total labor earnings would rise (O'Herlihy 1972, p. 278).

Table 13-9 contains the values of the GESL elasticities broken down by farm size. These are computed for the same arcs as the elasticities presented in table 13-7, above. The first column of elasticity values, referring to all farms, is identical to the GESL elasticities contained in table 13-7. The most significant feature of the elasticity values computed by farm size is the large difference between the values for small farms and those for medium and large farms. In all but two cases, the small-farm values are below the weighted averages for all farms. The values for medium and large farms are correspondingly above the average and, in general, are greater than unity. The implications for policy of this discrepancy are that a wage-rate increase would be beneficial for small farmers and harmful to medium and large farmers. There would be little substitution of machinery for labor on small farms, whereas the substitution would be great on medium and large farms.

The elasticities suggest that small farmers would experience some decline in employment, but that this would represent a smaller percentage decrease than the percentage increase in wages. Total small-farm wage earnings would rise. Following the same reasoning, total wage payments on medium and large farms would decline because machinery would be rapidly substituted for labor. Extrapolating further from these basic data on elasticities, one can conclude that raising wages tends to equalize the income distribution among farmers.

Table 13-9. *GESL Arc Elasticities of Factor Substitution,*
by Farm Size, Derived from the Wage-rate Sweep

Arc end points	Associated percentage of base wage	Elasticity by farm size			
		All	Small	Medium	Large
	Group 1: Arcs of increasing length centered on the base machinery-labor ratio				
E–F	90–110	0.574	0.313	2.367	0.915
D–G	80–120	0.645	0.200	2.933	2.561
C–H	70–130	0.545	0.138	2.613	2.105
B–I	50–150	0.506	0.222	1.677	1.533
A–J	0–200	0.593	0.576	0.652	0.749
	Group 2: Arcs extending upward from the base machinery-labor ratio				
X–F	100–110	0.566	0.720	−0.165	−0.321
X–G	100–120	0.782	0.242	2.987	4.012
X–H	100–130	0.733	0.119	4.990	3.566
X–I	100–150	0.742	0.300	4.823	2.393
X–J	100–200	1.386	1.157	4.712	2.055
	Group 3: Arcs extending downward from the base machinery-labor ratio				
X–E	100–90	0.580	−0.062	4.240	2.288
X–D	100–80	0.533	0.162	3.138	1.242
X–C	100–70	0.404	0.147	2.163	1.042
X–B	100–50	0.340	0.155	1.247	1.218
X–A	100–0	0.220	0.244	0.324	0.083

Source: Data from general-equilibrium wage-rate sweep (national produce price elasticities) in model's solution. The letters A through J are keyed to the points along the dashed line
in figure 13-5. (The GESLs for each of the three farm sizes are not depicted in figure 13-5.)

Day laborers suffer, however, under the wage increase because they are
displaced by machinery on the larger farms. Although day laborers are
paid at a commercial wage rate twice as high as the farmers' reservation
rate, they are employed only seasonally. Yearly day-labor earnings are
among the lowest of all the classes of agricultural labor.[30] Thus, the effects
on income distribution of a wage increase are equivocal: the poorest group
of farmers gains, but an even poorer class suffers. Table 13-10 summarizes
these results.

30. See Centro de Investigaciones Agrarias (1970, vol. 1, chapter 3). That study reports
that, of the five classes of agricultural labor considered—namely, day laborer, employee,
self-employed worker, large-scale entrepreneur, and domestic servant—the day laborer
receives by far the lowest income.

Table 13-10. *Distribution of Labor Income by Farm Size and Type of Labor at Selected Wage Rates*
(thousands of pesos)

Farm size	Base ($\bar{w}$) wage rate (1)	$1.2\bar{w}$ (2)	$1.3\bar{w}$ (3)	$1.5\bar{w}$ (4)	Percentage change from (1) to (4)
Small					
Payments to day labor	0	0	0	0	
Payments to farmers	21,524	24,753	26,275	28,511	32
Total farmer labor earnings	21,524	24,753	26,275	28,511	32
Medium					
Payments to day labor	907	570	313	273	−70
Total farmer labor earnings	2,286	2,572	2,594	2,611	14
Large					
Payments to day labor	1,803	1,268	976	855	−53
Total farmer labor earnings	1,092	1,304	1,495	1,824	67
Total					
Payments to day labor	2,710	1,838	1,289	1,128	−58
Total farmer labor earnings	24,902	28,629	30,328	32,946	32

Source: Data from general-equilibrium wage-rate sweep (national product price elasticities) in model's solution.

An Alternative Approach to Employment Creation and Income Redistribution

A major issue throughout development economics is how to gainfully employ the ever-growing labor force in most developing countries. Industrialization has not been as successful as had been hoped in the early 1960s. Difficulties in expanding the industrial sector or in using sufficiently labor-intensive techniques (or a combination of these) have resulted in a lower rate of labor absorption in industry than expected in many countries. Thus, agriculture is being examined as a possible source of increased employment as well as a source of export earnings, capital for industry, food for the country, and other not necessarily consistent objectives.

This chapter offers some insight into employment in the agricultural sector. The perspective is from a small geographical area that is not particularly representative of the country as a whole. The problems of employment creation which are uncovered in this study, however, will be present to a greater or lesser degree in most irrigated areas of Mexico. These problems may be addressed by the "factor-proportions hypotheses" of Eckaus (1963). Although these hypotheses concern the problem of underemployment in an economy as a whole, they may be usefully applied to underdevelopment in a subsector of an economy—namely, in this case, irrigated agriculture. In fact, Eckaus posits that the factor-proportions problem is not as prevalent in agriculture as in industry, and in a comparative sense he is probably right. Nonetheless, it does exist in agriculture, and may be shown to be particularly severe on small farms.

The factor-proportions hypotheses stem from two possible explanations of unemployment in less developed areas:

> The first type assumes that available technology would permit full use of the working force at some set of relative prices and finds the source of unemployment in various types of "imperfections" in the price system. The second type suggests that there are limitations in the existing technology or the structure of demand which lead to a redundancy of labour in densely populated, underdeveloped areas. (Eckaus 1963, p. 350.)

The first type of explanation may be called the "market-imperfections hypothesis." In the present context, this hypothesis implies that wages are artificially pegged above their equilibrium level. The way to achieve full employment in this case is to allow wages to fall to their equilibrium level. Another possibility is to shift out the curve for derived labor demand by raising the price of substitute inputs. As has been demonstrated above, the market-imperfections hypothesis does not explain the unemployment of small farmers in the Tula district. High unemployment rates are encountered even when a market equilibrium is simulated under conditions of costless labor.

Eckaus calls the second of the two explanations for unemployment contained in the extract above the "technological-restraints hypothesis" (Eckaus 1963, p. 354).[31] Included in this hypothesis are three elements: few alternative techniques of production; inappropriate factor endowments for these production techniques; and a structure of product demand that leads to production of goods with low labor content. This hypothesis is borne out in the case of Tula. Although there are as many as nine degrees of substitution between labor and machinery per crop, the range of choice is still too restricted. There is a lack of even more labor-intensive techniques at the labor-intensive end of the isoquant. Given the available

---

31. Eckaus treats this hypothesis separately from the market-imperfections hypothesis. He then combines the two to form the factor-proportions hypothesis. Because the technological-restraints hypothesis seems to be the more significant of the two subhypotheses, the term "factor-proportions hypothesis" will be used here to denote the explanation of unemployment as "limited opportunities for technical substitution of factors and inappropriate factor endowments."

Table 13-11. *Employment and Income Levels when Vegetable Production Is Increased*

	Farm size							
	Small		Medium		Large		All	
Employment, wages and income, and land use	Value	Percent change[a]	Value	Percent change[a]	Value	Percent change[a]	Value	Percent change[a]
Employment (man-years)								
1. Farmers	11,187	8	656	0	313	0	12,156	8
2. Unemployment rate	55	−4	0	0	0	0	53	−3
3. Day labor	0	0	258	19	921	113	1,179	81
4. Total employment	11,187	8	914	5	1,229	65	13,330	12
Wages and income (thousands of pesos)								
Wage payments								
5. Day-labor wage bill	0	0	1,077	19	3,841	113	4,918	81
Farmer income								
6. Total farmer labor income	23,331	8	2,372	4	1,134	4	26,837	8
7. Labor income/farmer	935	8	3,616	4	4,198	20	1,035	8
8. Other rents accruing to farmers	29,566	146	2,002	8	2,227	208	33,795	181
9. Total farmer income	52,897	58	4,374	91	3,361	208	60,632	64
10. Total income/farmer	2,119	58	6,668	91	10,738	208	2,338	64
11. Total income	52,897	58	5,451	71	7,202	149	65,550	65
Land use								
12. Utilization rate (percent)	91	1	85	2	65	21	86	5

Source: Data from simulation of upward shift of vegetable demand curves in model's solution.

a. Percentage change is measured from corresponding base-solution values as presented in tables 13-3 and 13-4.

techniques, the endowment of farmer labor on small farms is too large for all of it to be fully employed. Finally, it has been shown how the product demand structure permits little substitution of labor-intensive crops for capital-intensive crops. The national product price elasticities used in specifying the crop demand curves yield highly inelastic demand functions.[32]

The factor-proportions problem may be ameliorated in several ways. One is to introduce more labor-intensive cultivation techniques. Oxen and small, garden tractors are alternative sources of traction that require a greater labor input than mules and large tractors, respectively. If technical input-output data on these forms of traction were available, they could be included in a respecified model, and the potential reduction of unemployment associated with their use could be calculated. A second way to lessen the factor-proportions problem is to change the product demand structure so that more land is devoted to crops with a high labor content. This may be accomplished by subsidizing the production of labor-intensive crops, taxing relatively capital-intensive crops, or both. Assuming that national tastes in consumption do not change, the government would be plagued by surpluses of the former and deficits of the latter crops. The logical next step would be to try to change national consumption tastes to initiate a massive foreign trade effort to export the former and import the latter.

Mexico has been successful to some extent with the foreign trade alternative. Many vegetables are now exported to the United States, particularly those grown in winter (U.S. Department of Agriculture 1969). Vegetables require relatively labor-intensive techniques of production, and the six vegetables grown in Tula —garlic, onions, dry chile, green chile, tomatoes, and cucumbers—are among the most labor-intensive crops cultivated in Mexican irrigation districts.

An experiment with the model in which demand for these vegetables is increased demonstrates the possibilities of resolving the factor-proportions problem by means of changing the structure of demand. The demand curves for each of these crops were shifted upward such that the overall production of vegetables increases by 54 percent.[33] The effects of employment and income of this strategy are listed in table 13-11. Signifi-

32. The national elasticities are indeed appropriate in the case of variations in wage and interest rates, for such variations would not be confined to the Tula district alone if they were to take place. Hence, there would be induced supply response elsewhere in the sector. This chain of reasoning leads to the conclusion that the national elasticities are proper even in a local model. See chapters 11 and 15 of this volume for a complete discussion.

33. This percentage increase represents the change in a Laspeyres quantity index. The upward shift in the demand curves may represent a change in tastes or a specific subsidy to purchasers of x pesos per ton of crop bought. These subsidies represent an average of 59 percent of the base vegetable price.

cantly, the major effect on employment is not on small farms but on medium and large farms. Farmer self-hiring on small farms rises 8 percent, whereas day-labor employment on medium and large farms increases 19 and 113 percent, respectively. The unemployment rate on small farms decreases four percentage points from the 59 percent level of the base period.

The relatively low increase in employment on small farms is attributable to the scarcity of land. Vegetable production may be increased only by taking land away from the cultivation of other crops. This is shown by the small increase in the land utilization rate from 90 to 91 percent on small farms. Essentially, land is being used at full capacity on small farms in the base solution. On medium farms, the situation is similar but not as severe. On large farms, where land is underutilized in the base solution, the situation is very different: vegetable production may be increased on large farms by bringing idle land into production rather than by displacing other crops.

A policy of increasing vegetable demand helps to alleviate the unemployment problem on small farms to a small extent. It has a greater effect on day-labor hiring. But its greatest impact of all, in percentage change, is on farmer income on medium and large farms. Because of additional rents earned on these farms, incomes rise 91 and 208 percent, respectively. Thus, a policy of increasing demand for labor-intensive crops will be modestly successful if the professed goal is the reduction of unemployment among small farmers. It may, on balance, be undesirable if great weight is attached to equalizing income distribution, or at least to not increasing the disparity.

References

Ahluwalia, Montek S. 1974. "Income Inequality: Some Dimensions of the Problem." In Hollis Chenery and others. *Redistribution with Growth*. London: Oxford University Press.

Allen, R. G. D. 1962. *Mathematical Analysis for Economists*. London: Macmillan.

Centro de Investigaciones Agrarias. 1970. "Tenencia de la tierra, población, y empleo." In *Estructura agraria y desarrollo agrícola en México*. Vol. 1. Mexico City.

Eckaus, R. S. 1963. "The Factor Proportions Problem in Underdeveloped Areas." *American Economic Review* (September 1955). Reprinted in A.N. Agarwala and S.P. Singh. *The Economics of Underdevelopment*. New York: Oxford University Press, pp. 348–78.

Hicks, John R. 1963. *The Theory of Wages*. London: Macmillan.

Howell, Hunt. 1974. "Machinery-Labor Substitution in an Irrigated Agricultural Area of Central Mexico." Ph.D. dissertation. University of Pennsylvania.

O'Herlihy, C. St. John. 1972. "Capital/Labor Substitution and the Developing Countries: A Problem of Measurement." *Bulletin of the Oxford Institute of Economic Statistics*. Vol. 34, no. 3 (August), pp. 269–80.

Robinson, Joan 1969. *Economics of Imperfect Competition*. New York: St. Martin's Press.

Secretaría de Recursos Hidráulicos (SRH). 1966–69. *Estadística agrícola*, volumes for agricultural cycles 1966–67, 1967–68, and 1968–69. Mexico City.

U.S. Department of Agriculture. 1969. *Supplying U.S. Markets with Fresh Winter Produce: Capabilities of U.S. and Mexican Production Areas*. Washington, D.C.: U.S. Government Printing Office.

14

Farmers' Response to Rural Development Projects in Puebla

CARLOS A. BENITO

THE OBJECT OF THIS CHAPTER is to develop a model representing a farm family's response to modernization projects in rural development such as Plan Puebla Project in Mexico (CIMMYT 1974a, 1974b). Plan Puebla is a strategy for rapidly increasing yields of maize among smallholders. Operationally, it is a package of technological recommendations and organizational practices reducible to the following components: (1) technical information about the optimal combination of fertilizer to produce maize, (2) a schedule for fertilizer application, (3) guidance concerning plant density, (4) a suggested procedure for facilitating access to the fertilizer market, and (5) a procedure for obtaining access to the financial capital market. The first three constitute a technological package geared to increase maize yields. The last two components are also linked to each other because credits are given to participants for the specific purpose of buying fertilizers; that is, credit and fertilizer are "tied" sales. The two organizational elements of the project facilitate the adoption of the technological component.

An important question related to modernization projects in rural development is: what are the factors explaining the rate of *adoption* of the recommended technical practices? Attracted by this question, researchers from various fields of the social sciences have advanced explanations about farmers' innovativeness in general or farmers' adoption of particular new and recommended techniques.

A survey of the literature indicates that farmers' innovativeness or adoption of technological recommendations can be explained as a function of three major factors: the fundamental behavioral rule of farm families (motivational view), the admissible set of opportunities for the farm family (structuralist view), and the dynamics of social group formation

Note: This chapter originally appeared as Benito (1976). Permission of the original publisher to use material here, with some editorial changes, is gratefully acknowledged. I am indebted to Alain de Janvry, Edgardo Moscardi, and Donald Winkelmann for a number of ideas set forth in this paper. I have also benefited from the comments of William Nickel and Peter Hazell.

and participation (organizational view). Consequently, adoption models can be classified by the relative importance given to these three factors. A first group of models emphasizes the importance of farmers' "norms" with respect to economic activities. It includes discussions on the nature of their objective function—profit maximizer (Schultz's poor but efficient farmer) versus family utility satisfier (Chayanov's peasant economy; Schultz 1964; Chayanov 1966; Lipton 1968; Nakajima 1969; Winkelmann 1972). It also embraces considerations of the farmers' behavior under uncertainty and the determination of this behavior according to wealth or other socioeconomic characteristics such as safety-first rules (CIMMYT 1974b; Moscardi 1974, 1971; Roumasset 1973), expected utility approaches, and stochastic dominance approaches (O'Mara 1972; chapters 9 and 10 of this volume).

Overview

Models that either emphasize the structural opportunities (Hymer and Resnick 1969; Sen 1966) or social group formation of farmers have received less attention by economists, although they have been intensively studied by social anthropologists, rural sociologists (Cancian 1967; Galjart 1971; Gartrell, Wilkening, and Presser 1973), and specialists in communication (Rogers 1969, 1962). This chapter is an integrative approach to the rural economy under the paradigm of a *choice* model (Benito 1973). The model gives more relative importance to the *structural* characteristics of the rural economy but also examines the *organizational* process generated by modernization projects in rural development. Within this volume, the chapter's linear programming model is the only one that incorporates social variables; that is, variables representing a farmer's use of his time in pursuits other than consumption and direct production.

The basic hypothesis of this chapter is that, besides the risky nature of farming activities and the behavior of smallholders under uncertainty, a major factor explaining different rates of technique adoption among farmers is the different degree of socioeconomic development of the farmers. At a given point in time, the state of development of a farm family can be represented by its endowments of human capital, physical capital, and organizational power. The process by which this development takes place depends both on the fundamental behavioral rule of the family and the set of opportunities offered by the socioeconomic structure. This chapter assumes a choice model in representing a farmer's motivation within a socioeconomic structure of the *minifundio* kind.

The remainder of this chapter is organized as follows.

The second section ("Structure of the Rural Family Economy") develops an explanatory model of the behavior of a rural family within an economic structure of the *minifundio* (smallholding) kind. The agricultural family is taken as a homogeneous unit; therefore, the problem of the intrafamily allocation of time is factored out in the model.[1] The model explains the allocation of time between agricultural and labor market activities, the development process of the farm economy through time, and the financing process.

The third section ("Activities of a Modernization Project") expands the model of the farm family by specifying the new set of activities that peasants are exposed to through modernization projects in rural development—for example, information gathering and organizational activities. These two new activities generate new agronomic knowledge (human capital) and facilitate the access to input and credit markets (organizational power). The organizational power is generated when small farmers form a "solidarity group" with the assistance of the modernization project team. The process of social interaction leading to the formation of a group is specified.

The fourth section ("Smallholders' Motivations and Additional Constraints") completes the model by specifying the fundamental behavioral rule of the family (a welfare function) as well as the human time and income constraints.

The fifth section ("Farmers' Response to a Rural Modernization Project") analyzes some of the decision rules corresponding to the model of the agricultural economy described in the previous sections. The decision rule regarding the allocation of time between agricultural and labor market activities is studied. The decision rules regarding the formation of, and participation through, a solidarity group for those who adopt all or part of the project package are also investigated. Finally, for those smallholders who allocate all or part of their time to agricultural activities, the decision to use agronomic inputs (mainly fertilizers and seeds) is examined. Likewise, the cost-benefit relation of additional time spent in agricultural activity on owned plots and its potential effects on the adoption of project recommendations is analyzed.

The sixth section ("Empirical Model of Smallholders' Response") presents some preliminary empirical results on farmers' responses to Plan Puebla that were derived from a linear specification of the model described in earlier in the chapter. Simulations of farm size, family size, on-the-job experience, financial fund variables, conditions in the labor market, and various patterns of human time allocation are determined. A summary is

---

1. Intrafamily allocation of time is a relevant aspect of farm economies—in particular, the distribution of functions between men and women and between adults and children. For the sake of analytical simplicity, this problem is factored out here.

given in the seventh section, and some conclusions and implications for policy are offered in the eighth section.

The last section of the text is a "Mathematical Appendix" that presents the specification and solution of the optimal control problem utilized in the model [equations (14.63) through (14.95)].

Structure of the Rural Family Economy in the Puebla Area

The average rural family in the Puebla area comprises six members. Although landholdings average 2.5 hectares, there are farmers (10 percent) who farm units of 0.5 or less hectares, as well as farmers (10 percent) who operate 5 or more hectares of land. Ninety-two percent of the farmers have effective possession of land, although they usually have several plots differing in soil type and distance from the farmstead (67.8 percent of the farmers possess four or more plots).

Schooling of farm operators averages 2.2 years, although 30 percent of this group are either illiterate or self-taught, and only 7 percent have six or more years of schooling. Physical mobility of farmers is limited—62 percent of them leave the village rarely if ever. Nevertheless, they have contact with ideas from outside of the village, principally through radio (CIMMYT 1974a). Given the economic structure of the Puebla social formation, a family can allocate its working time either to agricultural activities on its own plot or to labor market activities. Therefore, many rural residents can be characterized as part-time farmers or semiproletarian workers. In turn, agricultural time can be distributed among farming activities, learning and gathering information about agricultural practices, and organization of and participation in solidarity groups. These last two activities are introduced by modernization projects in rural development, for example, the Plan Puebla (CIMMYT 1974b).

Agronomic practices and yields

Important cropping systems under rainfed conditions in the Puebla area are maize alone, the maize/pole-bean combination, bunce beans alone, maize interplanted in orchards, and scarlet-runner beans. Maize monoculture is the most important in the project area (CIMMYT 1974b).

The local agronomic practices—here referred to as "traditional"—are the result of centuries of interaction of the farmers with their environment. Plan Puebla exposes peasants to new agronomic practices—here referred to as "modern." The set of feasible agronomic practices in Plan Puebla is represented by the following production function:

(14.1) $$Y(t) = s_a(t) \pi_a(t) g_a[x_a(t), C_a(t), C_k(t), u_a(t)],$$

where

> Y = agricultural production
>
> s_a = proportion of human time allocated by the peasant family to agricultural activities
>
> π_a = proportion of time for agricultural activities specifically allocated to farming
>
> $g_a(\cdot)$ = production function (agricultural output per unit of human time)
>
> x_a = vector of agronomic inputs (seeds, plants, fertilizer, and insecticides)
>
> C_a = vector of durable means of production (land size and quality, well, animals, implements, and so forth)
>
> C_k = index of knowledge of agricultural practices
>
> u_a = stochastic variable reflecting agronomic risks.

To complete the description of the agronomic technologies, it is necessary to specify the characteristics of these technologies. The function $g_a(\cdot)$ expresses agricultural productivity per unit of labor time:

(14.2)
$$\frac{\partial g_a}{\partial x_a} > 0 \quad \frac{\partial g_a}{\partial C_a} > 0 \quad \frac{\partial g_a}{\partial C_k} > 0$$

$$\frac{\partial^2 g_a}{\partial x_a^2} \leq 0 \quad \frac{\partial^2 g_a}{\partial C_a^2} \leq 0 \quad \frac{\partial^2 g_a}{\partial C_k^2} \leq 0.$$

The human capital variable C_k is introduced in an explicit form since one of the major aims of Plan Puebla is to increase maize yields by increasing farmers' knowledge concerning plant densities and alternative combinations and the timing of fertilizer application. In figure 14-1, the most efficient production frontier, $Y\{C_k^M > C_k^T\}$, where M indicates knowledge of "modern" agronomic practices, is presented with a "less" technical-efficient curve, $Y\{C_k^T\}$, associated with "traditional" (T) knowledge of agronomic practices. A displacement of the production function can result from a more efficient agronomic management—that is, from embodied (as human capital) technical change.[2]

The production function (14.1) is assumed to be linear and homogeneous in human time. The use of human time within the farm, however, is limited by the level of other resources, mainly land size, C_a. Plan Puebla's agronomic recommendations can be described as a land-saving and labor-using technology. In fact, it comprises practices

---

2. In this sense change is *not* of the "green revolution" technical kind, since it is not based on the adoption of new seed varieties. This approach is similar to Welch's (1970) and Schultz's (1972) "allocation ability."

Figure 14-1. *Agricultural Production Function for Puebla Area*

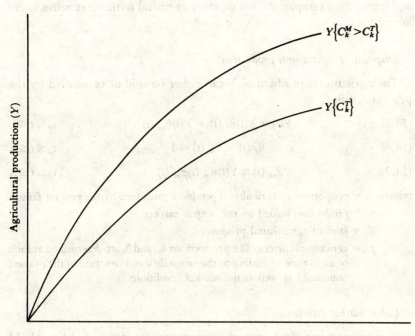

Note: C_k^M = index of knowledge of "modern" (technological) agricultural practices; C_k^T = index of knowledge of "traditional" (empirical) agricultural practices. C_k^M represents a higher level of knowledge (in some abstract units) than C_k^T.

demanding the use of more man-hours per hectare and a more efficient management of resources:

$$x_a\,(t) \geq \ell_x\,[C_k\,(t)]$$

(14.3)

$$\frac{\partial \ell_x}{\partial C_k} > 0 \quad \frac{\partial^2 \ell_x}{\partial C_k^2} \leq 0,$$

where $\ell_x(\cdot)$ describes the complementarity between new knowledge and modern agronomic inputs; for example, knowledge of modern agronomic practices implies the use of more chemical fertilizer. Furthermore,

$$s_a\,(t) \cdot \pi_a\,(t) \geq \ell_s\,[x_a\,(t)]$$

(14.4)

$$\frac{\partial \ell_s}{\partial x_a} > 0 \quad \frac{\partial^2 \ell_s}{\partial x_a^2} \leq 0,$$

where $\ell_s(\cdot)$ describes the complementarity between agronomic inputs and labor; for example, the use of more chemical fertilizer requires more labor.

Disposal of agricultural production

The agricultural production Y can either be sold or consumed by the peasant family:

(14.5) $$Y(t) \geq Y(t)\theta_c(t) + Y(t)\theta_m(t) \qquad t_0 \leq t \leq t_1$$

(14.6) $$\theta_c(t) + \theta_m(t) = 1 \qquad t_0 \leq t \leq t_1$$

(14.7) $$Z_a(t) \leq Y(t)\theta_m(t)p_m(t), \qquad t_0 \leq t \leq t_1$$

where θ_c = proportion of agricultural products consumed by the peasant family

θ_m = proportion traded on the output market

Z_a = sales of agricultural products

p_m = agricultural prices. The proportions θ_m and θ_c are determined at each point in time according to the rationality and structure of the peasant household as well as the market conditions.[3]

Labor market activities

Earning power and labor market opportunities of the rural household are represented by:

(14.8)
$$Z_w(t) = s_w(t)g_w[C_w(t)] \qquad t_0 \leq t \leq t_1$$

$$\frac{\partial g_w}{\partial C_w} > 0 \quad \frac{\partial^2 g_w}{\partial C_w^2} \leq 0$$

(14.9)
$$s_w(t) \leq g_e[C_w(t)] \qquad t_0 \leq t \leq t_1$$

$$\frac{\partial g_e}{\partial C_w} > 0 \quad \frac{\partial^2 g_e}{\partial C_w^2} \leq 0,$$

3. A more complete model can represent the storage process of agricultural outputs. In this case,

$$\dot{C}_k(t) = Y(t) - \gamma_c(t) - \gamma_m(t)$$
$$\gamma_c(t) = Y(t)\,\theta_c(t)$$
$$\gamma_m(t) = Y(t)\,\theta_m(t)$$
$$C_k(t) = C_k(t_0) + \int_{t_0}^{t} \dot{C}k(\tau)\,d\tau,$$

where C_k is the stock of agricultural products; $\dot{C}_k$ is the net storage in time t; and t_0 is the initial time.

where Z_w = labor earnings

s_w = supply of labor

$g_w(\cdot)$ = wage function

$g_e(\cdot)$ = employment opportunity (demand for labor) function

C_w = index of on-the-job experience.

Here labor earnings are defined as the wage rate times the quantity of labor supplied. The wage rate is an endogenous variable in the long run because workers can increase it by investing in on-the-job experience. On-the-job experience, as defined here, includes skills and knowledge of a job, labor market information, and seniority.[4]

The employment function $g_e(\cdot)$ indicates that the farmer's opportunities in the labor market are limited, although in the long run his opportunities can be enhanced through investment in on-the-job experience.[5] The employment function is a reduced-form expression whose parameters are determined by the aggregate supply of and demand for labor within a particular economic structure.

Consumption activities

The consumption of the rural family is represented by:[6]

$$(14.10) \qquad Z_c(t) = g_c\left[Y(t)\theta_c(t), x_c(t)\right], \qquad\qquad t_0 \leq t \leq t_1$$

where Z_c is the index of family consumption, $g_c(\cdot)$ is the consumption-transformation function, and x_c is the amount of other goods purchased in the market.

The consumption activities of the family shall always generate a minimum level of nutrition sufficient to maintain its members alive and productive; that is, the smallholders' economy is subject to a survival constraint

---

4. A more complete explanation of the wage rate will also make it a function of education, past mobility, health, and social status. The influence of nutrition on productivity can be very important in poor rural areas. In this case, the wage function will be $g_w[C_w(t), C_n(t)]$, where C_n is health (human) capital (Grossman 1972).

5. This is an alternative "deterministic" representation to the "stochastic" representation (Todaro 1969) of equation (14.3). Todaro's model explicitly discusses the total supply of and demand for labor while assuming homogeneous workers. This model takes as given the reduced form of the labor market model but allows for heterogeneous workers by way of $g_d(C_w)$. A more complete representation will include health, schooling, and partial mobility as explanatory factors.

6. Consumption and other nonmarket activities are also time consuming. For simplicity here, it is assumed that the allocation of time between working and home activities has already been determined.

(14.11) $$Z_c(t) \geq \underline{Z}_c, \qquad\qquad t_0 \leq t \leq t_1$$

where $\underline{Z}_c$ is a minimum necessary level of consumption.[7]

Development and financing processes

The state of economic development of a farm family at a time t is defined by $C(t) = [C_a(t), C_w(t), C_k(t), C_o(t)]$ where physical capital C_a and human capital C_w were the major, or perhaps the only, state variables before the existence of Plan Puebla. A development project enhances or gives more relevance to technological knowledge, C_k, and promotes the development of modern institutions (that is, the institutional power, C_o, of solidarities).

The development processes of physical and human capital are represented by

(14.12) $$\dot{C}_a(t) = J(t) - \delta_a C_a(t)$$

and

$$\dot{C}_w(t) = f_w[Z_w(t)] - \delta_w C_w(t)$$

(14.13)

$$\frac{\partial f_w}{\partial Z_w} > 0 \quad \frac{\partial^2 f_w}{\partial Z_w^2} \leq 0,$$

where C_a and J are net and gross accumulation of physical capital, respectively;[8] $\dot{C}_w$ and $f_w(\cdot)$ are net and gross accumulation of on-the-job experience, respectively; and δ_a and δ_w are geometric rates of depreciation (see Benito 1973 and Weiss 1971).

The processes of learning and gathering agronomic information and of organizing modern institutions are studied in the next section of this chapter.

Farm families have three major sources for funding their activities: personal savings; loans from "traditional" institutions (relatives, friends, priests, moneylenders, and the like); and loans from modern institutions (banks and other credit agencies). The farmer has limited saving capacity and usually seeks traditional sources of funds to cushion the disastrous effects of agronomic risk or illness of family members. (In 1967 only 4.4 percent of farm operators in the Puebla area had credit from moneylenders; CIMMYT 1974a.) Although farmers' use of credit from

---

7. In a model that includes explicitly the investment process in health, the survival constraint can be expressed as $C_n(t) \geq \underline{C}_n(t)$, where $t_0 \leq t \leq t_1$ and $\underline{C}_n(t)$ is the minimum level of health (Grossman 1972).

8. Accumulation of physical capital is defined here as the result of the input-consuming process alone. However, and in the case of smallholder economy in particular, the amount of the family's time that is allocated to their production may also be important.

banks and other agencies is very limited—in 1967 only 7.6 percent of farm operators had credit given by a bank or other credit institutions (CIMMYT 1974*a*)—these institutions represent a potential source for financing the future development-oriented activities of farm families.

The process of farmer indebtedness (Hochman, Hochman, and Razin 1973) can be represented by

$$(14.14) \qquad \dot{B}(t) = r \cdot B(t) - M(t),$$

where B = accumulated financial liabilities or debt

$\dot{B}$ = rate of change of B over time

r = average rate of interest

$M(t)$ = net payments to creditors.[9]

For agriculturalists of the Puebla area, the major transactional costs of obtaining credits are generated by the bureaucratic process of having access to credit markets. Even when the average rate of interest can be taken as independent of the amount of credits B, the farmer's problem is how to increase B. This requires an ability on the part of the farmer to deal with modern bureaucracy as well as the existence of a financial guarantee. This ability and guarantee can be increased by the farmer's joining credit groups, or *grupos solidarios*.

Activities of a Modernization Project in Rural Development: The Plan Puebla

The main activities of the rural family having been described and formally represented, it is now necessary to describe and represent the new subset of activities added by a modernizing rural development project such as the one in Puebla.

Plan Puebla promotes the adoption of modern agronomic practices among smallholders and provides both information and organizational help. Farmers receive information about new agronomic practices and expected increases in yields and net income. Major communication media employed are radio and pamphlets, village meetings, demonstration of new practices in the field, exchanges or excursions of farmer groups during the crop-growing season, and demonstrations at harvest time (CIMMYT 1974*b*).

Farmers organize themselves for buying needed agronomic inputs (seeds, fertilizers, and insecticides) and for obtaining credit from banks. An institutional procedure for this is the formation of credit groups.

9. *M* will be negative when credits received are higher than payments to creditors.

Paralleling the above description of Plan Puebla, a model for explaining the rate of adoption of Plan Puebla's recommendations should take into account the process of change in the *information structure* among farmers and the process of their *group formation* (toward relating marginal, traditional economic agents with the modern capitalistic institutions such as input markets, output markets, and credit markets). The informational structure of Puebla farmers is changed through a learning process that includes both receiving new knowledge through communication media and learning-by-doing (experimenting in their own economic unit) (Rogers 1962). All these learning activities are time-consuming.[10]

The process of group formation is also promoted by the Plan Puebla team and involves a selection process among farmers as well as a functional procedure. First the results (benefits) generated by the project, and then the organizational process (group formation) that smallholders have to be involved in to facilitate their access to modern institutions, will be presented.

Effects of Plan Puebla

The acquisition of information and new skills, as well as the development of the entrepreneurial capacity for dealing with bureaucratic institutions, represent an investment in human capital for the farm households. This acquisition and development can be represented, respectively, by

$$(14.15) \qquad C_k(t) = Z_k(t) - \delta_k C_k(t) \qquad\qquad t_0 \leq t \leq t_1$$

and

$$Z_k(t) = s_a(t)\pi_k(t)g_k[D(t), C_k(t), C_o(t)] \qquad t_0 \leq t \leq t_i$$

(14.16)

$$\frac{\partial g_k}{\partial D} > 0 \quad \frac{\partial g_k}{\partial C_k} > 0 \quad \frac{\partial g_k}{\partial C_o} > 0,$$

where π_k = proportion of agricultural time allocated to learning and information gathering

$g_k(\cdot)$ = learning function

D = index of technical assistance of the Puebla project team to the rural family

C_o = index of organizational level (or institutional power expressed in some abstract units) as a result of the formation of a solidarity group.

---

10. Because agricultural production is always a risky, stochastic phenomenon, the informational structure of farmers is not only expressed in farmers' knowledge of alternative practices and yields but also in farmers' perceptions of risk or beliefs regarding possible states of nature.

The organizational level C_o also determines the farmer's access to the input credit markets:

$$x_a(t) \leq G_x[C_o(t), D(t)] \qquad t_0 \leq t \leq t_1$$

(14.17)

$$\frac{\partial x_a}{\partial C_o} > 0 \quad \frac{\partial^2 x_a}{\partial C_o^2} \leq 0 \quad \frac{\partial x_a}{\partial D} > 0 \quad \frac{\partial^2 x_a}{\partial D^2} \leq 0,$$

where $G_x(\cdot)$ is a marketing function. The condition (14.17) indicates that, at a time t, a peasant is able to acquire at the most $G_x(\cdot)$ amount of agronomic inputs. The value of $G_x(\cdot)$, however, is not a constant and can be extended by having more institutional power, C_o, and more technical assistance, D, from the Puebla project team.

Similarly, access to the credit market can be represented by:

$$B(t) \leq G_b[C_o(t), D(t)] \qquad t_0 \leq t \leq t_1$$

(14.18)

$$\frac{\partial G_b}{\partial C_o} > 0 \quad \frac{\partial^2 G_b}{\partial C_o^2} \leq 0 \quad \frac{\partial G_b}{\partial D} > 0 \quad \frac{\partial^2 G_b}{\partial D^2} \leq 0,$$

where $G_b(\cdot)$ is a financing function. The condition (14.18) also indicates the maximum amount of debts, B, that the credit institutions will grant to an individual. This maximum G_b can also be expanded by means of institutional power, C_o, and the technical assistance of the Puebla project team.[11]

Functions $G_x(\cdot)$ and $G_b(\cdot)$ are crucial elements in representing the modernization process in the institutional vector of a community.

The social interaction and solidarity formation of farmers

The results and process of community organization are studied here. The process of community organization refers to the process of social interaction that determines the formation of a credit group.

The following analysis of this process is inspired by the treatment of individual choices and social processes suggested by Roberts and Holdren (1972). The contribution of farmer involvement to the institutional power of a credit group can be represented by

11. One of the activities of the Puebla project team has been to ensure that services provided by agricultural agencies are adequate to the needs of small producers. The project coordinator keeps the representatives of agricultural agencies (credit, distribution of inputs, crop insurance, and marketing agencies) informed of project activities and the needs of the farmers (CIMMYT 1974b).

$$Z_o\,(t) = s_a\,(t)\pi_o\,(t)g_o[D(t)] \qquad\qquad t_0 \le t \le t_1$$

(14.19)

$$\frac{\partial g_o}{\partial D} > 0 \quad \frac{\partial^2 g_o}{\partial D_o^2} \le 0;$$

similarly, for the other J individuals who can get involved in the credit group,

$$Z_o^j\,(t) = s_a^j\,(t)\pi_o^j\,(t)g_o^j\,[D^j\,(t)] \qquad\qquad \begin{matrix} t_0 \le t \le t_1 \\ j = 1, \ldots , J \end{matrix}$$

(14.20)

$$\frac{\partial g_o^j}{\partial D^j} > 0 \quad \frac{\partial^2 g_o}{\partial D^{j2}} \le 0,$$

where Z_o^j is the organizational contribution (in some abstract unit) of the jth individual *if* he joins the group; π_o^j is the proportion of agricultural time allocated to community organization and relations with other institutions; and $g_o^j(\,\cdot\,)$ is the "organization" generation function.

It is assumed that institutional power C_o is an additive result of the farmers' organizational contributions generated in forming the credit group:[12]

(14.21) $$C_o\,(t) = Z_o\,(t) + \sum_j Z_o^j\,(t) - \delta_o\,C_o\,(t), \qquad\qquad \begin{matrix} t_0 \le t \le t_1 \\ j = 1, \ldots , J \end{matrix}$$

where C_o is the net increase in institutional power (by way of organizational improvements), and δ_o is a rate of the deterioration of institutional power. The farmers' ability in generating institutional power is obviously conditioned by the assistance provided by community organizers (that is, the Puebla project team); this is represented by a functional dependence $g_o^j\,(D^j)$ between Z_o^j and D^j.

The specification of the psychological and social interaction process leading to the formation of a solidarity has to be based in a socioanthropological interpretation of rural values and social structure.

As perceived by the farmer, the institutional power C_o, generated by the existence of the credit group, shall maintain or increase other farmers' welfare in order to motivate them to form the solidarity and to accept him as a member. That is, $F_o^j\,(C_o) \ge \bar{F}_o^j$ for all $j \ne i = 1, \ldots , J$, where F_o^j is an index of family welfare. The distribution of $\{\bar{F}_o^j \,|\, j \ne i = 1, \ldots , J\}$ is a representation of the distribution of social status or class structure within

---

12. Multiplicative or other forms are also possible. Additivity is assumed here for analytical convenience.

the community. This social structure cannot be radically changed without generating threats against the farmers that will stop the institutional evolution.

The welfare F_o^j of any other farmer j also is determined through an economic process as described by equations (14.1) through (14.20). It is assumed that the farmer knows the distribution of social status $\bar{F}_o^j$ that he should respect. This is a plausible assumption because the group members are relatives or long-time neighbors of the same community. The analytical representation of such a problem, however, becomes very complicated. An analytically convenient simplification is the following:

$$(14.22) \qquad F_o^j (C_o) \geq \bar{F}_o^j \qquad j = 1, \ldots, J$$
$$t_0 \leq t \leq t_1$$

and

$$(14.23) \qquad s_a^j (t) \pi_o^j (t) \leq \bar{s}^j, \qquad j = 1, \ldots, J$$
$$t_0 \leq t \geq t_1$$

where $F_o^j (C_o)$ is the "reduced" form of other farmers' welfare as a function of social power alone; $\bar{F}_o^j$ is the index of the social status to be maintained; and $\bar{s}^j$ indicates resource (human time) constraints.

Smallholders' Motivations and Additional Constraints

Given the economic structure of the peasant family as described above, the level and composition of its activities will be determined by the meanings it assigns to the results of these activities. These meanings in general correspond to a value structure the family associates with specific institutions. Abstracting from the problem of the symbolic representation of meanings, as well as the change in the organization of knowledge and cognitive styles generated by modernization, it is assumed that the rural family behaves *as if*

$$(14.24) \qquad \max W_o = \int_{t_0}^{t_1} F [Z_c (t)] \exp [- \tau (t - t_0)] \, dt,$$

where W_o is an index of welfare, and $F(\cdot)$ is an instantaneous concave welfare function.[13]

---

13. The instantaneous concave welfare function $F(\cdot)$ is assumed to satisfy the following conditions: $[dF (Z_c)]/dZ_c = F' (Z_c) > 0$; $0 < Z_c < \infty$; $[d^2 F (Z_c)]/dZ_c^2 = F'' (Z_c) \leq 0$; $\lim F' (Z_c) = \infty$; $Z_c \to 0$; $\lim F' (Z_c) = 0$; $Z_c \to \infty$.

A more complete representation of this will also include the results of on-the-farm and on-the-job activities as arguments. Perhaps a complete representation of the farmers' ranking systems could be $\max W_o = \int_{t_0}^{t_1} F [Z_c, Z_a, Z_w, Z_o] \exp [- \tau (t) (t - t_0)] \, dt$.

(*Note continues on the following page.*)

Finally, the working and organizational activities of the family are subject to a human time constraint

$$(14.25) \qquad s - s_a(t) - s_w(t) \geq 0 \qquad\qquad t_0 \leq t \leq t_1$$
$$s = 1,$$

where s is the total time availability of the family,

$$(14.26) \qquad \pi_a(t) + \pi_k(t) + \pi_o(t) = 1, \qquad\qquad t_0 \leq t \leq t_1$$

and to a monetary budget constraint

$$(14.27) \qquad \int_{t_0}^{t_1} [Z_a(t) + Z_w(t) - \sum_i x_i(t)p_i(t) - Jp_j - M(t)]$$
$$\exp[\tau(t - t_0)] \, dt \geq 0,$$

where, as before,

Z_a = agricultural sales
Z_w = wage earnings
x_i = the amount of the ith agronomic input
p_i = the price of the ith input
$M(t)$ = net payments to creditors
J = gross capital formation vector
p_j = price of the jth capital good.

Farmers' Response to a Rural Modernization Project

The major interest in specifying the structure and rationality of a rural family economy is to explain the determinants of its observed economic behavior. This section studies (1) the farmer's decision criteria with respect to allocation of his time between agricultural and labor market activities (under circumstances of unlimited and limited working opportunities); (2) the process of forming a credit group; (3) the decision rule for using agronomic inputs, an indicator of peasants' adoption of modern agronomic practices. In the course of developing these points, this section also sheds light on the household process of accumulating physical, human, and institutional capital.

Optimal allocation of human time is analyzed under two conditions:

This formulation considers the nonpecuniary, or psychological, benefits of agricultural, wage-labor, and organizing activities. The welfare effects of Z_a, Z_w, and Z_o will describe the nature (for example, traditional or modern) of the cultural patterns of the family. This subject has been of great interest to cultural anthropologists.

when opportunities to work on the farm or in the labor market are not limited [that is, equations (14.2) and (14.9) are not binding constraints] and when opportunities are limited [that is, (14.2) or (14.9) (or both) is (are) binding constraints]. The first case is relevant for farmers who work medium-size or large farms and who have invested in human capital sufficiently to have unrestricted working opportunities; the second case is considered in three versions. Both cases are discussed below.

Farmers' choice between agricultural and labor market activities under unrestricted working opportunities

At a given time t, a peasant will allocate his labor time to agricultural activities or to labor market activities according to the following criteria:[14]

(14.28)
$$\begin{cases} s_a = 1 \\ s_w = 0 \end{cases} \Bigg| \text{if } b_a > b_w$$

$$\begin{cases} s_a = 0 \\ s_w = 1 \end{cases} \Bigg| \text{if } b_a < b_o \,.$$

That is, in each time period t, the farmer will participate either in agricultural activities ($s_a = 1$, $s_w = 0$) or in labor market activities ($s_a = 0$, $s_w = 1$), depending on his subjective economic calculus of the relative contribution (b_a, b_w) of these activities to the family welfare.

Synthetic measures of the contribution to welfare of agricultural activities are defined by[15]

(14.29)
$$b_a = \pi_a b_{aa} + \pi_k b_{ak} + \pi_o b_{ao},$$

where b_{aa} is the direct effect of on-the-farm activities; b_{ak} is the household development effect (by way of new knowledge and information about agricultural practices acquired through Plan Puebla); and b_{ao} is the community development effect (by way of new organizational ability promoted also by Plan Puebla). These three effects are weighted by the proportion of agricultural time allocated to farming (π_a), learning (πk), and organizing (π_o). That is, the welfare contribution of agricultural time s_a also depends on its optimal allocation among farming, learning, and organizing. The direct effect of farming is measured by

(14.30)
$$b_{aa} = b_{aa}^m + b_{aa}^c.$$

---

14. From appendix equation (14.89) when $g_f(\cdot) = g_e(\cdot) = s$.

15. From appendix equation (14.86) combined with equations (14.65), (14.66), and (14.68) through (14.70).

The term b_{aa}^m represents the earning contributions resulting from the proportion θ_m of agricultural production Y during period t sold in the market at a price p_m:

$$(14.31) \qquad \frac{b_{aa}^m}{\lambda} = p_m(Y\theta_m).$$

The contribution of farming by way of sales, b_{aa}^m, is divided by the marginal utility of money λ in order to transform it into monetary value.

The term b_{aa}^c in equation (14.30) represents the consumption satisfaction derived from the proportion $\theta_c = (1 - \theta_m)$ of the same agricultural production Y used by the farm family:

$$(14.32) \qquad \frac{b_{aa}^c}{\lambda} = \left[\frac{1}{\lambda} \left(\varepsilon + \frac{\partial F}{\partial Z_c} \right) \frac{\partial g_c}{\partial Y\theta_c} \right] (Y\theta_c).$$

The welfare generated by one unit of farm production consumed at home, $Y\theta_c$, is measured in monetary terms by the bracket of the right-hand term of equation (14.32). According to the habits of the family, the product $Y\theta_c$ is transformed into consumption $g_c(\cdot)$. The marginal effect over consumption, $\partial g_c / \partial Y\theta_c$, is then weighted by the marginal satisfaction $[\varepsilon + (\partial F / \partial Z_c)] / \lambda$ that it generates. This marginal satisfaction includes two elements: when the family is at the survival level (that is, $Z_c = \underline{Z}_c$), besides the direct marginal satisfaction $\partial F / \partial Z_c$ of consumption activities, the farmer also assigns to them a survival marginal value ε. But if the family is above the survival level (that is, $Z_c > \underline{Z}_c$), the marginal welfare of survival is zero.

In the case where there exists strict complementarity between farming time $s_a \pi_a$ and agronomic inputs x_a, as determined by a known technology C_a,[16] the benefits attributed to farming time are computed together with the marginal benefits generated by the associated agronomic inputs x_a. In this case, equation (14.30) becomes

$$(14.33) \qquad b_{aa} = b_{aa}^m + b_{aa}^c + b_{aa}^s$$

$$(14.34) \qquad \frac{b_{aa}^s}{\lambda} = \frac{\psi_{as}}{\lambda},$$

where ψ_{as} is the marginal utility jointly generated by farming time and agronomic inputs.

The second term of equation (14.29), or the development effect of the

16. In this case the condition (14.3) becomes a binding constraint: $s_a \pi_a > \ell_s (x_a)$. If $s_a \pi_a > \ell_s (x_a)$, this indicates the existence of a surplus of family time and, therefore, ψ_{as}, the marginal utility generated from both farming time and agronomic inputs, will have zero value.

agricultural time $s_a\pi_k$ allocated to learning activities promoted by the modernization project, is

$$(14.35) \qquad \frac{b_{ak}}{\lambda} = \frac{q_k}{\lambda} g_k(\cdot),$$

where the right-hand side measures the marginal monetary value of investment in new agricultural knowledge and information. The expression q_k/λ is the monetary demand price of human capital, and $g_k(\cdot)$ is the knowledge and information generated because of the family's participation in Plan Puebla in period t. The demand price of human capital is the present value of the future flow of farm earnings and consumption generated by a new (abstract physical) unit of skills and information. Under conditions of optimal behavior of the rural family, it equals the marginal cost of "producing" new knowledge:

$$(14.36) \qquad \frac{q_k}{\lambda} = \frac{(\mu/\lambda)\pi_k}{Z_k},$$

where μ/λ is the marginal cost of the family's time.[17]

The last term of equation (14.29) is the development effect of agricultural time $s_a\pi_o$ dedicated to organizational activities promoted by the modernization project,

$$(14.37) \qquad \frac{b_{ao}}{\lambda} = \left(\frac{q_o}{\lambda}\right) g_o(\cdot),$$

that is, the marginal monetary value of participation in a credit group in period t. The expression q_o/λ is the monetary demand price of institutional power, and $g_o(\cdot)$ is the farmer's contribution (in some physical abstract unit) to the institutional power of the solidarity. The demand price of institutional power is the present value of the future flow of earnings and consumption generated by adding a new unit of institutional involvement. Under conditions of optimal behavior, it equals the unit cost of generating institutional power for a group of farmers:

$$(14.38) \qquad \frac{q_o}{\lambda} = \frac{(\mu/\lambda)\pi_o + \sum_j (\mu^j/\lambda)\pi_o^j}{Z_o - \sum_j Z_o^j}.$$

When the analysis of equation (14.28) is completed, the welfare contribution of labor market activities, as perceived by the farm family, is defined by[18]

17. Consequently, b_{ak}/λ equals the opportunity cost of human time times the proportion π_k allocated to learning activities: $b_{ak}/\lambda = (q_k/\lambda)g(\cdot) = (\mu/\lambda)\pi_k$.
18. From appendix equation (14.87) combined with (14.67).

(14.39)
$$b_w = b_{wm} + b_{wk},$$

where

(14.40)
$$\frac{b_{wm}}{\lambda} = g_w(\cdot),$$

that is, the *earning* contributions as measured by the market wage rate $g_w(\cdot)$.

In addition, when labor activities include a learning process (of both skills and market information), they also have a *developmental* contribution—as measured by

(14.41)
$$\frac{b_{wk}}{\lambda} = \frac{q_w}{\lambda}[(\partial f_w / \partial Z_w) g_w(\cdot)],$$

where $(\partial f_w / \partial Z_w) g_w(\cdot)$ is now learning (in abstract physical units) and q_w / λ is the stock demand price of human capital. In equilibrium, the demand price of on-the-job experience shall be equal to its unit cost[19]

(14.42)
$$\frac{g_w}{\lambda} = \frac{(\mu / \lambda) - [g_w(\cdot)]}{Z_{kw}},$$

where Z_{kw} is new knowledge (in abstract physical units) from labor market experience. In other words, the value of investment in human capital $(q_w / \lambda) Z_{kw}$ in time t is equal to the forgone income; that is, the opportunity cost of time μ / λ less the wage rate $g_w(\cdot)$.

Farmers' choice between agricultural and labor market activities under limited working opportunities

Three major cases of limiting working activities can be considered: (1) $g_f < s$ and $g_e > s$; (2) $g_f > s$ and $g_e < s$; and (3) $g_f < s$ and $g_e < s$. In case (1), the size of the farm is relatively small in relation to the available family time and the known technology C_k, but it has unlimited opportunities in the labor market at a given wage rate. Case (2) is the converse situation; and case (3) is a combination where there exist limited opportunities, both on the farm and in the labor market.

In case (1), the benefits of farming time are computed net of the opportunity cost of farm size $\xi_a \lambda$. Thus, equation (14.33) becomes

(14.43)
$$b_{aa} = b_{aa}^m + b_{aa}^c + b_{aa}^s - b_{aa}^f$$

(14.44)
$$\frac{b_{aa}^f}{\lambda} = \frac{\xi_a}{\lambda}.$$

19. From appendix equations (14.73) and (14.67).

In case (2), the benefits of labor market opportunities are calculated net of the opportunity cost of employment ξ_w / λ. Thus equations (14.40) and (14.42) become

$$(14.45) \qquad \frac{b_{wm}}{\lambda} = g_w(\cdot) - \frac{\xi_w}{\lambda}$$

$$(14.46) \qquad \frac{g_w}{\lambda} = \frac{(\mu/\lambda) - [g_w(\cdot) - (\xi_w/\lambda)]}{Z_{kw}}.$$

Under limited opportunities to work on the farm or in the labor market (or both), the general rule (14.28) of allocation of human time is modified.[20] For $g_f(\cdot) \leq s$ and $g_e(\cdot) \leq s$,

$$(14.47) \qquad \begin{cases} s_a = g_f(\cdot)\, \pi_a^{-1} \\ s_w = \min\{(s - s_a),\, g_e(\cdot)\} \end{cases} \Bigg| \text{if } b_a > b_w$$

$$\begin{cases} s_w = g_w(\cdot) \\ s_a = \min\{(s - s_w),\, g_f(\cdot)\} \end{cases} \Bigg| \text{if } b_a < b_w.$$

Condition (14.28) becomes a particular case of (14.47) when $g_f(\cdot) = g_e(\cdot) = s$. Condition (14.47) is a better representation of reality because most smallholders work both on their farms and in the labor market.

Criteria for forming a solidarity group

A peasant household's decision to participate in a credit group is governed by two rules: a subjective comparison between benefits and costs and a social or interpersonal comparison of own costs with others' costs. The first was already discussed and is implicit in equations (14.27) in general and in equations (14.37) and (14.38) in particular. The second, or criterion of social behavior is that a farm family will allocate a proportion π_o of agricultural time for community involvement in such a way that the unit cost of its organizational contribution equals the marginal cost of every other farmer entering the credit group; that is,

$$(14.48) \qquad \frac{\pi_o\, \mu}{Z_o} = \frac{\pi_o^1\, \mu^1}{Z_o^1} = \ldots = \frac{\pi_o^J\, \mu}{Z_o^J}$$

---

20. From appendix equation (14.89).

for those $j = 1, \ldots, J$ with $F_o^j \geq \bar{F}_o^j$. This is a behavioral rule with substantive counterparts in social life, and this condition can be fulfilled in many situations. One particular case is when farmers are a homogeneous class; that is, when the value of their human time μ and their individual organizational contributions are the same:

$$\mu = \mu^1 = \ldots = \mu^J$$

(14.49)
$$g_o(\cdot) = g_o^1 (\cdot) = \ldots = g^J(\cdot)$$

$$\pi = \pi_o^1 = \ldots \pi_o^J.$$

The equality between organizational contributions $g^j(\cdot)$ can take place in two ways: when the organizational *abilities* of contributors are equal and when the organizational help of the project team is the same for every farmer:

$$g_o = g_o^1(\cdot) = \ldots = g_o^J(\cdot)$$

(14.50)
$$D = D^1 = \ldots = D^J.$$

This will be the case for a completely homogeneous group of farmers.

Another case with more empirical relevance is when the marginal values of farmers' time are equal but their organizational abilities are different:

(14.51)
$$g_o(\bar{D}) \neq g_o^1(\bar{D}) \neq \ldots \neq g_o^J(\bar{D}).$$

Then, if the organizational help of the project team is different for each farmer,

$$g_o(D) = g_o^1(D^1) = \ldots = g_o^J(D^J)$$

(14.52)
$$D \neq D^1 \neq \ldots \neq D^J.$$

For example, if the organizational ability of a natural leader d is the highest among the group,

$$g_o^d(\bar{D}) \geq g_o(\bar{D})$$

(14.53)
$$g_o^d(\bar{D}) \geq g_o^j(\bar{D}), \qquad\qquad j = 1, \ldots, J$$

then the organizing help, D^j, dedicated to other peasants has to be greater:

$$D > \bar{D}$$

(14.54)
$$D^j > D. \qquad\qquad j = 1, \ldots, J.$$

However, even if farmers are not of equal ability, condition (14.48) still is socially viable. For example, if the value of human time μ of a given family is higher than it is for other families, the family's organizational contribution Z_o can also be higher so as to have a unit cost similar to others. Individuals whose time value is relatively higher are likely to be those with higher levels of innate ability and endowments of human, physical, and institutional capital; also, farmers with these higher endowments are likely to have more organizational capacity than poorer farmers.

Use of agronomic recommendations

The rate of use of agronomic inputs x_a is determined by the following behavioral rule:

$$(14.55) \quad \left[\theta_m p_m + \left(\varepsilon + \frac{\partial F}{\partial Z_c}\right)\frac{1}{\lambda}\left(\frac{\partial g_c}{\partial Y\theta_c}\right)\theta_c\right]s_a\pi_a\left(\frac{\partial g_a}{\partial x_a}\right) = p_a + \frac{\psi_{ox}}{\lambda},$$

where the left-hand side measures the benefits (in monetary terms) of the last unit of agronomic inputs and the right-hand side measures its associated cost (also in monetary terms).

The physical marginal product of agronomic inputs $s_a\pi_a(\partial g_a/\partial x_a)$ is distributed in proportions θ_m and θ_c between market sales (which generate earnings) and family consumption (which generate satisfactions). Earnings per unit of output are measured by $\theta_m p_m$ within the left-hand bracket. Consumption satisfaction is measured by $(\partial g_c/\partial Y\theta_o)\theta_c$ (objective consumption) times the marginal family welfare of consumption. The marginal welfare is measured by $[\varepsilon + (\partial F/\partial Z_c)]$, deflated by the marginal utility of money λ to make it comparable with the earnings contributions. The marginal welfare of consumption is made of two elements: a direct marginal welfare effect $\partial F/\partial Z_c$, derived from consuming, and a survival welfare effect ε perceived only for those families whose economy provides only for a survival level of consumption (that is, $Z_c = \underline{Z}_c$). If the family is above the survival level (that is, $Z_c > \underline{Z}_c$), the marginal utility of survival ε will be zero. That is, families at the survival level will perceive more benefits from agricultural production than would families who are above such level; therefore, those at the survival level will have a higher propensity to innovate. This formal conclusion is the theoretical foundation for the observed phenomenon mentioned by some specialists (Cancian 1967; Peters 1973): very poor peasants have a higher propensity to adopt the recommendations of modernizing rural development projects than do medium-size farmers.

The right-hand side of equation (14.55) represents the two cost elements associated with the use of agronomic inputs x_a for a farmer exposed to Plan Puebla recommendations: the first cost component is the market price p_a of agronomic inputs; the second is the transaction cost ψ_{ox}/λ generated by the process of having access to input markets. For a farmer

Table 14-1. Linear Programming Tableau for Model of a Puebla Family's Farming Behavior

Concept	Value	Type	Technology Traditional C_a^t (1)	Technology Modern C_a^m (2)	Total output Y (3)	x_n (4)	x_{ph} (5)	x_p (6)	x_f (7)	Supply labor (1) a_{w1} (8)	Learning ΔC_{w1} (9)	Supply labor (2) a_{w2} (10)	Project participation (organizing) a_o (11)	Perceived deviations Traditional C_a^t (12)	Perceived deviations Modern C_a^m (13)
Objective function	Maximize		0	0	p_a	$-p_n$	$-p_{ph}$	$-p_p$	$-p_f$	w_1	q_{w1}	w_2	0	0	0
1. Maize output (kilograms)	O	=	γ_t	γ_m	-1										
2. Family time (days per year)	s	$\geq$	a_{ts}	a_{ms}						1		1	1		
3. Land (hectares)	C_a	$\geq$	a_{tl}	a_{ml}											
4. Nitrogen (kilograms)	0	=	a_{tn}	a_{mn}		-1									
5. Phosphate (kilograms)	0	=	a_{tph}	a_{mph}			-1								
6. Seeds (thousands)	0	=	a_{tp}	a_{mp}				-1							
7. Fixed cost per hectare	0	=	a_{tf}	a_{mf}					-1						
8. On-the-job experience	0	=									-1				
9. Minimum supply to job 1 (permanent job)	g^1_{min}	$\geq$								1					
10. Maximum supply to job 2 (temporary job)	g^2_{max}	$\geq$										1			
11. Credit availability	B	$\geq$				p_n	p_{ph}	p_p					$-b$		
12. Safety-first rule	$\underline{Z}_c$	$\geq$			p_a	$-p_n$	$-p_{ph}$	$-p_p$	$-p_f$	w_1		w_2		$-\eta\bar\sigma_{at}$	$-\eta\bar\sigma_{am}$
13. Combination constraint, $\bar\sigma_t$	0	=	1											-1	
14. Combination constraint, $\bar\sigma_m$	0	=		1											-1

who has no limited access to input markets, the marginal transaction cost ψ_{ox}/λ will be zero; therefore, this consideration will not influence his decisions. Accessibility to input markets $G_x(\cdot)$, as well as its associated cost ψ_{ox}/λ, will be determined by the institutional power C_o (for example, whether or not the farmer is a member of a solidarity) and by the direct organization help D provided by Plan Puebla extension agents.[21]

Empirical Model of Smallholders' Response to Modernization Projects in the Puebla Area

A linear model representing the behavior of a farm family in the Puebla area is developed and solved in this section. On the basis of previous empirical studies and data collections, as well as on a priori personal knowledge of the Puebla area and Plan Puebla administration, various initial states of development are postulated, and the associated farmers' activity levels are estimated.

A linear programming model of Puebla's farmer economics

A complete linear programming tableau is formulated in table 14-1 that *approximates* a solution to the control problem implied by the model given in the second through fourth sections ("Structure of the Rural Family Economy" through "Smallholders' Motivations") of the chapter and further elaborated in the chapter's mathematical appendix.

FARMERS' WELFARE FUNCTION. The objective function is linear with full income as its only argument.[22]

(14.56) $$\max H = F(\cdot) + q_{w1}\Delta C_{w1},$$

where

(14.57) $$F(\cdot) = Yp_a - \sum_i x_i\, p_i + s_{w1}w_1 + s_{w2}w_2$$

and the second term of the right-hand side of equation (14.56) is the demand value of investment in on-the-job experience.

FARMING ACTIVITIES. Maize production can be generated by means of two activities—a traditional agronomic practice labeled with the subindex

21. Other important determinants that are not explicitly introduced as arguments of $G_x(\cdot)$ are education, nutrition, and physical capital. Their effects will be reflected in the shift parameters of this function.

22. The objective function (14.56) is basically the Hamiltonian function of the mathematical appendix—that is, the farm family maximizes "full income" or earnings plus future discounted income generated by human capital.

t and a modern practice recommended by Plan Puebla and identified by the subindex *m*.

LABOR MARKET ACTIVITIES. There exist two major job opportunities, depending on the previous level of investment in human capital. The first, job 1, is basically a permanent kind of job; the second, job 2, is temporary in nature. Job 1 pays a higher wage rate than job 2 and, in addition, generates on-the-job experience. The demand price q_{w1} of one unit of investment in on-the-job experience is an exogenous variable in the linear programming model. This modification transforms a multistage control problem into a simpler, one-stage programming case. An additional simplification is made: specifically, investment in human capital is measured by the time worked in job 1—that is, $\Delta C_{w1} = s_{w1}$. Constraint 8 in the stub of the tableau (table 14–1) indicates that earnings (column 8) and human capital (column 9) are outcomes of joint production activities. Finally, the allocation of time to labor activities is subject to constraints imposed by labor-demand conditions: the permanent job 1 is subject to a minimum constraint, $g_{\min}^1$ (row 9) and the temporary job 2 is subject to a maximum constraint, $g_{\max}^2$ (row 10).

ORGANIZATIONAL ACTIVITIES. It is assumed that the farmer has been exposed to Plan Puebla and that he knows the nature and expected results of the recommended modern technology. He must then decide whether or not to participate in a solidarity group in order to have access to the credit and fertilizer markets. The value B of constraint 11 in table 14–1 is a measure of the farmer's own funds for fertilizer purchases. The funds that he can borrow from banks or from other credit institutions depend on the time that he allocates to organizational activities, s_o (column 11). The level of $s_o^\star$, determined by the optimal solution of the linear programming model, will be a proxy for the farmer's participation in Plan Puebla.

Table 14–2 shows the input-output coefficients of maize production, and table 14–3 the price parameters of the objective functions.

Subsistence constraint, risky earnings,
and behavior under uncertainty

Both net farming income and labor earnings are results of risky activities. Agronomic risk affecting yields per hectare is the major source of variance in net farm income because maize prices are relatively stable over time.[23] Annual variations in yields can result from variable rainfall patterns

---

23. The Mexican government maintains a policy of price support for maize. Producers can sell their maize production to CONASUPO (Compañía Nacional de Subsistencia Popular) at given official prices.

Table 14-2. *Estimated Input-Output Coefficients of Maize Production, Puebla Areas I–IV, 1971–72*

Variable	Symbol	Traditional $(i=t)$	Modern $(i=m)$
Family labor (days)	a_{is}	41	53
Land (hectares)	a_{ie}	1	1
Nitrogen (kilograms)	a_{in}	22	115
Phosphate (kilograms)	a_{iph}	15	40
Seeds (thousands)	a_{ip}	33	60
Fixed inputs (hectares)	a_{if}	1	1
Yields (kilograms)	y_i	2,000	3,500

Note: Coefficients are expressed in units of the variable per hectare per year.
Source: CIMMYT (1974*b*) and Villa Issa (1974).

Table 14-3. *Estimated Price Parameters of the Objective Functions, 1971–72*

Variable	Symbol	Pesos
Maize price (per kilogram)	p_a	0.92
Nitrogen price (per kilogram)	p_n	4.48
Phosphate price (per kilogram)	p_{ph}	3.17
Seed price (per thousand)	p_p	0.30
Fixed costs (per hectare)	p_f	35.0
Daily wage rate		
Job 1	w_1	36.0
Job 2	w_2	25.0
Demand price for days of on-the-job experience (job 1)	q_{w1}	5.0

Sources: CIMMYT, (1974*b*) and Villa Issa (1974).

and uncertain availability of fertilizers. Since the implementation of Plan Puebla, however, the dominant source of risk in the area has been vagaries in rainfall. Variations in labor earnings are relatively more important for jobs of a temporary nature, for which both the number of days that a peasant can work and the wage rate are nonfixed parameters.

The presence of risk in production activities, as well as proximity to the survival income level, conditions the farmer's selection of activities. Moscardi (1974) has argued that, in the case of the Puebla area, farmer behavior under uncertainty is best represented by safety-first rules.[24] In agreement with this argument, the linear programming model described above is transformed into a stochastic model. Telser's (1955–56) approach is followed and modified for implementation in the linear programming algo-

24. See also Kataoka (1963).

rithm. It is assumed that the farm family maximizes expected full income subject to a subsistence constraint:

$$(14.58) \qquad \max \hat{H} = \hat{F}(\cdot) + \hat{q}_{w1} \Delta C_{w1},$$

subject to

$$(14.59) \qquad P\left[\hat{F}(\cdot) \geq \underline{Z}_c\right] \geq (1 - \alpha),$$

where circumflexed ("hatted") symbols indicate expected values. Condition (14.59) is read as the probability of the expected current income $F(\cdot)$ being as large as the survival consumption level (in monetary terms). The (subjective) probability of disaster is measured by α.

The certainty equivalence of the above formulation is max $\hat{H}$ subject to

$$(14.60) \qquad \left[\hat{F}(\cdot) - \eta_\alpha \, \tilde{\sigma}_{\hat{F}}\right] \geq Z_c,$$
$$\eta_\alpha = A_\alpha^{-1}$$

where A_α is the cumulative distribution function of total income and $\tilde{\sigma}_{\hat{F}}$ is the perceived standard deviation of total income.

Farm income is the only stochastic variable considered in this model; based on a priori information, it is assumed that yields from traditional farming and modern farming are closely related because both are subject to the same ecological conditions. For analytical simplicity, the correlation coefficient of both is assumed to be positive and unity.[25] These assumptions, which are plausible for the case of the Puebla area, transform a nonlinear programming model into the much simpler case of a linear programming model.

Therefore, the expression (14.60) can be written as

$$(14.61) \qquad \left[\hat{F}(\cdot) - \eta_\alpha \left(C_a^t \tilde{\sigma}_{at} + C_a^m \tilde{\sigma}_{am}\right)\right] \geq Z_c$$

where $\tilde{\sigma}_a$ is the set of perceived standard deviations of net farming income per hectare. The condition (14.61) is represented by row 12 in table 14-1. This version of the safety-first criterion differs from other approaches in that the objective function does not include the standard deviation as an argument.

Empirical studies (CIMMYT 1974b) indicate that the observed distributions of maize yields in the Puebla area are normal, similar to other natural phenomena. Therefore, it is assumed that the peasant knows both the form of the distribution (that is, normality) and its two first moments (the mean and the standard deviation). In this case the value of η is estimated from a standard normal table. If the farmer does not know the form of the

---

25. See the final section of the appendix ("Solution when correlation coefficients are unity"), equations (14.90) through (14.95), for an understanding of the implications of correlation coefficients equal to $+1$.

probability distributions of incomes but only their mean and variance (see Moscardi 1974, p. 44), using the Chebychev inequality it is found (Telser 1955–56, p. 2) that:

$$(14.62) \qquad \eta_\alpha = \frac{1}{\sqrt{\alpha}}.$$

An additional feature of the present approach is the difference between the perceived (or subjective) distribution of probabilities and the observed (or objective) distribution of probabilities for each activity—that is, $\tilde{\sigma} = \beta\sigma$, where σ is standard deviation "observed" by researchers; $\tilde{\sigma}$ is standard deviation "perceived" by the farmer; and β is the degree of information of the farmer.[26] In other words, define:

$$[\beta = 1] = \text{complete information}$$

$$\begin{bmatrix} \beta > 1 \\ \\ \beta < 1 \end{bmatrix} = \text{incomplete information.}$$

In the empirical analysis of this study, it is assumed that the perception of the standard deviation of yields from traditional farming coincides with the observed. The justification for this assumption is that a farmer has gathered sufficient information through long experience to know what is the actual standard deviation. On the basis of previous empirical studies (Villa Issa 1974) and personal observations, however, it is assumed that the perceived standard deviation of maize yields from the modern technology is higher than the deviation observed in the agronomic experiments conducted by the Plan Puebla team.

A solution of the linearized model

The·model attempts to represent the economy of an average farm family of regions I–IV in the Puebla area. Table 14-4 describes the state of development and the economic opportunities of this family.

THE CASE OF AN AVERAGE FARM FAMILY. The family consists of six members whose available working time in a year is a maximum of 450 days, the working time of 1.5 adults. The family operates a holding of 2.5 hectares and controls funds for fertilizer expenses limited to 450 pesos per year. The family's investment in human capital (school years and trade skills) is relatively low and, therefore, the family has access only to jobs of a temporary nature (that is, job 2). Thus, the maximum expected demand for total family labor services cannot exceed 280 days per year. The

---

26. For a general treatment of this information problem, see Gould (1974).

Table 14-4. *Estimated Constraints Characterizing the Economic Opportunities of an "Average" Puebla Farm Family, 1971–72*

Variable	Symbol	Value
Holding size (hectares)	C_a	2.5
Family size (members)	S	6
Family working time (days)	s	450
Minimum demand for job 1 (days)	g^1_{min}	300
Maximum demand for job 2 (days)	g^2_{max}	280
Family-owned funds (pesos)	B	450
Subsistence income (pesos)	Z_c	10,000

Sources: Based on CIMMYT (1974*b*); Centro de Investigaciones Agrarias (1970).

family's subsistence consumption is estimated at 10,000 pesos per year. Table 14-5 presents alternative cases of the peasant perception of risk and his behavior in the face of uncertainty. The basic case of reference (case 0) is one in which the peasant perceives the true risk of traditional farming, perceives the risk of modern farming as twice the observed risk, and perceives no risk in wage rates but rather an awareness of the maximum job opportunities (row 10 in table 14-1). In addition, the family's probability of disaster, α, is established at 15 percent, implying that $\eta_\alpha = 1$.[27]

Once an optimal solution is obtained for the linear programming model of the average farm family under this specification, the quantitative changes in adoption rates are investigated for three alternative cases: the farmer correctly perceives the risk inherent in modern technologies (case 1); the initial state of development of the family is different from that of the average case; that is, different size of landholdings (case 2); and, for a given family (with permanent job opportunities), the displacement of the equilibrium is studied when the wage rate changes (case 3). In other words, in case 2 new linear programming solutions are obtained for different initial conditions, whereas in case 3 sensitivity analysis is used.

Table 14-6 presents the results of the optimal solution of case 0 of the linear programming model. The average peasant family will adopt the modern technology but only partially—40 percent of the land will be cultivated in accordance with Plan Puebla's recommendations, and the remainder under traditional methods. The rationale for this decision is

27. η_α was obtained from a standard normal table. Notice that the level of minimum consumption Z_c and the probability of disaster α are likely to be positively correlated; that is, a lower level of subsistence income (for example, at the biological level) is going to be associated with a lower probability α.

Table 14-5. *Estimated Parameters of Risk and Uncertainty, 1971–72*

| | Risk parameter | | Uncertainty |
Variable	Observed	Perceived	parameter
Standard deviation (in pesos) of:			
Net income from traditional farming	330	330	
Net income from modern farming	760	1,520	
Accepted probability of disaster			0.15
η_α			1

Sources: For observed risk parameters, CIMMYT (1974*b*); for perceived risk parameters, Villa Issa (1974).

that, given the risk perceptions and behavior of the farmer when he is confronted with uncertainty, his survival income becomes a binding constraint. To satisfy this constraint, the peasant opts for (technical) diversification. An alternative possibility is that the peasant will utilize a technique that represents a combination of the traditional and modern practices.

Even when the peasant adopts the new technology partially, his own financial resources will not be sufficient to cover the recommended expenditures for fertilizer. Therefore, he joins a credit group, allocating a full five days to organizational activities. This can be interpreted as an indicator of his decision to participate in Plan Puebla since, in fact, credit will not be a linear function of the number of days devoted to organizational duties.

In addition to agricultural activities, the family members will annually work 280 days—the maximum time demanded by the labor requirement of job 2. Therefore, the unemployment of the family members will be approximately 12 percent of the total available time.

With this combination of activities, the family will generate a total net income of 12,000 pesos per year, with 42 percent attributable to farming activities and 58 percent to extra farm activities.

The combination of technologies adopted by the peasant permits the production of 2,592 kilograms of maize per hectare; that is, 74 percent of the output possible if only modern technology were used.

THE CASE OF DIFFERENT PERCEPTIONS OF AGRONOMIC RISK (CASE 1). If the farmer's perception of risk coincides with the observed risk of modern agronomic practices (for example, $\tilde{\sigma}_{am} = \sigma_{am} = 760$), his adoption rate will be higher than in the former case (that is, he will cultivate his entire holdings under the modern technology). The first three columns of numbers in table 14-7 presents the set of optimal activities and results corresponding to this new formulation. A different pattern of allocation of

Table 14-6. *Optimal Linear Programming Solutions of the "Average" Puebla Farm Family's Activities, 1971–72*

Variable	Total for 2.5 hectares	Percent	Per hectare
	Solution (Case 0)		
Allocation of human time (days)			
Farming			
Traditional	62		
Modern	52		
Subtotal	114	25	46
Labor market			
Job 1			
Job 2	280	62	
Organizing	5	1	
Unemployment	51	12	
Total	450	100	
Allocation of land (hectares)			
Traditional	1.5	60	
Modern	1.0	40	
Total	2.5	100	
Use of agronomic inputs			
Nitrogen (kilograms)	147		59
Phosphate (kilograms)	62		25
Plants (thousands)	109		44
Use of funds (pesos)			
Private funds	450		
Credit from banks	405		
Total	855		342
Maize output (kilograms)	6,481		2,592
Income (pesos)			
Net farm income	5,000	42	2,000
Labor earnings	7,000	58	
Total	12,000	100	

Note: Blanks indicate "not applicable."

resources is generated because survival income no longer is a binding constraint. Even in the case of disaster (as anticipated by the farmer), the total net monetary income will be 14 percent higher than the survival income.

In this case the unemployment rate of the family is reduced from 12 percent to nearly 5 percent because the family allocates more time to farming activities and organizational activities. Net farm income will, therefore, increase 21 percent, and total family income will increase 10 percent relative to the initial formulation.

THE CASE OF LARGER HOLDINGS (CASE 2). If the family possesses 3.5 hectares of land instead of 2.5 hectares, and the other conditions presented above are maintained, the rate of adoption of the modern technology will increase. To satisfy the survival constraint, 86 percent of the land will be cultivated under modern technology, while the remaining 14 percent will still be operated with the traditional technology. The new results are given in the columns for case 2 of table 14-7.

Under these circumstances, there will be no unemployment. Also, more time will be allocated to farming because of the increased size of the holding, the use of the modern labor-intensive technology, and a considerable amount of time devoted to organizational activities.

THE CASE OF BETTER EMPLOYMENT OPPORTUNITIES (CASE 3). A case of a family whose embodied human capital is higher than for the average case is studied here. Because of the schooling level, on-the-job experience, and health, its members can work in permanent kinds of jobs. The wage rate w_1 of job 1 is estimated at 36 pesos per day, whereas the demand price of on-the-job experience q_{w1} is estimated at 5 pesos per unit (days) of learning by doing. If the job no longer generates new knowledge, the case can be reinterpreted as if the wage rate were 41 pesos per day.

A new optimal solution of the linear programming model is obtained. The results are written in the columns for case 3 of table 14-7. Because of the relative higher level of human capital, the family not only is paid a higher wage rate but also can employ all the available working time now used in agricultural activities. In other words, the family's unemployment rate drops to zero.

The survival constraint is not binding, and the family cultivates all its land with the modern technology and participates 3 percent of the time in organizational activities in order to obtain credit. The allocation of time to, and the output results from, agricultural activities of this case are the same as in case 1 in table 14-7.

The supply of labor response to wage rate changes was studied by means of sensitivity analysis. Variations of the wage rate within the range of 20 pesos and 36.50 pesos will not affect the quantity of labor supplied to job 1. But since the observed average wage rate of permanent jobs is about 36 pesos, individuals with slightly better job opportunities (that is, with wage rates equal to or higher than 36.50 pesos) will have fewer incentives to work on the farm and, other things being equal, to participate in the modernization project. This result challenges the effectiveness of labor-using technologies per se.

THE CASE OF LOWER SUBSISTENCE CONSUMPTION. The subsistence level of consumption attributed to the average rural family throughout the examples above shall not be considered a survival consumption determined by biological factors alone. Rather, it is also determined by biological and cultural factors. Historically, it represents the average net income of a

Table 14-7. Optimal Linear Programming Solutions for Cases Other than the "Average" Puebla Farm Family, 1971–72

	Solution								
	Case 1			Case 2			Case 3		
Variable	Total for 2.5 hectares	Percent	Per hectare	Total for 3.5 hectares	Percent	Per hectare	Total for 2.5 hectares	Percent	Per hectare
Allocation of human time (days)									
Farming									
Traditional	0			22			0		
Modern	133			157			133		
Subtotal	133	30	53	179	40	53	133	30	53
Labor market									
Job 1	280	62		252	56		303	67	
Job 2	14	3		19	4		14	3	
Organizing	23	5		0	0		0	0	
Unemployment	0	0		0	0		0	0	
Total	450	100		450	100		450	100	
Allocation of land (hectares)									
Traditional	0	0		0.5	14		0	0	
Modern	2.5	100		3.0	86		2.5	100	
Total	2.5	100		3.5	100		2.5	100	
Use of agronomic inputs									
Nitrogen (kilograms)	287		115	352		101	287		115
Phosphate (kilograms)	100		40	127		36	100		40
Plants (thousands)	150		60	195		56	150		60
Use of funds (pesos)									
Private funds	450			450			450		
Credit from banks	1,155		642	1,530		565	1,155		642
Total	1,605			1,980			1,605		
Maize output (kilograms)	8,750		3,500	11,442		3,270	8,750		3,500
Income (pesos)									
Net farm income	6,325	47	2,530	8,384	57	2,395	6,325	34	2,530
Labor earnings	7,000	53		6,296	43		10,910	58	
Investment in human capital	0	0		0	0		1,515	8	
Total	13,325	100		14,680	100		18,750	100	

Note: Blanks indicate "not applicable."

family that possesses the given characteristics. This is why the probability of disaster that is considered as acceptable by the family (see table 14-5) is relatively high ($\alpha = 0.15$). If the level of subsistence consumption were lower—for example, 7,500 pesos or even 5,000 pesos—it is likely that the probability of disaster would also be lower than $\alpha = 0.15$.

Table 14-8 presents a sensitivity analysis of the linear programming solutions with respect to the subsistence income Z_c for each case studied above. In all cases but case 1, lower levels in this constraint (for a probability of disaster $\alpha = 0.15$) will change the optimal solution. The solution for case 1 is very interesting for, when farmers have perfect information (that is, they perceive the time risk of modern farming), the solution will be the same even for levels of survival consumption (that is, $Z_c \geq 4,424$ pesos).

To complete the example, table 14-9 shows the allocation pattern of case 4, in which the subsistence income is set at 7,500 pesos a year. In this case, probability of disaster is assumed at $\alpha = 0.05$ and, therefore, $\eta_\alpha = 1.64$; that is, the relative cost imputed to risky activities is higher than in the latter cases. All other variables and parameters are the same as in case 0 in table 14-6.

Summary

This chapter has developed both a theoretical and a linearized empirical model for explaining the economic behavior of a farming family within economic structures of the minifundio kind. More specifically, the model has attempted to represent a farmer's response to a modernizing rural development project.

The following are the major conclusions drawn from the study:

• According to the theoretical representation and the empirical exemplification of a farm family's economy, different initial states of development of these familial units determine varying optimal combinations of

Table 14-8. *Sensitivity Analysis of the Linear Programming Solutions with Respect to Subsistence Consumption Z_c in Puebla, 1971–72*

Case	Subsistence consumption (pesos)		Case	Subsistence consumption (pesos)	
	Lower bound	Upper bound		Lower bound	Upper bound
0	9,525	10,256	2	9,585	11,190
1	4,424	11,425	3	10,924	12,268

Source: Sensibility analysis of the linear optimal programming solutions.

Table 14-9. *Optimal Linear Programming Solutions of the "Average" Puebla Farm Family's Activities (for Sensitivity Tests on Subsistence Consumption and α), 1971–72*

Variable	Total for 2.5 hectares	Percent	Per hectare
Allocation of human time (days)			
Farming			
Traditional	15		
Modern	113		
Subtotal	128	28	51
Labor market			
Job 1			
Job 2	280	62	
Organizing	12	3	
Unemployment	30	7	
Total	450	100	
Allocation of land (hectares)			
Traditional	0.4	16	
Modern	2.1	84	
Total	2.5	100	
Use of agronomic inputs			
Nitrogen (kilograms)	252		101
Phosphate (kilograms)	91		36
Plants (thousands)	140		56
Use of funds (pesos)			
Private funds	450		
Credit from banks	968		
Total	1,418		567
Maize output (kilograms)	8,145		3,258
Income (pesos)			
Net farm income	5,995	46	2,398
Labor earnings	7,000	54	
Investment in human capital			
Total	12,995	100	

Note: A subsistence-level income of 7,500 pesos a year is assumed (probability of disaster α = 0.05). Blanks indicate "not applicable."

farming and nonfarming activities and varying combinations of traditional and modern agronomic practices. Differences in adoption rates for agronomic techniques among peasants can be explained by a detailed empirical examination of the distribution of farmers by age, schooling, family size, on-the-job experience, ability to relate with institutions, saving capability, minimum subsistence income, and so forth. Therefore, more emphasis should be placed on the study of *variances* than on the analysis of *mean* values of which empirical studies in the past have tended to concentrate.

• The level of adoption of modern agronomic practices is determined by considerations of added costs and benefits, the components of which are of a specific nature: (1) additional costs include the price of new inputs, the opportunity cost of learning new agronomic practices and gathering information, and the transaction costs of having access to input and credit markets by way of community development; (2) in minifundio farming that is integrated with labor market economies, the benefits of modern agronomic practices are not only weighed against the benefits derived from traditional farming but also against earning opportunities in the extra farm-labor market.

• For farm families whose accumulated human and physical capital is so low that they are able to generate only a survival level of income, new activities are also evaluated according to their marginal survival utility. This provides a theoretical explanation for the observed case, in which farmers with relatively small holdings demonstrate a higher propensity to adopt modern recommended practices.

• Empirical investigation of risk and uncertainty should be able (1) to differentiate observed risk (for example, by means of controlled agronomic experiments such as Plan Puebla) from perceived agronomic risk; (2) to investigate perceived risk in the labor market; and (3) to differentiate between perception of agronomic risk (that is, the subjective distribution of probabilities) and behavior in response to factors of uncertainty (that is, survival probabilities and minimum subsistence income in the case of safety-first rules or utility parameters in the case of expected utility approaches).

• A simple linearized model, such as the one developed in the empirical section of the chapter, presents a useful approach for explaining—and perhaps forecasting—rates of adoption of new technology. Some contributions of this simple model for project design are that it establishes, with a fair degree of accuracy, what is a minimum set of data that should be collected in any empirical study before designing a rural modernization project and that it can forecast the likely pattern of adoption by the farmers of a given area.

• The additional empirical research to improve the explanatory and forecasting power of the above model will require socioanthropological case studies, econometric estimations, and budgeting studies. Socioanthropological case studies will help in extending the model to account for (1) seasonality or intrayear allocation of human time and its relations to the labor supply and the unemployment periods; (2) allocation of human time over the family life cycle; (3) intrafamily allocation of human time or distribution of activities by sex and age; (4) decisions on fertility and family size within the farm family; and (5) nutrition activities in the family. A complete study of the welfare effects of modernization projects at the household level could account for their effects over all the mentioned dimensions of the rural economy. Econometric estimation is necessary for two major purposes: First, econometric studies would express the relations between subsistence consumption Z_c, the state of development of the rural family, the probability of disaster α, and the perceived risk.[28] The values of these variables used in the present study were obtained from a priori information, but a more complete study requires an econometric approach. Second, econometric studies would test the adoption patterns explained by the peasant family model discussed above.[29] Finally, budgeting studies are needed to estimate input–output coefficients of agricultural activities other than maize production.

• As the empirical part of this chapter has exemplified, a personal knowledge of farmers and their leaders—and the exchange of ideas with agronomists, sociologists, and communication specialists familiar with the study area—can also provide valuable sources of information for building a model of the farm family. This is possible when the research is guided by an explicit model of political economy (to understand the structure of the rural economy and its function within the social system) and a specific model of the allocation of human time.

Conclusions and Implications for Policy

Differences in the endowment of human capital, physical capital, and organizational power among farmers determine the differences in opportunity costs of human time, transaction costs, and behavior in the face of risky events. The particular combination of these differences determines the observed distribution of agronomic adoption rates among rural households.

The structural condition of the rural economy studied in this chapter challenges the simplistic view that labor-intensive technologies per se will rapidly increase agricultural production and improve farmers' welfare. If the objective of a rural development policy is to increase agricultural production in the *minifundista* sector, the following steps are necessary. At the level of agronomic research, the generation of less risky technologies is required—for example, adaptability of the new high-yielding crop varieties to different environmental conditions and development of improved varieties and practices for crop systems. At the level of design and implementation of modernization projects, the organizational component

28. The research project of Edgardo Moscardi (1974) is oriented in this direction.

29. Various econometric estimations have already been done in master's theses of the Postgraduate School of Chapingo. Their findings, in general, support the conclusions of the model of this chapter (see Villa Issa 1974, and Moscardi 1971).

should have as much importance as the information (extension) component—Plan Puebla is a good example of the complementarity between organization and communication in dissemination processes. At the adoption level, an adaptive strategy should be offered. For example, one might begin with a package of land-saving and labor-using technologies, such as that which characterizes the "green revolution," so as to generate an accumulation process within the household, and then introduce more capital-using technologies. When doing so, two complementary approaches can be followed: to promote the development and adoption of intermediate technologies that are not labor displacing and can be produced within the rural sector, and to continue the process of community organizing by developing intermediate organizations (as, for example, in a few cases of Plan Puebla in which solidarity groups bought tractors that were then rented to individual households).

Economists can provide valuable assistance in the design and evaluation of these kinds of programs if their research is grounded in a model of political economy that takes into account the overall opportunity set of rural households within each specific socioeconomic structure and a structural form model of the allocation of human time at the household level. This simple application of economics promises to be a necessary substitute for reduced-form representations and unimodal explanations of the adoption problem.

Further theoretical and empirical research will require an expansion of the model to study the complexities of crop production systems (for example, maize-bean combinations) including the seasonal and intrafamily allocation of human time (for example, women-men and adults-children). Also required will be the development of a multistage model to investigate the changes of farmers' perception of risk over time, β, because of learning experiences as well as changes in the subjective probability of disaster, α_1, as resource constraints change.

Mathematical Appendix

The farming family can be viewed as an economic agency trying to solve an optimal control problem. In order to discover the optimal decision rules of the family, the Maximum Principle, combined with Kuhn-Tucker conditions, is applied (see Hestenes 1966; Hochman, Hochman, and Razin 1973).

Definition of variables

Activity variables:

Y = agricultural production
Z_a = sales of agricultural products
Z_w = earnings from the labor market
Z_c = family consumption (physical units)
Z_k = flow of agronomic knowledge (in abstract units)
Z_o = flow of organizational contribution (in abstract units)
Z_{kw} = flow of on-the-job experience (in abstract units).

Control variables:

s_a = human time allocated to agricultural activities
s_w = human time allocated to labor market activities
π_a = proportion of agricultural time allocated to farming
π_k = proportion of agricultural time allocated to learning agronomic practices
π_o = proportion of agricultural time allocated to organization
x_a = flow of agronomic inputs (seeds, fertilizers, insecticides)
x_c = flow of consumption goods bought in the market
θ_m = proportion of agricultural output sold in the market
θ_c = proportion of agricultural output consumed by the family
J = flow of investment in physical capital (in abstract units)
M = net payments to creditors.

State variables:

C_a = stock of physical capital (in abstract units)
C_k = stock of knowledge of agronomic practices (in abstract units)
C_w = stock of on-the-job experience (in abstract units)
C_o = stock of organizational power of the solidarity group (in abstract units)
B = debt.

Implicit values and prices:

ξ_a = marginal utility of on-the-farm employment opportunity

ξ_w = marginal utility of labor market employment opportunities

ψ_{ax} = marginal utility of efficient allocation of agronomic inputs

ψ_{as} = marginal utility of efficient allocation of farming time

ψ_{ox} = marginal transaction cost (in units) in the input market

ψ_{ob} = marginal transaction cost (in units) in the credit market

χ = marginal utility of agricultural output

ζ_a = marginal utility of agricultural sales

ζ_w = marginal utility of labor earnings

ζ_c = marginal utility of consumption

ζ_k = marginal utility of agronomic knowledge

ζ_o = marginal utility of organizational contribution

ε = marginal utility of survival

μ = marginal utility of human time

ϕ = marginal utility of one unit of agricultural output

λ = marginal utility of money

γ_{ow}^j = marginal cost (in units) of social cooperation

q_a = stock demand price of physical capital

q_w = stock demand price of on-the-job experience

q_k = stock demand price of agronomic knowledge

q_o = stock demand price of organizational power

q_b = stock demand price of credits.

Exogenous variables and parameters:

p_m = price of agricultural products

p_a = price of agronomic inputs

p_c = price of market consumption goods

p_j = price of investment goods

D = level of organizing activities of the Puebla project team

$\underline{Z}_c$ = survival level of consumption

s = total available human time.

Solution of the optimal control problem

At each t $(t_0 \le t \le t_1)$, there exists a Lagrangian defined by:

$$\max \mathscr{L} \cdot \exp\left[-\tau\,(t-t_0)\right] = \mathscr{H}$$
$$+ \chi\,[s_a\pi_a g_a\,(x_a,\,C_a,\,C_k) - Y]$$
$$+ \psi_{ax}\,[x_a - \ell_x\,(C_k)] + \psi_{as}\,[s_a\,\pi_a - \ell s\,(x_a)]$$
$$+ \zeta_a\,[Y\theta_m p_m - Z_a] + \zeta_w\,[s_w g_w\,(C_w) - Z_w]$$
$$+ \xi_w\,[g_e\,(C_w) - s_w]$$

(14.63)
$$+ \zeta_c\,[g_c\,(Y\theta_c,\,x_c) - Z_c] + \varepsilon\,[Z_c - \underline{Z}_c]$$
$$+ \phi\,Y[1 - \theta_c - \theta_m]$$
$$+ \mu\,[s - s_a\,(\pi_a + \pi_k + \pi_o) - s_w]$$
$$+ \lambda[Z_a + Z_w - \sum_i x_i p_i - Jp_j - M]$$
$$+ \zeta_k\,[s_a\pi_k g_k\,(D,\,C_k,\,C_o) - Z_k]$$
$$+ \zeta_o\,[s_a\pi_o g_o\,(D) - Z_o]$$
$$+ \sum_j \zeta_o^j\left[s_a^j\pi_o^j g_o^j\,(D^j) - Z_o^j\right]$$
$$+ \sum_j \gamma_{ow}^j\left[F_o^j\,(C_o) - \bar{F}_o^j\right] + \sum_j \mu^j\left[\bar{s}^j - s_a^j\pi_o^j\right]$$
$$+ \psi_{ox}\,[G_x\,(C_o,\,D) - x_a] + \psi_{ob}\,[G_b\,(C_o,\,D) - B],$$

where $\mathscr{H}$ is the Hamiltonian function:

(14.64) $\mathscr{H} = F\,(Z_c) + q_a\,[J - \delta_a C_a] + q_w\,[f_w\,(Z_w) - \delta_w C_w]$

$$+ q_k\,[Z_k - \delta_k C_k] + q_o\left[Z_o + \sum_j Z_o^j - \delta_o C_o\right] + q_b\,[rB - M].$$

The necessary conditions for an optimal behavior are given by the following.

Allocation among activities:

(14.65) $\dfrac{\partial L}{\partial Y} = -\chi + \zeta_a\theta_m p_m + \zeta_c\left(\dfrac{\partial g_c}{\partial Y\theta_c}\right)\theta_c = 0$

(14.66) $\dfrac{\partial L}{\partial Z_a} = -\zeta_a + \lambda = 0$

(14.67) $\dfrac{\partial L}{\partial Z_w} = -\zeta_w + \lambda + q_w\dfrac{\partial f_w}{\partial Z_w} = 0$

(14.68) $\dfrac{\partial L}{\partial Z_c} = -\zeta_c + \varepsilon + \dfrac{\partial F}{\partial Z_c} = 0$

(14.69) $\dfrac{\partial L}{\partial Z_k} = -\zeta_k + q_k = 0$

$$(14.70) \qquad \frac{\partial L}{\partial Z_o} = -\zeta_o + q_o = 0$$

$$(14.71) \qquad \frac{\partial L}{\partial Z_o^j} = q_o - \zeta_o^j = 0 \qquad\qquad j = 1, \ldots, J$$

Allocation of human time:

$$(14.72) \qquad \frac{\partial L}{\partial s_a} = \chi \pi_a g_a(\cdot) + \psi_{as} \pi_a - \mu + \zeta_k \pi_k g_k(\cdot) + \zeta_o \pi_o g_o(\cdot) = 0$$

$$(14.73) \qquad \frac{\partial L}{\partial s_w} = \zeta_w g_w (C_w) - \xi_w - \mu = 0$$

Allocation of agronomic inputs:

$$(14.74) \qquad \frac{\partial L}{\partial x_a} = \chi s_a \pi_a \left(\frac{\partial g_a}{\partial x_a}\right) + \psi_{ax} - \psi_{as}\left(\frac{\partial \ell_s}{\partial x_a}\right) - \lambda p_a - \psi_{ox} = 0$$

$$(14.75) \qquad \frac{\partial L}{\partial x_c} = \zeta_c \left(\frac{\partial g_c}{\partial x_c}\right) - \lambda p_c = 0$$

Use of agricultural output:

$$(14.76) \qquad \frac{\partial L}{\partial \theta_m} = \zeta_a Y p_m - \phi Y = 0$$

$$(14.77) \qquad \frac{\partial L}{\partial \theta_c} = \zeta_c \left(\frac{\partial g_c}{\partial Y \theta_c}\right) Y - \phi Y = 0$$

Demand for credit:

$$(14.78) \qquad \frac{\partial L}{\partial M} = q_b - \lambda = 0$$

$$\lambda = q_b.$$

In addition, the following conditions must be satisfied; they require that, along the optimal life development path, there are no gains from the passage of time as such (the overdot denotes d/dt):

$$(14.79) \qquad \dot{q}_a = q_a (\tau + \delta_a) - \chi s_a \pi_a \left(\frac{\partial g_a}{\partial C_a}\right)$$

$$(14.80) \qquad \dot{q}_w = q_w (\tau + \delta_w) - \zeta_w s_w \left(\frac{\partial g_w}{\partial C_w}\right) - \xi_w \left(\frac{\partial g_e}{\partial C_w}\right)$$

$$(14.81) \qquad \dot{q}_k = (\tau + \delta_k) q_k - \chi s_a \pi_a \left(\frac{\partial g_a}{\partial C_k}\right) + \psi_{ax} \left(\frac{\partial \ell_x}{\partial C_k}\right) - \zeta_k s_a \pi_k \left(\frac{\partial g_k}{\partial C_k}\right)$$

$$(14.82) \qquad \dot{q}_o = q_o \left(\tau + \delta_o\right) - \zeta_k s_a \pi_k \left(\frac{\partial g_k}{\partial C_o}\right) - \left(\sum_j \gamma_{ow}^j\right) \left(\frac{\partial F_o^j}{\partial C_o}\right)$$

$$- \psi_{ox} \left(\frac{\partial G_x}{\partial C_o}\right) - \psi_{ob} \left(\frac{\partial G_b}{\partial C_o}\right)$$

$$(14.83) \qquad \dot{q}_b = q_b \left(\tau - r\right) + \psi_{ob}.$$

Finally, the complementary Kuhn-Tucker conditions for each one of the constraints have also to be satisfied.

Sufficient conditions are also satisfied because of the concavity assumption imposed on the Hamiltonian function.

The time constraints (14.25) and (14.26) and the associated conditions (14.72) and (14.73) define a bounded maximization problem. In addition, these control variables enter the Lagrangian function in a linear form. Therefore, the solution has to be of the "bang-bang" type. Rewriting the Lagrangian (14.63) as

$$(14.84) \quad \max \left\{ \mathcal{L} \cdot \exp\left[\tau \left(t - t_0\right)\right] - R - \sum_j b_d^{'j} s_o^j \right\} = b_a' s_a + b_w' s_w$$

$$(14.85) \qquad \begin{cases} b_a' = b_a - \mu \\ b_w' = b_w - \mu \\ b_o^{'j} = b_o^{'j} - \mu^j \qquad\qquad j = 1, \ldots, J \end{cases}$$

$$(14.86) \qquad b_a = \pi_a \left[\chi g_a(\cdot) + \psi_{as}\right] + \pi_k \zeta_k g_k(\cdot) + \pi_o \zeta_a g_o(\cdot)$$

$$(14.87) \qquad b_w = \zeta_w g_w(\cdot) - \xi_w$$

$$(14.88) \qquad b_o^j = \pi_o^j \zeta_o^j g_o^j \left(D^j\right) \qquad\qquad j = 1, \ldots, J$$

and R includes all other terms in (14.63) and (14.64).

For given values of $x_i, J, M, s_a^j \pi_o^j, C, B, \chi, \xi, \psi, \zeta, \varepsilon, \phi, \mu, \lambda, \gamma,$ and q, the maximization of (14.84) reduces to:

$$(14.89) \qquad \begin{cases} \begin{cases} s_a = g_f(\cdot)\, \pi_a^{-1} \\ s_w = \min\left\{(s - s_a), g_e(\cdot)\right\} \end{cases} \Bigg| \text{if } b_a > b_w \quad t_0 \le t \le t_1 \\[2em] \begin{cases} s_w = g_w(\cdot) \\ s_a = \min\left(s - s_w\right), g_f(\cdot)\} \end{cases} \Bigg| \text{if } b_a < b_w \quad t_0 \le t \le t_1. \end{cases}$$

If $b_a = b_w$, the maximum is not unique, and the solution with respect to s_a and s_w is arbitrary.

Solution when correlation coefficients are unity

When all the correlation coefficients of a stochastic programming model are unity, the problem reduces to a case of a linear programming problem (see Sinha 1963, p. 11).

Since a correlation coefficient, ρ, always lies between -1 and $+1$ (that is, $|\rho| \leq 1$), we have

$$(14.90) \qquad \left(x' \, \Sigma \, x\right)^{1/2} \leq \sum_{j=1}^{n} \sigma_{jj} \, x_j,$$

where the equality holds only when all the correlation coefficients are $+1$.

For the case that

$$(14.91) \qquad \rho = \frac{\sigma_{tm}}{\sqrt{\sigma_t^2 \sigma_m^2}} = 1$$

$$(14.92) \qquad \sigma_{tm} = \sigma_t \sigma_m \cdot \, ,$$

define:

$$(14.93) \qquad x = \left(C_a^t C_a^m\right)$$

$$(14.94) \qquad \sigma_{\hat{F}}^2 = x' \, \Sigma \, x = \left(C_a^t C_a^m\right) \begin{bmatrix} \sigma_t^2 & \sigma_{tm} \\ \sigma_{mt} & \sigma_m^2 \end{bmatrix} \begin{bmatrix} C_a^t \\ C_a^m \end{bmatrix}$$

Inserting (14.92) into (14.94) and operating, then

$$(14.95) \qquad \begin{aligned} \sigma_{\hat{F}}^2 &= x' \, \Sigma \, x = \left(C_a^t \sigma_t + C_a^m \sigma_m\right)^2 \\ \sigma_{\hat{F}} &= C_a^t \sigma_t + C_a^m \sigma_m. \end{aligned}$$

References

Benito, C. A. 1976. "Peasants' Response to Rural Development Projects in Puebla." *American Journal of Agricultural Economics*, vol. 58 (May).

———. 1973. *The Choice Model of Human Resource Development—A Methodological Note*. Davis: University of California.

Cancian, F. 1967. "Stratification and Risk-taking: A Theory Tested on Agricultural Innovation." *American Sociological Review*, vol. 32, no. 6 (December), pp. 912–27.

Centro de Investigaciones Agrarias. 1970. *Estructura agraria y desarrollo agrícola en México*. Tomo II. Mexico City.

Chayanov, A. V. 1966. *Theory of Peasant Economy.* Edited by D. Thorner and others. Homewood, Ill.: American Economic Association.

CIMMYT (Centro Internacional para el Mejoramiento de Maíz y Trigo, International Maize and Wheat Improvement Center). 1974a. *El Projecto Puebla, 1967–69.* Mexico City.

———. 1974b. "The Puebla Project: Seven Years of Experience, 1967–73." Mexico City. Processed.

del Campo, M. 1972. *La capacidad del agricultor para adoptar tecnología: Caso de estudio para el P. Puebla.* Chapingo, Mexico: Colegio del Postgraduados.

Galjart, B. 1971. "Rural Development and Sociological Concepts: A Critique." *Rural Sociology,* vol. 36, no. 1 (March), pp. 31–41.

Gartrell, J. W., E. A. Wilkening, and H. A. Presser. 1973. "Curvilinear and Linear Models Relating Status and Innovative Behavior: A Reassessment." *Rural Sociology,* vol. 38, no. 4 (Winter), pp. 391–411.

Gould, John P. 1974. "Risk, Stochastic Preference, and the Value of Information." *Journal of Economic Theory,* vol. 8, no. 1 (May), pp. 64–84.

Grossman, M. 1972. "On the Concept of Health Capital and the Demand for Health." *Journal of Political Economy,* vol. 80, no. 2 (March–April), pp. 223–55.

Hestenes, M. R. 1966. *Calculus of Variations and Optimal Control Theory.* New York: John Wiley and Sons, 1966.

Hochman, Eithan, Oded Hochman, and Assaf Razin. 1973. "Demand for Investment in Productive and Financial Capital." *European Economic Review,* vol. 4, no. 3 (April), pp. 67–83.

Hymer, S., and S. Resnick. 1969. "A Model of an Agrarian Economy with Nonagricultural Activities." *American Economic Review,* vol. 59, no. 4 (September), pp. 493–506.

Kataoka, Shinji. 1963. "A Stochastic Programming Model." *Econometrica,* vol. 31, nos. 1-2 (January-April), pp. 181–96.

Lipton, M. 1968. "The Theory of the Optimizing Peasant." *The Journal of Development Studies,* vol. 4, no. 3 (April), pp. 327–51.

Moscardi, E. R. 1974. "A Behavioral Model of Adoption of New Technologies among Smallholding Farmers." Berkeley: University of California. Processed.

———. 1971. *Riesgo y Transferencia de Tecnología: Estudio para el Caso de P. Puebla.* Chapingo, Mexico: Colgeio de Postgraduados.

Nakajima, C. 1969. "Subsistence and Commercial Family Farms: Some Theoretical Models of Subjective Equilibrium." In *Subsistence Agriculture and Economic Development.* Edited by C. R. Wharton, Jr. Chicago: Aldine Publishing Co.

O'Mara, G. 1972. "A Decision-theoretic View of an Agricultural Diffusion Process." Presented at the Purdue Workshop on Empirical Studies of Small-farm Agriculture in Developing Nations. West Lafayette, Indiana, 1972.

Peters, J. B. "Factors Hindering More Widespread Participation in Puebla Project." Puebla, Mexico. Processed.

Pontryagin, L. S., B. G. Boltyanskii, R. V. Gamkrelidge, and R. F. Mischenko.

1962. *The Mathematical Theory of Optimal Processes.* Translated by K. N. Trirogoff. New York: Wiley/Interscience.

Roberts, Blaine, and Bob R. Holdren. 1972. *Theory of Social Process: An Economic Analysis.* Ames: Iowa State University Press.

Rogers, Everett M. 1969. *Modernization among Peasants—The Impact of Communication.* New York: Holt, Rinehart and Winston.

――――. *Diffusion of Innovations.* 1962. New York: Free Press.

Roumasset, J. 1973. *Estimating the Risk of Fertilization: Rice Production in the Philippines.* University of California, Davis, Department of Economics Working Paper, no. 32.

Schultz, Theodore W. 1972. "The Increasing Economic Value of Human Time." *American Journal of Agricultural Economics*, vol. 54, no. 5 (December), pp. 843–50.

――――. 1964. *Transforming Traditional Agriculture.* New Haven: Yale University Press.

Sen, A. K. 1966. "Peasants and Dualism with or without Surplus Labor." *Journal of Political Economy*, vol. 74, no. 5 (October), pp. 425–56.

Sinha, S. M. 1963. "Stochastic Programming." Ph.D. dissertation. University of California, Berkeley, Department of Statistics.

Telser, L. G. 1955–56. "Safety First and Hedging." *Review of Economic Studies*, vol. 23, no. 1, pp. 1-16.

Todaro, M. P. 1969. "A Model of Labor Migration and Urban Unemployment in Less Developed Countries." *American Economic Review*, vol. 59, no. 1 (March), pp. 138–48.

Villa Issa, Luis A. 1974. "Adopción de tecnología nueva en zonas de temporal: El efecto del factor incertidumbre." M.A. thesis. Colegio de Postgraduados, Escuela Nacional de Agricultura, Chapingo, Mexico.

Welch, Finis. 1970. "Education in Production." *Journal of Political Economy*, vol. 78, no. 1 (January), pp. 35–59.

Weiss, Y. 1971. "Learning by Doing and Occupational Specialization." *Journal of Economic Theory*, vol. 3, no. 2 (June), pp. 189–98.

Winkelmann, Donald. 1972. *The Traditional Farmer: Maximization and Mechanization.* Paris: Organisation for Economic Co-operation and Development (OECD).

15

Procedures for Treating Interdependence in the Appraisal of Irrigation Projects

Luz María Bassoco, Roger D. Norton,
and José S. Silos

This chapter discusses the appraisal of irrigation investment projects at the micro-level—more specifically, at the level of an irrigation district—in relation to the kinds of agricultural district models used throughout this volume.

Overview

The chapter comprises seven different examples of the application of optimization models to public policy decisions in an environment in which decisionmakers are concerned with the simultaneous use of several instruments of economic policy. There are two overall themes in the discussion. The first one concerns the types of interdependence that affect the evaluation of irrigation investment projects. The second theme concerns the use of linear programming as a tool for capturing this interdependence and for reflecting alternative goals in project evaluation. Further details on the models used herein may be found in chapters 2, 4, and 16, and in Bassoco and others (1973), Bassoco, Norton, and Silos (1972), and Duloy, Kutcher, and Norton (1973).

The interdependence discussed here affects in some cases only the benefit side, and in other cases both the benefit and cost sides of a project evaluation. One example of the former is the set of complex relations between an investment program and other instruments of government policy, such as guaranteed prices. The stream of benefits to a proposed irrigation project is affected significantly by the government's decision on whether to guarantee the prices of crops grown with the aid of irrigation in

Note: This chapter first appeared as "Appraisal of Irrigation Projects and Related Policies and Investments," *Water Resources Research*, vol. 10, no. 6 (December 1974), pp. 1071–79; © American Geophysical Union. Permission of the original publisher to use material here, with some editorial changes, is gratefully acknowledged.

that area. Benefits are also affected by the availability of short-term credit, the price of chemical inputs, and other factors that can be influenced by public policy.

In all, this chapter deals with three kinds of interdependence: that between investment projects and other policy instruments, that among different kinds of investment projects, and that between local and sector-level decisions. All three kinds have been analyzed by means of linear programming models of irrigated agricultural producing areas—generally with parametric programming experiments performed on these models.

Generation of additional agricultural employment has been selected as illustration of an alternative policy goal, and linear programming analyses of employment possibilities are presented. Unlike most mathematical programming treatments of alternative goals, these models do not maximize a function of employment, but rather simulate producers' responses under different specified public policies. This approach ensures the possibility of implementation for a program outlined with the model.

The application of linear and dynamic programming to irrigation planning is by now common practice. To cite some examples of previous work, Soltani-Mohammadi (1972) used linear programming to analyze the choice of irrigation technique given the cropping patterns, and with the same method Rogers and Smith (1970) explicitly allowed for the interdependence between cropping patterns and selection of investment project. With strongly restricted variability in the cropping pattern, Gisser (1970) also used linear programming to estimate the demand for water as its marginal value product in agricultural activities. Cummings and Winkelmann (1970) made use of dynamic programming to discuss the optimal rate of release of stored water and its relation to the determination of the cropping pattern. Dudley, Howell, and Musgrave (1971a) have used stochastic dynamic programming to study alternative rates of release of stored water during successive ten-day stages in the growing season, given uncertainty about weather. They treated the case of one crop on the basis of detailed, experimental biological data on the effects of various moisture conditions on roots. In a companion paper (1971b), they solved for the optimal acreage planted, again for one crop, in the face of stochastic water supplies and demands. Both the Rogers and Smith study (1970) and a paper by Young and Bredehoeft (1972) were concerned with joint management of surface water and groundwater resources, and in this respect they treated interdependence among different kinds of investment projects. None of the studies, however has addressed the interdependence between investments in water and in other resources (for example, agricultural research outlays) or the interdependence between investment on the one hand and product pricing and other noninvestment policies on the other.

To make this study as relevant as possible to policy, care has been taken to construe the results of the analysis in a potentially applicable way. This chapter is organized around seven analytic experiments, and for each experiment italicized examples are provided of prescriptions that may be logically inferred from the analysis.

Two producing areas have been selected for study: the Bajío region in Mexico's Central Plateau and the Río Colorado district in the Northwest. Both areas are endowed with irrigation water—from tubewells as well as river control systems—but they offer interesting contrasts in production patterns. El Bajío is an older producing area of irrigated farms about 8 hectares in average size on which relatively labor-intensive techniques are used. There is considerable rainfed cultivation in the area around the irrigation zone, and these are also represented in the model. The Río Colorado irrigation district is a newer producing area and is located in an arid region in which no cultivation is possible without irrigation. The typical farm there is more than twice as large as its counterpart in El Bajío, and mechanization is more extensive. The mechanization is influenced not only by farm size but also by interregional differences in the cost of farm labor. The population density is much less in the Northwest than in the Central Plateau; this, combined with the proximity of the Río Colorado district to the United States border, has resulted in a farm labor wage that is more than twice as high as in El Bajío.

The Bajío area, as defined for the model, comprises 432,000 cultivated hectares in the states of Guanajuato and Michoacán—of which 360,000 hectares are rainfed, 112,500 hectares are irrigated by gravity-fed water, and 60,000 hectares are irrigated with pumped water from tubewells. The production pattern in the rainfed area is based on maize, beans, sorghum, and chickpeas. In the irrigated zones, a greater variety of crops is produced, including wheat, barley, tomatoes, alfalfa, garlic, and strawberries.

The average annual net income per producer in El Bajío, as estimated by solution of the linear programming model, was 12,400 pesos in the early 1970s. (From the mid-1950s up to the time of this study, the exchange rate in Mexico had been fixed at 12.5 pesos to the U.S. dollar.) This amount varies in accordance with the size of the plot and whether it is irrigated or not. The large irrigated farms that average 23 hectares in size generate about 55,000 pesos of annual net income, whereas at the other extreme a small rainfed farm of 7 hectares generates only 5,800 pesos.

The Río Colorado irrigation district embraces 203,000 hectares, of which about 120,000 hectares are irrigated by gravity-fed water from the Colorado River, whose use is regulated by an international treaty, and about 80,000 hectares are irrigated by pumping from deep wells.

The production patterns show that cotton and wheat are the most important crops grown in the area. Other crops are barley, alfalfa, safflower, oats, garlic, sorghum, and maize. Cotton traditionally has been the single most important crop in the region. Because of salinity problems and infestation of pests, however, the cotton yields and the number of hectares planted in cotton have decreased. Wheat has become the most important crop because of its higher yields and its guaranteed sale price to the government's marketing agency.

The annual net income per producer in the Río Colorado area, as estimated by the linear programming model, amounted to 25,880 pesos. This income is obtained from a representative farm of about 18 hectares. In both the BAJIO and RIO COLORADO models, casual livestock raising and tree crop operations are excluded. They represent relatively insignificant portions of the production on farms raising annual crops.

A more detailed investigation of investment choices in the Río Colorado area is reported in the next chapter.

Definitions and Approaches

The first type of interdependence, between investment and other policy instruments, was mentioned above with reference to guaranteed prices, short-term credit availability, and so forth. The second kind refers to complementarity or substitutability among investment projects. For example, an appraisal of a potential project in canal lining may yield a given estimate of its benefits; but if land is leveled simultaneously, the returns to canal lining may decline. This would occur because both kinds of projects yield increases in effective water availability by reducing conduction losses and waste and because the marginal productivity of water per hectare declines as more water is made available. In general, a proper assessment of the returns to irrigation investment requires specification of *all* the alternative potential irrigation investments in the same locale.

The third kind of interdependence is the relation between investments in a particular district and sector-wide investment programs. Because many producing areas compete for the same markets, investment (and expansion of production) in one area may affect market prices and hence may affect returns to investment in other areas. For this to occur, it is necessary only that the district's share of national supply be significant in at least one crop. ("Significance" in this context depends in part on the magnitudes of the price elasticities of demand for the products. Five percent of national production can be a significant share for products facing inelastic demands.) It can also occur when a few districts are used as examples to analyze a hypothetical policy that could be applied to the majority of the districts in the sector once approved. In this case, the supply response of one district could effectively represent the response of a large part of the sector.

In brief, then, this chapter calls attention to three elements of a proper project appraisal: other government policies in the district must be specified; the set of investment choices must be fully specified; and it may be necessary for district-level investment to be evaluated as part of an overall sector program.

Some of these questions, such as the effects of variations in guaranteed price policies, are often avoided by assuming shadow prices for inputs and outputs. Of course, there are many ways of determining shadow prices, and this diversity has led to a well-documented and lengthy controversy among economists. Implicitly or explicitly, a set of shadow prices refers to a desired or forecast long-run equilibrium. But implementation of a project that appears to be supported by a shadow-price evaluation may require extensive fiscal schemes. For example, if the shadow wage of labor is assumed to be zero in the project evaluation, farmers who have to pay the actual wage to hired labor may receive low or even negative net income in the project area. In reality, the fiscal measures required to implement a shadow-priced project may not be feasible. In this chapter, fiscal feasibility is ensured by using shadow prices *only* when the market forces or policy instruments required to induce those price levels are identified and included in the model.

The approach to pricing adopted here essentially depends on the degree of fixity of the good or resource. Toward this end, the models discussed below distinguish among three levels of spatial mobility for goods and factors: local, sectoral, and national. Local resources are land, irrigation water, and the labor of farmers and their families. Sectoral resources include, for example, fertilizers, draft animals, and farm machinery. National resources include day labor and short-term credit. For the analysis, the prices of national and sectoral resources are given exogenously. The sectoral resource prices may be set at alternative levels in parametric solutions because they fall within the purview of sectoral policies. National resource prices are taken as immutable because they are presumably opportunity costs determined by marginal productivities in other sectors. Local resource prices are determined, with certain qualifications, endogenously within the models.

In the case of product markets, depending on the circumstances, the district is assumed to be a price taker or to face downward-sloping product demand functions. In the former case, product prices may be taken to be either completely fixed or variable (as a representation of alternative levels of guaranteed prices).

In the latter case, when demand curves are downward sloping, a modified version of the Samuelson (1952) procedure is used to guarantee that the model replicates the competitive equilibrium on product markets. Chapter 3 of this volume presents a discussion of the procedure, and Takayama and Judge (1971) have made a full development of the quadratic programming case. It is in this sense that the models discussed here are simulation models: they simulate the behavior of a competitive market, given downward-sloping demand curves and profit-maximizing behavior on the part of producers. Simulations are conducted repeatedly under different hypothetical policy packages to explore the market response to possible policy interventions.

In other words, at the sector-wide level, products or sector-wide inputs are assigned shadow prices that are different from prevailing market prices only if it is possible to identify either the supply-demand behavior or specific policy instruments that have the potential of making the shadow prices real. For example, if a product is protected by a tariff, and its price therefore lies above the world market price, an experiment would be conducted using the world market price to show the effect on investment returns of removing that tariff. Further experiments with other price levels would also be conducted to simulate the situation under which the tariff remains unchanged or is altered but not removed. The reader who is disturbed by our acceptance of existing market distortions is referred to Baumol and Bradford (1970), who point out that, *in general*, prices must deviate from marginal costs in order to ensure efficient resource allocation. When, out of millions of prices, many are distorted, moving a few prices toward the international level (or some computed level) is no guarantee of moving toward a Pareto optimum.

At the local level, the following assumptions are made for irrigation water: farmers pay the actual fee for gravity-fed water or actual pumping costs, but water is allocated over crops optimally—that is, according to its marginal productivity in each crop and class of soil. These two assumptions need not be inconsistent, given that water allocation is governed by a district-wide management committee, even though the actual price of water is below its marginal productivity. Thus, the water price (tax) becomes a policy instrument that can be varied in alternative solutions. Land is assumed to be allocated over crops according to its marginal productivity in alternative uses.

The labor of farmers and their families is priced at least at a monthly reservation wage, which is set at half the market wage (see chapter 2). In months when farmer labor is fully utilized, the reservation wage may rise as high as the day-labor wage because of the element of economic rent accruing to the use of farmer labor. This reservation wage accounts for only one-fifth to one-third of the total net income of a farm family, and the

rest derives from economic rent to the land and water rights plus the rent
to the family's labor that accrues over and above the reservation wage.
The reservation wage clearly is a kind of opportunity cost. It is a *short-term*
(in this case, monthly) opportunity cost that the farmer demands as a
minimal acceptable return to his labor before undertaking an agricultural
task, in the knowledge that in the longer run he will receive substantially
higher returns in the form of rent to fixed resources. His *medium-term* (for
example, annual) opportunity cost, translated to a monthly basis, would
be substantially higher. This discrepancy between the short-run and long-
run opportunity wage reflects the farmers' lack of perfect job mobility in
the short run: the decision to leave the farm is a major one. In particular, it
would be more costly to leave in the middle of the crop year than at the end
of the crop year. Various simulation experiments were made to evaluate
the reservation wage level in reference to its effect on the cropping pattern
and the labor hiring pattern; these experiments have indicated that about
half of the rural market wage (day-labor wage) is an appropriate level of
reservation wage for irrigated agriculture in Mexico. It appears to be
lower for nonirrigated agriculture, as explained in chapter 2. The level of
the reservation wage is clearly important, for example, in the case of the
decision regarding acceptance of an agricultural innovation that promises
higher returns per hectare but also involves more labor by the farmer
himself.

Finally, it is stressed again that the models used here are behavioral
simulation models, although optimization is the mathematical tool used.
The models simulate the responses of constrained profit-maximizing
producers to alternative government programs. This does *not* imply that
the goal of public policy is maximal producers' profits. When the policy
goal is, say, higher employment or foreign exchange earnings, a model
such as these is used to find out, for example, which programs induce the
farmers to use those crops and techniques that are more labor intensive or
to grow more export crops. The approach adopted is to explore systematically the responsiveness of irrigated agriculture to various prespecified
policies so that a more rational selection can be made from among them.

Examples of Interdependence in Project Evaluation

Given the existing production technology set—that is, the known
possible practices for fertilization, pest control, and the like—the economic returns to additional water in a given locality with a fixed land
endowment are determined by three factors: the biological effects of water
on each crop, the prices of the crops, and the cropping pattern or composition of crops cultivated. The dimensions of the first factor can be estab-

lished through agricultural research based on the soils and climate of the particular area in question. The second factor is a matter of market forecasting and government interventions in product markets. The third is determined by the profit-maximizing behavior of producers. When relative prices of crops change, producers change their cropping patterns. The models presented here are designed to capture the functioning of the second and third factors, given the first.

The RIO COLORADO model contains eight alternative crops, each of which may be cultivated under two alternative degree of mechanization. Up to four alternative planting dates are specified for each crop; in all there are over 200 cropping activities in the model. This set of cropping activities is defined for each of four zones that had been established by the former Mexican Ministry of Water Resources (Secretaría de Recursos Hidráulicos, SRH) according to the degree of efficiency in water use. (Various factors determine efficiency in water use. For example, other things being equal, farms farther from the dam will incur more water losses through evaporation and seepage as the water travels longer distances through canals.)

The cropping activities are constrained by the monthly availability of water, land, and the labor of farmers and their families. Other inputs are assumed to be available in perfectly elastic supply at the market prices. These include day labor, agrochemicals, services of farm machinery and draft animals, improved seed, short-term credit, and miscellaneous cost items. Water is priced according to whether it is drawn from tubewells or from the dam, and land is not priced a priori, although the dual solution reports its marginal productivity on a monthly basis. The monthly specification is important to the allocational decision because the monthly cropping schedules differ among crops (and among planting dates for the same crop). Some crop combinations permit double-cropping in a twelve- to eighteen-month period, and others do not. Products in the RIO COLORADO model are assumed to be sold against infinitely elastic demand schedules, with the exception of some high-value crops for which marketing bounds are imposed. The price may be either a free-market price or a government-supported price. A more complete description of the RIO COLORADO model is contained in Bassoco, Norton, and Silos (1972) and in the next chapter of this volume.

Experiment 1

As experiment 1, the RIO COLORADO model was solved under a wide range of assumptions regarding the support price of wheat—from 780 pesos to 930 pesos per metric ton, with other crop prices (for barley, safflower, and cotton, for instance) held constant. In each case, the model

Table 15-1. *RIO COLORADO Solutions under Different Support Prices of Wheat*

Item	Support price (pesos per metric ton)								
	780	800	820	840	860	880	900	913ᵃ	930
Production (tons)									
Wheat	0	21.3	38.7	154.1	223.8	223.8	223.8	284.4	357.0
Barley	212.8	194.4	179.7	66.6	0	0	0	0	0
Safflower	1.2	0	0	0	0	0	0	0	0
Cotton	184.0	184.0	184.0	184.0	184.0	184.0	184.0	159.3	128.8
Employment (thousands of man-years)	33.4	33.4	33.9	32.1	32.1	32.3	32.3	31.3	27.5
Marginal product of gravity-fed water (pesos per 10,000 cubic meters)	797	802	824	863	922	981	1,040	1,077	1,091
Annual district income of farmers and day laborers (millions of pesos)	391.2	391.2	391.2	387.7	390.8	395.5	400.0	388.0	374.7

Note: Only the support price of wheat was varied; prices of other crops were held constant.
a. At the time these experiments were made, 913 pesos per metric ton was the actual support price of wheat.

determined the (private) marginal value product of water, in pesos per 10,000 cubic meters. This experiment addresses the interdependence of price supports and irrigation investment. Table 15-1 shows the principal results. In the table the marginal product of gravity-fed water is valued according to contributions to producers' profits. The table also records the level of district income, which includes three components: producers' own-wage payments; producers' profits (beyond their reservation wage); and wages paid to day laborers.

Several conclusions can be drawn from table 15-1. First, regarding the price supports alone, the cropping pattern changes significantly over the range of wheat prices studied (see also figure 15-1). Substitution between wheat and cotton, however, does not begin until the wheat price is raised above 900 pesos. (Below this wheat price, cotton stays at an upper bound determined by the extent of soils suitable for cotton.) Employment declines irregularly as the cropping pattern changes, but when the wheat-cotton substitution begins, employment declines monotonically and noticeably. This occurs because cotton is much more labor intensive than wheat. Total district income also moves irregularly, since the number of day laborers hired depends on the cropping pattern. On the whole, the price support for wheat is quite effective in increasing the wheat supply, but it has negative effects on employment and uncertain effects on regional income.

Regarding irrigation, the (private) marginal value of water increases markedly with the support price of wheat. The elasticity, or percentage responsiveness of the value of water to changes in the wheat price, is about 1.9 over this range of prices. These particular marginal values of water, however, have a strictly limited meaning: they refer to the annual contribution to producers' profits of an additional 10,000 cubic meters of water. They do not measure the contribution of water to total regional income or to other policy goals such as employment.

In other words, the marginal water values constitute measures of the annual benefits in relation to a conventional criterion for private rate of return. The quantitative implication of experiment 1 for investment programs is the following: *if a conventional criterion for rate of return is used to justify Río Colorado irrigation projects, at a support price of wheat of 930 pesos projects are justified that cost 37 percent more per unit of water supplied than those that are justified at 780 pesos for wheat; a 19 percent increase in the wheat price permits a 37 percent increase in the cost of supplying water.* The relevant calculation is as follows: the marginal value product of water is 1,091 pesos at a wheat price of 930 pesos, and it is 796 pesos at a wheat price of 780 pesos (1,091/796 = 1.37).

But a conventional criterion for rate of return may not be the desired criterion for judging irrigation projects. Employment effects, for exam-

Figure 15-1. *Product Supply Levels in Río Colorado as Functions of the Price of Wheat*

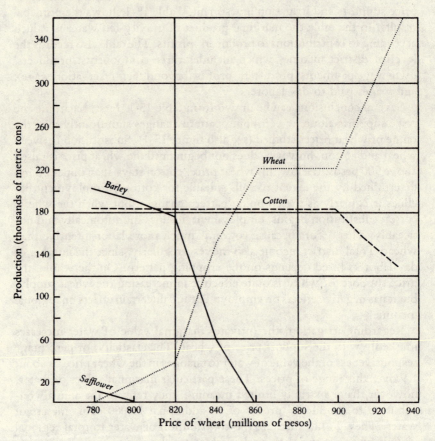

ple, may be more important from the viewpoint of policymakers. In the next section, comparisons of projects by their employment effects are made, but first additional examples are offered regarding the use of linear programming to evaluate projects purely by criteria of private returns.

Experiment 2

A frequent concern in agriculture is how to define the relation between agricultural research and extension on the one hand and other programs on the other. A brief example of measuring this particular relation with linear programming is also offered for the Río Colorado district as experiment 2. Again, the benefit side, and not the cost side of the program is analyzed. In some parts of the irrigation district, cotton yields are declining, so a sample solution was conducted under the assumption that cotton yields were 13 percent lower, averaged over the entire district. From an extrapolation from recent history and without increased efforts in agricultural research, it appears that this 13 percent decline in yields could well occur within four to five years. At a wheat support price of 913 pesos, this yield reduction lowered the marginal product of gravity-fed water from 1,077 pesos to 794 pesos, a reduction of 26 percent (table 15-1). *In other words, the returns to investment were reduced by 26 percent by a decrease in cotton yields of 13 percent.*

Because the benefits of expenditures on agricultural research have a large stochastic element, it is difficult to translate these results into precise operational implications. For this particular case, however, the following can be said about experiment 2: *if in the judgment of experts allocation of 10 to 15 percent of the investment budget to research and extension would be likely to halt further decline in cotton yields, then such an allocation would be easily justified by the increase in the returns to the total investment package in the district.* Strictly speaking, if this effect on cotton yields could be guaranteed with certainty by allocating no more than 26 percent of the investment budget to research, then such an allocation would be justified. Yet, because of the difficulty of ever predicting research results with certainty, the italicized wording is likely to be more meaningful for decisionmaking. This is a clear example of interdependence (complementarity) between different kinds of investment expenditures. In actuality, such a budgetary allocation would augment the current research effort in the Río Colorado area severalfold, and specialists feel that it would very likely arrest the decline in cotton yields. In other words, the current allocation of investment funds to different uses in the district may not equalize expected returns among those uses.

Experiment 3

More generally, if several investment activities are fully specified, a linear programming model can be used to establish a joint ranking among them and to trace out a schedule of marginal efficiency of capital (MEC). This brings the cost side of project appraisal explicitly into the picture. As experiment 3, the MEC schedule was traced out with the model of El Bajío on the basis of the following kinds of alternative activities for irrigation investment: leveling land, lining irrigation channels with concrete, digging tubewells on rainfed farms, and sinking tubewells on already irrigated land. Several alternative activities in canal lining were distinguished according to the width of the channel to be lined.

The BAJIO model is structurally similar to that of RIO COLORADO. However, in place of the four water-efficiency zones, the following distinctions are made: irrigated and nonirrigated, and within each, large and small farms; within irrigated, tubewell and gravity-fed irrigation; and within each irrigated category, leveled and nonleveled land. Also, in addition to the constraint set specified for RIO COLORADO, BAJIO contains management constraints. Farmers are divided into three levels of management efficiency on the basis of sample field data, and constraints are placed on the availability of management skills at the upper two levels. For a complete description of the model and a presentation of some numerical results, see Bassoco and others (1973).

In experiment 3, the model was solved for the levels and types of investment (as well as cropping patterns and water allocation) at different rates of return, and the schedule of marginal returns to capital shown in figure 15-2 was traced out. The schedule shows that *the justifiable amount of investment more than doubles when the criterion for rate of return is reduced from 24 percent to 12 percent.*

Experiment 3 also revealed that, for this particular area, *digging wells on rainfed farms turned out to be the most profitable investment, followed in order, by land leveling, digging wells on irrigated farms, and canal lining.* Canal lining did not enter the solution at any of the interest rates used, so its rate of return is estimated by the model as being less than 12 percent. This kind of result is useful for establishing both the size and composition of the investment program for a given area and is explored further in chapter 16. Details of this computer experiment are presented in Duloy, Kutcher, and Norton (1973).

Indirect Project Evaluation

In the foregoing discussion, explicit investment projects have been considered for the BAJIO model but not for the RIO COLORADO model. By indicating the possible consequences of investments (and other public policies) without making the cost side explicit, the RIO COLORADO model has been used indirectly for project evaluation. In an ideal procedure, this step would be followed by an engineering and economic evaluation of potential project costs, and then benefit-cost comparisons would be made. Because it is often the case that complete cost information is lacking when potential projects are initially screened, the analysis with the model helps assess, first, whether it is worthwhile to make a full engineering appraisal of the cost side and, later (when cost data are available), whether the project is economically justifiable.

In experiments 1 and 2 above, the RIO COLORADO model was applied in this indirect sense to wheat pricing options (or pricing plus investment

Figure 15-2. *Schedule for Marginal Efficiency of Capital (MEC)*
in El Bajío Given by the Linear Programming Model
for Investment in Water Resources

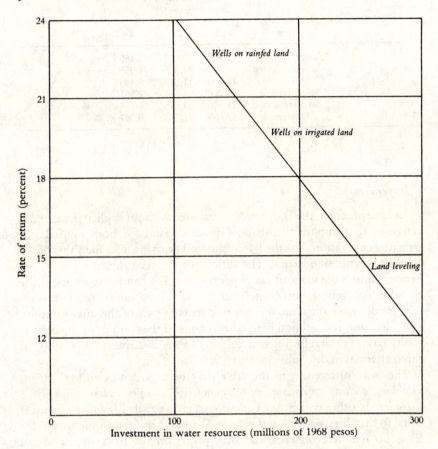

options) and to possible expenditures on research and extension. In experiment 5 below, this indirect approach is extended by varying the amount of irrigation water assumed to be available in the Río Colorado district to show the consequences of irrigation investments without yet specifying costs. In that experiment, both economic efficiency (private returns) and equity (employment) are analyzed as alternative objectives of public policy. But first another brief experiment with BAJIO is presented as a further illustration of indirect project evaluation under the efficiency goal.

Table 15-2. *Computed Prices of Land in El Bajío*
(1968 pesos per hectare)

Type of farm	Annual discount rate (percent)		
	12	15	18
Large rainfed	3,960	3,060	2,460
Large irrigated	19,150	14,875	11,958
Difference	15,290	11,815	9,498
Small rainfed	4,268	3,642	2,651
Small irrigated	21,956	14,875	11,958
Difference	17,688	11,233	9,307

Experiment 4

As experiment 4, the BAJIO model was used without explicit investment activities to compute the shadow price of land for both rainfed and irrigated cultivation. For the latter, the land price includes the value of the associated irrigation water. The difference between the two prices, per hectare, shows the value of water per hectare. The computation was made by taking an infinite discounted sum of annual economic rents to land. The model gave the annual economic rents as part of the dual solution. (The discount rate as used here corresponds to the concept of the opportunity cost of capital for the landholder.) Table 15-2 gives these results for three alternative discount rates.

The price differences in the table may be regarded as an *upper limit, according to efficiency criteria, on the permissible cost per hectare of supplying irrigation.* In other words, the benefit-cost ratio equals 1.0 when the cost is equal to this difference in land values. The linear programming model for El Bajío has determined these land prices on the basis of the set of cultivable crops, available technologies that transform inputs into outputs, prices of purchased inputs, and product prices. Given input and output prices, the model has selected the cropping patterns and technologies that maximize producers' profits. The model's primal solution in fact closely paralleled the actual situation. The selection of crops and techniques implies a certain return (economic rent) to land, which is the basis for the figures in table 15-2. The results for the 18 percent discount rate correspond closely to prices that prevailed in 1968, the model's base year. (A drawback of this approach in practice is that the differences in table 15-2 show the resource values only at the margin. The values may decline substantially with investments in more land, water, or both.)

Table 15-3. *Effects of Making More Irrigation Water Available in the Río Colorado District*

Item	Base-solution value	Percentage increase in water				
		10	15	20	25	30
Farmers' incomes (millions of pesos)	305.0	335.4 (+10.0)	341.4 (+11.9)	362.5 (+18.9)	374.2 (+22.7)	385.8 (+26.5)
District income (millions of pesos)	388.0	425.4 (+9.6)	434.7 (+12.0)	458.5 (+18.2)	471.6 (+21.5)	484.7 (+24.9)
Employment (thousands of man-years)	31.60	34.14 (+8.0)	35.59 (+12.6)	36.21 (+14.6)	36.55 (+15.7)	36.89 (+16.7)
Shadow price of gravity-fed water (pesos per 10,000 cubic meters)	1,077	1,052 (-2.3)	937 (-13.0)	937 (-13.0)	737 (-31.6)	737 (-31.6)

Note: Figures in parentheses denote percentage change.

Experiment 5

Indirect evaluation techniques can also be used when economic efficiency is not the only goal of public policy. In experiment 5, for the case of Río Colorado, several solutions were conducted for successively larger water endowments, so that the effects of irrigation investment packages of different sizes could be represented. Table 15-3 summarizes the results. As could be expected, the marginal value of water ("shadow price" in the stub of the table) declines noticeably as more water is made available on a fixed land endowment (Gisser 1970).

The shadow price of water shown in the last row of the table again refers to the private returns (increments in producers' profits) associated with the additional water supplies. If this were the criterion for defining project benefits, then to maintain the benefit-cost ratio at or above 1.0 the project costs (per 10,000 cubic meters) could not exceed the shadow prices reported in table 15-3. The shadow price declines by 13 percent as the water availability is increased by 15 percent, and by 32 percent as water availability is increased by 30 percent.

Farmers' incomes, district income, and employment all increase with additional water, but at a decreasing rate. For example, the water elasticity of employment (defined as the percentage increase of employment divided by the percentage increase of water availability) is 0.80 as water is increased to 110 percent of the base amount; it drops to 0.50 as water is further increased to 120 percent of the base amount; and it further drops to 0.22 as water is increased to 130 percent of the base amount. Rapidly diminishing returns in employment generation are reached after the water supply is increased to 115 percent of the base amount. In contrast, the water elasticity of farm income remains in the range of 0.80–1.00. Clearly, it is up to policymakers to decide where the cutoff point is for the investment, but experiment 5 shows that, *in the case of Río Colorado, if employment concerns are paramount, then the cutoff point is lower than if efficiency concerns predominate.* We do not feel that it is possible to know explicitly the policymakers' preference function as regards employment and efficiency (or other multiple goals), but we feel that the decisionmaking process can be facilitated by showing the program consequences for each goal. A similar attitude is put forth by Freeman and Haveman (1970). More precisely, the policymakers themselves normally would not have articulated precise tradeoffs among aggregate goals, but they may react with unambiguous choices when confronted with detailed alternatives for particular cases.

District versus Sector

To gain perspective on these results from the viewpoint of formulating sectoral investment programs, it is important to compare several producing areas. This is done in the following paragraphs, but a distinction must first be made among three ways of providing additional water. Extra water can be provided to an existing irrigation district without increasing the land area under cultivation in the district; it can be supplied on land that is already cultivated, but only on the basis of rainfall; or it can be supplied in areas that were formerly uncultivated because of insufficient rainfall. (Equivalently, water control projects can be established in areas that are uncultivated because there is too much rainfall and natural runoff.) In the last case, new farms are created. In the second case, rainfed farms are given water. In the first case, the amount of water per hectare is increased on existing irrigated farms.

Clearly, the employment-generating effects of water are different in the three cases. They also depend on the crops cultivated and on the farm size, which influences the degree of mechanization in cultivation techniques. Nevertheless, some generalizations can be made. Experiment 6 consists of comparing the employment aspects of the solutions reported earlier.

Experiment 6

Take first the case of creating new irrigation districts in arid zones. With the existing endowments of land and water, the RIO COLORADO model gives an estimate 2.5 man-months of annual employment per hectare cultivated. (This is employment at a marginal productivity equal to or greater than the prevailing daily wage of 42.5 pesos for hired laborers and equal to or greater than a daily wage of 21.25 pesos for farmers.) Given the average farm size in the area, this is about 32.7 man-months per farm. Since farmers work in crop cultivation less than twelve full months a year because of the seasonality of agriculture, and day laborers much less, 32.7 man-months means more than three jobs. These figures would presumably apply to newly created farms if the land and water endowments were increased proportionately, maintaining at the margin the same farm size. Also for Río Colorado, experiment 5 shows that increasing only the water on existing irrigated land by 20 percent increases employment by 0.5 man-months per hectare.

For the case of adding water to zones that are already cultivated, but with rainfall only, it is informative to look at the results for El Bajío. The employment rates per hectare shown in table 15-4 were calculated with the model. As the table indicates, supplying irrigation to large rainfed farms in El Bajío increases annual labor absorption by about 3.0 man-months per hectare, and by about 7.0 man-months per hectare on small farms. These figures are higher than the corresponding ones for Río Colorado. Increasing both land and water proportionately increases employment by 4.3 to 8.7 man-months per hectare in El Bajío, whereas the same steps in Río Colorado would increase employment by only 2.5 man-months per hectare. The differences are because of differences in cropping patterns and in farm size. Small-scale farms have a comparative advantage in using labor-

Table 15-4. *Annual Employment Rates in El Bajío*

Type of farm	Annual employment per hectare (man-months)	Annual employment per farm (man-months)
Large rainfed	1.3	15.5
Small rainfed	1.7	7.5
Large irrigated	4.3	53.3
Small irrigated	8.7	17.9

Source: BAJIO model.

intensive techniques of cultivation and in growing labor-intensive crops such as fruits and vegetables.

Although the various effects cannot be separated easily, it is legitimate to say that, *given prevailing farm sizes and cropping patterns, irrigation investment increases employment per hectare in El Bajío at least twice as much as it does in Río Colorado.* Because El Bajío is close to the Mexico City market, it may be said to have a permanent advantage over Río Colorado in the production of high-value fruits and vegetables. The differences between the two areas as regards both cropping patterns and farm size are likely to persist for decades at least.

Thus, the lessons of experiment 6 may be summarized as follows: although irrigation investment creates additional employment in Río Colorado, *a comparison with other parts of the sector reveals that an employment-oriented program must assign priority to areas that are similar to El Bajío in farm size and cropping patterns.* Similar comparisons should be made for many producing areas before completing the formulation of the sectoral investment program. (As of this writing, some further comparisons have been made, and they confirm that the comparisons between El Bajío and Río Colorado may be generalized to the entire Central Plateau versus the entire Northwest.)

This is one example of looking at an investment decision from the viewpoint of the locality (district) as against that of the sector. There are other ways to illustrate the district-sector relation. In fact, the discussion above of the interdependence between the sectoral level of price support for wheat and the returns to a local investment program is one example. Earlier research with the BAJIO model (Duloy, Kutcher, and Norton 1973) probed the sector-district relation more fully by means of formal linkages between the district model and a sector-wide model. Here a brief summary of that work is presented as the final example, experiment 7.

Experiment 7

If a district or region is large enough to account for about 5 percent or more of total production, it appears that there can be significant interdependence between the investment program in that area and the program for the entire sector. With this share of the total value of production in all crops, a district is likely to account for 10 to 20 percent or more of the production of at least one individual crop. If there is such a relation, *analyzing the investment program for the district alone, without taking into account the rest of the sector, will almost always result in an overestimate of the returns to producers.* (See chapter 11.)

Investment in the district under study expands sectoral production levels and tends to decrease agricultural prices relative to nonagricultural prices. This occurs because of the generally price-inelastic nature of the demand for agricultural products. Although demand certainly grows over time with income growth, it is also true that, other things being equal, more production decreases prices, at least relative to the economy-wide price level. Some obvious exceptions to this statement occur for crops exported at a constant price or for import-substitute crops. Nevertheless, the statement holds for the vast bulk of agricultural cases.

The Bajío area produces a significant share of the national supply of some crops, so an analysis with the district model alone was found to lead to a substantial overestimate—by as much as 50 percent—of the amount of investment justifiable at a given criterion for the rate of return, in comparison with the case in which the district model was solved as part of the sector model.

It was also found that the solutions for the district alone could be substantially improved without formal linkage to the sector model *if* the sector model were used as a price forecasting device prior to setting up solutions of the district model. Prices for the sector model then became inputs to the district model. In effect, this permits evaluation of the district's investment program as part of the investment program for the entire sector (which has been approximately estimated). Since a single district's investment program can significantly influence sectoral price levels, a fortiori this is true of the entire sectoral investment program.

What are the operational implications of these findings? One possibility is that a sector model that includes details on particular producing areas should be constructed and continuously revised, but even a large sector model can be addressed to detailed project evaluation questions only in a few selected areas. A more practical approach would be to use, as a tool for forecasting sector-wide price and production, a somewhat aggregative model that would take into account approximate overall investment magnitudes. In a second stage, then, an individual's district program could be assessed in a framework of an approximately consistent sector-wide investment program, so that the district-sector linkage would be captured, even if imperfectly. If the second-stage analyses yielded a sum of district investments that differed significantly from the initial sectoral estimates, then further iterations of the procedure might be required.

Conclusions

Several kinds of interdependence have been outlined that are crucial to the evaluation of an agricultural irrigation project. Parametric linear programming has been shown to be a useful means of capturing this interdependence, even in the absence of full project cost information. The models, however, were not structured to maximize various policy objectives arbitrarily, but rather to simulate producer's production responses to alternative, specific interventions of public policy. It has been found that, when a model does contain project specification, it is important that all related forms of investment in the area be specified. It is also important to apply parametric programming to noninvestment policies that may affect the area's production response. Finally, the district–sector relation can be significant even when the district accounts for only five percent of national agricultural production. At present, a two-stage procedure involving sectoral then local analyses seems most practicable.

References

Bassoco, Luz María, John H. Duloy, Roger D. Norton, and Donald L. Winkelmann. 1973. "A Programming Model of an Agricultural District." In *Multi-Level Planning: Case Studies in Mexico*. Edited by Louis M. Goreux and Alan S. Manne. Amsterdam/New York: North-Holland/American Elsevier, pp. 401–16.

Bassoco, Luz María, Roger D. Norton, and José S. Silos. 1972. "Programación de la producción y uso de los recursos en el Valle de Mexicali, B.C." Mexico City: Secretaria de la Presidencia. Processed.

Baumol, W. J., and D. F. Bradford. 1970. "Optimal Departures from Marginal Cost Pricing." *American Economic Review*, vol. 60, pp. 265–83.

Cummings, R. G., and Donald L. Winkelmann. 1970. "Water Resource Management in Arid Environs." *Water Resources Research*, vol. 6, pp. 1559–98.

Dudley, N. J., D. T. Howell, and W. F. Musgrave. 1971a. "Optimal Intraseasonal Irrigation Water Allocation." *Water Resources Research*, vol. 7, pp. 770–88.

———. 1971b. "Irrigation Planning 2: Choosing Optimal Acreages within a Season." *Water Resources Research*, vol. 7, pp. 1051–63.

Duloy, John H., Gary P. Kutcher, and Roger D. Norton. 1973. "Investment and Employment Alternatives in the Agricultural District Model." In Goreux and Manne, pp. 417–34.

Freeman, A. M., III, and R. H. Haveman. 1970. "Benefit-Cost Analysis and Multiple Objectives: Current Issues in Water Resources Planning." *Water Resources Research*, vol. 6, pp. 1533–39.

Gisser, M. 1970. "Linear Programming Models for Estimating the Agricultural Demand Function for Imported Water in the Pecos River Basin." *Water Resources Research*, vol. 6, pp. 1025–32.

Rogers, P., and D. V. Smith. 1970. "The Integrated Use of Ground and Surface Water in Irrigation Project Planning." *American Journal of Agricultural Economics*, vol. 52, pp. 13–24.

Samuelson, Paul A. 1952. "Spatial Price Equilibrium and Linear Programming." *American Economic Review*, vol. 42, pp. 283–303.

Soltani-Mohammadi, G. R. 1972. "Problems of Choosing Irrigation Techniques in a Developing Country." *Water Resources Research*, vol. 8, pp. 1–6.

Takayama, T., and G. G. Judge. 1971. *Spatial and Temporal Price and Allocation Models.* Amsterdam: North-Holland.

Young, R. A., and J. D. Bredehoeft. 1972. "Digital Computer Simulation for Solving Management Problems of Conjunctive Groundwater and Surface Water Systems." *Water Resources Research*, vol. 8, pp. 533–56.

16

Incorporating Policy Guidelines in the Design of Agricultural Investment Projects

Luz María Bassoco, Anthony Mutsaers,
Roger D. Norton, and José S. Silos

This chapter reports a method for project selection and appraisal that was developed during the course of the World Bank–Mexican government collaboration on development of operational policy planning techniques for agriculture. In a sense, it takes up where the preceding chapter left off: analysis of the question of interdependence between the sectoral and local levels of decisionmaking. Illustrative results are presented for the Río Colorado irrigation district in northwestern Mexico.

Overview

The discussion offered here addresses the following four concerns that are frequently encountered in agricultural project appraisal (to varying degrees, the same concerns arise in nonagricultural sectors). The first concern is how to reflect adequately the national (or sectoral) policy guidelines in each project's selection and appraisal so that the program of investments, which added up for the entire sector, is consistent with prior guidelines for sectoral policy. The practice of establishing a shadow price of capital (or labor, or both)—for use in the cost computations for all projects in the sector—provides consistency in one dimension. But in agriculture, it may be as important to gear the investment program to the forecast (or desired) intersectoral terms of trade. Given that the total investment program influences sector output levels, and hence affects prices,[1] the target level of the terms of trade can be an important determi-

Note: The authors wish to acknowledge the helpful comments of Wilfred V. Candler, John H. Duloy, and Gerald T. O'Mara and the computational assistance of Armando Calvo, Roberto Canovas, and Hunt Howell.

1. In this context, it is relative prices, especially agricultural versus nonagricultural prices, that are of interest. The well-known price inelasticity of demand for most agricultural products tends to make agricultural prices and incomes quite sensitive to variations in output levels.

nant of the size and composition of the investment program. Thus, in this chapter reflecting the sectoral policy viewpoint in the appraisal of projects means formal incorporation of both the targeted terms of trade and the cost of primary factors in the appraisal procedure—that is, price levels of both outputs and scarce factors.

This concern for price consistency is simply the corollary of a concern for adequate national production levels. An agricultural program designed to substitute domestic supplies for food imports, for example, is likely to be accompanied by significant changes in the domestic price ratio between foods and nonfoods.

The second principal concern addressed in this chapter relates to the process of project design (sometimes termed "project identification" or "project selection") that precedes the appraisal stage in the project cycle. The problem is that the identification itself, by greatly narrowing the range of alternatives to be evaluated, may strongly influence the final evaluation. In the case of agriculture, the project often consists of a package comprising several different small-scale investments. In putting together the package in certain proportions, it is necessary to rely heavily on elements of judgment,[2] and the best combinations may therefore be inadvertently excluded before the final appraisal. The solution proposed here is to make the selection and appraisal stages simultaneous, so that all potential alternatives are formally evaluated without having to perform separate analyses for each potential package.

The third principal concern is the need for simple procedures for numerically measuring corollary effects of the project such as employment generation and foreign exchange earnings, so that the project can also be screened on those grounds.

Finally, the fourth concern is perhaps a peculiarly agricultural one: the need to reflect in the analysis the significant interdependence that is observed between annual production decisions (cropping patterns) and investment decisions. By changing relative factor endowments (such as the ratios among land, irrigation supplies, and labor) the investment project often changes the desired crop mix. In traditional project appraisal, the crop mix is forecast in advance of deciding the scale and content of the investment, and hence this interdependence is overlooked.

The proposed method turns out to be easy to implement; a few man-weeks sufficed for the complete selection and appraisal analysis for a rather complex Mexican irrigation project involving fifty-eight distinct investment choices. The method makes use of mathematical programming, and the farm production choices are represented in process-analysis format.

2. Engineering considerations usually enter formally at the selection stage, but as far as economic aspects are concerned, the project selection stage is usually judgmental.

The selected terms of trade (for the entire bundle of agricultural outputs) are expressed as an annual rate of change of agricultural prices, relative to other prices, over the investment planning horizon.

It should be emphasized that we are not concerned here with making agricultural price forecasts per se, but rather with the problem of representing the mutual interdependence of prices and investment projects. After all, if a particular investment project is justified under current prices, then others will be too; if taken together they may significantly affect prices. Harberger (1974, pp. 18–19) has made this point cogently:

> One often hears projects justified, in practice, on the basis that even if they are not profitable today, they will become profitable in the future because of the growth of demand. There can indeed be such a justification for particular projects, but when this is the case it is more subtle than many people think. Almost any investment made today would become profitable with time if no competing investments were made in the future. . . . the "profitability" of today's investments should be estimated on the assumption that all "profitable" future investments will also be made. . . . Here, of necessity, the project analyst himself has to estimate an expected time path of the price—not on the assumption that his project stands alone, nor on the assumption that future projects will be held up in order to "protect" the profitability of his current project, but on the much more rigorous assumption that future investments will be made on their own merits. . . . it must be realized that the expected price path here means more than just a guess about future prices—it means rather a guess as to the prices that will be generated . . . by the continuous application in the future of valid investment decision rules.

In essence, this chapter offers formal but simple-to-use procedures for carrying out Harberger's suggested approach numerically. For simple situations of one crop and few investment choices, the formal procedure would not be required, but most actual investment situations are more complex, particularly when definition of the composition of the investment package is seen to be part of the problem.

The remainder of the chapter is organized as follows. The next section recapitulates some recent advances in the use of linear programming as a tool of practical analysis for agriculture. The third section ["The Static Cropping Model (MEXICALI)"], describes the static (one-period) version of the model used in this study, emphasizing its representation of farmers' annual cropping and input use decisions. (The static model is structurally similar to those reported in chapters 2, 8, 11, 12, and 13.) In the fourth section ("The Model and Criteria for Project Selection"), the concepts of project benefits and rates of return as incorporated in the model are discussed. The fifth section ["The Investment Version of the Model (TECATE)"] then introduces the full dynamic specification of the model, taking into account project lifetimes, technological change, and targeted terms of trade for some future date. This completes preparation of the necessary analytic tools for a practical application, the subject of the sixth section ("Numerical Results"), in which the model is applied to the problem of project identification and evaluation for a case involving, as mentioned, fifty-eight different potential subprojects to be screened and combined into an investment package. Some concluding remarks are made in the final section.

Linear Programming as a Tool of Analysis

In recent years, linear and nonlinear programming have become accepted as useful tools in the design of large-scale water-control projects,[3] but they have not been exploited widely for the analysis of small-scale projects or projects that comprise a collection of small-scale investments in irrigation or other aspects of agriculture. There are two basic reasons for the slow acceptance of these methods in this area: the potential for using models for project *identification* as well as evaluation has not been fully realized; and in the past, the typical linear programming model gave solutions that were too extreme to be realistic and were too specialized in the more attractive production and marketing activities.[4]

The first obstacle can be overcome by taking fuller advantage of linear programming's potential to reflect numerous aspects of reality simultaneously. This chapter offers suggestions and numerical examples along these lines. The second obstacle can now be overcome more satisfactorily than in the past, thanks to some recent methodological advances that are susceptible to inclusion in project evaluation models. One particularly useful advance has been Hazell's demonstration of how to incorporate risk—as perceived by producers—into a linear programming model of farm cropping decisions.[5] The great practical advantage of his innovation

3. For two of the hundreds of examples of such applications, see the discussions in Cummings and Winkelmann (1970), and Rogers and Smith (1970).
4. The existing nonlinear programming algorithms are not powerful enough computationally to handle models that include a low of real-world detail, so it has generally proven more efficient to utilize linear programming algorithms and to approximate nonlinear behavior by means of piecewise linear functions. One of the earliest efforts to exploit linear programming in this way was that of Adelman and Sparrow (1966). Linear programming is, of course, not applicable where economies of scale are present.
5. See Hazell (1971) and chapters 7 and 8 of this volume. Extensions of the theory to aggregate market models are made in chapter 7.

Figure 16-1. *Schematic Tableau of MEXICALI Model*

	Cropping variables	Demand variables	Input variables	Labor and water variables	Risk variables	RHS
Objective function		Crop prices (+)	Input costs (−)		Risk penalty (−)	(max)
Commodity balances	Crop yields (+)	Crop sales (−)				= 0
Input balances	Input use (+)		Input supplies (−)			= 0
Risk identities	Historical revenue fluctuations (+ and −)				Risk summation (−)	≤ 0
Land constraints	Land use (+)					≤ R₁
Labor and water balances	Labor and water use (+)			Labor and water supplies (−)		= 0
Labor and water constraints				(+ and −)		≤ R₂

is that it captures uncertainty regarding both prices and yields in a linear model: it permits representation of the essential aspects of a variance-covariance matrix of gross revenues by crop.

Another recent contribution that is relevant is the procedure devised by Duloy and Norton (chapter 3 of this volume) for the incorporation of complex price- and income-elastic demand structures into linear models. Empirical tests have shown that this step contributes as much as, or more than, the risk parameters to diversifying in a realistic manner the cropping pattern selected by a model. As is shown in a later section of this paper (and in chapter 11), under normal circumstances it is reasonable to specify a significant downward slope to the price-elastic demand curves (as opposed to horizontal demand curves) facing the producers in a given locality or region. It becomes especially important to do so if the sectoral viewpoint is to be included in the project evaluation.

Taken together, these developments permit construction of a noticeably more realistic agricultural programming model—without imposing arbitrary upper or lower bounds on production of specific crops—than was previously thought possible by practitioners in the field. Improvements in the realism of linear programming models usually are based in part, as are the ones reported here, on the incorporation of nonlinear functions through piecewise linear segmentation.

The Static Cropping Model (MEXICALI)

Because the MEXICALI model bears a family resemblance to other models reported in this volume, the description given here is brief. It is called MEXICALI after the principal town in the agricultural district represented, and the dynamic version applied to this study, TECATE, is named after a smaller town in the same vicinity. MEXICALI is an extended version of the RIO COLORADO submodel of CHAC (chapter 2). A version of MEXICALI was used for the experiments reported in the preceding chapter.

The basic principle of model specification is complete separation of production activities, product demands, factor supplies, risk activities, and investment activities. This makes it possible to specify downward-sloping demand and upward-sloping supply curves, and it makes it much easier to perform sensitivity analyses. For example, a change in the price of water is achieved by changing the value of one price coefficient, whereas had we specified net income coefficients for each cropping activity, it would have been necessary to recalculate and then alter in the matrix some 231 income coefficients in order to achieve the same result.

The structure of MEXICALI is illustrated by figure 16-1, in which columns represent variables and rows represent equations or inequalities.

486   BASSOCO, MUTSAERS, NORTON, SILOS

The blocks are submatrixes of a tableau of detached coefficients, in the usual manner of linear programming presentations.

Ten crops are represented in the production activities part of the matrix, with alternatives regarding planting dates and the degree of mechanization. Crop yields from these production activities enter into commodity balances, where equality between production and sales is imposed. Similarly, purchased inputs used in the production process are registered in input balance rows that ensure that the appropriate quantities are supplied and that the cost of supplying them is charged to the objective function.

Land is not priced in the primal structure but is bounded. Restraints are specified in fortnightly periods to capture crop rotation possibilities. Labor is undifferentiated (except for tractor drivers) at the point where it is used in field tasks, but on the labor supply side, farm family labor is distinguished from hired labor. Consistent with the earlier CHAC studies, farm family labor is assumed to have a conception of its own opportunity cost, which is reflected in a reservation wage. The latter is set at half the market wage for hired labor.

Water is both bounded and priced, at the prevailing level of administrative water charges. These charges tend to lie well below the marginal productivity of water.

The treatment of risk directly follows the procedures of Hazell and Scandizzo in chapter 7 of this book. In essence, farmers are assumed to place a negative value on uncertainty; for crops with a relatively high variance of gross revenue (yield × price), they demand a high average (expected) return as compensation. In other words, they operate according to an (E, V) utility function.[6]

An objective function such as the one in CHAC is adopted but is modified by a negative risk term. The deterministic part of the objective function corresponds to maximization of the sum of producer and consumer surpluses, which guarantees that the model will reproduce the competitive market equilibrium, in prices and quantities, for each crop.

In all, there are 198 rows in the static version, and they may be summarized as follows:

Twenty-four biweekly land constraints

Twenty-six water constraints (monthly and annual, for pump and gravity-fed water)

Two balances for water pricing

Twelve monthly restrictions for water management that govern the rate of change of water releases from the dam from period to period

6. That function is transformed into an (E, σ) function over the relevant range of values, for ease of handling in the model.

Twelve monthly balances for labor

One annual farmer constraint

One accounting row for total day labor

One accounting row for wages

One accounting row for total costs

One accounting row for total employment

Two balances for chemical and seed inputs

One balance for short-term credit inputs

One balance for long-term capital

Two pricing balances for use of machinery and draft animal services

Twenty-four accounting rows for use of machinery and draft animals

Eleven balances for risk computation

Nine commodity balances for outputs

Nine special constraints for demand functions

Nine accounting balances for district-wide degrees of mechanization

One accounting balance for sector income

One accounting balance for producer's profit

Two revenue counters, at endogenous and at exogenous prices

One objective function

Forty-four miscellaneous accounting balances.

A few major *economic* characteristics of the model structure are worth noting. First, as discussed, the demand structure drives the model to the competitive equilibrium point on product markets. Second, producers maximize profits minus a measure of uncertainty in both yields and prices. Third, purchased inputs, including short-term credit, are available in perfectly elastic supply at exogenous prices. Fourth, land is allocated according to its endogenous shadow price by fortnight; that is, according to its seasonal opportunity cost within the district. Fifth, water is allocated according to the same principle, subject to the special restrictions for water management. These restrictions limit the rate at which water applications may be increased from period to period. Water is also priced at prevailing administrative quotas, but, since its marginal productivity is higher than this price, its allocation is governed by the endogenous shadow prices. Sixth, the labor supply function has two steps: in the first one, family labor is hired at half the prevailing day-labor wage, and in the second step day labor is hired from an infinitely elastic supply function.

Regarding the wages for family labor, the reservation wage, it is clear that family members are willing to work on their own farm at tasks with a marginal productivity less than the market wage. Yet it is also clear that in Mexico they do not work for zero returns. Finding the appropriate level between zero and the market wage thus becomes an empirical question.

The pricing of labor is of some importance for project evaluation, and it has been discussed at some length in chapters 2 and 15.

For input pricing, it is possible, for example, to explore the effects of trade policies by experimenting with, say, different levels of the fertilizer price that would be brought about by different tariff levels. In this case, the input would be shadow-priced exogenously—as distinct from the case of land and water, which are assigned endogenous shadow prices by the model.[7]

As a final point of the structure of MEXICALI, the question of the specification of demand functions is taken up. The usefulness of this kind of programming model for the analysis of investment projects depends in large part on its ability to simulate farmers' production choices under varying resource endowments and pricing policies. But when the simulations are designed, the response of other producers in the sector must be taken into account. Two alternative approaches are available. The first is to say that the choices analyzed with the model are purely local choices, and the rest of the sector's supply response is therefore zero with respect to those choices. An example of this situation would be the introduction of a new plant variety that is suitable only to the local soil conditions. In this case, the appropriate *demand* elasticity facing local producers is the national price elasticity divided by the local share of production. As the local area becomes smaller, this elasticity approaches infinity, so that in the extreme case of the individual farm unit producing an infinitesimal share of national output, we have the atomistic price-taker situation.

The circumstances are very different, however, if the model is being used to analyze choices that are not purely local. For example, if the interest rate on short-term credit were to be changed locally, it almost certainly would be changed elsewhere in the sector, either through the functioning of credit markets or as a reflection of a national policy decision. Similar comments hold for the prices of fertilizer and other inputs. In these cases, a simulated local production response to the price change would be accompanied by responses in other parts of the sector. If the local area were representative of the sector in the sense that its supply response functions had roughly the same elasticity as in the sector as a whole, then the appropriate *demand* elasticities facing local producers are the national elasticities.[8]

The same reasoning would hold for investment appraisal, since invest-

---

7. Family labor in the model sometimes receives a shadow price that is higher than the reservation wage. This occurs in peak labor periods, and the shadow price may not go higher than the day-labor wage.

8. If possible, adjustments should be made to the extent that the local area is not representative in this sense. To make such adjustments would require knowing in advance supply elasticities throughout the sector. In the absence of such information, the assumption of representativeness may be made.

ments are taking place simultaneously in various regions. If it is profitable to invest in expansions of productive capacity locally, it is also likely to be profitable to do so elsewhere. This brings the argument back to Harberger's point, and so for MEXICALI and TECATE the national demand elasticities are used—applied, of course, to local supply quantities.

The Model and Criteria for Project Selection

A linear programming model is governed by its objective function. It utilizes additional resources only as long as their marginal contribution to the objective function is at least as great as their marginal cost; in fact, it drives to the point where marginal benefits and costs are equalized. How can this characteristic be used as a tool for project appraisal? There are two aspects to this question. One is the nature of the appraisal criterion (private rate of return, social rate of return, employment, or the like). The other is the formal (mathematical) identity between the concept of internal rate of return (IRR) and the rate of return calculated by the model. This latter aspect is taken up first, and the former is pursued subsequently.

Unlike econometric models, a mathematical programming model is an explicit instrument of choice: it *selects* from among alternative actions that are laid out. For investment analysis, the model has been set up to choose among alternative potential investment projects. To do so, the model simulates the consequences for farm income, cropping patterns, employment, and so forth of each possible investment package; it also applies the prespecified investment criterion in each case. It scans and evaluates all possible alternatives studied in a single solution. The ways in which a model is set up to do this are discussed at greater length below; here we wish to discuss the nature of the calculated rate of return.

The model's results pertain to a single representative year at some specified point in the future. Hence it equalizes the annual benefit flow and annualized capital costs. For this result to be useful, we must make an important assumption: that the expected benefit and cost streams are uniform over time, or, if not, that it is possible to convert them to equivalent forms that are uniform over time.[9] This assumption effectively permits representation of a dynamic problem by means of comparative statics. Given this assumption, the model chooses investments[10] such that

---

9. Some adjustments—to take account of start-up periods and the like—are required for analysis of some projects that generate nonuniform streams of consequences over time. For projects in which nonuniformities are attributable only to stochastic variables, however, there are no difficulties, for here we refer to *expected* values.

10. For the present discussion, investment activities in the model can be defined as column vectors that relax the restrictions on resource availability (at a cost per unit of incremental resources).

(16.1) $B_t = rI,$ all $t \in T$

where B_t is the flow of incremental benefits accruing to the investment project at time t, r is the annual price of capital, I is the total amount of investment selected, and T is the project lifetime. The interest rate r is varied under alternative solutions to trace out a schedule of investment amounts that are worthwhile at each value of r; this yields the marginal efficiency of capital (MEC) schedule.

Therefore, the discounted stream of benefits is equal to

(16.2) $$B^* = \sum_{t=1}^{T} \frac{B_t}{(1+r)^t} = rI \sum_{t=1}^{T} \frac{1}{(1+r)^t}$$

From the nature of convergent geometric series, it follows that, as T approaches infinity, the benefit stream converges to a specific value:

(16.3) $$B^* \to rI \frac{1}{r} = I.$$ as $T \to \infty$

In other words, in the limit the discounted benefit stream equals the value of the total investment, which is the defining characteristic of the IRR. Hence, the alternative model values of r are in fact alternative values of the IRR. The procedure of working with the model is: to set the IRR at one value, say 8 percent, and observe the model's choice of investment; then to set it at another value, perhaps 10 percent; then to repeat the procedure. These steps are followed until all IRR values of concern are examined. For each step, the amount and composition of investment consistent with that IRR value are calculated by the model. For this study, the IRR has been set in succeeding runs at eight values: 0, 4, 8, 12, 16, 20, 24, and 36 percent.

Two questions may be asked about equation (16.3). First, suppose that all investment does not occur instantaneously in the first year. This is in fact the case for the Río Colorado investment package analyzed in this chapter. For each IRR value, the present values of investment cost have been calculated with that value and so entered in the model. In other words, the model solution fulfills the following condition:

(16.4) $$B^* = \sum_{t=1}^{T} \frac{B_t}{(1+r)^t} \to \sum_{t=1}^{T} \frac{I_t}{(1+r)^t} = I^*,$$ as $T \to \infty$

which is the more general statement of the equation which defines the IRR.

The second question concerns the rate of convergence. Only in the limit do we have a solution that exactly corresponds to the IRR criterion. It is readily shown, however, that the degree of approximation for finite project lifetimes is quite small. The exact statement for finite lifetimes is:

(16.5) $$B^* = rS_T \sum_{t=1}^{T} \frac{I_t}{(1+r)^t} = rS_T I^*,$$

where S_T is the Tth partial sum of a geometric series:

(16.6) $$S_T = \sum_{t=1}^{T} \left(\frac{1}{1+r}\right)^t = \frac{1 - \left(\frac{1}{1+r}\right)^{T-1}}{r}$$

Therefore, equation (16.5) becomes

(16.7) $$B^* = \left[1 - \left(\frac{1}{1+r}\right)^{T-1}\right] I^* = kI^*$$

To give illustrative numerical values, for a project lifetime of thirty-three years, $k = 0.9734$ at $r = 0.12$; and $k = 0.9913$ at $r = 0.16$. The factor k measures the degree by which the method approximates the discounted benefit and cost streams for finite lifetimes; k approaches unity as the project lifetime increases and as the interest rate increases. For low IRR values, the distortion could become significant if the lifetime were not long enough; in these cases, it is recommended that the project planning horizon be extended by assuming renewal (replication) of the project at the termination of its lifetime. By this procedure, the approximation error can be kept down to acceptable levels.[11]

The foregoing has shown that the model generates investment levels (and compositions) that correspond to each alternative value of the IRR. The question of the *type* of IRR, however, has been left open: it could be defined with respect to private returns, social returns, returns in terms of additional employment, and so on. The type of returns treated follows from the definition of the objective function in the model.

Earlier, the issue of definition of benefits was raised. Traditional project appraisals go by the private rate of return, which is equivalent to maximizing the incremental stream of private (producers') profits deriving from investing in the project. In some cases, the concept is modified to include benefits that accrue to other groups in the society besides producers, and thus the criterion for decision becomes a social rate of return. At times a multiple set of criteria is specified: the additional goals may include variables such as employment, income distribution, and foreign exchange earnings. It may be sought to maximize one of these other goals, but more often they are given secondary consideration: either they are called upon to decide among projects that are equal on grounds of rates of return, or else some minimum acceptable level is established for values of the secondary variables, and then all projects passing that threshold are judged

---

11. Alternatively, the rate-of-return axis could be rescaled to compensate for the distortion.

according to rates of return.[12] Project identification is properly viewed as a two-level problem: it is composed of the *behavioral* problem of representing how decentralized decisionmakers (chiefly producers) will react to policy instruments, and the *policy* problem of choosing among behaviorally feasible alternatives.[13] It is only at the level of the policy problem that social weights can be entered into the objective function. At the behavioral level, the objective function must take on a market-simulating character. That is, it must drive the model to simulate as faithfully as possible the reactions of producers to policy measures, such as pricing and investment levels.[14] The behavioral simulation problem can be solved many times and different assumptions about the values taken on by policy instruments. Then the outputs of these behavioral solutions can be used, in a subsequent stage, as inputs to the definition of the problem of policy choice.

Normally, the first stage—simulation of likely reactions—is the more difficult, especially in agriculture, where crop supply functions can be highly interdependent. Therefore, the power of the model should be applied to doing as good a simulation as possible, for then selection rules are fairly easily applied ex post on the basis of tables of estimated consequences for many variables. MEXICALI and TECATE are simulation models in the economic sense.

In constructing a simulation model, the first question that arises is the kind of behavior to be simulated. For the case of TECATE, and this is likely to be the same in many other agricultural situation studies, there are two principal behavioral assumptions: that producers maximize expected profits minus a measure of risk, and that the equilibrium market form for products competitive (atomistic) and not monopolistic or semi-monopolistic.[15]

The allocation of resources in TECATE is governed by these rules of behavior, subject to limitations on the availability of technologies and fixed resources and subject to decisions on public pricing policy or rules regarding shadow pricing of particular inputs and outputs. For a given future year, the model is solved under these assumptions for each of the several alternative rates of return, and the consequent resource allocation

12. We do not wish to suggest that employment criteria *should* be relegated to a secondary role; rather, we are describing typical practices.

13. This distinction is spelled out in considerable detail in Candler and Norton (1977).

14. In the absence of an explicit two-level analysis, however, shadow prices that are in principle derived from a social preference function could be used for the exogenously priced goods and factors in the behavioral simulation model. In the rest of this chapter, we assume that such shadow pricing has been carried out as appropriate.

15. Descriptions of modified competitive market behavior to incorporate features such as production for own consumption are readily added to these models; see chapter 5 of this volume for a discussion of such characteristics of the sector model CHAC.

decisions are tabulated. The sequence of investment decisions under different interest rates forms the MEC schedule, as pointed out above, according to an IRR criterion.

Because a competitive market is assumed, however, the benefit stream B^* represents more than just profits to producers: it represents a particular measure of social welfare, the sum of Marshallian surpluses.[16] Hence the IRR that is measured may be called a limited kind of internal *social* rate of return. Alternatively, it is not necessary to assign normative content to the objective function: it may simply be said that this is the MEC schedule that would be the result of the workings of atomistic product markets and the actions of profit-maximizing, risk-averse producers under various possible interest rates. Either interpretation is viable.

The market simulation interpretation, however, may be preferred when the other kinds of benefits (for example, employment) are brought into the decision. The investment decision may ultimately be based in part on the project's effects on production of exportable crops, on employment, on the regional income distribution, and so forth. The model is designed to simulate these other effects and thus to provide a quantitative basis on which the project selection can be made ex post according to a decision rule with multiple criteria, where employment and other goals can be given weights of any magnitude.

Strictly speaking, TECATE performs the task of project identification—that is, narrowing down a virtually infinite array of investment packages to a relatively few—by following an IRR criterion, as defined above, at selected values of r. The final project choice, but not the identification, may then be made according to alternative goals.

The Investment Version of the Model (TECATE)

Establishing the investment version of the model basically involves three steps: incorporating the alternative investment activities in the choice set; shifting the price-elastic demand functions rightward over time to permit evaluation of the stream of future benefits; and normalizing the parameters of the model to account for differing project lifetimes among the components of the investment package. As will be seen, it is in the second step that the policy guidelines on the terms of trade are taken into account.

The various investment options for the Río Colorado district can be grouped into four categories: lining irrigation canals with concrete, level-

---

16. That this is true may be seen by the fact that in order to *simulate* a competitive market's functioning, the model maximizes the sum of Marshallian surpluses. See chapter 3.

ing irrigable land, digging new wells, and desalinization. The physical effects of these activities are as follows: canal lining reduces water losses in transmission; land leveling both saves water and increases crop yields per hectare; wells increase the supply of nonsaline water; and desalinization increases yields and arrests the process of land abandonment (thus saving land, in comparison with what occurs in the absence of investment). Desalinization requires investment in drains, diversion of water for the land-washing process, and diversion of cultivable land away from cropping activities for the duration of the washing process.

Space limitations prohibit a completely detailed description of the investment activities, but their characteristics are summarized to provide a feeling for the degree of project detail that may be captured in a linear programming model.

TECATE contains twenty-two canal-lining activities, each corresponding to a certain subregion and to a certain canal width, and each is permitted up to a maximum number of kilometers. There are eighteen land-leveling activities, two for each of the nine crops. The twofold distinction refers to the type of water saved—well (pump) or dam (gravity-fed). Because the two types of water differ in cost, this is an important distinction. The crop distinctions are necessary because the yield savings may be sufficient to warrant land leveling for some crops but not for others. Incidentally, these kinds of distinctions, which ultimately refer to different degrees of economic benefits per unit of investment, are difficult to make in conventional project appraisal.

Well construction for the Río Colorado area refers not to expansion of total water availability, but rather to replacement of hopelessly saline wells with new ones in a less contaminated part of the district. Thus, the net impact of the investment in new wells may be expressed as an increase in crop yields per hectare, over the levels which would have obtained in the absence of the investment. For this reason, well construction is also made crop specific, to find out the crops for which it is profitable to undertake it. Similarly, there are nine crop-specific desalinization activities.

Since the model covers one cropping year, the general procedure for incorporating investment activities is to put into the model a column vector representing all the (recurring) annual costs and *physical* effects. After suitable adjustments to make the model represent some typical future year when the projects would be in full operation, the model is solved to obtain an *economic* evaluation of the benefits of physical effects such as water savings. As in any area, water has a declining marginal productivity per hectare as more is applied, so the returns to any one investment activity may depend on how many of the others are also undertaken—hence, the importance of evaluating all of them in a simultaneous framework.

After specifying the investment activities, a future year T must be selected as the representative year with respect to investment costs and benefits. This selection depends on project lifetimes. In broad terms, the model is then set up for solutions representing year T by incorporating the components of technological progress that may be independent of the contemplated investments and by rightward shifts in the price-elastic demand functions (to represent the growth of markets over time).

An example of technological progress in agriculture is the autonomous trend in yields per hectare, apart from increases arising from specific programs for desalinization and the like. Incorporating the technological progress effectively shifts the model's implicit supply functions rightward. The shift in the demand function then becomes determined by the targeted terms of trade, assuming that they can be controlled by appropriate sectoral policies.

More specifically, the procedure is as follows:

- Initial values are adopted for the sectoral terms of trade and the rate of return to investment in the sector. This is the counterpart to the usual procedure of choosing values of product prices and the interest rate for purposes of project appraisal.

- Through iterative solutions, a demand shift factor is found that gives a solution conforming to the targeted terms of trade, with investment costed at the target interest rate. This amounts to finding a local equivalent to the factor of national income growth, which, via income elasticities of demand, shifts all the static (price-elastic) demand curves rightward.

- Because the model now represents year T under the initial sectoral policy guidelines, via solutions under different interest rates, the consequences of altering those guidelines for the desired rate of return can now be evaluated. Given that national price elasticities of demand are used, this procedure takes into account Harberger's concern for including the contemporaneous effects of investments occurring elsewhere in the sector (see chapter 11).

Positioning the demand curves for year T also amounts to partitioning the future marketed quantities among producing areas. Seen in this light, the procedure provides an answer to the question about future markets for a given region—a question that often arises in the course of project appraisals. Areas with relatively inelastic supply possibilities will require relatively little shift in the demand functions to meet the targeted terms of trade, and hence they will be participating relatively little in the national expansion of marketing possibilities.

For the sectoral model CHAC, straightforward procedures have been devised that take into account exogenous estimates of national income and population growth and that apply income elasticities to those estimates in order to shift the demand functions. For a local model, however, the matter is somewhat different, for the aggregate growth rate of sectoral production is a composite of differential growth rates in the various producing regions of the country. A given rate of expansion of the national economy therefore has different implications for the markets for each producing locality. Concretely, an established producing region such as Río Colorado, although its productive capacity is continuously being increased by projects such as the one studied here, will not experience as rapid an output growth as new areas where land is being reclaimed for the forest or swamps or where new dams are being built. Therefore, the question becomes: What does a given expected real GNP growth rate at the national level (for example, 7 percent) mean for the markets of the Río Colorado products? How much should the demand curves be shifted over a ten-year period? Over twenty years?

In formal terms, the procedure for determining the amount of shift in the demand functions implicitly constitutes solution of a set of supply and demand equations for the year T. The supply equations are implicit in the linear programming model, but for purposes of this discussion we may write them explicitly, along with the demand equations:

$$(16.8) \qquad q_i = a_i \, y^{\eta i} \, \Pi_j \, p_j^{\varepsilon j}$$

$$(16.9) \qquad q_i = (b_i \, p_i^{\alpha i} \, \Pi_j \, p_j^{\alpha_{ij}}) \, f(I^{r*})$$

$$(16.10) \qquad p^* = \sum_i p_i \, q_i / \sum_i q_i,$$

where i is the goods index and

q_i = quantity demanded and supplied in equilibrium

p_i = price

y = "relevant income" as regards demand for goods produced in the Río Colorado district

ε_i = price elasticity of demand

η_i = income elasticity of demand

a_i = demand function constant

b_i = supply function shift parameter representing the cumulative effects of technological progress between the present and year T

α_i = own-price elasticity of supply

α_{ij} = cross-price elasticity of supply

I^{r*} = quantity of investment expressed as incremental production capacity, which is profitable at interest rate $r*$

p^* = sectoral target value of the price index for agricultural commodities.

Equation (16.8) is the set of demand functions, (16.9) the set of implicit supply functions, and (16.10) the requirement that, at an interest rate $r*$, the targeted terms of trade be met. The particular form of the demand and supply functions (16.8) and (16.9) is only illustrative; any well-behaved demand functions can be put into the model, and the supply functions are endogenous to the model. The function $f(I^{r*})$ shifts the price-elastic supply function rightward. There are $2n + 1$ equations for n goods, and they determine $2n$ prices and quantities and the value of income (y). In effect, the iterative sequence of solutions of the linear programming model provides a solution to equations (16.8) through (16.10), in multiple products. The value of y turned out to be equivalent to 3 percent annual income growth over the project planning period; this may be compared with an expected 7 percent national income growth. In other words, as suspected, Río Colorado will not expand its supply as rapidly as the sector as a whole over the next twenty years. This local equivalent value of y is then maintained constant for all other experiments in project selection performed with the model.

Two parameters significantly influence this local equivalent of y, which represents the degree of rightward shift in the price-elastic demand functions. One is b_i, the technological progress parameter, and the other is r^*, the interest rate, which influences the equilibrium amount of investment. For TECATE, on the basis of historical values the yields per hectare of all crops were assumed to grow at 1.75 percent per year.[17] The interest rate was set at 12 percent, the current market rate, to reflect expected supply cost conditions.

For finding the local equivalent value of y, recourse was made to a simple iterative process on the computer. The matrix generated by the SECGEN routine (see chapter 19) includes accounting rows for gross revenue both at base-year prices and at endogenous prices. Once having set the number of years in the planning horizon (coding it on one of the SECGEN instruction cards), the iterative process consists of making repetitive solutions under different values of a parameter representing the annual growth rate in y, until the point is reached where the two gross revenue variables take on equal values. In practice, this required about four to five solutions. This procedure does not fix individual crop prices, but rather only the index of all crop prices. It finds a rate market expansion (percentage growth in y) for the local area that is consistent with the forecast or desired national terms of trade.

It should also be recognized that the price index (terms of trade) is fixed only for $r = r^*$. At higher and lower values of r, the investment program takes on lower and higher magnitudes, and hence aggregate crop supply and prices respond correspondingly. This behavior is consistent with the sectoral viewpoint discussed previously. If public sector decisionmakers

17. If the data so warrant, the yield increases can be made to vary by crop.

Figure 16-2. *Procedure for Shifting the Demand Curve*

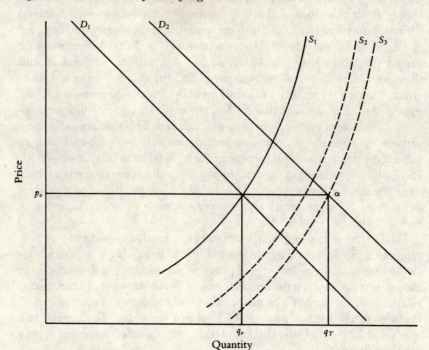

were to appraise all agricultural investments at a higher interest rate, then relatively few investments would take place, sectoral production would expand slowly, and agricultural prices would be likely to rise relative to nonagricultural prices.

To conclude this part of the discussion, the procedure for shifting the demand curve is illustrated in figure 16-2 for the simplified example of one aggregate product. The (unknown) base-period supply curve is S_1; the supply curve at the end of the planning horizon (time T) that results only from the cumulative effects of technological progress is S_2; and S_2 as modified by the investments that take place at $r = r^\star$ is shown by the curve S_3. The initial demand function, linearized with respect to prices, is D_1. The problem is then to find point α on the (unknown) curve S_3 and ensure that D_2, the demand function for time T, passes through that point.

The final remark regarding the investment version of the model concerns the choice of a particular year for T. In principle, any consistent normalization procedure could be used. In practice, it was simple to adjust the economic parameters for all project components to conform to the lifetime $L^\star$ of the component with the longest life. For a component with lifetime L, if it becomes part of the project, reinvestment is assumed to

Table 16-1. *Results of MEXICALI versus Actual Cropping Patterns*
(thousands of metric tons)

Crop	MEXICALI Results	Actual pattern			
		1968–69	1969–70	1970–71	1971–72
Garlic	1.4	1.1	0.5	0.5	2.1
Cotton	140.6	149.5	118.4	113.2	149.3
Alfalfa	171.9	150.2	185.9	180.0	210.0
Safflower	27.7	36.2	15.6	46.7	30.9
Barley	34.4	15.6	23.7	23.6	50.7
Sorghum	15.2	4.0	14.3	19.6	21.3
Maize	6.0	0	6.2	9.7	4.6
Wheat	172.9	148.0	239.6	190.3	171.6

Source: Actual figures are from Secretaría de Recursos Hidráulicos (SRH, 1968–72).

take place in year L in the amount $\bar{Q} = [(L^{\star}/L) - 1]\, Q$, where Q is the amount of initial investment in that component. And $\bar{Q}$ is then discounted to present value.

Numerical Results

The static version, MEXICALI, was based on data for the crop year 1970–71. For purposes of comparison, MEXICALI's cropping pattern is shown in table 16-1 alongside the actual cropping patterns for four crop years. MEXICALI estimates the marginal social productivity of pump water (from wells) at 327 pesos, or 6 percent greater than its administered price. Water from dams (gravity-fed water) is less productive because it must travel much further through canals; hence it incurs greater transmission losses in reaching the fields under cultivation. Nevertheless, gravity-fed water also has a true value that is higher than its cost: 229 pesos, or 13 percent higher than the cost. As is subsequently shown, these marginal productivities rise over time, especially in cases where little investment is undertaken to release the water bottleneck.

The model shows that the district provides 28,500 man-years of employment each year, more than half of which represents employment of hired laborers. There is no direct way to verify these figures, since agricultural employment statistics are not collected,[18] but they can be com-

18. The decennial censuses provide estimates of the labor force by sector, but that is not the same as actual employment. The divergence between the two concepts is particularly marked in agriculture, where there is substantial seasonal unemployment. The models MEXICALI and TECATE compute employment in man-years by adding up actual gainful occupation by man-month.

Table 16-2. *Basic Marginal Efficiency of Capital (MEC) Schedule in TECATE*

Interest rate (percent)	Present value of investment by category (millions of pesos)				
	Total investment	Land leveling	Desal- inization	Wells	Canal lining
0	1,658.9	597.2	130.4	124.0	807.3
4	1,452.0	454.0	110.9	107.4	774.7
8	1,037.4	364.9	73.3	93.0	505.3
12	773.5	314.8	42.1	84.0	322.6
16	554.3	246.2	39.3	79.5	187.3
20	457.7	230.8	37.3	73.8	115.8
24	391.8	189.2	34.5	70.9	115.8
36	202.3	170.1	32.2	0	0

Note: Investment figures are expressed in present values (1971) at the corresponding rate of interest for each solution. The base case used an interest rate of 12 percent.

pared usefully with rates of employment elsewhere in the sector. Río Colorado generates about 2.0 man-months of employment per hectare cultivated; this is significantly less than the corresponding figure for the small-scale irrigated vegetable farms in the Central Plateau of Mexico, but it is much more than the figure for nonirrigated agriculture.

Total agricultural net income in the district, including that of both farmers and day laborers, is estimated by MEXICALI to be about 223,200,000 pesos; or about 1,329 pesos per hectare cultivated. This is equivalent to 7,832 pesos per man-year worked. Both these figures are significantly higher than the national average.

The investment version, TECATE, was used to address a number of questions related to the amount of investment and its composition and the effects of data errors and changes in policy parameters. Table 16-2 and figure 16-3 show the basic MEC schedule for year T, 1991.[19] Eight points on this schedule were generated by varying the level of the IRR required of the project and by solving the model anew for each level. Shifting the criterion from 8 percent to 12 percent reduces the justifiable amount of investment by some 264 million pesos (21 million dollars). Over the entire range of 0 to 36 percent, the variation in investment amounts is 1,457 million pesos (117 million dollars). Thus, we are dealing with a wide range of choices in financial outlays, and the point chosen in that range is shown to be quite sensitive to the rate of social return demanded.

19. The longest project lifetime among the project components is twenty years, and the model's base year is 1971.

Figure 16-3. *Basic Marginal Efficiency of Capital (MEC) Schedule from TECATE*

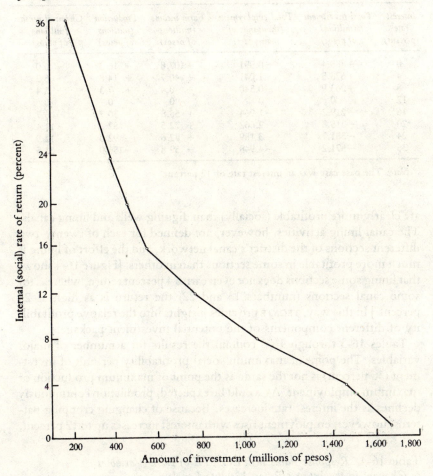

The composition of the investment package is also sensitive to the rate of return. In table 16-2, the fifty-eight categories of investment are grouped into four basic categories for illustrative purposes. It can be seen that canal lining, for example, accounts for 49 percent of expenditures at a 0 percent interest rate, 34 percent of expenditures at a 16 percent interest rate, and it falls to zero at a 36 percent interest rate. Land leveling, on the other hand, moves in the opposite direction: it goes from 36 percent of the package at no interest cost to 84 percent of the package at the highest interest cost.

Within hierarchies of project categories, land leveling and desalinization

Table 16-3 *Response Paths in TECATE of Basic Variables to Changes in the Interest Rate: Increments with Respect to the Base Case*

Interest rate (percent)	Total investment (millions of pesos)	Total employment (thousands of man-years)	Farm income (millions of pesos)	Production (millions of pesos)	Chemical inputs (millions of pesos)
0	+ 885.4	− 0.991	+ 107.8	+ 16.3	− 6.0
4	+ 678.5	− 1.041	+ 40.7	+ 14.8	− 6.2
8	+ 263.9	− 0.540	− 0.4	+ 7.3	− 2.4
12	0	0	0	0	0
16	− 219.2	− 1.964	+ 56.8	− 50.7	− 6.7
20	− 315.8	− 2.662	+ 72.5	− 82.2	− 11.4
24	− 381.7	− 3.189	+ 82.6	− 92.0	− 12.9
36	− 571.2	− 4.948	− 59.8	− 156.0	− 21.5

Note: The base case was an interest rate of 12 percent.

are clearly more profitable (socially) than digging wells and lining canals. The canal lining activities, however, are defined for each of twenty-two different sections of the district's canal network, and the effort of lining is much more profitable in some sections than in others. [Figure 16-4 shows that lining some sections does not even earn a 4 percent return, whereas for some canal sections (numbers 13 and 22) the return is as high as 24 percent.] In this way, TECATE provides insights into the relative profitability of different components of the potential investment package.

Tables 16-3 through 16-6 contain the results for a number of major variables. The point of maximum social profitability per unit of investment (36 percent) is not the same as the point of maximum production or maximum employment. As would be expected, production continuously declines as the interest rate increases. Because of changing cropping patterns, however, employment rises with interest increases up to 12 percent,

Table 16-4. *Response Paths in TECATE of Basic Variables to Changes in the Interest Rate: Absolute Levels*

Interest rate (percent)	Total investment (millions of pesos)	Total employment (thousands of man-years)	Farm income (millions of pesos)	Production (millions of pesos)	Chemical inputs (millions of pesos)
0	1,658.9	34.634	802.2	1,338.5	196.3
4	1,452.0	34.584	735.1	1,337.0	196.1
8	1,037.4	35.085	694.0	1,329.5	199.9
12	773.5	35.625	694.4	1,322.2	202.3
16	554.3	33.661	751.2	1,271.5	195.6
20	457.7	32.963	766.9	1,240.0	190.9
24	391.8	32.436	777.0	1,230.2	189.4
36	202.3	30.677	634.6	1,166.2	180.8

Table 16-5. *Response Paths in TECATE of Basic Variables to Changes in the Interest Rate: With Wage Rates 25 Percent Higher*

Interest rate (percent)	Total investment (millions of pesos)	Total employment (thousands of man-years)	Farm income (millions of pesos)	Production (millions of pesos)	Chemical inputs (millions of pesos)
0	1,658.2	32.074	780.0	1,338.3	196.2
4	1,414.0	31.921	716.4	1,333.7	195.8
8	1,041.9	30.650	740.8	1,292.3	189.8
12	822.9	30.463	714.9	1,278.8	190.2
16	489.7	30.512	763.8	1,245.5	190.8
20	457.5	30.515	741.4	1,245.5	190.8
24	383.1	29.752	768.8	1,221.4	187.1
36	202.8	28.015	809.9	1,161.5	179.1

and then it declines monotonically thereafter. Thus, in terms of total employment generated, the 12 percent package happens to be the best. To select according to employment criteria, it is necessary to review experience in other parts of the sector to derive benchmarks for judging employment performance. Local farm income (defined here as including incomes earned by day laborers) also moves irregularly over the MEC curve, but nevertheless is highest when the greatest amount of investment occurs.

Tables 16-5 and 16-6 also show, respectively, how the MEC curve is displaced if wage rates are 25 percent higher and if the physical benefits (water savings, yield increases, and the like) to all investment components are higher by 30 percent. The first variation illustrates what would happen if wages were to rise, relative to other prices in the economy, during the

Table 16-6. *Response Paths in TECATE of Basic Variables to Changes in the Interest Rate: With Higher Investment-benefit Coefficients*

Interest rate (percent)	Total investment (millions of pesos)	Total employment (thousands of man-years)	Farm income (millions of pesos)	Production (millions of pesos)	Chemical inputs (millions of pesos)
0	1,697.3	37.011	635.7	1,453.1	205.2
4	1,442.4	36.737	586.3	1,443.6	203.6
8	1,061.8	35.134	627.6	1,393.2	197.3
12	838.4	34.303	645.0	1,362.4	194.3
16	502.7	35.201	684.7	1,336.1	199.0
20	468.4	35.132	654.1	1,335.1	198.8
24	372.8	34.996	672.5	1,319.3	198.2
36	211.8	32.143	769.7	1,234.2	185.8

Note: Water savings, yield increases, and the like are higher by 30 percent than in the solutions reported in tables 16-3 and 16-4.

project planning horizon. The second variation constitutes a type of sensitivity analysis regarding possible data errors. The approach is to investigate with the model whether uncertainty regarding the physical effects of the investment is important to the financial amount of investment authorized. If so, it becomes important to narrow that range of uncertainty before the final project evaluation.

As regards the change in the wage rate, the overall effect is to lower the returns to capital. That the sum of investments justified at interest rates of 16 to 36 percent is lower, and that the sum of investments at zero to 12 percent is higher under the new wage level, confirms this effect. Clearly, some projects, but not all, have been pushed down to lower points on the MEC schedule. Because of the existence of interdependent crop supply functions, the effect of any such change in production costs is complex and often nonuniform. For this reason, the numerical simulations by the model are quite helpful. For example, it would be impossible to foresee a priori that the higher wage rate *reduces* the justified amount of investment by 65 million pesos at $r = 16$ percent and *increases* it by 49 million pesos at $r = 12$ percent. In retrospect, the explanation is straightforward: some investments that formerly earned 16 percent or more were knocked down to 12 percent by the wage increase, and yet others that have earned 12 percent were knocked down to lower levels of return. TECATE quantifies these various effects. The effects of a wage change are also registered on the production and farm income variables. To illustrate, at zero interest cost production is essentially unchanged, but income is lower under the higher wage rate. Clearly, the higher labor cost has forced farmers to move to more machine-intensive production techniques that have lower profit margins. Or even if the profit margins are not lower, the displacement of day labor shows up in lower levels of farm income. Yet at 36 percent interest the wage change results in lower production levels but higher income levels. Here the labor displacement effects are dominated by the pure wage effect, which gives higher incomes to those who have to work.

The point of these results is that simple conceptual models do not suffice to predict behavior, for in the real world there are many complementary and competitive effects acting simultaneously. Without a detailed numerical simulation, it is perhaps impossible to foresee which effect will dominate.

An irregular pattern of results is also found with respect to the change in physical benefits (table 16–3 again). Here the opposing effects are different. On the one hand, the higher benefits simply make the investment activities more productive. On the other hand, the higher productivity represents a rightward shift in the supply function; other things being equal, this shift drives prices and farm incomes down.[20] Again, at times one effect is dominant, and at times the other. It does emerge as important, however, to make a precise estimation of the physical benefits for each investment component. TECATE shows that the 30 percent change in unit benefits changes the justifiable amount of investment by 65 million pesos (5 million dollars) at $r = 12$ percent. Thus, the sensitivity tests indicate that a significant financial savings could be gained by investing a few hundred thousand dollars in refinements of the basic project data.

Figure 16–4 is a breakdown of parts of tables 16–2 and 16–3 through 16–6 by more detailed subproject categories. As mentioned, to facilitate their incorporation in the model it was necessary to define three types of investment (land leveling, desalinization, and well digging) according to the crop to be grown. In practice, of course, it is not possible to guarantee which crop will be grown on the leveled land, but from knowledge of each district's usual cropping patterns a good approximation may be made. This format has the advantage of helping to indicate how many hectares should be leveled, drained, and so forth and which crops justify the investment. For the fourth category of investment, canal lining, the breakdown was simply by sections as defined by the project engineers.

In the results, it turned out that all land leveling for all crops was highly profitable and that new wells were justified only for garlic and alfalfa zones. For desalinization, high returns are indicated for land dedicated to garlic, alfalfa, barley, and wheat. The model also computes directly the number of hectares devoted to each category of investment so that plans can be made on the basis of hectareage.

Table 16–7 shows the cropping patterns associated with each interest rate and, hence, with each potential investment package. There are a few changes from the base-year pattern, as may be seen by a comparison with table 16–1.[21] As the interest rate varies, some crops (corn, wheat) remain fairly stable, while others (safflower, forage barley) are quite variable. These shifts in cropping pattern are responsible for a change of about 15 percent in total revenue as measured in constant prices, over the range 0–36 percent in interest rates, and for a larger change in revenue measured in endogenous prices. Although not overwhelming, these variations are significant enough to indicate the importance of tying the cropping patterns to the investment package in the project appraisal.

An indirect measure of the project's aggregate input on the district is provided by TECATE's measures of the marginal productivity of water,

---

20. Recall that all price elasticities of demand for Río Colorado products are less than unity in absolute value.

21. Table 16–1 is expressed in metric tons and table 16–7 in hectares.

Figure 16-4. *Composition of TECATE Investment Package at Varying Interest Rates*

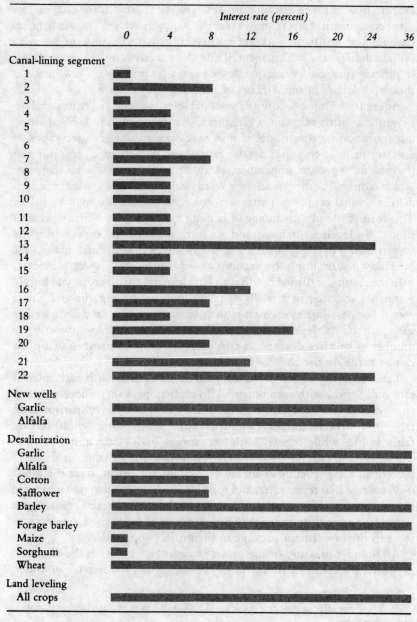

shown in table 16-8. Recall that in the base year the marginal productivity was of the order of 0.2 to 0.3 pesos per cubic meter. It remains in that range in 1991 (in constant prices) but it varies substantially according to the size of the investment package.

Conclusions

There is a growing literature on the incorporation of policy guidelines in project appraisal. Briefly, five approaches (or combinations thereof) have been attempted:

- Estimation of appropriate economy-wide shadow prices for goods and resources, by reference to international price levels and fiscal parameters (as in Little and Mirrlees 1968)
- Use of a linear programming model's dual solution to derive those shadow prices (as in Clark, Foxley, and Jul 1974)
- Use of an input-output model to derive guidelines for sectoral rates of capacity expansion (Lee 1969)
- Without a model, ranking of a project according to different criteria and experimenting with alternative weights for those criteria (McGaughey and Thorbecke 1972)
- Explicit linkage of price-endogenous project and sector models to internalize the shadow pricing (Duloy, Kutcher, and Norton 1973).

In agriculture, many goods are nontraded, and for many others the gap between c.i.f. and f.o.b. prices is quite large, so that the Little-Mirrlees procedure is not appropriate. Similarly, models that do not reflect market structures and the corresponding mechanisms for price determination cannot be expected to give useful guidance on product price levels. The Duloy-Kutcher-Norton linkage experiment concluded that the intersectoral terms of trade quite significantly affect the project returns in agriculture.

This chapter, therefore, represents an attempt to include output pricing effects on the investment decision, in the context of project selection as well as appraisal, with an appropriate representation of market structures. It is based on what appears to be a reasonable assumption: that the criterion for rate of return adopted for one locality will be used for all other regions.

In the context of multilevel planning systems, the procedure of this chapter may be viewed as follows: on the basis of historical evidence or plan guidelines, a tentative selection is made for values of r (the interest rate) and p (the terms of trade) for the sector as a whole. Then, with models such as TECATE explorations are made, for a few districts, of the consequences of varying r. The models reflect what would happen to

Table 16-7. Changes in Cropping Pattern along the Basic MEC Curve in TECATE

Interest rate (percent)	Total investment (millions of pesos)	Total hectares planted (thousands)	Hectares in each crop (thousands)								
			Garlic	Alfalfa	Cotton	Safflower	Forage barley	Grain barley	Maize	Sorghum	Wheat
0	1,658.9	204.547	0.213	15.222	77.883	21.407	0.294	14.404	3.340	6.184	65.600
4	1,452.0	204.156	0.213	15.222	77.883	21.104	0.285	14.323	3.340	6.184	65.600
8	1,037.4	204.847	0.213	15.222	80.547	19.481	0.278	14.125	3.318	6.063	65.600
12	773.5	206.478	0.214	14.879	82.383	20.318	0.270	13.845	3.318	6.063	65.185
16	554.3	197.078	0.214	14.192	78.790	15.450	0.247	13.286	3.297	5.823	64.771
20	457.7	193.560	0.213	14.192	77.198	14.551	0.248	13.286	3.275	5.823	64.771
24	391.8	189.855	0.213	13.849	77.200	13.006	0.232	13.006	3.276	5.703	69.356
36	202.3	180.025	0.240	14.591	73.438	6.868	0.217	12.447	3.233	5.463	63.527

Table 16-8. *Marginal Productivity of Water in TECATE*
(pesos per cubic meter)

Interest rate (percent)	Irrigation water		Interest rate (percent)	Irrigation water	
	Gravity-fed	Pump		Gravity-fed	Pump
Basic MEC schedule					
0	0.1435	0.2050	16	0.1700	0.2428
4	0.1353	0.1933	20	0.1697	0.2426
8	0.1378	0.1969	24	0.1792	0.2560
12	0.1433	0.2047	36	0.2042	0.2916
Higher wage rate case					
0	0.1275	0.1822	16	0.1558	0.2225
4	0.1258	0.1797	20	0.1536	0.2140
8	0.1412	0.2018	24	0.1622	0.2317
12	0.1473	0.2104	36	0.1822	0.2603
Higher benefits case					
0	0.1018	0.1454	16	0.1354	0.1935
4	0.0944	0.1348	20	0.1319	0.1884
8	0.1140	0.1628	24	0.1441	0.2059
12	0.1334	0.1906	36	0.1796	0.2566

Note: The difference between pump and gravity-fed water is a constant 0.7, which reflects the relatively lower efficacy of gravity-fed water in comparison with pump water.

prices sector-wide if a different value of r were selected throughout the sector's investment program. Subject to the important caveat that the sampled districts be sufficiently "representative" of the sector in their supply response, this procedure leads to a sector-wide value of p as the outcome of both the newly chosen value of r and the investments that are justified at that value of r.

Of course, in less formal terms the ultimate project decision is likely to be based on criteria other than r and p. In this respect, the procedure of this paper is offered as a numerical aid to decisionmaking because it quickly generates estimates of employment levels and other consequences of varying the price of capital—or the price of any other input or output.

References

Adelman, Irma, and F. T. Sparrow. 1966. "Experiments with Linear and Piece-wise Linear Dynamic Programming Models." In *Theory and Design of Economic Development*. Edited by Irma Adelman and E. Thorbecke. Baltimore, Md.: Johns Hopkins University Press.

Bassoco, Luz María, Roger D. Norton, and José S. Silos. 1972. "Programación de la producción y uso de los recursos en el Valle de Mexicali, B.C." Mexico City: Secretaria de la Presidencia. Processed.

Candler, Wilfred V. and Roger D. Norton. 1977. "Multi-Level Programming." Development Research Center, Discussion Paper, no. 20. Washington, D.C.: World Bank. Restricted circulation.

Clark, P. B., A. Foxley, Jr., and A. M. Jul. 1974. "Project Evaluation within a Macroeconomic Framework." In *Analysis of Development Problems*. Edited by R. S. Eckaus and P. N. Rosenstein-Rodan. Amsterdam: North-Holland.

Cummings, R. G., and Donald W. Winkelmann. 1970. "Water Resource Management in Arid Environs." *Water Resources Research*, vol. 6, pp. 1559–98.

Duloy, John H., Gary P. Kutcher, and Roger D. Norton. 1973. "Investment and Employment Alternatives in the Agricultural District Model." In *Multi-Level Planning: Case Studies in Mexico*. Edited by Louis M. Goreux and Alan S. Manne. Amsterdam/New York: North-Holland/American Elsevier. Pp. 417–34.

Harberger, Arnold. 1974. *Project Evaluation*. Chicago: Markham Publishing Co.

Hazell, Peter B. R.. 1971. "A Linear Alternative to Quadratic and Semivariance Programming for Farm Planning under Uncertainty." *American Journal of Agricultural Economics*, vol. 53 (February), pp. 53–62.

Lee, Hee Il. 1969. "Project Selection and Evaluation: Formulation of an Investment Program." In *Practical Approaches to Development Planning: Korea's Second Five-Year Plan*. Edited by Irma Adelman. Baltimore, Md.: Johns Hopkins University Press.

Little, Ian M. D., and James A. Mirrlees. 1968. *Manual of Industrial Project Analysis in Developing Countries*. Vol. 2. Paris: Organisation for Economic Co-operation and Development.

McGaughey, S. E., and E. Thorbecke. 1972. "Project Selection and Macroeconomic Objectives: A Methodology Applied to Peruvian Irrigation Projects." *American Journal of Agricultural Economics*, vol. 54 (February), pp. 32–40.

Rogers, P., and D. V. Smith. 1970. "The Integrated Use of Ground and Surface Water in Irrigation Project Planning." *American Journal of Agricultural Economics*, vol. 52, pp. 13–24.

Secretaría de Recursos Hidráulicos (SRH). 1968–72. *Estadística agrícola*. Ciclos 1968–69, 1969–70, 1970–71, 1971–72. Mexico City.

17

A Demonstration Model
for the Dairy Industry of La Laguna

Wilfred V. Candler

This CHAPTER provides an example of model building suspended in midcourse.[1] It shows a model structure, with "best guess" data, some of which are demonstrably inconsistent.

Overview

The project is particularly well documented (errors and all) exactly because the project had to be suspended when counterpart responsibilities in Mexico were changed. It was hoped that after a period alternative working arrangements could be made, and that the model could be taken up again and completed.

The project was started with the enthusiastic support of a young and able Mexican administrator. Relatively rapid progress was made, but when he was promoted to a still more important job outside the livestock area the project was unable to recapture the sense of policy direction with which he had initially motivated the development of the model.

The problem is a familiar one in countries well on their way in development: as the administrative and professional infrastructure rapidly grows, promotions remove promising administrators and researchers from the

1. *Editors' note*: At our request, Wilfred Candler compiled his notes on this half-finished exercise in model building. It was halted because promotions and opportunities to study abroad led to the complete dispersal of his five-person Mexican counterpart team. Because the aim had been to build a model that actually could be used in policy formation, there was no point in carrying through without counterparts to help specify model structures and policy options relevant to Mexico and to participate in the iterative process of developing a data set that ultimately could be defended as reasonable. Nevertheless, we felt Candler's notes would make a valuable contribution to this book precisely because they illuminate the *process* of building a usable model, in a way that the other chapters in this book cannot, inasmuch as they tend to focus either on concepts or on completed numerical models. If data and policy issues are to be confronted squarely, then this process inevitably involves going through some stages that will yield odd results and learning from them, as this chapter illustrates.

field; similarly, these people may expatriate, if only temporarily, for study abroad to further their training and careers. Both circumstances create a brain drain away from the grass-roots project level and discontinuities in project planning and implementation.

In addition to the intrinsic interest of the proposed model structure, this chapter also provides a timely reminder that not all modeling exercises may come to an immediate, clean, and satisfactory completion. In the normal course of events, this model would have been revised, relevant data inserted, and an apparently immaculate product exhibited to the world.

The original motivations for the present study included a desire to expand CHAC to include livestock as well as crops; to explore the methodology of modeling multiyear agricultural products; and to help institutionalize capabilities for constructing models similar to CHAC in other Mexican institutions besides the presidency. Work was also undertaken on models of other livestock subsectors apart from dairying, but for a variety of reasons documentation of these models was not carried through to the point at which even the structure of the models could be usefully presented.

The current draft model was constructed and documented with the following objectives: to illustrate the proposed model structure; to present starting-point estimates, in the hope that markedly improved data sources (or even more experienced opinion) would enable these estimates to be improved to the point where the model manages to capture the salient features of dairy production in La Laguna in north-central Mexico (as an aid to this process, some alternative sets of assumptions about the La Laguna dairy industry that have emerged from discussions of a rough draft of this documentation have been included); to provide the modeling experience needed to improve the speed with which subsequent dairy models can be constructed.

Precisely because it is incomplete, this chapter may give a better feeling for the process of model building than can be obtained from examination of a completed model. By the same token, it cannot illustrate the usefulness of model results, since it has not yet reached the point of containing consistent data.

Policy Questions

Five important policy questions motivated the desire to build an interregional and seasonal model of Mexican dairy production. The concern was to study:

- The economics of Mexican versus imported replacement stock, and to a lesser extent whether Mexican replacements should be raised on dairy farms or on specialized or extensive farms
- The extent to which seasonal adjustments in supply could be coordinated among different regions to give a smooth annual flow of milk for consumption
- The extent to which credit or other policies could be used to increase the rate of expansion of the dairy industry
- Locational problems as to whether whole milk production should be concentrated in consumption or forage areas, and where manufactured milk production should be located
- The extent to which improved processing and distribution systems could be used to reduce consumer prices, to raise producer prices, or both.

The model should be helpful in answering the first three concerns. It is unlikely to throw much light on the last, but if additional dairy models were constructed for other regions, these would allow the regional locational questions to be answered.

The Existing Model

The present model has 64 rows (including the objective function) and 154 columns (including the right-hand sides). Some of the restraints (such as quarterly water balances) were designed to facilitate eventual meshing of this model with CHAC (see chapters 2 and 5). A full list of the restraints and activities used is given in the chapter's appendix (table 17-17).

The model is for one production technology (stable-and-pasture) in one region (La Laguna) in the north-central part of the republic. One of the CHAC irrigation submodels corresponds to this area (see chapter 4). The dairy model's activities are defined for one "equilibrium" year, which is disaggregated for milk production into four quarters of equal length. A stock reconciliation matrix (rows 2–9 and rows 50–64, and columns 2–54), which incorporates stock sales and purchases, ensures that herd numbers are kept in balance.

Restraints

Four restraints ensure that seasonal (spring, summer, autumn, winter) balance is maintained for milk production (rows 12–15), land and water usage (rows 30–37), milk sales maxima (rows 44–47) and feed (rows 16–27). Varying degrees of seasonal uniformity in milk production can be enforced by changing the numerical restraint values on rows 12–15, and thus some of the consequences of uniformity, or lack thereof, can be explored.

Feed requirements are expressed per quarter in terms of protein, energy, and total dry matter (TDM). These requirements can be met by crops purchased in appropriate seasons. Seasonal price differentials have been applied to the cost of crops, but the price of milk is the same in all four quarters. Capital requirements are divided into long and short term (rows 38 and 39), whereby annual labor requirements are expressed in skilled and unskilled workers (rows 28 and 29). Maxima and minima on herd size (rows 48 and 49) have been added to prevent an unbounded solution being obtained when the milk market maxima are relaxed.

Activities

The activities used in the model are also listed in the appendix (the second part of table 17-17) in the order in which they appear in the matrix.

The selling activities appear in columns 2–5 for young stock. Male calves can be sold within two days of birth or when they are ten months old. Males are not retained for bulls, but rather these animals are purchased from specialized breeders. Female calves can be sold within two days of birth, at ten months old, or just before the first lactation (column 10). Cows can be sold after their lactations (columns 35–55). Milk sales, columns 69–72, are identified by quarter.

Purchases (except feeds) are identified in columns 8 and 9 and 58–68. Springing heifers (that is, animals ready to enter their first lactation) can be bought from the United States or elsewhere in Mexico (columns 8 and 9). Irrigation water can be bought quarterly. Labor time is divided into skilled and unskilled. The input of the farmer's labor is included as part of the hiring of skilled labor—that is, the model does not assume the existence of a pool of proprietor's labor that can be had for free. Capital can be borrowed long-term at successively higher interest rates (that is, 8, 10, and 11 percent), whereas short-term capital can be borrowed at 12 percent and 15 percent.

There are three expansion activities in columns 73–75. The first allows for the renting of land for crop production, and the others allow for the expansion of old, and building of new, dairy facilities. Activities 74 and 75 (expansion of buildings) have not been implemented in the current version of the model.

Replacement heifers (that is, females being raised to enter the milking herd or for sale), are raised on the home farm for the first ten months, but may then either be kept at home or grazed on another ranch until the thirtieth month (one month before calving). These two kinds of replacement activity appear as columns 6 and 7.

There are twenty lactation activities, columns 11–34. Lactations are identified by the season in which calving occurs. A cow that calves in the spring is available to calve the next spring or can be sold. The selling activities for cows that have been milked appear as columns 35–54. The biological lactation curve, first rising and then declining, is represented via the coefficient values in columns 11–34 (see table 17-1).

Cows may be inseminated naturally or artificially. These activities appear as columns 56 and 57. It should be noted that natural mating is calculated per bull (that is, forty-five matings annually), whereas artificial insemination is calculated per mating.

The remaining activities, columns 76–154, refer to different ways of feeding the dairy herd.

Columns 76–101 refer to crops grown specifically for dairy feed, and these activities are identified by season. Columns 102–127 refer to crops bought locally, whereas the final columns 128–154 refer to crops bought from outside the region. (The exact crops and units are listed in the appendix, table 17-17.)

In summary, young stock produced can be sold at once, at ten months old, or at thirty-one months old (just prior to entering the milking herd). Replacement stock can only be bought at thirty-one months old. Cows that enter the milking herd remain in the herd for that lactation, with appropriate allowances for death losses, sale of cows that repeatedly fail to come into calf, and slight culling for disease. Cows may be sold after any lactation, but after the sixth lactation all cows are sold for slaughter.

Production from (and feed requirements for) cows in the milking herd are identified by whether the cow calved in spring, summer, autumn, or winter. Cows may be mated naturally or by artificial insemination. Labor is available in two qualities, and capital is available at increasing prices from stepped supply functions. Feed may be purchased on farms, locally, or outside La Laguna.

This model structure has at least three features that may be considered undesirable:

- Replacement stock (and hence their feed requirements) are identified only by the year in which they enter the herd.
- That only three nutritive restraints are used in each quarter means that, at most, three crops will be fed in each quarter. This objection may be met by supplying feeds from upward-sloping (stepped) supply functions. At present, the model has seventy-eight feeding activities of which at most twelve can be basic.
- The equilibrium nature of the model means that the *time path* of developments in herd size cannot be studied. The model cannot accept an opening inventory of dairy cattle in 1972 and trace how this herd should be adjusted in each successive year.

If the path of development were of the essence, then a different model-

Table 17-1. Performance per Cow in Successive Lactations

Lacta-tion	Age at start (months)	Weight (kg)	Concep-tion (percent)	Calving (percent)	Death loss (percent)	Culled (percent)	Milk (kg)	Feed maintenance Protein (kg)	Energy (kcal)	TDN (kg)	Services needed to conceive
						Basic assumptions					
1	31	454	95	90	2	0	5,160	102.51	2,055		2.0
2	43	491	97	95	1	0	5,626	110.69	2,301		1.3
3	55	522	98	95	2	1	5,160	116.97	2,467		1.5
4	67	545	97	93	2	1	4,300	120.69	2,558		1.5
5	79	570	95	90	2	2	2,924	126.42	2,658		1.8
6	91	570	90	90	3	100	2,580	126.42	2,658		1.8
						Alternative assumptions					
1	28	500	90	83	2	16	4,380	36.09	21.54	2.45	2.0
2	42	550	90	95	1	20	5,037	39.70	23.69	2.70	1.3
3	56	550	90	95	2	20	5,547	39.70	23.69	2.70	1.5
4	70	550	90	95	2	22	5,824	39.70	23.69	2.70	1.5
5	84	550	90	95	3	22	4,987	39.70	23.69	2.70	1.8
6	98	550	90	95	3	100	4,248	39.70	23.69	2.70	1.8

Note: *Basic assumptions.* Feed requirements were: for first pregnancy (see Morrison 1963), 18 kilograms (kg) of protein and 367.5 kilocalories (kcal) of energy; for later pregnancies (see Morrison 1963), 22.88 kg of protein and 431.3 kcal of energy; for milk production (per kg of milk with 3 percent fat content), 0.0395 kg of protein and 0.55 kcal of energy. TDN (total digestible nutrients) requirements were not determined. *Alternative assumptions.* Data were obtained from discussions with people familiar with cattle-feeding practices in La Laguna; reconciliation of data under basic and alternative assumptions was not completed at termination of the study. Feed requirements under the alternative assumptions were: for maintenance and lactation, 12 percent of a cow's body weight in fresh alfalfa daily; for milk production, 0.25 kg of feed concentrates per kg of milk. Under both basic and alternative assumptions, 50 percent of the calves were male, 50 percent were female.

ing approach would be appropriate. At least two such approaches may be suggested:

- Link about seven of the present models together to give a model with, say, 500 rows and 1,000 columns. This would allow the age groups to be treated separately—and hence "aged," with milk yield, death losses, fertility, and so forth corresponding in each year to the actual age of the herd. Extra rows and columns would be needed to specify individual age groups and control their transfer from year to year.
- A second approach would be through recursive programming, in which only the present model is run, but after each year some calculations are done "off-line" to find herd composition. This is reflected in the feed requirements, yield, and death losses for the next one-year round.

As indicated below, the present model can be used to examine the overall effect of herd expansion at an average rate of, say, 0, 5, 10 percent or higher.

An alternative to specifying nutrient requirements would have been to specify amounts of maintenance and production rations by quarter, in which case we then would have activities to meet these requirements. The activities, however, would not be individual feeds, but suitable mixtures of feeds. This could ensure that a wide mix of forage crops would be used in place of the three that are likely to be used with the modeling actually done in this model. As mentioned below, feed requirements based on observed rations in La Laguna are suggested in this account, but they have not been implemented on the computer.

Assumptions and Calculations of Coefficients

Specific assumptions were used to calculate the coefficients used in this model. One set of assumptions refers to the performance of the cows in successive lactations. These basic assumptions and an alternative set are summarized in table 17-1. The "basic assumptions" section of the table is derived from accepted Mexican nutritional tables; the "alternative assumptions" section is based on feeding practice as observed by extension personnel in La Laguna. Work on the model terminated before these two sets of estimates could be reconciled, and thus before the estimates to be used could be finally agreed upon.

Annual feed requirements in the existing model

The annual protein and energy requirements listed in table 17-2 were interpolated from information in Morrison (1963, p. 658) according to the following system. If w is the weight of the dairy cow in kilograms and w^-,

Table 17-2. Calculation of Protein and Energy Requirements for Maintenance of Lactating Cows

Lactation	Weight (kg) w	Less Weight w^-	Less Daily d^-	More Weight w^+	More Daily d^+	Δ $d^+ - d^-$	δ_1 $w - w^-$	δ_2 $w^+ - w^-$	$\Delta\delta_1$	$\Delta\delta_1/\delta_2$	Requirement Daily d	Requirement Annually a
						Protein requirements						
1	454	453	0.28	476	0.30	0.02	1	23	0.02	0.00086	0.28026	102.5139
2	491	476	0.30	499	0.305	0.005	15	23	0.075	0.00326	0.30326	110.6899
3	522	521	0.32	544	0.33	0.01	1	23	0.01	0.00043	0.32043	116.96595
4	545	544	0.33	567	0.345	0.015	1	23	0.015	0.00065	0.33065	120.68725
5	570	567	0.345	589	0.355	0.01	3	23	0.03	0.00136	0.34636	126.42140
6	570	567	0.345	589	0.355	0.01	3	23	0.03	0.00136	0.34646	126.42140
						Energy requirements						
1	454	453	5.6	476	6.2	0.26	1	23	0.26	0.03	5.63	2056.11
2	491	476	6.1	499	6.4	0.3	15	23	4.5	0.20	6.30	2301.08
3	522	521	6.75	544	7.0	0.25	1	23	0.25	0.01	6.76	2467.71
4	545	544	7.0	567	7.25	0.25	1	23	0.25	0.01	7.01	2558.96
5	570	567	7.25	589	7.5	0.25	3	23	0.75	0.01	7.28	2658.15
6	570	567	7.25	589	7.5	0.25	3	23	0.75	0.03	7.28	2658.15

Note: $\Delta = d^+ - d^-$; $\delta_1 = w - w^-$; $\delta_2 = w^+ - w^-$; $d = d^+ + (\Delta\delta_1/\delta_2)$; $a = 365d$.
a. Calculation of requirements is as discussed in text and expressed in equation (17.1). See text for identification of variables.
Source: Morrison (1963, p. 658).

w^+ are the nearest weight entries in Morrison, respectively below and above w, and if d^-, d^+ are the corresponding daily protein maintenance requirements in kilograms, then the daily protein requirement for an animal of weight w for maintenance is calculated as:

$$(17.1) \qquad d = d^- + \frac{w - w^-}{w^+ - w^-}(d^+ - d^-).$$

Similar calculations may be made for energy requirements.

FOR MAINTENANCE. The annual protein (or energy) requirement for maintenance is simply:

$$(17.2) \qquad a = 365d.$$

These calculations are illustrated in table 17-2. The derivation of the total digestible nutrients (TDN) requirements used in the model is unknown to the author.

FOR GESTATION AND LACTATION. Morrison states that for the last seventy-five days of gestation additional protein and energy requirements by weight incurred as follows:

	Protein (kilograms)		Net energy (kilocalories)	
	Daily	*75 days*	*Daily*	*75 days*
<475 kilograms	0.24	18.0	4.9	367.5
>475 kilograms	0.305	22.875	5.75	431.25

Morrison also states that the protein requirement for lactation (per kilogram of 3 percent fat milk produced) is 0.395 kilogram, whereas the energy requirement is 0.55 kilocalorie per kilogram produced. Applying these multipliers to the annual production assumed gives us the protein and energy requirements for lactation listed in table 17-3.

FOR MILK PRODUCTION. In general, the annual feed requirements for a cow can be calculated as the simple sum of requirements for maintenance and milk production. Summing the maintenance and lactation feed requirements as given in tables 17-2 and 17-3 yields the total feed requirements per lactation, as calculated in table 17-4.

FOR CALVES. We took the daily feed requirements of calves weighing 136 kilograms (1.36 kilograms of TDM, 0.325 kilogram of digestible protein, and 5.0 kilocalories of energy) and multiplied by 300 days to get total feed requirements of 97.5 kilograms of digestible protein, 1,500 kilocalories of energy, and 408 kilograms of TDM. Dividing these numbers by 4 gave seasonal feed requirements of 24.37 kilograms of digestible protein, 375 kilocalories of energy, and 102 kilograms of TDM.

Table 17-3. *Protein and Energy Requirements for Lactation*

Lactation	Milk production (kg)	Protein (kg)	Energy (kcal)	Lactation	Milk production (kg)	Protein (kg)	Energy (kcal)
1	5,160	203.82	2,838	4	4,300	169.85	2,365
2	5,626	222.23	3,094	5	2,924	115.50	1,608
3	5,160	203.82	2,838	6	2,580	101.01	1,419

Annual feed requirements in the alternative model

As a result of discussions of the above model, an alternative approach to the estimation of feed requirements was proposed. Starting with the observed feeding pattern of 12 percent of body weight in fresh alfalfa daily for maintenance and gestation of milking cows, and 1 kilogram of concentrate (dry alfalfa meal) per 4 kilograms of milk for production, it is possible to argue "backward" to what the feed requirements (in metabolizable energy) and digestible protein requirements must be for maintenance and growth. This "backward" argument is now illustrated in detail.

FOR MAINTENANCE. The alternative model assumes that heifers weigh 500 kilograms and mature cows weigh 550 kilograms. Maintenance requirements of mature cows are 10 percent more than for heifers.[2] Thus, only maintenance requirements for heifers need be laid out in detail. Twelve percent of the body weight of a heifer weighing 500 kilograms is 60 kilograms of fresh alfalfa.

From the 1970 National Academy of Sciences' (1971) "Nutritional Feeding Standards for Dairy Cows" (table 4, p. 24), we find that fresh alfalfa comprises 27.2 percent of the TDM. Thus, dairy cows require 16.32 kilograms (60 kilograms × 27.2 percent) of dry matter from alfalfa per day, or its equivalent.

The same source indicates that each kilogram of dry matter from fresh alfalfa contains 2.21 megacalories (1,000 kilocalories) of net energy for milk production, and 0.15 kilogram of digestible protein. Thus, 16.32 kilograms of dry matter will contain:

36.09 = 16.32 × 2.21 megacalories net energy for maintenance

21.54 = 16.32 × 1.32 megacalories net energy for milk production

2.45 = 16.32 × 0.15 kilogram of digestible protein.

The maintenance requirements for mature cows in this alternative formulation have been set 10 percent higher than the above figures.

2. The generally accepted body-weight maintenance relation of $M = aW^{0.75}$ notwithstanding.

Table 17-4. *Total Annual Nutrient Requirements per Head for Dairy Cattle in La Laguna Model*

Lacta-tion	Protein (kg)				Energy (kcal)			
	Mainte-nance[a]	Gesta-tion[b]	Lacta-tion[c]	Total	Mainte-nance[a]	Gesta-tion[b]	Lacta-tion[c]	Total
1	102.51	21.58	203.82	324.33	2,055.11	367.5	2,838	5,260.61
2	110.69	22.88	222.23	355.80	2,301.08	431.25	3,094	5,826.33
3	116.97	22.88	203.82	343.67	2,467.71	431.25	2,838	5,736.96
4	120.69	22.88	169.85	313.42	2,558.96	431.25	2,365	5,355.21
5	126.42	22.88	115.50	264.80	2,658.15	431.25	1,608	4,697.40
6	126.42	22.88	101.91	251.21	2,658.15	431.25	1,419	4,508.40

a. See table 17-2.
b. See text table under "Annual feed requirements in the existing model."
c. See table 17-3.

FOR MILK PRODUCTION. The alternative model specifies 1 kilogram of dry alfalfa meal per 4 kilograms of milk production. Again using our "backward" argument, the National Academy of Sciences' (1971) "Nutritional Requirements for Dairy Cattle" indicates (p. 30) that 1 kilogram of dehydrated ground alfalfa (minimum 17 percent protein) provided 0.93 kilogram of dry matter. Each kilogram of dry matter provides 1.31 megacalories of net energy for maintenance, 0.69 megacalorie of net energy for milk production, and 0.15 kilogram of digestible protein. For each kilogram of milk, we feed 0.25 kilogram of alfalfa meal, or 0.2325 ($= 0.25 \times 0.93$) kilogram of dry matter. Hence we feed:

$0.305 = 0.2325 \times 1.31$ megacalories net energy for maintenance

$0.160 = 0.2325 \times 0.69$ megacalorie net energy for production

$0.035 = 0.2325 \times 0.15$ kilogram of digestible protein.

FOR CALVES. Calves are assumed to be fed 2.5 kilograms of concentrate (dried alfalfa) and 4 liters of milk for the first 4 weeks.

Herd composition

In using these assumptions it is further assumed that when a cow is available for a lactation it either conceives a calf (after up to two or three mating services) or is sold for slaughter. Of the cows that conceive, *all* come into milk, but only a percentage of live calves are born. All cows that conceive survive the entire lactation, but may die at the end of the lactation. If they do not die, they may still be sold (culled) for slaughter because of low levels of milk production. If we take the first line of table 17-5, the above procedure may be illustrated numerically:

Table 17-5. *Calculations of Herd Composition*

Lactation	Cows available	Nonconception		Cows in milk	Calving		Deaths		Cows that could be culled	Culling		Cows for next lactation
		Proportion	Sold		Proportion	Number born	Proportion	Lost		Proportion	Sold	
1	1	0.05	0.05	0.95	0.90	0.855	0.02	0.019	0.931	0	0	0.931
2	1	0.03	0.03	0.97	0.95	0.9215	0.02	0.0194	0.9606	0	0	0.9606
3	1	0.02	0.02	0.98	0.95	0.9310	0.02	0.0196	0.9604	0.01	0.0096	0.9508
4	1	0.03	0.03	0.97	0.93	0.9021	0.02	0.0194	0.9306	0.01	0.0093	0.9213
5	1	0.05	0.05	0.95	0.90	0.8550	0.03	0.0285	0.9215	0.02	0.0094	0.9121
6	1	0.10	0.10	0.90	0.90	0.81	0.03	0.027	0.873	1.0	0.873	0

- If a cow is available for the first lactation, it has a 5 percent probability of failing to conceive, thus 5 percent of the cows are sold for slaughter, and 95 percent complete the lactation.
- If 95 percent of the cows conceive, they have a 90 percent chance of having calves, hence the number of calves are born to 85.5 percent of the cows $(0.95 \times 0.90 = 0.855)$; half of the calves are males and half are females.
- If 95 percent of the cows complete the first lactation, then they have a 2 percent chance of dying—so 1.9 percent (0.95×0.02) of the cows die, and 93.1 percent $(0.95 - 0.029)$ are available for culling.
- If 93.1 percent of the cows are available for culling, none are sold the first year for slaughter, so 93.1 percent of them are available to enter the second lactation.

Thus, for the cows available to enter the first lactation, only 93.1 percent are available to enter the second lactation.

We have assumed that only the bare minimum of culling (that is, sales on the basis of low production) takes place. The major reason for culling is assumed to be failure of the cow to conceive rather than its low milk production. The model may decide to sell cows with less than six lactations, but if it does so it will be because the average cow of this age is not worth retaining. One useful extension of the model might be to include culling as a specific economic activity, with effects on production as minimal culling was carried out.

Herd Production Coefficients

As indicated above, a cow entering the herd for the first lactation is treated as a single activity. The calculation of total production, calves, mating requirements, and so forth for these activities are illustrated in table 17-6. Taking the first line of the table, we see that for every cow

Table 17-6. *Calculation of Production, Calves, Matings, and So Forth per Lactation*

| Lacta-tion | Cows avail-able | Number of mating services | | Cows in milk | Milk (kg) | | Protein maintenance (kg) | | Energy maintenance (kcal) | |
		Per animal	Total		Per animal	Total	Per animal	Total	Per animal	Total
1	1	2.0	2.0	0.95	5,160	4,902	102.51	97.38	2,055	1,952
2	1	1.3	1.3	0.97	5,626	5,457	110.69	107.37	2,301	2,232
3	1	1.5	1.5	0.98	5,160	5,057	116.97	114.83	2,467	2,418
4	1	1.5	1.5	0.97	4,300	4,171	120.69	117.07	2,558	2,481
5	1	1.8	1.8	0.95	2,934	2,787	126.42	120.10	2,658	2,525
6	1	1.8	1.8	0.90	2,580	2,322	126.42	113.78	2,658	2,392

entering the first lactation 2.0 mating services have been provided, but only 95 percent of the cows serviced actually conceive. *All* cows that conceive are assumed to come into milk, but, as shown in table 17-5, only a proportion of these milking cows produce a calf. The milk production per cow in milk is 5,160 kilograms, but only 95 percent of the available cows conceived, hence production *per cow mated* is only 4,902 kilograms. Similarly, the protein requirement per cow in milk is 102.51 kilograms, but since only 95 percent of cows mated actually conceived, the protein requirement *per cow mated* is only 97.38 kilograms, and the same proportion holds for energy and TDN.

Annual production model

The basic structure of the portion of the model for stock reconciliation can best be illustrated by first ignoring the resources used for production and the quarterly analysis of annual production. Such a simplified and aggregated representation of the model is given in table 17-7. The asterisks in the coded labels of table 17-7 indicate suppressed seasonal identifiers, which take appropriate values in the full model.

From Table 17-7 we see that male calves can either be sold at once (VRM), or at ten months old (VIM). Female calves can be sold at once (VRH), at ten months (VIAH), or they can be held as replacements raised on another ranch (REPR) or on the dairy farm (in the dairy area) itself (REPC).

The heifer calves held as replacements (REPR and REPC) produce springing heifers (ready to enter the milking herd for their first lactation). Because of a 7.85 percent death loss, only 0.9215 springing heifers are produced per heifer born. Springing heifers can also be obtained by buying them from elsewhere in Mexico (COMME) or from the United States (COMMU). Springing heifers can either enter the milking herd (ILAC*) or be sold to other regions of Mexico (VMEX). The revenue from sale of springing heifers is lower than for their purchase. This reflects the transport costs into and out of La Laguna (and, incidentally, avoids the possibility of an unbounded solution).

Lactating cows ILAC* to 5LAC* (and 6LAC*) use a cow from the appropriate age balance row 01LAC* to 56LAC* and contribute a fraction of a cow (allowing for death loss, and sale of cows for low production and infertility) into the corresponding balance row for the next lactation. These lactation balance rows also allow for the sale of cows at the end of each lactation. All cows that have completed their sixth lactation are sold (VVEI) for slaughter.

Lactating cows require protein, energy and TDN, and insemination services. They also require labor and capital (not shown) and supply milk and calves. Mating services can be provided by bulls (TORO) at the rate of

Table 17-7. *Simplified and Aggregated Representation of Dairy Production Portion of the La Laguna Dairy Model*

| Row description | Code | Calves | | Yearlings | | Heifers | Sales activities | | | | | | Milk |
		Male VRM	Female VRH	Male V1M	Female V1AH	VMEX	ex 1 VTRA1*	ex 2 VTRA2*	ex 3 VTRA3*	ex 4 VTRA4*	ex 5 VTRA5*	ex 6 VVEI*	VLE*
Male calves	VAM	1		1.45									
Female calves	VAH		1		1.45								
1st lactation	01LAC*					1							
2d lactation	12LAC*						1						
3d lactation	23LAC*							1					
4th lactation	34LAC*								1				
5th lactation	45LAC*									1			
6th lactation	56LAC*										1		
Cows for slaughter	VENV											1	
Milk production	LE***	30	30	240	240								1,000
Matings	SERTO												
Protein	A***P			317.40	317.40								
Energy	A***E			1,500	1,500								
TDN[a]	A***M			1,100	1,100								
Objective						5,950	5,250	4,500	3,750	3,000	3,000	3,000	1,440

Note: Asterisks to codes indicate suppressed seasonal identifiers, which take appropriate values in the full model.

a. The derivation of the TDN coefficients is not available to the author at present.

b. Artificial insemination.

(Table continues on the following page.)

525

Table 17-7 (continued)

Row description	Buy heifers Mex. COMME	Buy heifers U.S. COMMU*	Breeding Bulls TORO	Breeding A.I.[b] INSE	Replacements Ranch REPR	Replacements Home REPC	Lactating cows 1st 1LAC*	Lactating cows 2d 2LAC*	Lactating cows 3d 3LAC*	Lactating cows 4th 4LAC*	Lactating cows 5th 5LAC*	Lactating cows 6th 6LAC*
Male calves	-1						-0.4275	-0.461	-0.466	-0.451	-0.4275	-0.405
Female calves		-1					-0.4275	-0.461	-0.466	-0.451	-0.4275	-0.405
1st lactation	-0.9215	-0.9215			1	1	1					
2d lactation							-0.9310	1				
3d lactation								-0.9606	1			
4th lactation									-0.9508	1		
5th lactation										-0.9213	1	
6th lactation											-0.9121	1
Cows for slaughter							-0.05	-0.03	-0.0296	-0.0393	-0.0594	-0.8730
Milk production					240	240	-0.4902	-0.5457	-0.5057	-0.4171	-0.2787	-0.2322
Matings			-45	-1			2.0	1.3	1.5	1.5	1.8	1.8
Protein							324.33	355.80	343.67	313.42	264.80	251.21
Energy							5,260.61	5,826.33	5,736.96	5,355.21	4,697.40	4,508.40
TDN[a]												
Objective	6,000	6,050					-226	-226	-226	-226	-226	-226

Note: Asterisks to codes indicate suppressed seasonal identifiers, which take appropriate values in the full model.
a. The derivation of the TDN coefficients is not available to the author at present.
b. Artificial insemination.

forty-five per year, or by artificial insemination (INSE) on the basis of one mating service per insemination.

Disaggregation of milk production

As indicated above, the full model disaggregates the annual milk production data given in table 17-7 into four quarters. This requires the definition of twenty-four milk production activities ILACP, ILACV, ILACO, and ILACI (for first lactation calving in spring, summer, autumn, and winter, respectively) through 6LACP, 6LACV, 6LACO, and 6LACI (for sixth lactation starting in spring, summer, autumn, and winter, respectively) to replace activities in ILAC* through 6LAC* in table 17-7. The seasonal identifier in each case relates to the time of calving. Activity 5LACP, for example, refers to animals in the fifth lactation that calve in the spring. These animals produce milk in all four seasons, but least in the winter (that is, just before the next calving).

To keep track of seasonal milk production, the annual milk reconciliation restraint LE*** of table 17-7 is replaced by LEPRI, LEVER, LEOTO, and LEINV; and annual milk sales VLE* of table 17-7 is replaced by quarterly sales VLEP, VLEV, VLEO, and VLEI. In the same way, the annual lactation balance equations OILAC* and following have to be disaggregated to identify the seasonal calving date; the cow purchase activity COMME* (and so on) and cow sale activities VTRA* (and so on) have to be disaggregated to identify the season to which they apply.

Having defined the extra restraints and activities used in the disaggregation, we now need to consider the disaggregation process itself. Milk production per quarter is assumed to depend on: which lactation the cow is in; how many quarters since the last calving; and the season of the year.

Lactating yields have already been listed in table 17-1. The quarterly distribution of production following calving was taken to be as in table 17-8; seasonal effects are also shown.

Applying the production profile in table 17-8 to the lactation activities in table 17-1 gives the distributions of milk production in table 17-9. From table 17-9 we can calculate a production profile, as shown in table 17-10.

As indicated earlier, Morrison (1963, p. 658) says that each kilogram of 3 percent fat milk requires 0.0395 kilogram of protein and 0.55 kilocalorie of energy. Applying these factors to the seasonal milk production in table 17-10, and adding maintenance and gestation as indicated in table 17-4, yields the feed requirements specified in table 17-11. For example, a cow calving in the spring in the first quarter of its first lactation is assumed to yield 1,318 kilograms in that quarter (top left of table 17-10); applying this to Morrison's factors (above) yields for production:

Table 17-8. *Quarterly Distribution of Production in Relation to Calving and Seasonal Effects*

Quarter after calving	Percentage of total milk production	Season	Percentage effect on production
1st	24	Spring	112
2d	35	Summer	84
3d	26	Autumn	112
4th	15	Winter	92

$$53.06 = 1{,}318 \times 0.0395 \text{ kilograms of protein}$$
$$724.9 = 1{,}318 \times 0.55 \text{ kilocalories of energy}$$
$$1{,}252.0 = 1{,}318 \times 0.95 \text{ TDM.}$$

Annual protein and energy requirements for maintenance in the first lactation are given in table 17-4 as 102.51 kilograms of protein and 2,055.11 kilocalories of energy. Seasonal maintenance requirements are just a quarter of this, or 25.62 kilograms of protein and 513.78 kilocalories of energy. Adding maintenance and production for the first quarter, we obtain total requirements of:

$$77.68 = 25.62 + 52.06 \text{ kilograms of protein}$$
$$1{,}238.68 = 513.78 + 724.9 \text{ kilocalories of energy.}$$

In the first quarter (that is, the quarter before calving) we also add in the gestation requirements of 22.58 kilograms of protein and 367.5 kilocalories of energy.

Feed requirements for replacement stock

Calculation of feed requirements for replacement stock is less difficult than that for the lactating animals, since annual totals are calculated and *prorated* equally over all four seasons.

Table 17-9. *Production Indexes Relating Calving Date to Seasonal Production*

	Production index			
Calving	Spring	Summer	Autumn	Winter
Spring	26.88[a]	29.40	29.12	13.80
Summer	16.80	20.16[a]	39.20	23.92
Autumn	29.12	12.60	26.88[a]	32.30
Winter	38.92	21.84	16.80	22.08[a]

a. Cow calves in this quarter.

Table 17-10. *Seasonal Milk Production*
(kilograms)

Lactation	Code	Season				Total
		Spring	Summer	Autumn	Winter	
1 Spring	1LACP	1,318	1,441	1,427	676	4,863
Summer	1LACV	823	988	1,922	1,173	4,906
Autumn	1LACO	1,427	618	1,318	1,578	4,941
Winter	1LACI	1,922	1,071	824	1,082	4,898
2 Spring	2LACP	1,467	1,604	1,589	753	5,413
Summer	2LACV	917	1,100	2,139	1,305	5,461
Autumn	2LACO	1,589	688	1,467	1,757	5,501
Winter	2LACI	2,139	1,192	917	1,205	5,453
3 Spring	3LACP	1,359	1,487	1,473	698	5,017
Summer	3LACV	850	1,019	1,982	1,210	5,061
Autumn	3LACO	1,473	637	1,359	1,628	5,097
Winter	3LACI	1,982	1,104	850	1,117	5,053
4 Spring	4LACP	1,121	1,226	1,215	576	4,138
Summer	4LACV	701	841	1,635	998	4,174
Autumn	4LACO	1,215	526	1,121	1,343	4,204
Winter	4LACI	1,635	911	701	921	4,168
5 Spring	5LACP	749	819	812	385	2,765
Summer	5LACV	468	562	1,092	667	2,789
Autumn	5LACO	812	351	749	897	2,809
Winter	5LACI	1,092	609	468	615	2,785
6 Spring	6LACP	624	683	676	320	2,303
Summer	6LACV	390	468	910	555	2,324
Autumn	6LACO	676	293	624	748	2,341
Winter	6LACI	910	507	390	513	2,320

For the first two months of its life, a calf obtains almost all its feed requirements from milk. This is assumed to require 240 kilograms of milk per calf. Male calves ten months old are sold. Heifers ten months old may be sold or they may be held to become springing heifers (that is, about to calve) at thirty-one months. A thirty-one-month-old springing heifer is assumed to weigh 454 kilograms, or a gain of 227 kilograms in twenty-one months.

We assume that a thirty-one-month-old heifer weighs 454 kilograms. To approximate this animal's total feed requirements, we take thirty-one months at the ration for an animal weighing 226 kilograms (Morrison 1963, p. 659): 5.4 kilograms of TDM, 0.395 kilogram of digestible protein, and 6.65 kilocalories of net energy per day. For 930 days, this comes to a total of 154 kilograms of digestible protein, 3,909 kilocalories of net energy, and 2,108 kilograms of TDM. Dividing by 4, we get the quarterly requirements of 38.5 kilograms digestible protein, 977 kilocalories of net energy, and 527 kilograms of TDM.

For heifers raised off the farm, we assume the quarterly feed requirements of calves raised to ten months, plus the feed requirements (Morri-

Table 17-11. *Seasonal Feed Requirements*

Lactation	Code	Protein (kg)				Energy (kcal)				TDM (kg)[a]			
		Spring	Summer	Autumn	Winter	Spring	Summer	Autumn	Winter	Spring	Summer	Autumn	Winter
1 Spring	1LACP	77.67	82.55	82.01	73.93	1,238	1,306	1,299	1,253	2,102.65	2,226.15	2,211.90	2,106.40
Summer	1LACV	79.74	64.66	101.53	71.94	1,334	1,057	1,571	1,159	2,418.0	2,705.9	1,956.35	1,607.40
Autumn	1LACO	82.01	71.60	77.67	87.98	1,299	1,221	1,238	1,382	2,102.65	3,007.95	2,210.95	1,402.50
Winter	1LACI	101.53	67.92	79.74	68.38	1,571	1,103	1,334	1,109	1,867.05	2,705.9	1,500.57	1,607.7
2 Spring	2LACP	85.61	91.04	90.44	80.30	1,382	1,458	1,449	1,421	2,512.41	1,953.3	2,312.4	2,775.9
Summer	2LACV	86.77	71.13	112.17	79.23	1,511	1,180	1,752	1,293	3,166.35	2,970.75	2,596.45	2,970.75
Autumn	2LACO	90.44	77.71	85.61	97.08	1,449	1,385	1,382	1,542	2,431.15	2,873.75	2,431.15	2,042.6
Winter	2LACI	112.17	74.75	86.77	75.27	1,752	1,231	1,511	1,238	1,612.25	1,772.8	2,268.6	1,772.8
3 Spring	3LACP	82.94	87.97	87.41	79.69	1,365	1,435	1,427	1,432	2,210.70	1,891.05	2,210.70	2,910.10
Summer	3LACV	85.68	69.51	107.55	77.02	1,515	1,178	1,707	1,282	3,070.15	2,814.9	2,471.95	2,814.9
Autumn	3LACO	87.41	77.29	82.94	93.56	1,427	1,399	1,365	1,513	2,319.95	2,800.35	2,319.95	1,963.7
Winter	3LACI	107.55	72.87	85.68	73.35	1,707	1,224	1,515	1,231	2,319.95	2,800.35	2,319.95	1,963.7
4 Spring	4LACP	74.46	78.61	78.15	75.79	1,256	1,314	1,308	1,388	1,991.25	1,716.7	1,991.25	2,530.55
Summer	4LACV	80.73	63.39	94.76	69.58	1,456	1,102	1,539	1,188	2,829.8	2,494.75	2,208.8	2,494.75
Autumn	4LACO	78.15	73.81	74.46	83.22	1,308	1,360	1,256	1,378	2,083.4	2,605.6	2,083.4	1,786.05
Winter	4LACI	94.76	66.15	80.73	66.55	1,539	1,141	1,456	1,146	1,457.35	1,579.9	2,142.95	1,579.9
5 Spring	5LACP	61.20	63.97	63.66	69.68	1,077	1,115	1,111	1,307	1,639.75	1,453.55	1,639.75	2,241.75
Summer	5LACV	72.98	53.80	74.76	57.94	1,353	974	1,265	1,031	2,245.05	1,982.7	1,787.95	1,982.7
Autumn	5LACO	63.66	68.36	61.20	67.05	1,111	1,289	1,077	1,158	1,702.45	2,293.05	1,702.45	1,500.1
Winter	5LACI	74.76	55.65	72.98	55.91	1,265	999	1,353	1,003	1,276.85	1,360.45	1,978.6	1,360.45
6 Spring	6LACP	56.26	58.57	58.31	67.14	1,008	1,040	1,036	1,272	1,426.	1,482.	1,476.	1,132.
Summer	6LACV	69.89	50.10	67.56	53.54	1,310	922	1,165	970	1,197.	1,277.	1,699.	1,362.
Autumn	6LACO	58.31	66.04	56.26	61.14	1,036	1,257	1,008	1,076	1,476.	1,106.	1,426.	1,544.
Winter	6LACI	67.56	51.64	69.89	51.86	1,165	943	1,310	947	1,699.	1,315.	1,197.	1,320.

a. The derivation of these coefficients is not known. Note same TDM requirements for cows in third lactation, calving in autumn and winter, and other differences from systematic patterns.

son 1963, p. 659) for two months at a weight of 453 kilograms (that is, 9.05 kilograms TDM, 0.455 kilogram digestible protein, and 9.5 kilocalories of net energy per day). Two months is sixty days, so that each quarter should be credited with fifteen days at this feeding rate. Multiplying by 15, we get the figures for thirty to thirty-one months shown below, and hence the total feed requirements given:

	Quarterly requirements		
	10 months	*30–31 months*	*Total*
Digestible protein (kilograms)	24.37	6.83	31.20
Net energy (kilocalories)	375	142	517
TDM (kilograms)	102	136	238

Death losses in young stock

A death loss of 5 percent is assumed for calves raised to ten months, and another 2.85 percent death loss by the time they reach thirty-one months.

Feed requirements for bulls

Bulls are assumed to weigh 907 kilograms, so that their daily feed requirements (Morrison 1963, p. 659) are 10.5 kilograms of TDM, 0.615 kilogram of digestible protein, and 13.6 kilocalories of net energy. Multiplying by 90, we get the quarterly feed requirements: 5.535 kilograms of digestible protein, 122.4 kilocalories of net energy, and 94.5 kilograms of TDM.

Water Requirements

Seasonal water requirements have been assumed to be as shown in table 17-12; the basis for derivation is not known. These water requirements, however, vary only with season, not with level of milk production.

The nutrients provided by crops and their prices, yields, and costs of production are assumed to be the same regardless of the quarter in which they are used or whether they are produced on dairy farms, produced elsewhere in La Laguna, or purchased from outside the region. These assumptions about crop nutrients, prices, and the assumed yields for crops grown on dairy farms in La Laguna are given in table 17-13.

The crops grown on dairy farms are assumed to use land only in the season in which the crop is used, and to use no irrigation water. *This assumption is doubtless erroneous*, but we can hope that much better local cropping data can be obtained from CHAC for a later version of the dairy model. Table 17-13 suggests that the crop yields, nutritional contents, and

Table 17-12. *Seasonal Water Requirements*
(kiloliters)

Class of stock	Code	Season			
		Spring	Summer	Autumn	Winter
Replacement					
Male calves	v1M	0.33091	0.48259	0.2483	0.10473
Female calves	v1AH	0.33091	0.48259	0.2483	0.10473
Heifers on home farm	REPC	0.4947	0.7216	0.3712	0.1566
Heifers on ranch	REPR	2.068	3.016	1.552	0.6546
All lactations	1LACP to 6LACI	0.99259	1.44758	0.74481	0.3145

prices have not been estimated on a consistent basis. Thus, tables 17–4 and 17–13 together suggest that 1 kilogram of alfalfa has enough protein to feed a cow for ten years!

Labor Requirements

Labor needs are reconciled annually with respect to skilled and unskilled labor. It was assumed that one unskilled, and a tenth of a skilled dairy worker could look after twenty calves for their first ten months and ten heifers until the time they are ready to enter the herd. That is, 1.0 unskilled and 0.1 skilled workers can look after 410 "calf-months" per year. Thus, each calf-month requires $1/410 = 0.00243$ of an unskilled worker, and $0.1/410 = 0.00024$ of a skilled worker annually. The labor requirements for replacement stock of calves and heifers are as follows:

	Labor	
	Skilled	Unskilled
Calves to ten months	0.00243	0.02430
Heifers in dairy area	0.00510	0.05103
Heifers away	0.00049	0.00486

The labor requirements for a herd of 100 cows were taken to be: 0.9 of a skilled (managerial) worker, and 4.4 unskilled worker, or 0.009 and 0.044 of skilled and unskilled laborers per cow in milk.

Costs and Prices

The costs of purchased and home-produced feeds have already been listed in table 17-13.

Costs

Replacement heifers are assumed to cost 15 pesos per month for medicine, coming to 465 pesos for medicines.[3] In addition, replacements raised away from the home ranch need to be charged a rental, but because an appropriate figure is not available, no cost for replacements raised off the farm has yet been made. Dairy cows are charged 18 pesos per month for medicines and veterinary services, or 226 pesos annually.

A bull is assumed to cost 3,500 pesos annually, principally in depreciation of initial cost over his productive life. Inseminations cost 100 pesos each. Springing heifers cost 6,250 pesos if purchased from the United States, and 6000 pesos if purchased from other regions of Mexico. Water costs 50 pesos per unit regardless of season. Skilled labor costs 18,000 pesos annually, unskilled 9,920 pesos annually. Interest on long-term capital is 8 percent, on short-term capital 12 percent. There is provision for capital to be supplied from an upward-sloping supply curve, but this has not been implemented at present. Land rents for 2,000 pesos per hectare annually.

Prices

Milk sells for 1.46 pesos per kilogram. Day-old calves sell for 145.80, ten-month-old calves for 1,045.80 pesos. Surplus heifers can be sold for 5,950, old cows for 3,000 pesos. Cows that have completed less than six lactations can be sold after lactations at the following prices:

	Pesos
After 1	5,250
After 2	4,500
After 3	3,750
After 4	3,000
After 5	3,000

Capital Requirements and Herd-size Restraints

Since dairy production gives monthly income, short-term capital has been taken to be the amount to cover two months' operations. Labor, feed, and interest on long-term capital are charged against short-term capital as the expense is incurred. Only one-sixth of the annual cost is charged because we assume these payments have to be carried for only two of the twelve months. The only remaining charge for calves and

---

3. At an exchange rate of approximately 12.5 pesos to the U.S. dollar.

Table 17-13. Nutrients (per kilogram) and Yields of Crops Included in the La Laguna Model

Crop	Code	Protein (kg)	Energy (kcal)	TDM (kg)	Yield (metric tons per hectare)	Price in La Laguna (pesos per metric ton)	Price of imports (pesos per metric ton)	Cost of growing (pesos per metric ton)
Alfalfa	**AL*	34,000	0.3039	253,000	53.695	90	189	3,569
Barley	**C*	100,000	1.7643	894,000	0.66	691	1,094	1,531
Maize	***MG*	100,000	1.8062	870,000	1.456	829	939	1,800
Cottonseed meal		68,000	1.4868	924,000	n.g.	1,300	1,300	n.g.
Grain sorghum	**SG*	302,000	1.7159	89,100	4.09	634	625	1,832
Forage oats	**AF*	88,000	0.3127	216,000	9.0	70	100	1,080
Salvat	**S*	19,000	1.726	906,000	n.g.	1,900	1,900	n.g.
Forage sorghum	**SF*	424,000	0.3128	249,000	27.8	67	77	1,826
Silage sorghum	**SE*	8,000	0.2819	295,000	1.0	180	180	120
Forage sugarbeets	***RF*	15,000	0.2181	141,000	8.0	200	220	1,800
Protein								
Bermuda grass	**ZB*				40,000			2,000
Sudan grass	**ZS*				48,000			1,800
Energy								
Bermuda grass	**ZB*				5.374			2,000
Sudan grass	**ZS*				5.374			1,800
TDM								
Bermuda grass	**ZB*				5,000,000			2,000
Sudan grass	**ZS*				4,320,000			1,800

n.g. Not grown.
* Indicates suppressed source and seasonal identifiers.

replacements raised on the farm is 15 pesos a month for medicines. In the case of calves, this is just 30 pesos (2×15) per two months, but it is only charged for ten of the twelve months, so the annual rate is 25 pesos = (30×10)/12. For each heifer replacement, we have on hand a one-year-old, a two-year-old, and half the time we have a three-year-old. Hence total medicines for heifer replacements are: 75 pesos = ($2 \times 2.5 \times 15$) per two months.

The long-term capital requirements by class of stock were assumed as follows:

	Code	Requirements (pesos)
Replacements at home	REPC	11,801
All lactations	ILACP	8,383
	to	
	6LACV	

These coefficients were calculated from Silos (1970). The figure of 11,801 pesos should probably be recalculated.

For reasonable results, it was assumed that 300,000 hectares of land at most were available for dairy production in La Laguna, that maximum herd size was 56,100 animals, and that 45,080 liters of milk at most could be sold each quarter.

Changes in Model Needed to Give Positive Levels of Milk Production

The model as described above had an optimum solution to zero milk production. Examination of the matrix showed that alfalfa was the cheapest source of energy, but it cost 3,569 pesos per acre for 16.31 ($= 0.3039 \times 53.695$; see table 17-13) units of energy, or 218.82 pesos per unit; but about 1,300 units are required per quarter for lactating animals. This would amount to a cost of 11,378.64 pesos per animal annually. Revenue per animal yearly is only about 7,300 pesos, so that feed costs vastly exceed income.

Obviously we have here a major data error that cannot be corrected except by direct reference to Mexican conditions. Two guess corrections were made to see what the solution would be if farmers could afford to feed their animals. First, we transposed the energy and protein figures for all feeds; second, we reduced all feed costs to 1 percent of the level reported earlier in the documentation. This (understandably) gave us a "solution" of sorts, with milk production at positive levels. For what it is worth, the next section describes this solution.

A Solution with Milk Production
at Positive Levels

With the above alterations, the model solved to give "reasonable" results. Herd composition was as shown in table 17-14. The maximum allowed milk, 4,508,000 kilograms per quarter, was produced and sold. In addition, there was surplus milk in spring of 127,489 kilograms and surplus in fall of 132,972 kilograms (for a total of surplus milk of 260,461 kilograms a year).

Seasonal requirements for irrigation water were: spring, 56,956; summer, 83,067; fall, 42,738; and winter, 18,042. Labor utilization was: skilled labor, 481 workers; unskilled labor, 1,945 workers.

Capital borrowed (in pesos) was: long-term, 574,813,000; short-term, 507,608,000.

Feed used was as shown in table 17-15. TDM was in surplus in all seasons, and energy was in surplus in all seasons except the winter. The opportunity cost per kilogram in pesos of producer's milk by season was: spring, zero; summer, 0.114; autumn, zero; and winter, 0.096. This was contrasted with a price of 1.46 pesos per kilogram.

Table 17-14. *Herd Composition
with Milk Production at Positive Levels*

Class of stock	Number	Total
Calves		37,804
Male	18,902	
Female	18,902	
Steers sold off at 10 months		18,002
Females to calve in 31 months		17,418
Heifers sold	2,567	
Heifers entering herd in spring	8,269	
Heifers entering herd in fall	6,582	
Cows for second lactation		13,827
Calving in spring	7,699	
Calving in fall	6,128	
Cows for third lactation		13,282
Calving in spring	7,396	
Calving in fall	5,886	
Total cows in milk		44,527
Cows sold		
After third lactation	5,597	
Culled prior to third lactation	1,551	
Bulls		1,502

Table 17-15. *Feed Requirements*
with Milk Production at Positive Levels

Crop	Amount	Crop	Amount
Produced on farm (hectares)		Purchased locally (kg)	
Summer grain sorghum	265,617	Spring forage sorghum	14,578,000
Summer forage sorghum	300,000	Winter forage oats	21
Fall alfalfa	280,780		
Winter alfalfa	41		

Production limited by herd size

As a contrast, the model was also run with a market of 9,000,000 kilograms per quarter (in place of 4,508,000 kilograms per quarter). Though even this market limit was effective in winter, the fundamental restraint for this run was herd size. The shadow price on herd size (that is, the reward for being able to expand the herd by one cow) for this situation was 4,627 pesos. With an increased market, the model called for a younger herd. The maximum allowed milk, 9,000,000 kilograms per quarter, was produced in winter; there was no surplus milk produced. The seasonal pattern of milk production (in kilograms) was: spring, 8,272,881; summer, 3,677,583; autumn, 7,669,303; and winter, 9,000,000. In this case, though winter milk sales are doubled, there is a decline in summer milk production. The feeding regime again concentrated on farm-produced alfalfa and sorghum.

The main purpose of this sensitivity analysis is to test whether the model had the "right" type of response to changed market conditions. The shift toward less even production and a younger (more productive) herd is a plausible adjustment.

No salvage value for old cows

One of the most startling contrasts between the model answers and husbandry practices in La Laguna is the way in which the model recommends retaining cows for only three lactations. In case salvage values of old cows had been set too high, the model was run again with a limited milk market, but zero salvage value on old cows. The results of this run are summarized in table 17-16. The maximum allowed milk sales, 4,508,000 kilograms per quarter, was sold in each quarter except the summer, when sales were only 2,699,264 kilograms.

It is not entirely clear why summer milk production should be so uneconomical—but this result does not appear to be a technical error in model construction. Presumably, milk production is relatively uneco-

Table 17-16. *Herd Composition
with No Salvage Value for Old Cows*

Class of stock	Number	Total
Calves		35,794
Male	17,897	
Female	17,897	
Steers sold at 10 months		17,045
Females to calve at 31 months		16,492
Heifers sold	8,644	
Heifers entering herd in spring	2,023	
Heifers entering herd in fall	5,825	
Cows for second lactation		7,306
Calving in spring	1,883	
Calving in fall	5,423	
Cows for third lactation		7,018
Calving in spring	1,809	
Calving in fall	5,209	
Cows for fourth lactation		6,673
Calving in spring	1,720	
Calving in fall	4,953	
Cows for fifth lactation		6,148
Calving in spring	1,585	
Calving in fall	4,563	
Cows for sixth lactation		5,607
Calving in spring	1,445	
Calving in fall	4,162	
Bulls		1,445

nomic when cows can be sold only *before* entering the milking herd. A springing heifer is worth 5,950 pesos when sold before entering the herd, but worth nothing thereafter.

Although shadow prices on extra spring and fall sales are 1.46 and 1.438 pesos per kilograms, respectively, the shadow price on extra winter milk sales is only 0.17 peso. The model has also nearly reached the point where winter milk production is also unprofitable. In this last run the maximum herd size, 56,100 milking cows, was also used—in contrast with the first run, in which only 44,527 cows were in milk.

Summary

As indicated, the assumptions recorded in this documentation leave a lot to be desired. What we have produced, however, is a framework within which substantial inconsistencies in the assumptions can be identified.

The La Laguna dairy model is in no sense final at present. Yet we hope that readers will have found it of interest, since (understandably) few studies are published with data at this "first-cut" stage. Typically, obvious contradictions are removed prior to publication, and hence this "confrontation" stage of model construction—the point at which assumptions made from different starting points confront each other and the real world—tends to be underreported. (In the present case, removal of data inconsistencies would have produced a "better" documentation but would not have had other benefits.) As an example, there does not seem to be any reason why the milk yields and calving rates should not be about right. Yet the model persists in finding substantial numbers of heifers that could be sold outside the region, although in real life there are substantial imports. This is in itself a very interesting finding that stands up regardless of the model's current inability to reconcile feed consumption and supply. A number of explanations are possible for the gap between the model's prediction of substantial sales of surplus stock and the observed actual need to import heifers to the region:

- Possibly the dairy industry in La Laguna is growing much faster than allowed for in the model. If so, this affects predictions of future feed requirements of the industry and the level of future milk deficit in Mexico City. The latter is a primary policy consideration in Mexico and was one of the original motivations for the model.
- Possibly there are undiagnosed, major health problems with the La Laguna herd, such that death loss or calf mortality is much higher than would seem reasonable with sound husbandry practice. If this proves to be the case, it might occasion a highly effective veterinary health campaign.
- Possibly milk yields are much lower than allowed for in the model. Again, this suggests the scope for an effective extension campaign to improve feeding and yields.
- Possibly price ratios are such that it is cheaper to sell calves and import heifers; perhaps it is unprofitable to rear replacements. This would suggest the policy option of closing the border to imported dairy stock, thus forcing the industry to become self-sustaining.

Other possibilities exist, but this illustrates the model's capacity, even now, to stimulate significant policy questions.

The model structure also has the capacity to stimulate questions on feed utilization, ration composition, seasonality of milk supply, and so forth. Unfortunately, in the absence of some data reconciliation these kinds of questions cannot be addressed seriously.

Appendix. Restraints and Activities in Existing Model

The description given in text of the restraints and activities included in the existing La Laguna model is rendered in tabular form in table 17-17. Restraints are listed first, then activities. Activities are listed in the order in which they appear in the matrix.

Table 17-17. *Restraints and Activities in Draft Model of Dairy Production in La Laguna*

Restraint/ activity	Name	Code	Unit	Remarks
Restraint				
1	Objective function	OBJ	Million pesos	To be maximized
2	Male calves	VAM	Head[a]	Animals that have survived the first two days
3	Female calves	VAH	Head	Animals that have survived the first two days
4	Transfer 0 to 1	01LACP	Head	Heifer balance row
	Lactation (spring)[b]			
5	Transfer 1 to 2	12LACP	Head	Transfers a cow calving in spring from lactation 1 to lactation 2
6	Transfer 2 to 3	23LACP	Head	Transfers a cow calving in spring from lactation 2 to lactation 3
7	Transfer 3 to 4	34LACP	Head	Transfers a cow calving in spring from lactation 3 to lactation 4
8	Transfer 4 to 5	45LACP	Head	Transfers a cow calving in spring from lactation 4 to lactation 5
9	Transfer 5 to 6	GLAC	Head	Transfers a cow calving in spring from lactation 5 to lactation 6
10	Mating	SERTO	Matings	Balance equation to ensure enough bulls or semen for artificial insemination to allow herd to be mated.
11	Sales for slaughter	VENV	Head	Cows sold for slaughter (balance equation allows convenient change of price of cattle for slaughter)
	Milk			
12	Spring	LEPRI	Liter	⎫
13	Summer	LEVER	Liter	Seasonal milk balances
14	Autumn	LEOTO	Liter	
15	Winter	LEINV	Liter	⎭
16	Protein (spring)	APRIP	Kg	⎫
17	Energy	APRIE	Kcal	Feed balances (spring)
18	TDN	APRIM	Kg	⎭

540

19	Protein (summer)	AVERP	Kg	⎫
20	Energy	AVERE	Kcal	⎬ Feed balances (summer)
21	TDN	AVERM	Kg	⎭
22	Protein (autumn)	AOTOP	Kg	⎫
23	Energy	AOTOE	Kcal	⎬ Feed balances (autumn)
24	TDN	AOTOM	Kg	⎭
25	Protein (winter)	AINVP	Kg	⎫
26	Energy	AINVE	Kcal	⎬ Feed balances (winter)
27	TDN	AINVM	Kg	⎭
28	Labor (skilled)	MOC	Man-year	Skilled dairy labor
29	Labor (unskilled)	MONC	Man-year	Unskilled dairy labor
	Land			
30	Spring	TPRI	Ha	⎫
31	Summer	TVER	Ha	⎬ Seasonal land balances (for crops grown on dairy farms
32	Autumn	TOTO	Ha	⎪ for cattle feed)
33	Winter	TINV	Ha	⎭
	Water			
34	Spring	AGPRI	m³	⎫
35	Summer	AGVER	m³	⎬ Seasonal water balances
36	Autumn	AGOTO	m³	⎪
37	Winter	AGINV	m³	⎭
38	Capital (long)	CAPLP	Pesos	Long-term capital requirements
39	Capital (short)	CAPCP	Pesos	Short-term capital requirements (note that because of monthly payment for milk produced, only 2 months' short-term capital are assumed to be needed at any one time)

(Table continues on the following page.)

Table 17-17 (continued)

Restraint/ activity	Name	Code	Unit	Remarks
Restraint				
40	Long capital (max 1)	KMLP1	Pesos	Maximum available at lower (public) interest rate for long-term capital
41	Long capital (max 2)	KMLP2	Pesos	Maximum available on good security from private long-term capital market
42	Short capital (max)	KMCP	Pesos	Maximum available at lower (private) interest for short-term capital
43	Land (total)	PASBA	Ha	Balance equation allowing selection of permanent pasture to be used (not used in model reported here)
	Market maximum			
44	Spring	MATMP	Kg	⎫
45	Summer	MATMV	Kg	⎬ Seasonal maximums on milk that can be sold (note that this is an alternative restraint on herd size; see restraint 48)
46	Autumn	MATMO	Kg	
47	Winter	MATMI	Kg	⎭
	Herd limits			
48	Herd (max)	LMAX	Head	⎫ Sets maxima and minima on herd size
49	Herd (min)	LMIN	Head	⎭

	Summer			
50	Transfer 1–2 lact.	12LACV	Heac	
51	Transfer 2–3 lact.	23LACV	Heac	
52	Transfer 3–4 lact.	34LACV	Heac	
53	Transfer 4–5 lact.	45LACV	Heac	
54	Transfer 5–6 lact.	56LACV	Heac	
	Autumn			
55	Transfer 1–2 lact.	12LACO	Heac	
56	Transfer 2–3 lact.	23LACO	Heac	
57	Transfer 3–4 lact.	34LACO	Heac	Transfers cows from ith to jth lactation, identified by calving season
58	Transfer 4–5 lact.	45LACO	Heac	
59	Transfer 5–6 lact.	56LACO	Herd	
	Winter			
60	Transfer 1–2 lact.	12LACI	Heac	
61	Transfer 2–3 lact.	23LACI	Heac	
62	Transfer 3–4 lact.	45LACI	Heac	
64	Transfer 5–6 lact.	56LACI	Heac	

(*Table continues on the following page.*)

Table 17-17 (*continued*)

Restraint/activity	Name	Code	Unit	Remarks
Activity				
1	Restraint levels	RHS		Unit depends on row
	Young stock sales			
2	Male calves	VRM	Head	2-day-old male calves sold off
3	Yearling steers	V1M	Head	Male animals sold at 10 months old (no death loss in calculations so far)
4	Female calves	VRH	Head	2-day-old female calves sold off
5	Yearling heifers	V1AH	Head	Female heifers sold at 10 months old (no death loss in calculations so far)
	Replacement			
6	Off	REPR	Head	31-month-old heifers produced from calves in herd and, respectively, grazed off farm or on home farm (heifers grazed off farm are away from their 10th to 29th months)
7	Home	REPC	Head	
	Heifers bought			
8	Mexican	COMME	Head	31-month-old heifers ready to enter milking herd, brought from elsewhere in Mexico or United States, respectively
9	U.S.	COMEU	Head	
	Heifers sold[d]			
10	Springing[d]	VMEX	Head	31-month-old heifers ready to enter the milking herd, sold to enter another herd
	First lactation			
11	Spring	1LACP	Head	
12	Summer	1LACV	Head	
13	Autumn	1LACO	Head	
14	Winter	1LACI	Head	Allows milk production to be identified by lactation number and calving season
	Second lactation			
15	Spring	2LACP	Head	
16	Summer	2LACV	Head	
17	Autumn	2LACO	Head	
18	Winter	2LACI	Head	

	Third lactation			
19	Spring	3LACP	Head	
20	Summer	3LACV	Head	
21	Autumn	3LACO	Head	
22	Winter	3LACI	Head	
	Fourth lactation			
23	Spring	4LACP	Head	
24	Summer	4LACV	Head	
25	Autumn	4LACO	Head	
26	Winter	4LACI	Head	
	Fifth lactation			
27	Spring	5LACP	Head	
28	Summer	5LACV	Head	
29	Autumn	5LACO	Head	
30	Winter	5LACI	Head	
	Sixth lactation			
31	Spring	6LACP	Head	
32	Summer	6LACV	Head	
33	Autumn	6LACO	Head	
34	Winter	6LACI	Head	Allows milk production to be identified by lactation number and calving season
	Spring cows sold			
35	After 1 lact.	VTRA1P	Head	
36	After 2 lact.	VTRA2P	Head	
37	After 3 lact.	VTRA3P	Head	
38	After 4 lact.	VTRA4P	Head	
39	After 5 lact.	VTRA5P	Head	Sells a cow that has completed 1–5 lactations identified by season

(Table continues on the following page.)

545

Table 17-17 (*continued*)

Restraint/activity	Name	Code	Unit	Remarks
Activity				
	Summer cows sold			
40	After 1 lact.	VTRA1v	Head	
41	After 2 lact.	VTRA2v	Head	
42	After 3 lact.	VTRA3v	Head	
43	After 4 lact.	VTRA4v	Head	
44	After 5 lact.	VTRA5v	Head	
	Autumn sold			
45	After 1 lact.	VTRA1o	Head	
46	After 2 lact.	VTRA2o	Head	Sells a cow that has completed 1–5 lactations, identified
47	After 3 lact.	VTRA3o	Head	by season
48	After 4 lact.	VTRA4o	Head	
49	After 5 lact.	VTRA5o	Head	
	Winter cows sold			
50	After 1 lact.	VTRA1l	Head	
51	After 2 lact.	VTRA2l	Head	
52	After 3 lact.	VTRA3l	Head	
53	After 4 lact.	VTRA4l	Head	
54	After 5 lact.	VTRA5l	Head	
55	Old cows sold	VVEI	Head	Cow after 6th lactation sold for slaughter, or younger cow culled for low production
	Breeding			
56	Bulls	TORO	Head	Mating services from bulls and artificial insemination, respectively
57	Artificial	INSE	Matings	

546

No.		Code	Unit	Description
	Irrigation			
58	Spring	AGUP	m³ᵉ	
59	Summer	AGUV	m³ᵉ	Water provided seasonally for grazing
60	Autumn	AGUO	m³ᵉ	
61	Winter	AGUI	m³ᵉ	
	Employment			
62	Labor (skilled)	MOCA	Man-year	Hire of skilled and unskilled labor (no initial endowment of skilled
63	Labor (unskilled)	MONO	Man-year	labor assumed; operator input accounted for by skilled labor hire activity)
	Capital use			
64	Long (8%)	KLPPU	$1,000	
65	Long (10%)	KLPR1	$1,000	Long- and short-term capital borrowing at differing interest
66	Long (11%)	KLPR2	$1,000	rates (note that public capital is cheapest source of long-
67	Short (15%)	KCPPU	$1,000	term, and most expensive source of short-term capital)
68	Short (12%)	KCPPR	$1,000	
	Milk sold			
69	Spring	VLEP	Kgᶠ	
70	Summer	VLEV	Kgᶠ	Seasonal sale of milk
71	Autumn	VLEO	Kgᶠ	
72	Winter	VLEI	Kgᶠ	
	Expansion			
73	Land rental	RENTI	Ha	Rental for a year, when initial supply for dairying has been used
74	Expanding old facilities	IBVEJ	Head	Expands upper bound on herd size (but blocked out of present model)
75	Building new facilities	INNVE	Head	

(Table continues on the following page.)

547

Table 17-17 (continued)

Restraint/ activity	Name	Code	Unit	Remarks
Activity				
	Crop production on farm			
	Spring			
76	Alfalfa	PRALP	Ha	
77	Grain sorghum	PRSGP	Ha	
78	Maize	PRMGP	Ha	
79	Barley	PRCP	Ha	Crops grown explicitly for feed
80	Silage sorghum	PRSEP	Ha	
81	Bermuda grass	PRZBP	Ha	
82	Sudan grass	PRZSP	Ha	
	Summer			
83	Alfalfa	PRALV	Ha	
84	Grain sorghum	PRSGV	Ha	
85	Maize	PRMGV	Ha	
86	Forage sorghum	PRSFV	Ha	Crops grown explicitly for feed
87	Bermuda grass	PRZBV	Ha	
88	Sudan grass	PRZSV	Ha	
	Autumn			
89	Alfalfa	PRALO	Ha	
90	Grain sorghum	PRSGO	Ha	
91	Maize	PRMGO	Ha	
92	Barley	PRCO	Ha	
93	Silage sorghum	PRSEO	Ha	Crops grown explicitly for feed
94	Forage sugarbeets	PRRFO	Ha	
95	Forage oats	PRAFO	Ha	

	Winter		
96	Alfalfa	PRALI	Ha
97	Grain sorghum	PRSGI	Ha
98	Maize	PRMGI	Ha
99	Silage sorghum	PRSEI	Ha
100	Forage sugarbeets	PRRFI	Ha
101	Forage oats	PRAFI	Ha

} Crops grown explicitly for feed

	Local crop purchase		
	Spring		
102	Alfalfa	CZALP	Kg
103	Grain sorghum	CZSGP	Kg
104	Barley	CZCP	Kg
105	Maize	CZMGP	Kg
106	Cottonseed meal	CZHP	Kg
107	Salvat	CZSP	Kg
108	Forage sorghum	CZSFP	Kg
	Summer		
109	Alfalfa	CZALV	Kg
110	Grain sorghum	CZSGV	Kg
111	Maize	CZMGV	Kg
112	Cottonseed meal	CZHV	Kg
113	Salvat	CZSV	Kg

} Feed crops purchased locally

549

(*Table continues on the following page.*)

Table 17-17 (continued)

Restraint/ activity	Name	Code	Unit	Remarks
Activity				
	Local crop purchase			
	Autumn			
114	Alfalfa	CZALO	Kg	
115	Grain sorghum	CZSGO	Kg	
116	Maize	CZMGO	Kg	
117	Cottonseed meal	CZHO	Kg	
118	Salvat	CZSO	Kg	
119	Silage sorghum	CZSEO	Kg	
120	Forage oats	CZAFO	Kg	
	Winter			Feed crops purchased locally
121	Alfalfa	CZALI	Kg	
122	Grain sorghum	CZSGI	Kg	
123	Maize	CZMGI	Kg	
124	Cottonseed meal	CZHI	Kg	
125	Salvat	CZSI	Kg	
126	Silage sorghum	CZSEI	Kg	
127	Forage sugarbeets	CZRFI	Kg	
128	Forage oats	CZAFI	Kg	
	Outside crop purchase			
	Spring			
129	Alfalfa	CALP	Kg	
130	Grain sorghum	CSGP	Kg	
131	Maize	CMGP	Kg	Feed crops purchased outside La Laguna
132	Barley.	CCP	Kg	
133	Cottonseed meal	CHP	Kg	
134	Salvat	CSP	Kg	
135	Forage sorghum	CSFP	Kg	

550

	Summer			
136	Alfalfa	CALV	Kg	
137	Grain sorghum	CSGV	Kg	
138	Maize	CMGV	Kg	
139	Cottonseed meal	CHV	Kg	
140	Salvat	CSV	Kg	
	Autumn			
141	Alfalfa	CALO	Kg	
142	Grain sorghum	CSGO	Kg	
143	Maize	CMGO	Kg	
144	Cottonseed meal	CHO	Kg	Feed crops purchased outside La Laguna
145	Salvat	CSO	Kg	
146	Silage sorghum	CSEO	Kg	
	Winter			
147	Alfalfa	CALI	Kg	
148	Grain sorghum	CSGI	Kg	
149	Maize	CMGI	Kg	
150	Cottonseed meal	CHI	Kg	
151	Salvat	CSI	Kg	
152	Silage sorghum	CSEI	Kg	
153	Forage sugarbeets	CRFI	Kg	
154	Forage oats	CAFI	Kg	

Note: ha = hectares; m^3 = cubic meters; lact. = lactation(s).

a. Head = thousands of head of cattle.

b. Other lactation transfer controls (for summer, autumn, and winter) appear as restraints 50–64.

c. Thousands of m^3.

d. Springing (heifer) = ready to enter first lactation.

e. One hundred thousands of m^3.

f. Thousands of kg.

References

Morrison, Frank B. 1963. *Compendio de alimentación del ganado*. Translated by José Luis de la Loma. Mexico City: Unión Tipográfica, Edición Hispano-Americana.

National Academy of Sciences. 1971. "Nutritional Feeding Standards for Dairy Cows" and "Nutritional Requirements for Dairy Cattle." In *Nutrient Requirements of Dairy Cattle*, 4th ed. Washington, D.C.

Silos, José S. 1970. *Costos de producción de la leche en La Comarca Lagunera*. Técnica Pecuaria en México, no. 14 (January). Mexico City.

18

Joint Management of Water Resources in Irrigation and Lagoon Environments

R. G. CUMMINGS, H. C. LAMPE,
AND J. W. MCFARLAND

IN MANY ARID AREAS OF THE WORLD where rapid development of irrigated agriculture has taken place, there is growing awareness and concern for the physical and ecological effects of such development that were previously ignored or unanticipated. Examples include the problems associated with increased salt loads on downstream users brought about by the extensive development of irrigation along the upper reaches of the Colorado River (Howe and Orr 1974) and the problems associated with seawater intrusion in coastal aquifers as a result of the development of pump irrigation along coastal areas (Cummings 1971).

Overview

In this chapter, attention is focused on still another class of problems that may result from the diversion of surface waters for use in irrigation in arid coastal areas with productive lagoon environments. To put these problems in perspective, consider Mexico's coastal Northwest. This area contains a number of rivers that run down the slopes of nearby mountains and discharge into the sea. In many cases the rivers discharge directly into a coastal lagoon. In other cases, where major rivers are adjacent to a lagoon system, flooding in the rainy season provides a large part of the lagoon's fresh water. These freshwater infusions contribute to what has historically been a productive environment for the growth of the white and blue shrimp.

Over the past twenty years, however, there has been a concentrated drive by the Mexican government to develop irrigation districts in these coastal areas. Many of these new irrigation districts are formed for the

Note: The authors express their appreciation for the financial support for this work given by the Rhode Island Agricultural Experiment Station (contribution no. 1633) and Resources for the Future, Inc.

purpose of providing productive lands for landless peasants as a part of Mexico's continuing agrarian reform. Further, elaborate plans have been formulated for the transfer of large volumes of water from rivers along Mexico's west-central coastal areas for the purpose of developing (or maintaining, or both) irrigation in water-deficient areas in northwestern coastal areas (Cummings 1974).

With these ambitious programs for damming and storing water for irrigation or interbasin water transfers (or both), water is now being diverted for use in agriculture during those months in which the ecological environment of many lagoons is under its greatest stress (high levels of salinity; low volumes of water, nutrients, and dissolved oxygen).

Given existing capital investments for irrigated agriculture, along with the needs of communities that depend upon the harvest of shrimp from the lagoons in these areas, it is not surprising that the Mexican government is becoming increasingly concerned with the problems of allocating scarce water supplies between these two competing uses. Given also that in a poorly managed lagoon of some 75,000 acres an increase in freshwater flows of 100,000 acre-feet per month for four to five months could increase shrimp production by some 3.6 million kilos (at $2 a kilo, $7.2 million, or $14 to $18 per acre-foot of water), whereas net returns to irrigation waters in these areas range from $10 to $80 per acre foot (see Cummings 1974)—the need for some type of analytical apparatus that would allow for analyses of alternative allocations of water is apparent.

The purpose of this chapter is twofold. First, we wish to suggest a conceptual model that captures the major physical and economic relations relevant for the management problems described above. Although the "decision environment" for this presentation will be the coastal area in Mexico's Northwest, our model will hopefully be sufficiently general for application to a wide range of settings with differing physical, economic, and social-institutional characteristics. This will be the topic treated in the next section.

The third section gives attention to the economic structure of optimal decision rules that follow from the conceptual model presented in the second section and to the potential uses for the model. Particular attention is given here to the institutional setting for the irrigation-lagoon management problem as it exists in northwest Mexico, and to the use of our management model for allocating water resources within this setting. Concluding remarks are given in the last section. An appendix gives the derivation of equilibrium conditions for equations (18.17a–d).

Chapter 16 of this book deals to some extent with appropriate investment policies for coping with underground intrusions of salt water in a cropping zone, but apart from that example this chapter is the only one in the volume to explicitly treat the ecological side effects of agricultural activities. We have not carried our analysis as far as some of the other chapters in numerical implementation, but we feel it is important to make a beginning in this area by establishing one type of useful framework for quantitative analysis.

A Systems Model for Irrigation-Lagoon Control

In a general sense, consider a coastal lagoon in a relatively arid area for which the source of freshwater supplies is a river. River flows may be controlled by an upstream reservoir. Water is diverted from the river for use in irrigation districts in the coastal area between the lagoon and the reservoir.

This decision environment includes a number of interrelated systems: hydrologic, bioecologic, and economic. For expositional clarity, each system is developed separately in what follows; they are brought together in integrated form in the management model presented in the subsection "The model," below.

The hydrologic system

Let X_t be the volume of water stored in the reservoir at the beginning of period t ($t = 1, \ldots, T$). In this discussion, a period t is implicitly viewed as a month, and $T = 12$. For some applications, it may be convenient to view periods as weeks, irrigation periods, or the like. The nature of our analytical results, however, is unaffected by such a choice.

Let R_t and e_t be the release of water from the reservoir and water receipts to the reservoir during period t, respectively. Water receipts are clearly random variables and are treated deterministically here strictly for the purpose of simplifying the exposition. As suggested above, rainfall directly to the irrigation area or the lagoon (or both)—still another stochastic element—is not explicitly included in the model.

With a given initial reservoir stock X_1, the transition of reservoir stocks over time is assumed to be given by

$$(18.1) \qquad X_{t+1} = X_t + e_t - R_t - C_t. \qquad t = 1, 2, \ldots, T$$

In equation (18.1), reservoir stocks at the end of period t, X_{t+1}, equal initial stocks, X_t, plus water receipts to the reservoir, e_t, minus the periodic release, R_t, and *transfers* from the river system, C_t. We allow for transfers C_t for the purpose of generality, since it is common to find reservoirs serving as a source of water for municipal water supplies, for example, and possibly for interbasin water transfers (as is the case in northwestern Mexico).

This simplified formulation of "transfers" presupposes that withdrawals of water occur at or above the reservoir. In case of transfers occurring below the reservoir, C_t would be taken into account in the equation that allocates the release R_t, as in equation (18.2), below.

. River flows are assumed to be limited to the release R_t and, for simplicity, instream losses (evapotranspiration) are ignored. Such flows are either diverted for use in agriculture (a_t) or simply allowed to flow into the lagoon (b_t):

$$(18.2) \qquad a_t + b_t = R_t. \qquad t = 1, \ldots, T$$

The amount of fresh water entering the lagoon during any period t may exceed b_t, however, by an amount equal to return flows (which result from drainage and runoff) from a_t. Possible problems for the lagoon environment that are associated with the quality of return flows from agriculture are discussed below in the final section. Let return flows be a constant proportion α of a_t. Fresh water to the lagoon, f_t, is then given by

$$(18.3) \qquad f_t = b_t + \alpha a_t.$$

Equations (18.2) and (18.3) may be combined and written as

$$(18.4) \qquad a_t(1 - \alpha) + f_t = R_t. \qquad t = 1, \ldots, T$$

The bioecologic system

At the beginning of any period t we describe the ecological structure of the lagoon by the following set of state variables that are evaluated at the beginning of each period t. S_t is the average salinity level of water in the lagoon (parts per thousand). V_t is the volume of water in the lagoon (cubic meters). N_t is the amount of nutrients in the lagoon waters (parts per million). D_t measures dissolved oxygen (milligrams per liter).

Given initial values S_1, V_1, N_1, and D_1, the ecological state of the lagoon is assumed to change through time in the manner generally described by the following transition equations:

$$(18.5) \qquad S_{t+1} = S_t + s_t(f_t, w_t) \qquad \frac{\partial s_t}{\partial f_t} \leq 0, \frac{\partial s_t}{\partial w_t} \geq 0$$

$$(18.6) \qquad V_{t+1} = V_t + v_t(f_t, w_t) \qquad \frac{\partial v_t}{\partial f_t}, \frac{\partial v_t}{\partial w_t} \geq 0$$

$$(18.7) \qquad N_{t+1} = N_t + n_t(f_t, w_t) \qquad \frac{\partial n_t}{\partial f_t} \geq 0, \frac{\partial n_t}{\partial w_t} \leq 0$$

$$(18.8) \qquad D_{t+1} = D_t + d_t(f_t, w_t). \qquad \frac{\partial d_t}{\partial f_t}, \frac{\partial d_t}{\partial N_t} \geq 0, \frac{\partial d_t}{\partial w_t} \leq 0$$

In the systems (18.5)–(18.8), salinity is assumed to increase with saltwater inputs, w_t, and to decrease with freshwater inputs, f_t (see panel a of figure 18-1). Volume is directly related to f_t (panel c of figure 18-1) and w_t.

Nutrients in the lagoon, which are viewed here in a general sense to include all food available to shrimp, are assumed to be inversely related to salinity, and thus to vary directly with f_t and inversely with w_t. The principal source of nitrogen, phosphorus, and potassium to the lagoon is from fresh water, and the level of nutrients do indeed vary inversely with salinity. Further, nutrient levels for shrimp depend fundamentally on plant growth (microscopic or macroscopic) that in turn is a function of the water chemistry—particularly the levels of nitrogen, phosphorus, and potassium. We recognize, however, that there are cases where fresh water is not available to the lagoon, in which case the sea itself may supply these basic materials along with other nutrients to the shrimp. Shrimp feed directly upon some plants, the detritus to which plants contribute, and a variety of plankton, small animals, and other shrimp. In fact, shrimp will feed on most animal or plant material of the appropriate size. Of course, these are not the systems with which one is concerned in terms of control.

Dissolved oxygen is assumed to vary directly with freshwater inputs and nutrients, inversely with saltwater inputs. This representation of dissolved oxygen treats nutrients as a surrogate measure for a wide range of complex phenomena, some of which deplete oxygen and others of which increase it. For example, dissolved oxygen will vary with plant and animal growth respiration in the lagoon, other things being equal. The relation of dissolved oxygen to nutrient levels is hence a complex one for a variety of reasons. Animal growth depends upon nutrient supplies and oxygen, hence the rate of oxygen *depletion* will be directly related to animal biomass, and hence to nutrient levels. The decay of plant and animal material requires oxygen (biochemical oxygen demand, or BOD); the amount of material decaying at any time is directly related to the biomass of living plants and animals, which in turn is a function of nutrient materials. Photosynthetic activity generates oxygen and is a function of the availability of nitrogen, phosphorus, and potassium. On balance, however, we feel that the direct relation between dissolved oxygen and nutrients assumed here is acceptable, particularly for relatively shallow lagoons, which are the most likely to be subjected to control.

We assume that capital structures in the lagoon—such as barriers across the mouth of the lagoon and canals that provide additional access from the lagoon to the sea and vice versa—are fixed. Given these capital structures, saltwater flows to the lagoon, w_t, may be subject to control and are included as a variable for the purpose of generality. Little attention is given to w_t in our later analysis, however, given that such control occurs in only

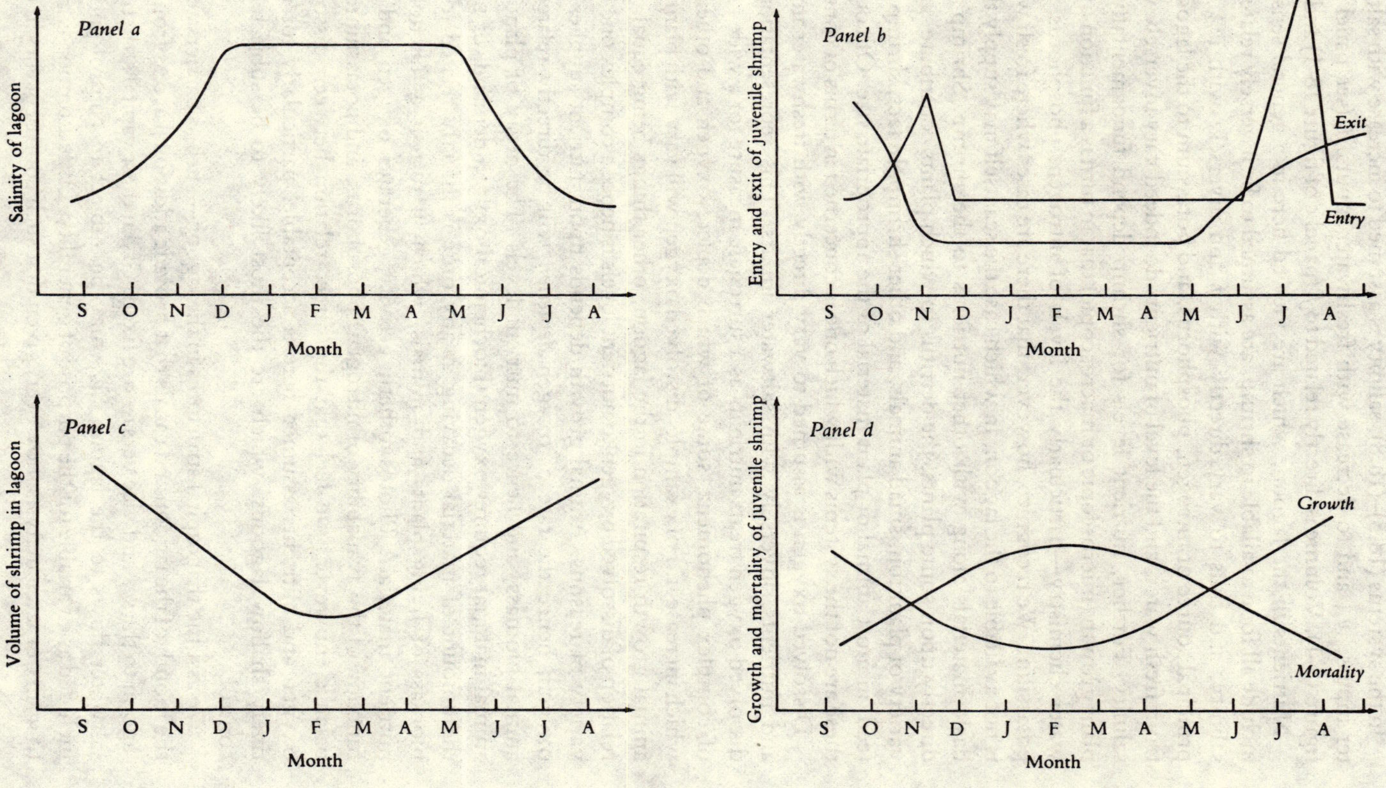

Figure 18-1. Bioecology of Lagoon Shrimp in Coastal Northwest Mexico

the rarest cases. In the more common case of an open lagoon, w_t is simply determined exogenously to the system. Tidal action would then be the primary determinant of w_t. In our conceptual framework, w_t would then most likely represent average conditions in the lagoon for the chosen decision period under consideration.

Finally, on its face equation (18.6) may appear to present conceptual problems in that no direct provision is made for outflows from the lagoon. However, fresh water additions, f_t, in a managed situation are not likely to be made at the rate at which V_t would increase dramatically. Indeed, shrimp are particularly susceptible to sudden *changes* in the lagoon environment, particularly in their early growth stages immediately following their entry into the lagoon. Most f_t additions would be made in such a way as to maintain volume in the face of evaporation losses and the like. Hence, for the general case outflows may be considered to be controlled, and volume assumed to be directly related to f_t and w_t. We acknowledge the exception, however, that even in this general case there may be critical, short periods of time that could require massive infusion of fresh water to the lagoons for a variety of reasons. These could occur particularly in the case of severe coastal storms in which breaches are made between the sea and the lagoon. As pointed out above, shrimp are more susceptible to sudden changes in the environment in the early stages immediately following entry, and any accidents during this period are critical because they would occur in agriculture immediately after planting.

The magnitude of the biomass of shrimp in the lagoon at the end of any period t, as well as the harvest of shrimp during t (H_t, developed below), depend upon three major relations: the entry of juvenile (postlarval) shrimp into the lagoon, the growth of individual shrimp, and mortality rates. These relations are depicted in figure 18-1 (panels a–d) as they are generally found in lagoons along the northwest coast of Mexico.

Shortly after the larval stage, juvenile shrimp move from the ocean toward the shore. Many of the juvenile shrimp find their way into the lagoon environment for reasons that are not completely understood. There is reason to believe, however, that their entry to lagoons is in some sense determined by the availability of nutrients, desirable levels of salinity and temperature, or a combination of these.

For notational simplicity, denote $\nabla_t = (S_t, V_t, N_t, D_t)$. It is assumed here that the entry of shrimp, E_t, is determined by ∇, is positively related to V_t, N_t, and D_t, and is inversely related to S_t:

$$(18.9) \qquad E_t = E_t(\nabla_t). \qquad \frac{\partial E_t}{\partial V_t}, \frac{\partial E_t}{\partial N_t}, \frac{\partial E_t}{\partial D_t} \geq 0, \frac{\partial E_t}{\partial S_t} \leq 0$$

Growth, G_t, and mortality, M_t, as they affect the biomass of shrimp, are assumed to be determined by the volume of the biomass, B_t, and the lagoon environment as described by ∇_t. As used here, the term "growth" is essentially the gross change in the biomass because of feeding [excluding entry, mortality, and so forth; see equation (18.13), below]. Growth is assumed to vary directly with B_t, V_t, N_t, and D_t. This growth-volume relation essentially assumes that space is required for feeding and growth and that water entering the system from land to increase volume is more rich in nutrients than the lagoons themselves. We assume an inverse relation between growth and salinity, recognizing that, although in arid environments where the competition for fresh water is likely to be very strong the inverse relation assumed for G and S is acceptable, there generally does exist a range over which G may vary directly with S (that is, some minimum level of salinity is required for shrimp growth).

Mortality is assumed to vary directly with B_t and S_t and inversely with V_t, N_t, and D_t:

$$(18.10) \qquad G_t = G_t(B_t, \nabla_t) \qquad \frac{\partial G}{\partial B_t}, \frac{\partial G}{\partial V_t}, \frac{\partial G}{\partial N_t}, \frac{\partial G}{\partial D_t} \geq 0, \frac{\partial G}{\partial S_t} \leq 0$$

$$(18.11) \qquad M_t = M_t(B_t, \nabla_t). \qquad \frac{\partial M_t}{\partial B_t}, \frac{\partial M_t}{\partial S_t} \geq 0, \frac{\partial M_t}{\partial V_t}, \frac{\partial M_t}{\partial N_t}, \frac{\partial M_t}{\partial D_t} \leq 0$$

The periodic harvest of shrimp, H_t, is assumed to take place passively, and is determined simply by the magnitude of the biomass and the ecological "state" of the lagoon, ∇_t, during each t. This will be the case when shrimp are captured by traps they encounter as they attempt to leave the lagoon at or near maturity, when the fishing effort is constant, or both. Such is the basic nature of shrimp harvesting from lagoons along Mexico's northwestern coast. Possible extensions of this analysis to include "effort" as a variable are discussed in the next subsection. We thus assume that

$$(18.12) \qquad H_t = H_t(B_t, \nabla_t). \qquad \frac{\partial H_t}{\partial B_t}, \frac{\partial H_t}{\partial V_t}, \frac{\partial H_t}{\partial N_t}, \frac{\partial H_t}{\partial D_t} \geq 0, \frac{\partial H}{\partial S} \leq 0$$

A transition equation for net changes in the biomass of shrimp may now be stated as follows:

$$(18.13) \qquad B_{t+1} = B_t + E_t(\nabla_t) + G_t(B_t, \nabla_t) - M_t(B_t, \nabla_t) - H_t(B_t, \nabla_t).$$

By equation (18.13), the biomass of shrimp at the end of period t, B_{t+1}, equals the initial stock, B_t, plus periodic growth, G_t, and new entrants, E_t, minus the harvest, H_t, and mortality losses, M_t.

The economic system

For given periodic volumes of water diverted from the river for use in the irrigation district, a_1, a_2, . . . , a_{12}, a wide range of management decisions are required concerning the number of hectares to commit for production, optimal crop selection, patterns of input use, and so forth. The literature abounds with alternative formats for laying out the management problem of optimizing water use in irrigation districts (see, for example, Burt and Stauber 1971). Most, if not all, of these production models for irrigation may be used to develop a straightforward functional relation between water use and net agricultural incomes of the form (Cummings 1974, chapter 4):

$$(18.14) \qquad A(a_1, a_2, \ldots, a_{12}). \qquad \frac{\partial A}{\partial a_t} \geq 0; t = 1, \ldots, T$$

For any value of $a_1, \ldots, a_{12}$, $\underline{A}$ measures net agricultural incomes that result from an *optimal* use of these waters—that is, from an optimal choice of crops, inputs, and the like, given crop prices and input costs. Thus, for a given value of the a's, $a_1^\star, a_2^\star, \ldots, a_{12}^\star$, the *value* $A(a_1^\star, \ldots, a_{12}^\star)$ subsumes crop patterns, input use, and the like; there is a one-to-one relation between values of $\underline{A}$ and the vector of optimally chosen variables in the irrigation district (Burt and Stauber 1971).

To keep the exposition as uncluttered as possible, we let equation (18.14) measure the benefits of water diversions from the river to the irrigation district and leave unspecified the *precise* operational framework (treated in other chapters of this volume) used for the suboptimization problem of generating $A(a_1, \ldots, a_{12})$. Although we argue that equation (18.14) is conceptually correct, we do recognize the many potential operational problems associated with the use of a periodic production or net benefit function for water as described in Burt and Stauber.

As argued above in the subsection "the bioecologic system," we have posited a passive harvesting process for shrimp wherein the periodic harvest of shrimp, H_t, is determined simply by the biomass of shrimp in each period t. Periodic net income from shrimp harvests may then be expressed as

$$(18.15) \qquad F_t(H_t) = F_t(B_t, \nabla_t). \qquad \frac{\partial F_t}{\partial B_t}, \frac{\partial F_t}{\partial V_t}, \frac{\partial F_t}{\partial N_t}, \frac{\partial F_t}{\partial D_t} \geq 0, \frac{\partial F_t}{\partial S_t}$$

$$\text{for all } t = 1, \ldots, 12$$

The relevant time path of prices for shrimp (as for agricultural prices) and all related costs are assumed to be known and subsumed in F_t.

We allow for the possibility that water may be extracted from the reservoir—our C_t in equation (18.1)—for a number of nonirrigation or lagoon-control purposes; for example, for supplies of potable water or interbasin water transfers to supplement water supplies in other basins. For whatever purpose the extracted water is to serve, we posit a relation of the form $K_t(C_t)$, which is assumed to measure net benefits associated with such extractions, where $\partial K_t/\partial C_t \geq 0$.

Finally, the annual income measures given by equations (18.14), (18.15), and $K_t(C_t)$ may seriously underestimate the value of reservoir water in arid areas in which a major function of the reservoir is that of providing for the interannual storage of water. In such cases, the reservoir stock at the end of the year (or horizon) T has a distinct value in relation to expected returns for future years of low water receipt (see, for example, Burt 1964). To avoid this limitation, we introduce a terminal value function $\psi(X_{T+1})$, $\partial\psi/\partial X_{T+1} \geq 0$, which is assumed to measure the value of terminal stocks of reservoir water. The terminal value function may be generated in a number of ways (for a discussion of this issue, see Burt and Cummings 1970, appendix).

The model

The optimization problems of interest here may now be stated as that of maximizing

$$A(a_1, a_2, \ldots, a_t) + \psi(X_{T+1}) + \sum_{t=1}^{T} \{F_t(B_t, \nabla_t) + K_t(C_t)\},$$

subject to the conditions and restrictions given by equations (18.1), (18.4) through (18.8), and (18.13). The Lagrangian function is

$$
\begin{aligned}
L = {}& A(a_1, a_2, \ldots, a_T) + \psi(X_{T+1}) + \sum_{t=1}^{T} \{F_t(B_t, \nabla_t) + K_t(C_t)\} \\
& - \theta_{t+1}\{X_{t+1} - X_t - e_t + C_t + R_t\} \\
& - \delta_{t+1}\{S_{t+1} - S_t - s_t (f_t, w_t)\} - \Gamma_{t+1}\{V_{t+1} - V_t - v_t (f_t, w_t)\} \\
& - \sigma_{t+1}\{N_{t+1} - N_t - n_t (f_t, w_t)\} \\
& - \xi_{t+1}\{D_{t+1} - D_t - d_t (f_t, w_t, N_t)\} \\
& - \beta_{t+1}\{B_{t+1} - B_t - E_t (\nabla_t) - G_t (B_t, \nabla_t) + M_t (B_t, \nabla_t) \\
& \qquad\qquad\qquad\qquad\qquad\qquad + H_t (B_t, \nabla_t)\} \\
& - \lambda_t\{a_t (1 - \alpha) + f_t - R_t\}.
\end{aligned}
$$

(18.16)

Necessary and sufficient conditions for a maximum of function (18.16) include the following, after rearranging terms:[1]

---

1. A derivation is given in the appendix; see equations (18.24a–j). For a more exhaustive treatment of necessary and sufficient conditions for an optimization problem of this general form, see Burt and Cummings (1971).

(18.17a) $\theta_t = \dfrac{\partial A}{\partial a_t} (1 - \alpha)^{-1}$

(18.17b) $= \dfrac{\partial K_t}{\partial C_t}$

(18.17c) $= \delta_{t+1} \dfrac{\partial s}{\partial f_t} + \Gamma_{t+1} \dfrac{\partial v_t}{\partial f_t} + \alpha_{t+1} \dfrac{\partial n_t}{\partial f_t} + \xi_{t+1} \dfrac{\partial d_t}{\partial f_t}$

(18.17d) $= \theta_{t+1} = \dfrac{\partial \psi}{\partial X_{T+1}}.$ for all $t = 1, \ldots, T$

In (18.17c), the terms θ_t, δ_{t+1}, Γ_{t+1}, α_{t+1}, and ξ_{t+1} are the Lagrangian multipliers associated with the trajectories of the state variables X_t, S_t, V_t, N_t, and D_t, respectively. The multiplier β_{t+1}, which is embedded in each of the multipliers listed above, is associated with the state variables B_t. Assuming strictly for simplicity that the terminal value for each of these state variables is zero, and defining

$$\nabla_{tS} \equiv \left(\frac{\partial E_t}{\partial S_t} + \frac{\partial G_t}{\partial S_t} - \frac{\partial M_t}{\partial S_t} - \frac{\partial H_t}{\partial S_t} \right)$$

$$\nabla_{tV} \equiv \left(\frac{\partial E_t}{\partial V_t} + \frac{\partial G_t}{\partial V_t} - \frac{\partial M_t}{\partial V_t} - \frac{\partial H_t}{\partial V_t} \right),$$

and similarly for ∇_{tN}, ∇_{tD}, the analytical structure for each of these multipliers is given by the following:

(18.18) $\beta_{t+1} = \displaystyle\sum_{\tau=t+1}^{T} \frac{\partial F_\tau}{\partial B_\tau} \prod_{i=t+1}^{\tau-1} \left(1 + \frac{\partial G_i}{\partial B_i} - \frac{\partial M_i}{\partial B_i} - \frac{\partial H_i}{\partial B_i} \right)$

(18.19) $\delta_{t+1} = \displaystyle\sum_{\tau=t+1}^{T} \left\{ \frac{\partial F_\tau}{\partial S_\tau} + \beta_{\tau+1} \nabla_{\tau S} \right\}$

(18.20) $\Gamma_{t+1} = \displaystyle\sum_{\tau=t+1}^{T} \left\{ \frac{\partial F_\tau}{\partial V_\tau} + \beta_{\tau+1} \nabla_{\tau V} \right\}$

(18.21) $\sigma_{t+1} = \displaystyle\sum_{\tau=t+1}^{T} \left\{ \frac{\partial F_\tau}{\partial N_\tau} + \xi_{\tau+1} \frac{\partial d_\tau}{\partial N_\tau} + \nabla_{\tau N} \right\}$

(18.22) $\xi_{t+1} = \displaystyle\sum_{\tau=t+1}^{T} \left\{ \frac{\partial F_\tau}{\partial D_\tau} + \beta_{\tau+1} \nabla_{\tau D} \right\}$

(18.23) $\theta_t = \theta_{t+1} = \dfrac{\partial \psi}{\partial X_{T+1}}.$ for all $t = 1, \ldots, T$

Management and Policy Issues Relevant
to Irrigation-Lagoon Control

The economic structure of decision rules for the optimal control of reservoir water is given by the set of equations (18.17a–d). The multiplier θ_t appears on the left-hand side of all equations. Examining the terms that appear in conditions (18.17), and θ_t is the periodic marginal value of water *in storage* (behind the reservoir). By (18.17d) the optimal allocation of reservoir water is one in which the marginal storage value of water in t equals the marginal storage value of water in all other periods and equals the marginal terminal storage value of water ($\partial \psi / \partial X_{T+1}$; that is, the marginal value of water in all future periods $t > T$).

On the right-hand side of equation (18.17a) is the periodic marginal value of water in irrigated agriculture net of return flows; on the right-hand side of (18.17b) is the marginal value of transferred water. In (18.17c) the terms $\partial s_t / \partial f_t$, $\partial v_t / \partial f_t$, $\partial n_t / \partial f_t$, and $\partial d_t / \partial f_t$ measure the *physical* (bioecologic) effects on the lagoon system of an incremental allocation of reservoir water to the lagoon in terms of associated changes in salinity, volume, nutrients, and dissolved oxygen, respectively. The multipliers δ_{t+1}, Γ_{t+1}, σ_{t+1}, and ξ_{t+1} are the marginal *values* associated with each respective physical change. Thus, for example, an increment of fresh water to the lagoon brings about a $\partial s_t / \partial f_t$ change in salinity, the *value* of which is $\delta_{t+1} (\partial s_t / \partial f_t)$. The right-hand side of (18.17c) is then seen to measure the periodic marginal value of freshwater to the lagoon.

Since the marginal value of water for agriculture (18.17a), transfers (18.17b), and the lagoon (18.17c) are *all* equated with the marginal storage value of water, the optimum periodic allocation of reservoir waters is seen to be one in which the marginal values of water earned in the agriculture, transfer, and lagoon sectors are equated.

Suppose now that we have empirically developed the bioecologic, hydrologic, and economic models required to generate operational statements for the system (18.1)–(18.15), in which case $a_1^\star, a_2^\star, \ldots, a_T^\star, f_1^\star, \ldots, f_T^\star, R_1^\star, \ldots, R_T^\star, C_1^\star, \ldots C_T^\star, X_{T+1}$, as well as operational arcs for the state variables, are obtained, as in the conditions given by equations (18.17). We may then ask the following questions: In relation to social-institutional structures, what are the assumptions implicit to this model—that is, under what kinds of social-institutional arrangements might the results from this model have *operational* significance for policy and under what arrangements might they not? In these latter instances, what kinds of modifications might be required in the model to accommodate alternative environments for institutional decisions?

As the model is stated, an appropriate institutional structure is one in which a central authority—public or private (as in a cooperative)—is responsible (in the broadest sense) for allocative decisions. Such is the case in Mexico, where the former Mexican Ministry of Water Resources (Secretaría de Recursos Hidráulicos, SRH; now the Ministry of Agriculture and Water Resources, SARH)—a relatively powerful ministry—has broad-ranging legal powers to control the use of Mexican water resources. In some areas (for example, the Costa de Hermosillo), the SRH and its successor agency exercise strict control over the use of water for irrigation; in the Costa de Hermosillo, water quotas are enforced with sanctions that include penalties equal to the *gross* value of water uses in excess of the quota.

With a central authority such as the SRH, the imposition of our optimal water allocations is tractable given that other societal criteria for water management—such as the distribution of income between sectors, the stability of income, the generation of employment or foreign exchange earnings (or both)—are embedded in the functions F, A, K, and ψ.

In most instances, however, and this is particularly true in Mexico, it is impractical (if not impossible) to develop a statement of the criterion function that will adequately encompass the many arguments relevant to social criteria for water resources management. Such a problem arises in the case where individual water rights have been legally established. Another example, common in many areas in northern Mexico, occurs when earlier colonization of newly formed irrigation districts has imposed an implicit commitment of minimum deliveries of water to the irrigation sector to sustain the livelihood of many small agricultural families therein (see Cummings 1974). An interesting phenomenon in this regard is the recent action of the Mexican government to permit the *ejido* farmers to organize into fishing cooperatives. In many cases, these farmers have rented their irrigated lands to entrepreneurs and hence are effectively unemployed. They are now becoming dependent upon the neighboring lagoons, which in effect compete with their own lands for scarce freshwater supplies, and this further places increased pressure on the lagoons for improved productivity.

In these as well as many other cases, substantial changes in the existing allocative pattern of water use may simply be untenable, considerations of efficiency notwithstanding. Even in such instances, however, the development of a management model such as the one suggested in this paper may be warranted for the purpose of generating tradeoffs associated with modest changes in allocative patterns.

For example, in the system described by equation (18.16) the only restriction on a_t is that of nonnegativity—that is, $a_t \geq 0$. As described above, reservoir waters are allocated so as to equate marginal value products from water in all uses, in which case it is conceivable that, in the optimal solution, a_t could be zero for all t. Denote basinwide benefits generated by the solution of (18.16) with $a_t \geq 0$ as $L(a_t \geq 0)$.

Suppose then that there is an institutionally determined lower bound on periodic water receipts to agriculture, $\hat{a}_t$. Clearly $L(a_t \geq 0) \geq L(a_t \geq \hat{a}_t)$, and the difference $L(a_t \geq 0) - L(a_t \geq \hat{a}_t)$ is a measure of the opportunity costs (in basinwide net incomes forgone) associated with the limit $\hat{a}_t$. One may well inquire then as to the rationale for this particular choice of $\hat{a}_t$—that is, why not impose a limit $a_t^* > \hat{a}_t$, or $a_t^* < \hat{a}_t$?

An economic evaluation concerning the choice of such limits requires information as to the *opportunity costs* associated with alternative choices for an $\hat{a}_t$. To this end, it is quite simple to add the restriction $a_t \geq a_t^*$ in our management model, and to generate opportunity cost measures $L(a_t \geq 0) - L(a_t \geq a^*)$ for a large number of arbitrarily chosen values for a^* in the neighborhood of $\hat{a}_t$. One would expect then to have information in the form depicted in figure 18-2. Should these opportunity costs be large for relatively small allocative changes, institutional structures that disallow large allocative changes may well admit more modest movements in the direction of Pareto efficiency.

Finally, an alternative approach that we wish to discuss briefly is the use of taxes for a decentralized system of water allocation. It is easily shown that the appropriate tax is θ_t, the periodic opportunity cost of water; with adequate means of measuring (metering) water inflows to the water-using sectors, a tax of θ_t per unit of water would result in the allocation described by equations (18.17) (with zero administrative costs).

A number of problems may arise with the use of taxes, however, that would apply equally well to a water-rights market system similar to that of Howe and Orr (1974). Taxes, or water rights, would generally be based on units of water used, and the tax (water price) would be equated with the marginal value earned by the water. Water infusions to the lagoon sector, however, represent the allocation of water to an unappropriated "pool" that serves a *biological* production process that is exploited (generally) by independent fishing units. The allocation of the tax over fishing units with varying efficiencies could impose a particularly thorny problem, with incentives for less efficient units to refuse to "buy" freshwater infusions and receive the more productive lagoon environment via freshwater "purchases" as an externality.

Instances in which such taxes would be relatively easy to administer are found where institutional arrangements of the type found in northern Mexico obtain, wherein the jurisdiction of a cooperative is clearly defined and the right to fish for shrimp resides in the cooperative, not with individual fishermen. Even here some problems arise in that a single lagoon may be shared by two or more cooperatives, although it is in

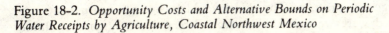

Figure 18-2. *Opportunity Costs and Alternative Bounds on Periodic Water Receipts by Agriculture, Coastal Northwest Mexico*

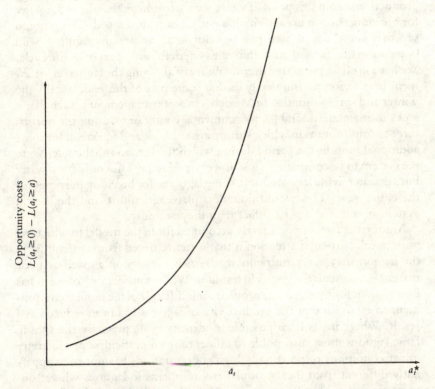

Note: L = opportunity cost of the lower bound a, measured in basin-wide income forgone; a = lower bound on periodic water diversions to irrigation; a_t = actual water diversions to agriculture in period t.

general possible to determine how much of the shrimp harvest is attributable to each cooperative. In other lagoons, however, there are no clear territorial rights, and in these fishermen from one cooperative may well be fishing alongside those from another. In Mexico, it can be generally stated that if one proceeds from north to south the territorial rights become less well established.

Conclusions

It should be made clear that the model, which presumes a certain fixity in the production patterns in agriculture, implies some rigidity in lagoon-management practices. The model would of course be enormously complicated were we to consider a completely closed system for shrimp production. From this point of view, we are fortunate that the technology for breeding shrimp in captivity has not yet been developed. This is not to say that some day it will not be. But here we are managing a wild population that is resident within our system for only part of its life cycle. We have in effect permitted the possibility of allowing the lagoon environment to deteriorate completely during some part of the year, usually the winter and spring months. In Mexico, current management is such that it seeks to maintain desirable environmental conditions during the normal three to four months in which shrimp are available and to extend for some additional months the period during which the lagoon is habitable. Were the system to be completely closed, we would require not only fresh water but saltwater in the lagoons, or in some lagoons, for breeding purposes for the entire year. This would doubtless place agriculture and the lagoon system in much sharper conflict than they are now.

Among the factors not directly accounted for in the model to which we wish to call particular attention is that water returned from agriculture to the lagoon may carry nutrients that enrich the lagoon as well as toxic materials, particularly pesticide residues. While considerable concern has been shown for the pesticide problem, it is difficult in the natural environment to establish that the productivity of lagoons has been reduced as a result. Where it has been possible to identify quite precisely the sites in large lagoons most susceptible to influence from pesticides, productivity under conditions of potentially high concentrations has not been apparently different from that in similar environments in lagoons where concentrations would be lower as a result of dilution. It is probably fortunate that, upon entry into the lagoon, shrimp have passed through their prejuvenile stages. If one generalizes from the effect of pesticides on other animals, it is likely that they would be most susceptible to pesticides, heavy metals, and other pollutants prior to the time that they enter the lagoon. Further, it is also probably fortunate that shrimp depart from the lagoon or attempt to leave the lagoon prior to achieving sexual maturity—hence the maturation process and breeding take place at sea. It is, of course, possible that the environments offshore are polluted in some areas, but it is fairly certain that a substantial dilution of toxic material takes place in the open ocean.

It is germane to ask, however, whether shrimp, as other animals, concentrate these toxic materials in their body tissues and hence affect their desirability for human consumption. From present evidence it would appear that toxic materials are indeed concentrated in the shrimp, but that this concentration takes place principally in the tissues of the cephalothorax, and these parts are not normally eaten by humans.

As to the enrichment of lagoons by residual nutrient phosphorus and potassium, it is not possible at this time to determine what fraction of the nutrients entering the lagoon in fresh water can be attributed to the natural nutrient content and what fraction is actually added by agriculture. There is no evidence at present to suggest that unusually high nutrient levels cause any significant change in the floral or faunal composition of the lagoon environment.

Appendix. Derivation of Equations (18.17)

A derivation of the equilibrium conditions (18.17a–d) is given as follows:

(18.24a)
$$\frac{\partial L}{\partial a_t} = \frac{\partial A}{\partial a_t} - (1 - \alpha)\lambda_t = 0$$

(18.24b)
$$\frac{\partial L}{\partial f_t} = \frac{\partial s_t}{\partial f_t}\delta_{t-1} + \frac{\partial v_t}{\partial f_t}\Gamma_{t+1} + \frac{\partial n_t}{\partial f_t}\alpha_{t+1} + \frac{\partial d_t}{\partial f_t}\xi_{t+1} - \lambda_t = 0$$

(18.24c)
$$\frac{\partial L}{\partial C_t} = \frac{\partial K_t}{\partial C_t} - \theta_{t+1} = 0$$

(18.24d)
$$\frac{\partial L}{\partial R_t} = -\theta_{t+1} + \lambda_t = 0$$

(18.24e)
$$\frac{\partial L}{\partial X_T} = -\theta_t + \theta_{t+1} = 0$$

(18.24e′)
$$\frac{\partial L}{\partial X_{T+1}} = \frac{\partial \psi}{\partial X_{T+1}} - \theta_{T+1} = 0$$

(18.24f)
$$\frac{\partial L}{\partial S_t} = \frac{\partial F_t}{\partial S_t} - \delta_t + \delta_{t+1} + \beta_{t+1}\left(\frac{\partial E_t}{\partial S_t} + \frac{\partial G_t}{\partial S_t} - \frac{\partial M_t}{\partial S_t} - \frac{\partial H_t}{\partial S_t}\right) = 0$$

(18.24g)
$$\frac{\partial L}{\partial V_t} = \frac{\partial F_t}{\partial V_t} - \Gamma_t + \Gamma_{t+1} + \beta_{t+1}\left(\frac{\partial E_t}{\partial V_t} + \frac{\partial G_t}{\partial V_t} - \frac{\partial M_t}{\partial V_t} - \frac{\partial H_t}{\partial V_t}\right) = 0$$

(18.24h)
$$\frac{\partial L}{\partial N_t} = \frac{\partial F_t}{\partial N_t} - \sigma_t + \sigma_{t+1} + \xi_{t+1}\frac{\partial d_t}{\partial N_t}$$
$$+ \beta_{t+1}\left(\frac{\partial E_t}{\partial N_t} + \frac{\partial G_t}{\partial N_t} - \frac{\partial M_t}{\partial N_t} - \frac{\partial H_t}{\partial N_t}\right) = 0$$

(18.24i)
$$\frac{\partial L}{\partial D_t} = \frac{\partial F_t}{\partial D_t} - \xi_t + \xi_{t+1} + \beta_{t+1}\left(\frac{\partial E_t}{\partial D_t} + \frac{\partial G_t}{\partial D_t} - \frac{\partial M_t}{\partial D_t} - \frac{\partial H_t}{\partial D_t}\right) = 0$$

$$(18.24\text{j}) \quad \frac{\partial L}{\partial B_t} = \frac{\partial F_t}{\partial B_t} - \beta_t + \beta_{t+1}\left(1 + \frac{\partial G_t}{\partial B_t} - \frac{\partial M_t}{\partial B_t} - \frac{\partial H_t}{\partial B_t}\right) = 0.$$

By equations (18.24c)–(18.24e′), we have

$$\theta_t = \theta_{t+1} \qquad \text{for all } t = 1, \dots, T,$$

$$\theta_{t+1} = \lambda_t, \; \theta_{t+1} = \frac{\partial K_t}{\partial C_t}, \text{ and } \theta_{T+1} = \frac{\partial \psi}{\partial X_{T+1}}.$$

Because $\lambda_t = (\partial A / \partial a_t)(1 - \alpha)^{-1}$ by equation (18.24a), and $\lambda_t = \theta_{t+1}$ $= \theta_t$ by (18.24d) and (18.24e), thus $\theta_t = (\partial A / \partial a_t)(1 - \alpha)^{-1}$ by equation (18.17a). Equations (18.17b) and (18.17d) follow directly from (18.24c), (18.24e), and (18.24e′). Equation (18.17c) is derived by solving (18.24b) for λ_t, and using $\theta_{t+1} = \theta_t$ by (18.24d) and (18.24e).

References

Burt, O. R. 1964. "The Economics of Conjunctive Use of Ground and Surface Water." *Hilgardia*, vol. 32, no. 2, pp. 31–111.

Burt, O. R., and R. G. Cummings. 1970. "Production and Investment in Natural Resource Industries." *American Economic Review*, vol. 9, no. 4 (December), pp. 576–90.

Burt, O. R., and M. S. Stauber. 1971. "Economic Analysis of Irrigation in Subhumid Climates." *American Journal of Agricultural Economics*, vol. 53, no. 1, pp. 33–46.

Cummings, R. G. 1971. "Optimum Exploitation of Groundwater Reserves with Saltwater Intrusion." *Water Resources Research*, vol. 7, no. 6 (December), pp. 1414–24.

————. 1974. *Interbasin Transfers of Water: A Case Study in Mexico*. Baltimore, Md.: Johns Hopkins University Press.

Howe, C. W., and D. V. Orr. 1974. "Economic Incentives for Salinity Reduction and Water Conservation in the Colorado River Basin." In *Salinity in Water Resources*. Edited by J. E. Flack and C. W. Howe. Boulder, Colo.: Merriman.

Part Four

Data Processing
for Agricultural Models

19

Computational Considerations
for Sectoral Programming Models

GARY P. KUTCHER AND ALEXANDER MEERAUS

THE TOPICS ADDRESSED in this volume range from policy uses of models, theoretical and methodological issues in model design, data requirements and sources, to specific applications of models in Mexican policy and projects. Most of the chapters, even those devoted to model applications, virtually ignore all computational aspects of actually building and using large-scale models—although most of the authors would agree that the computational aspect may have accounted for as much as half the resources required for their studies.

Overview

It is not our purpose here to provide explicit directions for building agricultural models. Aside from space limitations (an entire volume would be required), there are two other reasons that such a treatise is not desirable or feasible. First, the best computational approach is so highly dependent upon such a variety of factors as to preclude our suggesting precise methods for building agricultural models independently of the particular context. Second, the computational technology—not only computer hardware and software but the entire methodology of obtaining linear programming model results—is advancing at a pace so rapid that it would render a discussion along these lines obsolete.

Instead, our purpose is limited to a twofold objective: we wish to apprise potential model builders of the computational difficulties that are certain to arise and to offer some guidelines for efficient model building and use that may influence model design and data organization (as well as the direct computational work). The difference between proper and improper model management on the computer can spell the difference between a few days and several weeks in carrying out prescribed numerical experiments. This difference was in fact experienced with CHAC, between the early and later versions.

Computational Aspects in Perspective

Figure 19-1 shows a typical setting involving the use of mathematical modeling to assist in solving a real-world problem. By "model" we refer to a formal definition of the problem; that is, a complete algebraic statement or tableau and the associated data. The analyst desires to proceed directly to the box on the left labeled "solution," which refers to the final answer to the problem stated in the same language or notation as the model. In any applied study, the "analyst" box must be understood to contain also the policymaker, data collector, and other relevant individuals. Here we focus principally on the interaction of these individuals with the computer.

Consider the following example, taken fron Dantzig (1963). A company desires to supply its three warehouses from two canneries, with given inventories in each, to minimize the total shipping cost, which it needs to know. A formal definition of the problem using algebra and tables, and the optimal solution in the same notation as the problem's

Figure 19-1. *Use of Linear Programming Models in Problem Solving*

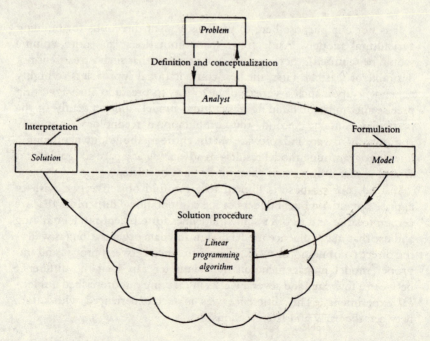

definition, are given in figure 19-2. If this were written for a general readership, only the model and solution as defined here would be of interest. How one proceeds through the procedure (the area below the broken line in figure 19-1) is of extreme practical importance and can often require more time and resources than the model formulation and solution analysis. In other words, the model must be transformed into a language or notation readable by a computer, it must be solved by a computer, and

Figure 19-2. *The "Cannery Problem" in the Format of Linear Programming: Algebraic Equations and Data Matrixes*

Problem

$$\min \Sigma_i \Sigma_j c_{ij} x_{ij}$$
$$\text{s.t.} \ \Sigma_j x_{ij} \le a_i$$
$$\Sigma_i x_{ij} \ge r_j,$$

where x_{ij} = shipment from cannery i to warehouse j (cases)
  r_j = requirement of warehouse j (cases)
  a_i = availability at cannery i (cases)
  c_{ij} = unit transport cost from cannery i to warehouse j (dollars per case)
  i = Seattle and San Diego
  j = New York, Chicago, and Kansas City.

Data and tableaux

Cannery (i)	Cases available (a_i)	Warehouse (j)	Cases required (r_j)
Seattle	350	New York	300
San Diego	650	Chicago	300
		Kansas City	300

From	Unit transport cost (c_{ij}) to		
	New York	Chicago	Kansas City
Seattle	2.5	1.7	1.8
San Diego	2.5	1.8	1.4

Solution

From	Shipments (x_{ij}) to		
	New York	Chicago	Kansas City
Seattle	0	300	0
San Diego	300	0	300

Total cost = **$1,680.**

Source: The basic problem is taken from Dantzig (1963).

the computer solution must then be translated back into the problem's original language or notation.

Historically, the critical computational element in this procedure has been the solution of the model by computer. Now, however, modern computer systems solve models of hundreds of equations in a matter of seconds, and models of some thousand equations are still within comfortable reach of today's linear programming technology. The advances in computer technology as applied to linear programming solution have been spectacular and have derived mainly from two sources. The first is the tremendous increase in capability of general purpose computer systems. Since 1955, the number of computations that can be performed for one dollar has, on the average, *doubled every year*. Second, similar advances have taken place in the development of algorithms and software systems for linear programming. These software advances have occurred in part because of hardware advances that permitted them; mostly they have been because the linear programming solution stage can be isolated from the rest of the modeling process via a standardized input format (mathematical programming system, or MPS) and an output requirement can be agreed upon. This isolation has allowed research and development resources to be focused on the "black box" of linear programming algorithms.

Although the standardization of MPS format[1] has permitted important advances in solution procedures, its rigidity and its scant relation to usual problem definitions have caused it to become a restrictive step in the process of obtaining solutions. Because a model must be translated into MPS format, a more complete definition of the term is warranted.

MPS Format

An MPS input file has at least three sections: the definition of the rows, the enumeration of the nonzero matrix coefficients, and the listing of the (nonzero) right-hand elements. Each of these sections is preceded by leader words that are, respectively, ROWS, COLUMNS, and RHS (right-hand side), all punched starting in column 1 of a data card. The very first card is a problem identification, with NAME punched in columns 1–4 and a user's given name punched starting in column 15. The very last card must be ENDATA, again starting in column 1.

Each row and column must have a unique name with a maximum of eight characters; this may be any combination of letters, numbers, and some special characters. All the row names must be placed in the ROWS

---

1. The format of IBM's MPS/360 (MPS for the IBM 360 computer series) has been adopted as the de facto industry standard and is referred to generically as "MPS format."

section with their corresponding type [the direction of the inequality (*L* or *G*), equality (*E*), or unconstrained (usually *N*)]. The type code is punched in column 2, and the row name is punched beginning in column 5. The elements of each column must be grouped together and are indentified by the column name (beginning in column 5), the row name (column 15), and the coefficient (in the field 25–34). The right-hand side, or constraint vector, is also given a name, and its elements are treated as the above matrix elements except that they are placed in the RHS section.

Figure 19-3 is the representation in MPS format of the cannery problem in figure 19-2. The algebraic statement and the data have been translated into MPS format as a prerequisite to solution. This format bears little resemblance to the algebraic statement, and is a clumsy and inefficient means of expressing the problem from the human viewpoint. It is, however, an efficient statement from the computer's viewpoint. And, since it is now standardized as the accepted input format for linear programming, the translation to MPS format must always be made in one manner or other.

Figure 19-4 is the linear programming algorithm's output of this solution.[2] Note that the output is in the same MPS notation, which bears little resemblance either to the notation of the original problem's statement or to the notation of the solution that the analyst desires (in the form of figure 19-2). Thus, an additional translation or mapping is required before the solution can be interpreted. This translation is, in fact, the inverse of the initial translation into MPS format.

Let us consider in more detail the initial mapping of the algebraic notation of the cannery problem into MPS matrix notation. First, each constraint and vector required a unique name. Complete flexibility is available up to the limitations described above. In this example, we chose to replace the *i* and *j* indexes with acronyms to facilitate the inverse mapping. In other words, by calling *i* = 1 SE, and *j* = 1 NY, we can more easily recall that activity "X.SENY" ships one unit from Seattle to New York. Of course, MPS format would have permitted us to name this activity "X.11" or even "11," but interpretation of the solution would then have required an additional translation step. That is, we would have had to refer to figure 19-2 to recall which source and destination had been assigned index 1.

In the relatively simple cannery problem there were no difficulties in assembly and manipulation of the data. This data mapping was direct because the data were organized in the form required. But suppose that

2. For this example we used APEX, a commercial linear programming package developed and supported by Control Data Corporation. This output is similar to that of most other commercial packages.

Figure 19-3. *Example of Mathematical Programming System (MPS) Format for Computer Entries*

NAME	DANTZIG	
ROWS		
N COST		
L A.SE		
L A.SD		
G R.NY		
G R.CH		
G R.KC		
COLUMNS		
X.SENY	COST	2.5
X.SENY	A.SE	1.0
X.SENY	R.NY	1.0
X.SECH	COST	1.7
X.SECH	A.SE	1.0
X.SECH	R.CH	1.0
X.SEKC	COST	1.8
X.SEKC	A.SE	1.0
X.SEKC	R.KC	1.0
X.SDNY	COST	2.5
X.SDNY	A.SD	1.0
X.SDNY	R.NY	1.0
X.SDCH	COST	1.8
X.SDCH	A.SD	1.0
X.SDCH	R.CH	1.0
X.SDKC	COST	1.4
X.SDKC	A.SD	1.0
X.SDKC	R.KC	1.0
RHS		
RHS	A.SE	350.
RHS	A.SD	650.
RHS	R.NY	300.
RHS	R.CH	300.
RHS	R.KC	300.
ENDATA		

Source: Representation of the cannery problem in figure 19-2.

instead of an enumeration of the coefficients c_{ij}, a functional form had been given for the costs that depend, say, on the distance, mode, season, commodity, and so on. In this case an additional transformation of the parameters of the functional form would have been required to arrive at the correct matrix elements.

Because this problem did not require manipulation of the data and was straightforward enough to allow us to keep the mapping (naming) conventions and structural relations in our memory, these transformations into and out of MPS format were simple enough to minimize the possibility of error. There was little scope for, say, transposing an index (for example, confusing Chicago and Kansas City), and the volume of data was small enough to allow quick visual verification. In the original CHAC (see chapters 2 and 4), however, there were approximately 5,000 row and column names to keep track of, and there were some 80,000 nonzero coefficients to manipulate into MPS format. The MPS statement of CHAC took up more than 1,000 pages of computer listing, and its linear programming output was more than 80 pages long!

Clearly, for processing at this level of size and complexity the computer must be used to automate at least parts of the translation procedures involved.

Matrix Generation

Certain aspects of this translation process may be facilitated through the use of computer programs. Once the problem is stated algebraically, the data assembled, and a naming convention adopted, a computer program referred to as a "matrix generator" may be employed to assemble the MPS-formatted matrix. A matrix generator is a *translator*, but we must be careful in defining exactly what it translates into the MPS matrix. One may, instead of proceeding directly from the algebraic representation of the problem into MPS format, proceed indirectly via an intermediate representation, which is the matrix generator. If this is undertaken, the computer may assume responsibility for the remaining stages.

Similarly, one must also be clear about what a computer program is. A computer program is simply a set of instructions written in what is called a "user's language." A user's language such as Cobol or Fortran is a hybrid between the human languages of English and algebra that is amenable to further translation (compilation) into computer language.

In using a matrix generator, we recognize that the computer is highly efficient in performing certain functions: it can manipulate vast amounts of data very rapidly, and it can assemble characters or strings of characters to arrive at the desired MPS format with virtual elimination of the possibil-

Figure 19-4. Example of Output from a Linear Programming Algorithm

APEX-I 1.014 PAGE 1

CONSTRAINTS

PRINT OPTION = COMPLETE OUTPUT
NAME = DANTZIG   OBJ = COST   RHS = RHS   BND =   VALUE OF OBJECTIVE = 1680.00000
DIR = MINIMIZE   COBJ =   CRHS =   RNG =   RPSOBJ = 1.0000   RPSRHS = 1.0000
RPCHOBJ = .0000   RPCHRHS = .0000

NUMBER	NAME	TYPE	STATUS	ROW ACTIVITY	SLACK	RHS LOWER	RHS UPPER	MARGINAL
1	COST	FR	SLACK	1680.00000	-1680.00000	-INF	+INF	.
2	A.SE	LE	SLACK	300.00000	50.00000	-INF	350.00000	.
3	A.SD	LE	SLACK	600.00000	50.00000	-INF	650.00000	.
4	R.NY	GE	BINDING	300.00000	.	300.00000	+INF	-2.50000
5	R.CH	GE	BINDING	300.00000	.	300.00000	+INF	-1.70000
6	R.KC	GE	BINDING	300.00000	.	300.00000	+INF	-1.40000

APEX-I 1.014 PAGE 1

COLUMNS

PRINT OPTION = COMPLETE OUTPUT
NAME = DANTZIG   OBJ = COST   RHS = RHS   BND =   VALUE OF OBJECTIVE = 1680.00000
DIR = MINIMIZE   COBJ =   CRHS =   RNG =   RPSOBJ = 1.0000   RPSRHS = 1.0000
RPCHOBJ = .0000   RPCHRHS = .0000

NUMBER	NAME	TYPE	STATUS	COL ACTIVITY	OBJ COEF	BND LOWER	BND UPPER	MARGINAL
1	X.SENY	PL **	LOWER	.	2.50000	.	+INF	.
2	X.SECH	PL	ACTIVE	300.00000	1.70000	.	+INF	.
3	X.SEKC	PL	LOWER	.	1.80000	.	+INF	.40000
4	X.SDNY	PL	ACTIVE	300.00000	2.50000	.	+INF	.
5	X.SDCH	PL	LOWER	.	1.80000	.	+INF	.10000
6	X.SDKC	PL	ACTIVE	300.00000	1.40000	.	+INF	.

Source: Solution of the cannery problem in figure 19-2.

ity of error. Of course, this holds true only if the set of instructions—the matrix generator—is a true representation of the algebraic statement of the model.

To understand more fully why this seemingly additional (and more complicated) translation step can be useful, consider the formation of the land constraints in CHAC through the use of Fortran instructions.

Assuming that the alphanumeric symbols for the twenty districts (for example, A, B, C, . . .), the four land classes (1, 2, 3, 4), and the twelve months (1, 2, ... , O, N, D) have been assigned earlier in the program, the Fortran statements become:

```
    DO 100  I = 1,20
    DO 100  J = 1,4
    DO 100 K = 1,12
100 WRITE(4,101)  DISTRICT(I), LCLASS(J), MONTH(K)
101 FORMAT(8H L   LAND, 3A1)
```

and these will write the land row names in MPS format:

```
L LANDA11
L LANDA12
      .
      .
      .
L LANDW4D
```

In this example, in which we substituted five Fortran statements to instruct the computer to write 960 ROW cards, the advantages are clear. In effect, these Fortran statements are a shorthand notation for the 960 MPS statements. If we did not make an error in these statements, we are assured of an accurate translation and of the computer's ability to assemble the varying symbols into an MPS-readable name set. But consider part of a model similar to CHAC in which nontrivial calculation of coefficients is required: that of the yields of the cropping activities, which may vary not only by crop and land class, but by district and, say, by time period. Suppose we are dealing with a five-year model, and the yields of each crop are expected to increase exogenously by 3 percent each year. If a 20×10 (twenty districts, ten crops) matrix Y is the input, and a vector RATIO = $(1.00, 0.85, 0.70, 0.55)$ indicates the yield in each land class relative to class 1, then we may generate the 4,000 MPS cards with the following handful of Fortran statements:

```
DO 200  I = 1,20
DO 200  J = 1,10
DO 200  K = 1,4
DO 200  N = 1,5
COEF = Y(I,J)*RATIO(K)*1.03**(N-1)
```

```
200  WRITE(4,201) DISTRICT(I),ICROP(J),LCLASS(K),N,ICROP(I),N,COEF
201  FORMAT(4X,4HCROP,3A1,I1,2X,1HQ,A1,I1,7X,F12.6)
```

which will generate a typical card

```
     CROPDM35   QM5              2.6547
```

that defines the matrix element associated with a cropping activity (each of which is prefixed CROP) in district D, for crop M, on land class 3, in year 5—which enters the row QM5, the output of crop M in year 5.[3] In this example we have taken advantage not only of the computer's ability to assemble strings of characters, but also its ability to perform relatively complex calculations. Most newer computers could execute these seven statements in a second or two, and the reader is invited to consider how long it would take to produce the 4,000 cards manually—including the tasks of calculating, coding, keypunching, and verifying each card. Because of its complexity, the entire model thus can be (and was) translated into MPS format using a matrix generator. That is, instead of writing out 80,000 MPS cards, we wrote about 1,000 instructions in Fortran that operated on about 400 compact data cards. By choosing this method of matrix generation, we of course completely altered the set of skills required. Instead of employing desk calculators and an army of coders and keypunchers, these inputs are now replaced by a programmer on whose shoulders lie the responsibilities of ensuring that the original problem is precisely represented in the matrix generator and that the data as accepted by the generator are correct. If these conditions are met, we may proceed to the solution and interpretation without further concern for error in any of our 80,000 coefficients.

The possible savings in manpower associated with achieving the initial MPS problem statement are clear. The full power of a matrix generator, though, arises when (as almost always is the case) variations in the model or its data (or both) are desired. Consider, for example, the labor coefficients in CHAC, which number in the thousands. Suppose hypothetically (but not impossibly) that it was discovered after the model had been built that the labor data were based on an assumed ten-hour workday, whereas in fact an eight-hour day was the measured norm. Without a matrix generator, these thousands of coefficients would have had to be scaled individually and the cards repunched and verified. With the generator, this correction could be made by adding or changing one instruction, and the thousands of cards thus altered in a matter of seconds. More important,

---

3. In this example we have ignored the MPS requirement that all the elements in a given vector be sequential. This ordering may be done with a sorting routine at the end of the matrix generation, or the Fortran statements may be so arranged as to yield a properly ordered list of elements.

proofreading the 400 data input cards for the matrix generator is an easier task than doing the same for the 80,000 individual CHAC coefficients. The simple process of "debugging" a manually constructed model can take months, and the debugging must be repeated for each significant change in model specification.

Although it is simple to demonstrate the power and desirability of employing matrix generators, the choice of how to proceed, in practice, is much more difficult, and subject to the consideration of many factors. In the case of CHAC, a special Fortran generator called SECGEN was written. It was special in the sense that its only purpose was to translate CHAC, which of course was developed and used initially in a research environment. Fortran was selected as the language for the single reason that the analyst writing SECGEN, who was also an economist, happened to be proficient in Fortran, and Fortran was usable on the computers both in Washington, D.C., and Mexico. Commercial packages for matrix generation were available in Washington but not in Mexico; thus, further consideration of their use was not possible.

The choice between a commercially available matrix-generation package and a specialized Fortran (or other user's language) generator depends on the type of study being undertaken and the skills of the person responsible for implementing the model. The most significant advantage of a commercial generating system is that it requires a particular approach to data organization, and the resulting translation is self-documenting. That is, another person who is proficient in the use of the given package can enter on the scene and understand precisely what the problem is as stated in say, MODGEN.[4] Thus, as algebra forces a unique problem statement, so does MODGEN— but in a form that permits direct access (via MPS format) to the solution algorithm. Of course, such a restrictive mode has a cost: to encompass the range of conceivable problem formulations, the program may be clumsy and costly. In contrast, a Fortran generator has virtually complete flexibility in the mode of data input and the way the algebra is translated. Fortran programs, however, are totally subject to the programmer's personal style of approaching and interpreting the problem. No two programmers would ever arrive at the same Fortran program to accomplish a task that was completely prespecified for input format and output requirements. Documentation is always a problem: no matter how carefully written and described a Fortran program might be, it always is difficult to transfer the usage of an existing program to other programmers. This difficulty arises from two sources. First, because Fortran is a language, much freedom is permitted the writer in expressing himself

4. MODGEN is a ficticious name we will employ for expository purposes to refer to the class of "table-driven" commercial generators (MAGEN, DATAFORM, GAMMA, MGRW).

individually and still accomplishing the desired result. Analogously, no two newspaper reporters would write the same article when viewing the same event, even though both are subject to the same stringent editorial requirements. Second, the use of a Fortran generator frees the programmer from the rigid program structure dictated by a commercial matrix-generating package. This additional freedom will likely induce vast differences in the way the input data, in particular, will be handled.

Data Organization and Preanalysis

Unlike industrial applications of linear programming, for example, which usually rely upon engineering data of a standard form, the data for agricultural models will vary widely according to the purpose of the study and will usually be derived from a hodgepodge of sources (agricultural censuses, farm management handbooks, the publications of specialized government agencies, partial surveys, meteorological reports, and the like). Such sources were employed in the original CHAC study, and some of the data tabulated from these sources are described in chapter 4 of this volume.

Regardless of the approach adopted for matrix generation, the data available for agricultural models are never in a form amenable to direct and efficient model generation. Some manipulation will always be necessary, but this manipulation has a cost that may influence the means of generation to be used.

It is conceivable that much of the data are available in tabular form and that the existing tables correspond roughly to submatrixes in the desired linear programming problem. If so, this would suggest that one of the table-driven commercial matrix generators could be employed with little modification or rearrangement of the data set. In other words, higher running costs of such a package would be offset by the savings in not having to process the input data further to make them amenable to an efficient (faster-executing) Fortran generator.[5]

If a Fortran generator is dictated for other considerations (such as not having a commercial generation package available), the choice remains whether to write the generator to accept the data in their existing format or to reorganize the data set to conform with efficient Fortran generation. This choice will depend in part on the computer storage available, the ability of the particular computer to access data in different forms of storage, and the language used. In addition, any computer program can be

5. For expository purposes, we continue to use "Fortran" to refer to a class of user's languages (Fortran, PL1, Basic, Cobol, and the like).

designed to be both shorter and faster if the programmer is given full choice regarding the way the input data are organized.

In the SECGEN experience, virtually none of the raw data was in computer accessible form (that is, on cards or tape), and thus the input data for any generator would have had to be coded and keypunched regardless of the approach taken. This consideration, plus the prospect of repeated generations of different versions of CHAC in its form for policy evaluation, led us to write the generator under the assumption that the data would be manually reorganized as required by the generator. A portion of these data is shown in figure 19-5, which the reader may contrast with the tables in chapter 4. The point is that the data, as required by SECGEN, are virtually meaningless unless read and interpreted by SECGEN. This in itself would be of no consequence to the analyst except that direct validation of the data input to the generator is impossible unless the analyst is intimately familiar with the generator.

To overcome this problem when the data set is reorganized, an additional preanalysis program may be used. At a minimum, such a program can serve two useful purposes. First, if it uses the identical set of "read" statements as the generator, the preanalysis program can detect certain input errors much more cheaply and quickly than can the generator. (Suppose a data error existed that would cause termination of the generator after 50,000 MPS cards had been generated; detection of this error by a preanalysis program could prevent an expensive mistake and, further, make it simpler to identify the data error's location. Second, preanalysis can unscramble the data and report them in a form understandable to the analyst.

Preanalysis programs may be put to much more important uses than merely ensuring that data will be read in as desired. In the case of mixed-integer programming models—in which reduction of the number of bivalent (0-1) variables is essential for keeping solution costs down—such programs may be used to determine, based on the a priori information in the data set, which variables are likely to be fixed at either zero or unity in the optimal solution (see Meeraus and Stoutjesdijk, forthcoming). In the case of agricultural models, simple economic theory may be applied in the preanalysis to determine if the model's structure and the data do indeed conform with accepted theory. For example, if we observe in the competitive environment of an agricultural sector that several crops are grown simultaneously in some narrowly delineated region, then apart from considerations of risk aversion, the gross margins from each of these crops should be approximately equal (otherwise the farmers would specialize in the most profitable). That this approximate condition is met in the model's representation of the environment could be determined by examining the simplex criteria (DJ's) from the linear programming solution, but it

Figure 19-5. *SECGEN Input, Partial Listing*

```
1 I CULMAY  16              1
ALV ARO AZU CAR CHV FRI GAR JIT JON MAI MEL PEP SAN SOR SOY TRI
     2     BC
147.274    255.409
127.143    62.622
.12        .026       .3707      .01685    .1       .1
.01-       .05
1.         1.36
17.65614.92 9.20141.8634.0172 5.232845.28782.16862.252193.145.220228.585
ALV  I     .0015      BC    2 9 8
MPN MPN
44.91      .603       .7         1.32288
.571       .192       .018
0.                    5.4
.252  .252  .252  .252  .252  .252  .252              .252  .252  .252
.5    .5    .5    .5    .5    .5    .5    .3    .3    .57   .6    .6
1.175 .875  .875  .875  .875  .875  .875  .775  .680  .95   .85   .85
1.2   .9    .9    .9    .9    .9    .9    .7    .82   1.07  .87   .87
```

Note: SECGEN is the matrix generator (written in Fortran) used in CHAC.

could be tested much more quickly and less costly if the preanalysis program "prices out" each activity in each region. That is, in addition to displaying the data in readable form, the program could calculate the net profit from each activity and permit the analyst to compare these before generating and solving the model. (See chapter 10 for model validation tests along these lines.)

Interpreting the Solution

As in the cannery example, the raw output from the linear programming solution package usually bears little resemblance to the solution as reported to the analyst's audience, and a further translation (inverse mapping from MPS format) is required. As in all of other aspects of computational considerations considered in this paper, tradeoffs are possible.

At one extreme, the model may be small enough and sufficiently simple so that, through careful consideration of the MPS naming procedure, the linear programming output is comprehensible and the solution fully interpretable. More often than not, though, the linear programming output will contain information of interest that is "embedded" within the solution and the input data. For example, a particular solution, although only yielding production patterns, in fact represents a pattern of income distribution among farmers in different regions and hired laborers, a pattern of implicit input demands (such as fertilizer), or even an agricultural foreign trade balance. Such output, which probably will not appear as a nicely labeled row or column in the linear programming output, is of course crucial to the economic evaluation of the model and its simulations.

Yet an additional program, called a "report writer," may also be of assistance and may spare the analyst from sorting through and interpreting the output, bringing to bear the necessary exogenous data, performing the manual calculations, and tabulating such information. Most current linear programming packages allow the option of writing the raw solution file onto some medium (tape, disk, drum) that can be read by a report writer. This program, which would also make the matrix generator's input file accessible, could then be used to combine the two information sources and the rules (for, say, calculating wage income) to produce tabular reports.

Certain tabulations—such as totaling fertilizer usage by district, farm type, and the like—may involve trivial calculations and be of only marginal interest. But others may be of crucial importance to the validation of the model. Suppose that an agricultural sector model is designed so that it is the aggregate of specific farm types. It could be virtually impossible to determine, from the aggregate solution, what the operations of the disaggregate farm types are without elaborate side calculations. In analyzing the aggregate output, it then might be mandatory to analyze the implicit outputs of individual farm types simultaneously. A report writer that displayed both the macro and micro output would then be of clear advantage.

Because they are additional programs (whether written in Fortran or as instructions for a commercial package having such options), report writers require essentially the same type of programming resource as matrix generators. Thus they have their own cost, which usually is proportionally greater than that of matrix generators because the task of selecting the required data from the solution file is quite tedious. In some cases, the combined use of appropriate naming procedures and *accounting* rows or columns (or both) may be used as a substitute. If the desired postsolution calculations are linear, then they may be directly included in the model so that a given row or column would automatically give the reported result. In a simple example of fertilizer use, if only the cost in monetary terms is included in the full model, and this cost is but a component of the total cost of each activity, additional rows may be added to the model to total the physical quantity by category of use. These "demands" are then simply read from the linear programming solution output. This procedure, of course, has a cost—larger models and higher solution times—that must be considered against manual calculation or the use of a report writer.

Conclusions

The recent dramatic advances in the efficiency of linear programming solutions have shifted the difficulty in applied modeling to the problem of translating models into—and back out of—MPS format. The resources required to solve the translation problem are almost always underestimated prior to the undertaking of a modeling exercise, much to the analyst's later chagrin.

It is almost always more efficient to employ a matrix generator to construct the MPS file than it is to attempt to construct it manually—not only because the computer is efficient in manipulating large amounts of numbers but because it is also efficient in assembling strings of characters into MPS row and column names. Both kinds of efficiency are useful not only for translating models into solvable form infinitely faster, but also for vastly reducing the chance of random error in the translation.

These conclusions hold for models used either in a research or operations environment. In the first case, many repeated generations of the model, either with data or structural improvements, are usually required. In the second case, the nature of operations implies repeated usage of the model, by individuals other than the original analyst, and usually with updated data. The writing of a matrix generator is thus an investment that has immediate payoffs in error reduction and continuing payoffs as the model is reused under different specifications.

The kind of matrix generator to be used depends on a variety of factors: the nature of the modeling exercise, the computer and software available, the skills of the personnel involved, the organization and physical form of the data, and the need or desire to transfer the model to different environments.

Just as critical in obtaining the solution to a model is the inverse translation from the raw linear programming out. A mirror of the matrix generator, a report writer, may be developed for this task. Alternatively, more meaningful naming conventions and the use of accounting relations built into the model may make this translation less difficult and time consuming.

These choices not only depend on the objectives of the study and the resources and skills available to it, but they are highly interrelated. Although few rules can be suggested, it is clear that careful planning is required for the computational aspects of linear programming.

References

Dantzig, George B. 1963. *Linear Programming and Extensions*. Santa Monica, Calif.: Rand Corporation.

Meeraus, Alexander, and Ardy Stoutjesdijk. Forthcoming. "The Solution Procedure for FERTILEA." In *Industrial Investment Analysis under Increasing Returns*. Edited by Ardy Stoutjesdijk and Larry Westphal.

Index

Chan, José Luis, 361

Chayanov, A. V., 413

Choice of technique theory, 252–57. *See also* Technique adoption

Climate, 84–85, 89; adverse effect on crops and, 167; irrigation projects and rainfall and, 474, 475

Comisión Coordinadora del Sector Agropecuario (COCOSA, later CONACOSA), 4, 8, 113

Commodities: balances (sectoral and district in CHAC), 48; demand and, 41; demand function and product groups and, 62, 63, 69–76

Compañía Nacional de Subsistancia Popular (CONASUPO), 107, 181–82; perishable products and, 198

Competition: exports (crop) and, 149, 151, 188; risk analysis and perfect, 205, 207–10; vegetable exports and, 353, 361, 364, 366, 371

Competitive market assumption test (PACIFICO), 330–32

Competitive market mechanism, 25; demand structure and, 58, 59, 62, 63–64, 79–80; maximizing objective function and, 58; risk and, 204; validation of CHAC and, 153, 155

Computation for programming, 16; data organization and, 584–87; mathematical modeling and, 574–76; mathematical programming system (MPS) and, 576–79, 588; matrix generation and, 579–84, 588–89; model building and, 573; preanalysis of data and, 585–87; solution interpretations and, 587–88, 589

Computers, 582; debugging of, 583; language of, 575, 576, 579–84; model management and, 573

CONASUPO, 107, 181–82, 198

Condos, Apostolos, 23n

Consumers, 63; policy planning and, 5; risk-averse, 311

Consumer surplus, 9; demand structure analysis and, 78

Consumption: choice of technique theory and, 252, 254; growth in agricultural sector and, 166; home, 118–19; maize-wheat, 152; Plan Puebla and, 419–20, 428, 433, 440, 443–45; price and producers', 28; regional model and, 319; subsistence, 6, 440, 443–45; technique adoption and, 259

Coordinating Commission for the Agricultural Sector (COCOSA, later CONACOSA), 4, 8, 113

Corn. *See* Maize

Costs: dairy model and, 533; exchange and domestic resource (as measure of comparative advantage), 147–51; irrigation projects and, 178–79, 478

Cotton, 33; irrigation projects and, 461, 467, 468–69; land cultivation and, 169–71; subsidies for inputs (PACIFICO) and, 345–46

Credit, 14, 178 n30, 195; algebraic statement in CHAC and, 51, 53; dairy industry and, 513; Plan Puebla and, 420–21; 423, 423–24, 426, 431, 436, 441; structure of, 197; technical production coefficients for, 97

Crop insurance, 45, 312

Cropping activities, 84; CHAC and, 28, 29, 30, 31, 33, 45; PACIFICO and, 322–23, 326, 346; rotation and, 97; vegetable exports and, 360, 370. *See also* Double cropping

Cropping model. *See* MEXICALI

Cropping patterns, 8, 11; CHAC and, 44, 45; changes in, 165; cultivation and, 169–71; data and regional, 87; investment and, 481, 505; irrigation and, 178, 231, 247; irrigation projects and, 459, 464–65, 467, 476; MEXICALI and, 499; Plan Puebla and, 415–18; price supports and, 143, 144, 146; production alternatives and regional, 86–87; sectoral dynamism and, 175; TOLLAN and, 381, 403; validation of CHAC and, 157

Crops: CHAC and ranking of, 10, 147–52; change in composition of, 180–81; export, 185–89; feed, 515, 531–32; MEXICALI and, 486; omission of long-cycle (CHAC model), 159; PACIFICO validation and, 330–32; short-cycle (CHAC model), 6, 25; short-cycle (Mexico), 322; subsidies for labor-intensive, 375, 409; substitution and, 184; TOLLAN and, 376, 377, 392, 393, 403

Cultivation, 30, 121; analysis of, 166–69; calendars for, 87–88, 99; employment and, 189–90, 191, 192; incentives for, 126–27; increase in, 165; irrigation and, 319; land programs and, 194; sectoral dynamism and, 175–77; tenure and, 197; TOLLAN and, 376

The full range of World Bank publications, both free and for sale, is described in the *Catalog of World Bank Publications*; the continuing research program is outlined in *World Bank Research Program: Abstracts of Current Studies*. Both booklets are updated annually; the most recent edition of each is available without charge from World Bank Publications, Department B, 1818 H Street, N.W., Washington, D.C. 20433, U.S.A.